Tools and Methods for the Improvement of Quality

The Irwin Series in Quantitative Analysis for Business
Consulting Editor   Robert B. Fetter   Yale University

Tools and Methods for the Improvement of Quality

Howard Gitlow
University of Miami

Shelly Gitlow
Miami Consulting Group, Inc.

Alan Oppenheim
Montclair State College

Rosa Oppenheim
Rutgers The State University of New Jersey

1989

IRWIN

Homewood, IL   60430
Boston, MA   02116

Various excerpts have been reprinted from *Quality, Productivity and Competitive Position* by W. E. Deming by permission of MIT and W. E. Deming. Published by MIT, Center for Advanced Engineering Study, Cambridge, MA 02139. Copyright 1982 by W. E. Deming.

Various excerpts have also been reprinted from *Out of the Crisis* by W. E. Deming by permission of MIT and W. E. Deming. Published by MIT, Center for Advanced Engineering Study, Cambridge, MA 02139. Copyright 1986 by W. E. Deming.

Sponsoring editor: Richard T. Hercher, Jr.
Project editor: Rita McMullen
Production manager: Bette Ittersagen
Compositor: Bi-Comp, Incorporated
Typeface: 10/12 Times Roman
Printer: R.R. Donnelley & Sons Company

LIBRARY OF CONGRESS
Library of Congress Cataloging-in-Publication Data

Tools and methods for the improvement of quality / Howard Gitlow . . .
  [et al.].
     p.   cm.
  Includes index.
  ISBN 0-256-05680-3
  1. Quality control—Statistical methods. I. Gitlow, Howard S.
TS156.T587 1989
658.5′62—dc19

88–12923
CIP

8 9 0 DO 5 4 3

Dedicated to the never-ending improvement of the species
Ali Gitlow
Adam and David Oppenheim

Preface

In recent years there has been a growing interest in the area of quality improvement. We believe that a comprehensive approach to this issue must consider Dr. W. Edwards Deming's philosophy for the improvement of quality, productivity, and competitive position.

Unlike other currently available publications, this book:

1. Focuses on the Deming philosophy and ties all discussions of statistical topics to the Deming philosophy.
2. Distinguishes between enumerative statistical studies (studies of fixed populations whose purpose is to take action on the population) and analytical statistical studies (studies of processes whose purpose is to take actions that will improve the process).
3. Deals with how to improve a stable process (a process that has a known capability to produce and is predictable).
4. Has a detailed discussion on diagnosing and analyzing control chart patterns.
5. Focuses attention on modern inspection policies (Deming's *kp rules*), as opposed to traditional acceptance sampling plans.
6. Is the only statistical quality control book to cover quality improvement stories (an effective format for presenting and working on statistical process improvement efforts).
7. Has many examples and mini-case studies which aid the reader in understanding and appreciating the topics covered.

This book can be used for three different levels of education. The basic level covers: Chapter 1 (Fundamentals of Quality), Chapter 2 (Fundamentals of Statistical Studies), Chapter 3 (Documenting and Defining a Process), Chapter 5 (Basic Statistics), Chapter 8 (Stabilizing and Improving a Process with Control Charts), Chapter 9 (Attribute Control Charts), Chapter 10 (Variables Control Charts), Chapter 12 (Diagnosing a Process), and Chapter 14 (Process Capability and Improvement Studies). The intermediate level covers all of the chapters in the basic level plus: Chapter 4 (Basic Probability), Chapter 6 (Probability Models), Chapter 7 (Sampling Distributions), Chapter 11 (Out-Of-Control Patterns), and Chapter 13 (Specifications). The advanced level covers all of the chapters in the basic and intermediate levels plus: Chapter 15 (Taguchi Methods—Quality Improvement in Product and Process Design), Chapter 16 (Inspection Policy), Chapter 17 (Deming's 14 Points and the Reduction of Variation), and Chapter 18 (Some Current Thinking about Statistical Studies and Practice).

We would like to acknowledge and thank Dr. W. Edwards Deming for his philosophy and guidance in our personal studies of quality improvement and statistics. Also, we would like to thank Robert F. Hart (University of Wisconsin, Oshkosh), Chandra Das (University of Northern Iowa, Cedar Falls), Donald Holmes (Stochos Incorporated), Peter John (University of Texas, Austin), Sudhakar Deshmukh (Northwestern University, Kellogg Graduate School), Erwin Saniga (University of Delaware, Newark), Theresa Sandifer (Kimberly-Clark Corporation), Jeffrey Galbraith (Greenfield Community College, Greenfield), Richard Hercher Jr. (Richard D. Irwin, Inc.), and Rita McMullen (Richard D. Irwin, Inc.) for their help in writing and editing this book. In the final analysis, we, the authors, accept total responsibility for the information contained in this book.

Most important, we want to thank our parents, Abraham and Beatrice Gitlow, Judy Dimmerman, Norman and Sylvia Oppenheim, and Aaron and Esther Blitzer, for their unwavering support and confidence.

We sincerely hope that you, the reader, find this book to be a valuable aid in your studies of quality improvement and statistics. Best of luck in your studies.

Howard Gitlow

Shelly Gitlow

Alan Oppenheim

Rosa Oppenheim

Contents

Control Charts: *Defect Detection. Defect Prevention: Attribute Control Charts. Never-Ending Improvement: Variables Control Charts.* An Example of an Attribute Control Chart. An Example of a Variables Control Chart. Two Possible Mistakes in Using Control Charts: *Overadjustment. Underadjustment.* Two Uses of Control Charts: *Evaluating the Past. Evaluating the Present.* Some Out-of-Control Evidence. Collecting Data: Rational Subgrouping: *Arrangement One. Arrangement Two. Arrangement Three.*

PART VII
Foundations of Quality Revisited *531*

Tools and Methods for the Improvement of Quality

PART I

Foundations of Quality

This text is based on concepts presented in this section. Chapter 1 discusses the three types of quality and the losses society incurs from the lack of quality in goods and services. Chapter 1 also provides a brief history of quality and focuses on W. Edwards Deming's 14 points for management, key to quality improvement in any service or manufacturing organization. Chapter 2 analyzes the basic distinction between enumerative and analytic statistical studies and discusses the significance of each type of statistical study as regards quality improvement efforts. Chapter 3 details the procedures used to document a process. When a process has been properly documented and defined, and characteristics to be studied have been operationalized, management can begin process improvement efforts using data, not guesswork and opinion.

The material presented in this section is critical to any effort towards quality and process improvement. Any improvement efforts must be preceded by a clear understanding of quality and the statistical studies required to improve quality. The remainder of this book focuses on the tools and methods needed to accomplish these goals.

Chapter 1

Fundamentals of Quality

DEFINITION OF QUALITY

Quality is a judgment by customers or users of a product or service; it is the extent to which the customers or users believe the product or service surpasses their needs and expectations. For example, a customer who purchases an automobile has certain expectations, one of which is that the automobile engine will start when it is turned on. If the engine fails to start, the customer's expectation will not have been met; and the customer will perceive the quality of the car as poor. On the other hand, if a worker on an assembly line consistently receives usable parts in a timely manner from the worker before him on the line, that worker's needs will be met and he will perceive the quality of those parts as good.

Quality also encompasses the never-ending improvement of a firm's *extended process*. This term refers to the expansion of the organization to include suppliers, customers, investors, employees, and the community. These are all integral parts of the firm's extended process. Figure 1.1 is a pictorial view of the extended process.

The extended process begins by communicating the needs of the consumer to the organization. It is important to realize that customer satisfaction is the ultimate goal of an organization; hence, communication of customer needs is critical to the functioning of the extended process. Viewing the customer as the most important element in the extended process means that a firm must have a continuing process to determine how its products and/or services are performing and what new characteristics would increase customer satisfaction.

At the other end of the extended process are the firm's suppliers. The firm communicates the needs and expectations of its customers to its suppliers so that the suppliers can help improve customer satisfaction. Firms and suppliers work together to produce quality products/services and pursue extended process improvement.

FIGURE 1.1   The Extended Process

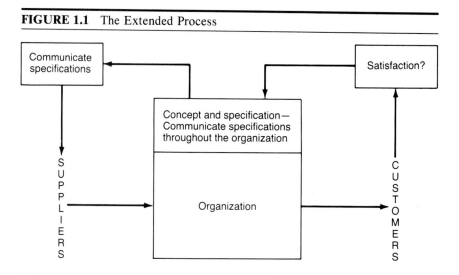

TYPES OF QUALITY

Managers desiring to improve quality in the extended process must consider three types of quality[1]:

a. Design/Redesign.
b. Conformance.
c. Performance.

Quality of Design/Redesign

Quality of design begins with consumer research and service/sales call analysis and leads to the determination of a product/service concept that satisfies the consumer. Specifications are then constructed for the product/service concept, as shown in Figure 1.2. Customers can exist for the firm internally as well as externally. For example, shipping is a customer for production.

The process of developing a product/service concept involves establishing and nurturing an interface between marketing personnel, service personnel, and design engineering personnel. In extended process terms, design engineering is one of marketing's customers, and vice versa. There may not be separate departments in small organizations; nonetheless, the above interfaces are important if an organization is to continuously meet consumer needs.

FIGURE 1.2 Design/Redesign

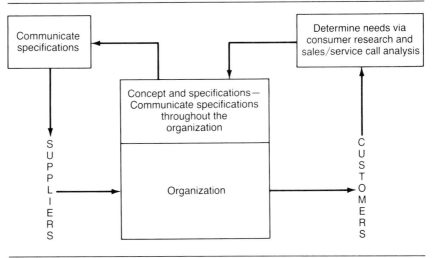

Consumer Research.   Continuous and never-ending improvement of an organization's product/service concept requires that consumer research and sales/service call analysis be an ongoing effort. *Consumer research* is a collection of procedures implemented to clarify customer needs, both present and future. Consumer research procedures range from nonscientific studies to scientific studies. An example of consumer research is a study of the reasons why dog food purchasers buy a particular brand of dog food. The goal of the investigation is to determine the customer's needs and redesign the dog food product around those needs: for example, alter the can size, make the can resealable, or modify the composition of the dog food. The study should be ongoing so the firm will always be in touch with changing customer needs.

Consumer research can be performed within an organization. For example, employees are the customers of some management policy decisions; hence, employee surveys are a form of consumer research that could lead to improved management policy.

Sales/Service Call Analysis.   *Sales call analysis* involves the systematic collection and evaluation of information concerning present and future customer needs, which is collected during sales interactions with customers. Sales call analysis helps determine customer needs by analyzing the questions and concerns customers express about products or services at the time of purchase; such analysis is an important window into the customer's needs. An example of sales call analysis is a formal investigation into salesman-customer interactions at a personal computer distributorship in order to modify and improve the selling protocol.

Service call analysis is the systematic investigation into the problems customers/users have with the performance of the product/service. Service call analysis provides an opportunity to understand which product/service features must be changed to surpass the customer's present and future expectations. An example of service call analysis is the formal collection of information from field repairmen by Sony Corporation concerning problems customers are encountering with 1986 Sony KV 1920 television sets. The basic source document for the service call analysis data is the service ticket, which indicates the problem and the work done to solve it. This information is collected, and overtime indicates problems ostensibly requiring specification changes—for example, redesigning the television tuner or reducing the time between customer request for service and the completed service call.

Service call analysis can also be performed internally in an organization. For example, an area supervisor might examine the problems the next operation encounters using the parts/service/forms his area delivers to the next operation. The purpose of the analysis could be to find out what the supervisor must do to pursue process improvement within his own area.

The continuous and never-ending conduct of consumer research and sales/service call analysis for product/services design or redesign is the goal of quality-of-design studies.

Quality of Conformance

Quality of conformance is the extent to which a firm and its suppliers are able to surpass the design specifications required to serve the needs of the customer, as illustrated in Figure 1.3.

Once the organization has determined product/service specifications via quality-of-design surveys, it must constantly strive to surpass those specifications so that customers receive products/services that perform properly the first time and every time during the product/service's life cycle. The ultimate goal of process improvement efforts is to create products/services of sufficiently high quality that consumers (both external and internal) brag about them. Statistical methods are extremely helpful in accomplishing this task. (Note: Some readers may question the concern with *surpassing* specifications rather than simply *meeting* them. Chapters 8 and 15 analyze the rationale for this statement.)

Quality of Performance

Quality of performance is the determination, through consumer research and sales and service call analysis, of how the firm's products or services

FIGURE 1.3 Conformance

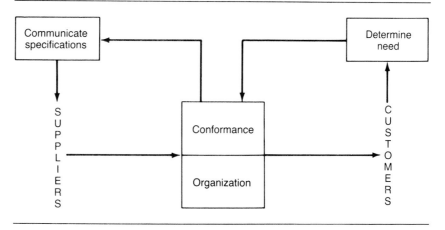

perform in the marketplace, as shown in Figure 1.4. The determination includes analysis of after-sales service, maintenance, reliability, and logistical support, as well as investigation into why consumers do not purchase the company's products/services.

The continual flow of information generated by quality-of-performance surveys clears the fog that exists between consumer research and sales/service call analysis, and the construction of product/service/job specifications. Design engineers must work with marketing people, for example, to determine the specifications (product/service characteristics) for a product/service concept that affects consumer satisfaction.

FIGURE 1.4 Performance

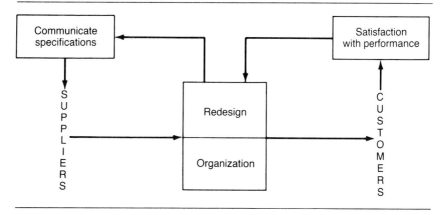

Once these characteristics are known and have been operationally defined, consumers can be grouped into market segments in accordance with their desired product/service characteristics. Product/service characteristics and price determine whether a consumer will initially enter a market segment; hence, product/service characteristics and price determine market size. A consumer will repurchase or brag about a product/service based on its performance. Performance determines the success of a product/service within a market segment; therefore, performance determines market share within a market segment.

Sources of Loss in Quality.   Two sources of loss in quality should be detected in a quality-of-performance study. First, a loss in quality occurs when a process generates products or services whose *quality characteristics package* (specifications) deviates from the needs of the individual (or group of individuals in a market segment). This type of loss can be remedied by increasing the number of market segments and tailoring the product to the consumer's requirements; that is, by increasing the number of "nominal" settings. For example, shirt neck sizes may be marketed in tenths of an inch rather than in halves of a inch; or Velcro may be used in shirt collars instead of buttons. This segmentation strategy minimizes the resultant loss in quality when the nominal levels of a product's quality characteristics package deviate from the needs of an individual (or group of individuals) in a market segment. This action could result from a quality-of-performance study.

A loss in quality also results when a process generates goods or services whose quality characteristics are not uniform (that is, there is high unit-to-unit variation). Such variation causes a product to function inconsistently for a given market segment.

Managers must balance the costs of having many market segments with the benefits of high consumer satisfaction resulting from small deviations between an individual consumer's needs and the product characteristic package for that consumer's market segment. Also, managers must continually strive to reduce variation in product characteristics for all market segments.

These two sources of quality loss must be detected in the quality-of-performance stage of the extended process. This information is then fed into the quality-of-design stage (for design engineers and marketing personnel) and the quality-of-conformance stage of the extended process.

THE HISTORY OF QUALITY

Issues of quality have existed since tribal chiefs, kings, and pharaohs ruled. An example of a quality issue in ancient times is found in the Code of Hammurabi, dating from 2150 B.C. Item 229 states: "If a builder has built a house for a man, and his work is not strong, and the house falls in and kills the householder, that builder shall be slain." Phoenician inspectors eliminated any repeated violations of quality standards by chopping off the hand of the maker of the defective product. Inspectors accepted or rejected products and enforced government specifications. The emphasis was on equity of trade and complaint handling. In ancient Egypt (approximately 1450 B.C.), inspectors checked the squareness of stone blocks with a string as the stonecutter watched. This method was also used by the Aztecs in Central America.

During the 13th century, apprenticeships and guilds developed. The craftsmen were both trainers and inspectors. They knew their trades, their products, and their customers; and they built quality into their goods. They took pride in their work and in training others to do quality work. The government set and provided standards (e.g., weights and measures); and, in most cases, an individual could inspect all the products and establish a single quality standard. This idyllic state of quality could have thrived and lasted if the world had remained small and localized; however, as the world became more populated, more products were needed.

During the 19th century the modern industrial system emerged. In the United States, Frederick Taylor pioneered scientific management, removing work planning from the purview of workers and foremen and placing it in the hands of industrial engineers. The 20th century ushered in a technical era that enabled the masses to avail themselves of products previously reserved for only the wealthy. Henry Ford introduced the moving assembly line into the manufacturing environment of Ford Motor Company. Assembly line production broke down complex operations so they could be performed by unskilled labor, resulting in the manufacture of highly technical products at low cost. As part of this process, an inspection operation was instituted to separate good and bad products; quality remained under the purview of manufacturing.

It soon became apparent, however, that the production manager's priority was meeting manufacturing deadlines—not product quality. Managers knew they would lose their jobs if they didn't meet production demands, whereas they would only be reprimanded if quality was poor. Upper management eventually realized that quality was suffering as a result of this system, so a separate position of "chief inspector" was created.

Between 1920 and 1940, industrial technology changed rapidly. The Bell System and Western Electric (its manufacturing arm) led the way in quality control by instituting an Inspection Engineering Department to deal with problems resulting from defects in their products and lack of interdepartmental coordination. Bell employees George Edwards and Walter Shewhart were pioneers and leaders in the field.

According to Edwards, quality control exists:

When successive articles of commerce have their characteristics more nearly like its fellows' and more nearly approximating the designer's intent, than would be the case if the application were not made. To me, any procedure, statistical or otherwise, which has the results I have just mentioned, is quality control, and any procedure which does not have these results is not quality control.[2]

Edwards coined the term *Quality Assurance* and advocated quality as part of management's responsibility:

This approach recognizes that good quality is not accidental and that it does not result from mere wishful thinking, that it results rather from the planned and interlocked activities of all the organizational parts of the company, that it enters into design, engineering, technical and quality planning specification, production layouts, standards, both workmanship and personnel, and even into training and fostered point of view of administrative, supervisory, and production personnel. This approach means placing one of the officers of the company in charge of the quality control program in a position at the same level as the controller or as the other managers in the operation. Its objective would be elimination of the hunch factors that at present so largely determine the product quality of too many companies. It puts a man at the head of the quality control program in a position to establish and make effective a company-wide policy with respect to quality, to direct the actions to be taken where it is necessary and to place responsibility where it belongs in each instance.[3]

In 1924, mathematician Walter Shewhart introduced the concept of *statistical quality control*, which provided a method for economically controlling quality in mass production environments. Shewhart was concerned with many aspects of quality control. In his book of lectures at the graduate school of the U.S. Department of Agriculture, for example, he asks the reader to write several letters *A* as carefully as possible, then observe them for variations. No matter how carefully one forms the letters, it seems that variations always occur—a simple yet powerful example of variation in a process. Although Shewhart's primary interest was statistical methods, he was very aware of principles of management and behavioral science.

World War II quickened the pace of quality technology. The need to improve the quality of products being manufactured resulted in increased study of quality control technology and more sharing of information. In 1946 the American Society for Quality Control (ASQC) was formed, and George Edwards was elected its president. He commented at the time: "Quality is going to assume a more and more important place alongside competition in cost and sales price, and the company which fails to work out some arrangement for securing effective quality control is bound, ultimately, to find itself faced with a kind of competition it can no longer meet successfully." In this environment, basic quality concepts expanded rapidly. Many companies implemented vendor certification programs.

Quality assurance professionals developed failure analysis techniques to problem-solve, quality engineers became involved in early product design stages, and companies began to test products' environmental performance. As World War II ended, however, progress in quality control began to wane. Many companies viewed it as a wartime effort and therefore unnecessary in the booming postwar market.[4]

In 1950, W. Edwards Deming, a statistician who had worked at the Bell System with George Edwards and Walter Shewhart, was invited by JUSE (the Union of Japanese Scientists and Engineers) to speak to Japan's leading industrialists. They were concerned with rebuilding Japan after the war, breaking into foreign markets, and improving Japan's reputation for producing poor-quality goods. Deming convinced them, despite their reservations, that by instituting his methods Japanese quality could become the best in the world. The industrialists took Deming's teachings to heart; and Japanese quality, productivity, and competitive position improved and strengthened tremendously. He was awarded the Second Order Medal of the Sacred Treasure by Emperor Hirohito for his contributions to Japan's economy. The coveted Deming Prizes are awarded each year in Japan to the company that has achieved the greatest gain in quality and to an individual for developments in statistical theory.[5] Prize-winning Japanese companies include Nissan, Toyota, Hitachi, and Nippon Steel. In 1985, Texas Instruments became the first U.S. company to receive the Deming prize.

The Deming methods are spreading throughout the United States. Companies such as the Nashua Corporation, the Ford Motor Company, and General Motors are incorporating his principles and trying to obtain the benefits of continuous process improvement. Deming's clients have included railways, telephone companies, consumer researchers, hospitals, law firms, government agencies, and university research organizations. He has written extensively on statistics and management and is a professor emeritus at New York University Graduate School of Business Administration.

In 1951, Armand V. Feigenbaum's book, *Total Quality Control,* was published; and it advanced the concept of quality control in all areas of business, from design to sales. Up until this time, quality efforts involved primarily corrective activities, not prevention. The Korean War sparked increased emphasis on reliability and end-product testing. Not all of the additional testing enabled firms to meet their quality and reliability objectives, so quality awareness and improvement programs began to emerge in manufacturing and engineering areas. Service Industry Quality Assurance (SQA) also began to focus on the use of quality methods in hotels, banks, government, and other service systems. By the end of the 1960s, quality programs had spread throughout most of America's major corporations. But American industry was still enjoying the top position in world markets as Europe and Japan continued to rebuild.

Foreign competition began to threaten U.S. companies in the 1970s. The quality of Japanese products such as automobiles and televisions began to surpass U.S.–made goods. Consumers became more sophisticated and began to consider price and quality over the long-term life of a product in purchase decisions. The combination of increased consumer interest in quality and foreign competition forced U.S. management to become more concerned with quality. The late 1970s and 1980s have been marked by a striving for quality in all aspects of businesses and service organizations, including areas of finance, sales, personnel, maintenance, management, manufacturing, and service. The focus is on the entire system, not just the manufacturing line. Reduced productivity, high costs, strikes, and high unemployment have caused management to turn to quality improvement as the means to organizational survival. Some of the quality leaders in the United States are Armand Feigenbaum, Joseph Juran, and W. Edwards Deming.[6] In this book we focus largely on the methods of W. Edwards Deming. His holistic approach allows for consistent, long-term improvement of quality in an atmosphere that encourages growth and development of employees, the organization's most valuable resource.

THE DEMING VIEW OF THE RELATIONSHIP BETWEEN QUALITY AND PRODUCTIVITY

Why should an organization try to improve quality? If a firm wants to increase its profits, why not raise productivity? For years, W. Edwards Deming has worked to change organizations that operate under the philosophy that increasing productivity will increase profits. The following example illustrates the folly of such thinking.[7]

For the past 10 years the Universal Company has produced 100 widgets per hour, 20 percent of which are defective. The Board of Directors now demands that top management increase productivity by 20 percent. The directive goes out to the employees, who are told that instead of producing 100 widgets per hour, the company must produce 120. The responsibility for producing more widgets falls on the employees, creating stress, frustration, and fear. They try to meet the new demands but must cut corners to do so. The pressure to raise productivity creates a defect rate of 25 percent and increases production to only 104 units, yielding 78 good widgets, fewer than the original 80, as shown in Figure 1.5(a).

Stressing productivity often produces exactly the opposite of the effect desired. The following example demonstrates a new way of looking at productivity and quality.

The Dynamic Factory produces 100 widgets per hour with 20 percent defective. Top management is continually trying to improve quality, thereby increasing productivity. Top management realizes that Dynamic

---

FIGURE 1.5   Productivity vs. Quality Approach to Improvement

(a)   Universal Company Output

	Before Demand for 20% Productivity Increase	After Demand for 20% Productivity Increase
	(Defect rate = 20%)	(Defect rate = 25%)
Widgets Produced	100	104*
Widgets Defective	20	26
Good Widgets	80	78

* Only reached 104, not required 120, but defect rate rose from 20 percent to 25 percent. More widgets were produced; but more were defective, yielding less productivity.

(b)   Dynamic Factory Output

	Before Improvement		After Improvement
	(Defect rate = 20%)		(Defect rate = 10%)
Units Produced	100	→	100
Units Defective	20	Process	10
Good Units	80	Improvement	90
		→	

---

is making 20 percent defective units, which translates into 20 percent of the total cost being spent to make bad units. If Dynamic's managers can improve the process, they can transfer resources from the production of defective units to the manufacture of additional good products. Management can improve the process by making some changes at no additional cost, so only 10 percent of the output is defective. This results in an increase in productivity, as shown in Figure 1.5(b). Management's ability to improve the process results in a decrease in defective units, yielding an increase in good units, quality, and productivity.

Benefits of Improving Quality

Deming's way of looking at the relationship between quality and productivity stresses improving quality to increase productivity. Several benefits result:

1. Productivity rises (in our Dynamic Factory example, from 80 good units in the 100 produced to 90 good units in the 100 produced).
2. Quality improves (from 80 percent good units to 90 percent good units).

3. Cost per good unit decreases.
4. Price can be cut.
5. Workers' morale improves because they are not seen as the problem. This last aspect leads to further benefits:
   1. Less employee absence.
   2. Less burnout.
   3. More interest in the job.
   4. Motivation to improve work.

In sum, stressing *productivity* means sacrificing quality and possibly decreasing output. Employee morale plunges, costs rise, customers are unhappy, and stockholders become concerned. On the other hand, stressing *quality* can produce all the desired results: less rework, greater productivity, lower unit cost, price flexibility, improved competitive position, increased demand, larger profits, more jobs, and more secure jobs. Customers get high quality at a low price, vendors get predictable long-term sources of business, and investors get profits.

THE QUALITY ENVIRONMENT

To improve quality, management must demonstrate its commitment to quality and accept responsibility for improving it. The "quality environment" encourages teamwork, communication, joint problem-solving, trust, security, pride of workmanship, and never-ending improvement. A true cooperative spirit prevails in this type of atmosphere, as teamwork is a prerequisite for the firm to function and constantly improve the extended process. The corporate culture changes so that workers are no longer afraid to point out problems in the system, and management is actively involved in the never-ending improvement of the extended process with the workers. Workers and management learn to speak the same language, the language of statistics and process control. Workers are responsible for communicating to management the information they have about the system so that management can act. Never-ending improvement of the process (not just within the organization, but of all aspects of the extended process, including vendors, customers, the community, and investors) eventually leads to higher quality, reduced costs, and greater profitability.[8]

Fourteen Points for Managing Never-Ending Improvement of the Extended Process

Dr. W. Edwards Deming has outlined his philosophy of management in his *Fourteen Points*. Acceptance and understanding of the 14 points lead to a commitment by management and provide a framework for action.

The following brief overview of these points is designed to provide a philosophical basis for effectively using statistical quality control. (See Deming's *Out of the Crisis*[9] for a full discussion of his methods for improving quality, productivity, and competitive position, and Gitlow and Gitlow's *The Deming Guide to Quality and Competitive Position*.[10])

Deming's methods incorporate the use of statistical tools and behavioral techniques for changing the corporate climate; both are integral to improving quality. The points are presented separately and can create improvement individually; nevertheless, it is the synergistic implementation of all of the points that will improve quality in a never-ending fashion. The 14 points are listed below and followed by a brief discussion.

1. Create constancy of purpose toward improvement of product and service with a plan to become competitive, stay in business, and provide jobs.
2. Adopt the new philosophy. We are in a new economic age. We can no longer live with commonly accepted levels of delays, mistakes, defective materials, and defective workmanship.
3. Cease dependence on mass inspection. Require, instead, statistical evidence that quality is built in to eliminate the need for inspection on a mass basis.
4. End the practice of awarding business on the basis of price tag. Instead, depend on meaningful measures of quality, along with price. Move toward a single supplier for any one item, based on a long-term relationship of loyalty and trust.
5. Improve constantly and forever the system of production and service, to improve quality and productivity, and thus constantly decrease costs.
6. Institute modern methods of training.
7. Institute modern methods of supervision.
8. Drive out fear, so that everyone may work effectively for the company.
9. Break down organizational barriers—everyone must work as a team to foresee and solve problems.
10. Eliminate arbitrary numerical goals, posters, and slogans for the workforce which seek new levels of productivity without providing methods.
11. Eliminate work standards and numerical quotas.
12. Remove barriers that rob employees of their pride of workmanship.
13. Institute a vigorous program of education and training.
14. Create a structure which will push the prior 13 points every day.

Point 1: Constancy of Purpose. Long-term perspective and constancy of purpose are necessary ingredients in never-ending improvement. An organization's mission statement, operating philosophy, and goals should provide a framework for consistent action on a day-to-day and long-term basis. Concern for the problems of today and tomorrow gives management the foresight to allocate resources for innovation, redesign, training, education, and research to improve quality.

Creating an environment that encourages everyone in the extended process to cooperate on continually improving quality and meeting customer needs will allow for 1) increased productivity, 2) better competitive position, 3) a reasonable return for stockholders, 4) secure employment, and 5) continued existence (to name a few of the possible benefits).

Point 2: Quality Consciousness. The foundation of the Deming philosophy is quality. Adopting the new philosophy means promoting a quality consciousness—rejecting commonly accepted levels of defects, rework, shoddy workmanship, and poor service. The costs to companies for rework, waste, and redundancy—including costs for material, manpower, capital equipment, facility space, warranty, retesting, reinspection, shipping, customer dissatisfaction, and schedule disruptions—are staggering. Accepting the above as "business as usual" has seriously eroded America's competitive position in world markets.

Defects are not free. Somebody makes them and gets paid for doing so. If a substantial portion of the work force corrects defects, then the company is paying to correct defects as well as make them. Most companies are in a state of *defect detection*; this is the lowest form of quality consciousness. They expect to make defects and assume that through inspection they will find them and do something about them. A more advanced state of quality consciousness is one of *defect prevention,* improving the process so that defects are not made. Once in the position of defect prevention, a firm can work on never-ending improvement of the process. This requires moving the process toward the desired level of performance and reducing unit-to-unit variation so that the process operates in an ever-decreasing range within specification tolerance.

Customer satisfaction is the key element to quality and the new philosophy. Providing customers goods and services that surpass their expectations and needs at a price they are willing to pay is paramount. Satisfying customers will lead to increased profits in the long run. A price tag cannot be put upon the advantages of a satisfied customer extolling the virtues of a company's products and/or services.

Point 3: End Dependence on Mass Inspection. Mass inspection is essentially checking goods with no consideration of how to make them better, improve the process, or achieve higher quality. Some have always believed that if product inspections are carried out properly, quality will improve. In reality, inspection neither improves nor guarantees quality. Mass inspection at any stage in the extended process fails to make a clean separation of good and bad—for mass inspection occurs too late. Nor is quality improved by after-the-fact inspection; the defective items have already been produced.

Inaccuracy is a major problem with 100 percent inspection. Play the role of inspector for a few moments and count the number of *f*s in the following passage:

> FINISHED FILES ARE THE RESULT OF YEARS OF SCI-ENTIFIC STUDY COMBINED WITH THE EXPERIENCE OF MANY YEARS.

Did you find six *f*s? When this exercise is done in a group, more than one third of the participants typically come up with the wrong answer. This demonstrates the fallibility of 100 percent inspection.

Deming advocates a plan that minimizes the total cost of incoming materials and final product. Simply stated, the rule is an inspect-all-or-none rule. Its logical foundation has statistical evidence of quality as its base. The rule for minimizing the total cost of incoming materials and final product is referred to as the *kp rule;* Chapter 16 focuses on inspection rules in the Deming philosophy.

Point 4: Quality in Supplied Goods. Many organizations purchase solely on the basis of price, without adequately measuring quality. Purchasing agents are encumbered by managerial policy that dictates buying from the lowest bidder as standard operating procedure. Without an adequate measure of quality, however, price is meaningless. The following example illustrates this point.

Three vendors submit bids to a firm for a large quantity of a certain part. Vendor A charges $12 per unit, vendor B charges $11 per unit, and vendor C charges $10 per unit. If the purchasing agent considers only the price tag, vendor C will get the contract. However, if the purchasing agent considers quality and price, another vendor might get the contract.

Suppose vendor A has been pursuing never-ending improvement of quality. He has improved his production process so that only one unit per million is defective. Consequently, the effective price the firm would pay for A's good product is:

$$\frac{\$12.00}{(1 - 0.000001)} = \$12.00 \text{ per unit.}$$

Vendor B has also been pursuing process improvement using statistical process control, but not as long as vendor A. He has reduced his average defect rate to 10 percent. The 10 percent figure has been stable over some period of time. The effective price the firm would pay for B's good product is:

$$\frac{\$11.00}{(1 - 0.10)} = \$12.22 \text{ per unit.}$$

Vendor A is the better buy; however, both vendors have the organizational capacity to improve because they are in control of their processes, even if vendor B is temporarily at a disadvantage.

Vendor C has no records showing the capability of his process to produce a good product. Therefore, the effective price the firm would pay for C's good product is unknown:

$$\frac{\$10.00}{(1 - ?)} = \text{Unknown per unit.}$$

It is likely that vendor C's defect rate is higher than A's or B's because his process is not in control. Vendor C's material may get prohibitively expensive if we consider the costs of using nonconforming material and the subsequent rework. Vendor A's price becomes more attractive because total cost includes the purchase cost plus the cost to put the material into production.

Performing the job of the purchasing agent requires the ability to judge quality, which necessitates education in statistics and process control. It also involves understanding the problems encountered with materials purchased as they move through the extended process. Purchasing agents interact with other employees, customers, and vendors and have to develop the skills necessary to determine satisfaction or dissatisfaction and feed this information back into the never-ending process of improvement.

Deming also stresses the need to move toward single-source suppliers instead of maintaining multiple sources for each item purchased. Quality is promoted by encouraging long-term single source relationships between buyers and vendors that are based on statistical evidence of quality. In single-source relationships based on trust, vendors will modify their processes to meet revised quality-of-design specifications and work with buyers to incorporate customer feedback into the improvement of their processes. This type of relationship allows for open negotiation of the arrangements required to meet the needs of the buyer and vendor— and ultimately, the customer.

Point 5: Process Improvement.   Management is responsible for the entire system and its various processes. This responsibility includes the design of the product or service, the measurement of the amount of trouble with the product or service, and the assignment of responsibility for action to remove the cause of the problem.

Process improvement is aided by statistical methods, which must be used by everyone in an organization. This book presents the statistical methods that are necessary to constantly improve processes. The chapters that follow provide an in-depth treatment of statistical tools for process improvement. Although using statistical methods is crucial to never-ending improvement, implementation must be gradual. First, the environment of the organization should demonstrate the new commitment to quality, a long-term perspective, and a growing trust between management and labor. This is a gradual process that takes time, generally a year or more. Statistical training is necessary for all employees—management, hourly employees, and everyone else in the organization.

The *Deming cycle*[11] is a method that can aid management in the pursuit of continuous and never-ending process improvement. It is a derivative of the scientific method aimed at processes. The Deming cycle was originally called the *Shewhart cycle* after its founder, W. A. Shewhart; in 1950, the Japanese renamed it the Deming cycle.

The Deming cycle is composed of four basic stages: a *plan* stage, a *do* stage, a *check* stage, and an *act* stage. Hence, the Deming cycle is sometimes referred to as the PDCA cycle (*Plan-Do-Check-Act* cycle). A group develops a plan (*plan*); they then implement the plan on a small scale or trial basis (*do*); they then monitor the effects of the trial plan (*check*); and then they take appropriate actions (*act*). These actions can lead to a new or revised plan (process modifications), so the PDCA cycle continues forever in an uphill cycle of never-ending improvement. The Deming (PDCA) cycle is shown in Figure 1.6.

The Deming cycle operates by recognizing that problems (opportunities for improvement) in a process are determined by the difference between customer (internal and/or external) needs and process performance. If the difference is large, customer dissatisfaction may be high; but there is great opportunity for improvement. If the difference is small, the consequent opportunity for improvement is diminished.[12] However, it is always economical to continually attempt to decrease the difference between customer needs and process performance, as discussed later in this book. This can be accomplished through the four PDCA stages.[13]

Stage 1: Plan.   The collection of data about process variables is critical when determining a plan of action for what must be accomplished to

FIGURE 1.6   The Deming (PDCA) Cycle

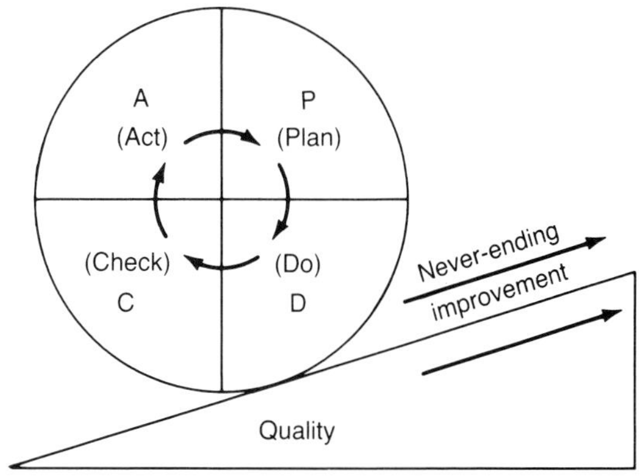

decrease the difference between customer needs and process performance. Earlier, we discussed the three types of quality that must be understood in the never-ending improvement of the extended process. Data concerning customer needs are collected via *quality-of-performance* studies. These same data are evaluated and transformed into variables, which may be acted upon during *quality-of-design* studies. Data concerning process capability are collected via *quality-of-conformance* studies. These process data should be collected on variables for which process improvement action can be taken.

A plan must be developed to determine the effect(s) of manipulating process variables upon the difference between process performance and customer needs. The plan must then be tested using *quality-of-design, conformance,* and *performance* studies.

Stage 2: Do.   The plan established in Stage 1 is set into motion on a trial basis in the Do stage. The trial plan should be conducted in a laboratory, production setting, office setting, or on a small scale with customers (both internal and external). The results of these tests will lead to a concrete plan for manipulating the process variables to decrease the difference between customer needs and process performance.

Stage 3: Check.   The trial plan, which was set into motion in Stage 2, must be monitored (checked) to answer two questions. First, are the manipulated process variables behaving according to the plan and causing a decrease in the difference between customer needs and process performance? And second, are the downstream effects of the plan creating problems or improvements? The results of statistical studies in this Check stage lead to the Act stage.

Stage 4: Act.   The purpose of the Act stage is to implement the modifications to the plan discovered in the Check stage or to implement the process improvements. If, at the Act stage, we learn that the manipulated process variables have not diminished the difference between customer needs and process performance, the PDCA cycle returns to the plan stage to search for other process variables that may decrease the difference. However, if at the Act stage we learn that the manipulated process variables have produced the desired results, the Act stage will either lead back to the Plan stage to determine the optimal levels at which to set the manipulated process variables or will lead to the process improvements. These process improvements lead to a narrowing of the difference between customer needs and process performance. Hence, the PDCA cycle continues forever in the never-ending improvement of the extended process.

The Act stage is accomplished in a three-part process. First, the organization must engage in education so that everyone understands the relationship between the manipulated variables and the proposed decrease in the difference between customer needs and process performance. Second, training is required so that everyone understands who the plan will affect and can modify their jobs accordingly. Once parts 1 and 2 are accomplished, the plan can be set into motion as the third part of the Act stage.

All procedures should be standardized so that everyone conducts a given procedure using the same format, thereby increasing communication and decreasing the possibility of error.

Point 6: Training. Training is a process, an ongoing, integrated approach to employee growth and development. Employees are the most important asset of an organization, and the company's long-term commitment to them includes proper training in the organization's philosophy, goals, and operating principles; and in how to perform their jobs, where *job* is broadly defined to include an understanding of the organization's product or service and the quality characteristics associated with that product or service. Employees must understand operational definitions, specifications, and the extended process.

Training is a part of everyone's job and should include formal class-work, experiential work, and instructional materials. Evaluating when training is complete is accomplished through the use of statistical methods; by using statistical methods, trainers can see whether their personnel have learned all they can from a given training program.

Proper training that gives workers a share in the philosophy and goals of the organization, an understanding of their jobs, specific procedures to do their jobs correctly, and a method of evaluating when training is completed will improve quality. Everyone knows his/her job and is in statistical control pursuing never-ending improvement. Further benefits of proper training for workers are security, pride, decrease in stress, and higher morale. This improves the organizational climate and promotes better working relationships.

Point 7: Supervision. Supervision is a critical link between top management and employees. Management first has to demonstrate that it understands its responsibility for process improvement, cares about improvement of quality, and knows how to guide the organization toward the goal of never-ending improvement. Management does this by removing, on statistical signal, inherent defects, bad incoming parts, machines not maintained, fuzzy operational definitions, or tools not designed for the job. Chapters 8 through 14 deal with this aspect in detail.

Supervision must be a supportive, positive endeavor that encourages learning, development, problem-solving, trust, and change by advancing training, removing barriers, fostering pride in work, showing workers how they fit into the extended process, and stressing quality. Supervisors are *coaches,* trained in statistical methods so that they can help employees pursue never-ending improvement. They can train employees properly and know when training is completed. Supervisors encourage two-way communication, meet on a regularly scheduled basis with workers to provide feedback and listen to their concerns, promote teamwork, and thereby improve the workers' ability to do their jobs.

Point 8: Eliminate Fear.   Fear in organizations profoundly affects those working in the organization and the functioning of the organization itself. On an individual level, fear can cause physical and physiological disorders such as a rise in blood pressure or an increase in heart rate. Behavior changes, emotional problems, and physical ailments result from fear and stress generated in work situations, as do drug and alcohol abuse, absenteeism, and burnout. These maladies impact heavily on any organization. An employee participating in a climate of fear experiences poor morale, poor productivity, stifling of creativity, reluctance to take risks, poor interpersonal relationships, and reduced motivation to work for the best interests of the company. The economic loss to the company is immeasurable.

Management has a moral obligation to ensure the physical and emotional health of its employees for their well-being and that of the organization. Many managers use their power to create fear because they believe that the way to motivate employees is through coercion—but this fear impedes employee performance and is counterproductive. (Fear may emanate from lack of job security, possibility of physical harm, ignorance of company goals, shortcomings in hiring and training, poor supervision, lack of operational definitions, failure to meet quotas, blame for the problems of the system, and faulty inspection procedures. Management has control over these elements and is responsible for changing the organizational climate.) Driving out fear is significant in creating the quality environment, for fearful, anxious employees cannot participate in improving quality and productivity.

Point 9: Eliminate Barriers.   That barriers exist in organizations is a fact of corporate life. Organizations are not created with barriers; team spirit, unity, and cooperation are the initial cries of a newly-formed company. However, these attitudes quickly disappear as people's roles become functional and as problems in communication, competition, and fears arise. Barriers impede the smooth flow of the extended process, and everyone suffers, most notably the customer; rework and costs increase, and quality and customer satisfaction decrease.

Barriers exist in many places within the extended process. Competition, personal grudges, different views of a problem, and different priorities all lead to barriers between departments or areas within departments. Barriers also exist between levels in the hierarchy. Poor communication between employees and supervisors, supervisors and middle managers, and middle and upper management is common. There are also barriers between the firm and its vendors, the firm and its customers, union and management, and the firm and the community.

Breaking down barriers involves changing attitudes so that people identify with the organization's unifying goals instead of specialized department goals. They should see the company as part of an extended process, not as an isolated entity, and should cooperate as a team, rather than work for individual gains. Open communication and confronting the barriers are the first steps to breaking them down. Creating teams that work effectively is an aid in removing barriers; teams can be created either within or across departmental lines. Training in teamwork is critical.

Point 10: Eliminate Numerical Goals and Slogans. It has become common practice to set numerical goals for people in organizations; having a goal is believed to motivate the individual to achieve, and to clarify organizational expectations. (Unfortunately, it usually has the opposite effect.) Goals generally are arbitrarily set by someone for someone else, as in the case of a sales manager asking for a five percent increase in sales from the sales force. If the sales manager does not suggest new ways to achieve that goal, the plea is meaningless. Slogans and posters such as "Do it right the first time," "Safety is job one," "Increase return on net assets 3 percent next year," "Zero defects" do not help anyone improve performance—for they represent management's wishes for a desired result instead of action statements for the workers. Workers who are measured against goals they do not know how to achieve, and are judged by a management that will punish them for not meeting those goals, are tense, afraid, and resentful—certainly not motivated. If management wants to motivate people by hanging up posters, the emphasis should be on the progress that *management* is making in never-ending improvement. Statistical methods demonstrating this are appropriate mechanisms for communicating management's commitment to the Deming philosophy. These methods will be discussed in Chapters 8 through 14.

Point 11: Eliminate Work Standards and Quotas. *Work standards, measured day work,* and *piecework* are names given to a practice in U.S. industry that has had devastating effects on quality, productivity, and competitive position. A work standard is a specified level of performance determined by someone other than the person actually performing the task. Work standards and quotas consider only quantity—not quality—so they are completely at odds with the Deming philosophy.

The effects of setting quotas are negative. Quotas do not provide a road map for improvement, and they prohibit good supervision and training. Workers are blamed for problems in the system that are beyond their control and, in effect, are encouraged to produce defective items to meet the quota—with dire consequences. Workers are robbed of their pride. Although employees want to produce high quality goods and feel positively about themselves and their jobs, management will not let them.

Additional critical problems result from setting quotas too high or too low. Quotas set too high result in increased pressure on workers and leads to the production of more defective units. Worker morale and motivation decrease significantly because the system encourages the making of defective units. Quotas set too low are also damaging. For example, workers who have met their quota may spend the last hour of the day doing nothing, which destroys morale.

As work standards are generally established through union negotiation and have nothing to do with the process capability, they often are inaccurate. Changes in process capability are ignored, and the standards consequently do not reflect the potential of the current system. Workers may deliberately perform below the standard because they know that once they achieve it, it will be raised. Management frequently employs work standards in budgeting, planning, and scheduling, but the information obtained is invalid; these crucial activities would improve greatly if they were based on the process capability as determined by statistical methods.

Point 12: Restore Pride.   Many organizations in the United States do not use workers to their fullest potential, robbing them of their pride of workmanship and treating them as commodities. This loss of pride is an obstacle to achieving competitive advantage. Pride provides the impetus to perform better and to improve quality—for the worker's self-esteem; for the company; and, ultimately, for the customer. People enjoy taking pride in their work, but very few are able to do so because of poor management. Managers pay too little attention to the problems of workers, who in turn become disenfranchised, uninvolved, and underutilized.

Other factors contribute to the loss of pride of workmanship. Often, for example, employees do not understand the company's mission, so are confused and do not identify with the organization. Moreover, they are often forced to act as automatons, unable to think or use their skills. Employees likewise may be continually blamed for the system's problems. Products may be hastily designed and inadequately tested, which translates into production of low-quality merchandise—''junk.'' Inadequate supervision and training foster fear, incompetence, and production of defective units—and bad feelings about workmanship. Faulty equipment, materials, and methods hinder workers' performance and lessen positive feelings about work and the organization.

Regardless of the source of loss of pride in workmanship, organizations that help employees reclaim pride in work will reap tremendous benefits. Maximizing the potential of the work force and creating loyalty, excitement, interest, and team cooperation are crucial to never-ending improvement.

Point 13: Education and Training.   Education and training are essential to clarifying new jobs and responsibilities, readying employees for future jobs, and preventing burnout. Education must be provided to ensure a smooth transition as jobs incorporate the use of statistical methods. Training and retraining should be geared to prepare employees for changes in procedures, materials, machines, techniques, quality characteristics, and operational definitions.

As products and services are improved, the organization must determine innovative ways to meet customer needs. Management must develop new products and services and invest in appropriate job retraining so employees can qualify for the new job opportunities created.

Education and retraining can help prevent employee burnout as new information and involvement in process improvement stimulate job interest. Employees who no longer look to their jobs as a source of fulfillment may feel renewed enthusiasm as a result of education and training. It sends the message that management wants them to grow and develop.

Training in the Deming philosophy is the first step in adopting the Deming methods. Everyone in the organization should receive this training, beginning with top management. Statistical education at all levels is also necessary to prepare employees to implement these methods. Training in fields related to the employee's current job, personal improvement, and retraining for the jobs of the future should be made available.

Point 14: Organizing for Quality. An organization's top management must commit to the organizational transformation. Setting the change process in motion involves the recognition that problems exist as well as a desire to create a new organizational environment, one in which the never-ending improvement of quality is the primary goal. Top management has to begin by creating a critical mass of people in the organization who understand the philosophy and want to change the corporate culture. Organizing for quality involves setting up a structure that incorporates statistical methods and never-ending improvement in every aspect of the organization. The structure of the company is very important and has to support the integrated implementation of the 14 points. Statistical guidance, either internal or external, is necessary for success in this endeavor. It takes time, often several years, before an organization begins to realize the major benefits of using these methods.

QUALITY IN SERVICE

The Census shows that 86 percent of Americans work in service organizations or service functions in manufacturing organizations. As so many people work in service in the United States, improvement in our standard of living is highly dependent on better quality and productivity in the service sector.[14]

Service organizations and service functions in manufacturing organizations require never-ending improvement of their extended processes. Inefficient service increases price to the customer and lowers quality.[15]

A denominator common to manufacturing and any service organization is that mistakes and defects are costly. The further a mistake goes without correction, the greater the cost to correct it. The cost of a defect that reaches the consumer or recipient may be the costliest of all.[16]

The list in Figure 1.7 highlights some actual administrative (service) applications of analytical studies in the Ford Motor Company.[17]

The principles and methods for process improvement are the same in service and manufacturing: the 14 points apply equally to both sectors of the U.S. economy.

FIGURE 1.7   Service Applications in the Ford Motor Company

Organization	*Application*
Central Laboratory	Time to process request of customer
	Errors in laboratory (based on audits)
Power train and chassis engineering	Time for supplier to notify company of failure
	Number of failures per month
Ford parts and service division	Errors in filling orders for dealer
Accounting	Time to process travel expense
Ford tractor operations engineering	Time to process engineering changes
Manufacturing staff, manufacturing engineering and systems	Time to review report on productivity sent from various Ford locations
Computer graphics	Variation in time of usage of discs
Product engineering office	Number of computer sign-on calls that gave busy signal
	Number of times that files (cabinets) were used for information
Product development, comptroller's office	Run chart on number of revisions in word processing
	Wasted man-hours resulting from late start of meetings
Comptroller's office	Errors in accounts payable, resulting in late payment of invoices to suppliers
Saline plant	Costs from errors in scheduling
Purchasing staff, transportation, and traffic office	Transit time by rail from manufacture of part to assembly plants
Transmission and chassis division	Errors in shipments of components to assembly plants (quantity, wrong parts)

SUMMARY

Quality, a judgment by customers or users of a product or service, also encompasses the never-ending improvement of a firm's extended process—the internal processes along with those associated with customers, suppliers, investors, employees, and the community. Three types of quality (design/redesign, conformance, and performance) are integral to improvement of the extended process.

Issues of quality have existed since tribal chiefs, kings, and pharaohs ruled. The modern history of quality is marked by great advances between 1920 and the 1950s by George Edwards, Walter Shewhart, W. Edwards Deming, Armand Feigenbaum, and Joseph Juran. The 1970s and 80s have witnessed foreign competition regularly threatening U.S. companies. The response has involved renewed emphasis on quality control. W. Edwards Deming, Joseph Juran, and Armand Feigenbaum are among the leaders in this area; this book focuses on the philosophy and methods of W. Edwards Deming.

Top management, responsible for the never-ending improvement of quality, must understand the three types of quality, the relationship between quality and productivity, and the benefits of improving quality. The quality environment of a firm is critical and stresses teamwork and communication. Deming's 14 points for management provide a guide to creating and establishing the quality environment through behavioral change and using statistical methods to continually improve the process. Subsequent chapters focus on the statistical methods and techniques that facilitate quality improvement in service and manufacturing organizations.

EXERCISES

1.1 Discuss the concept of "extended process." How should this concept affect an employee's view of a job?

1.2 Define: *quality of design, quality of conformance,* and *quality of performance.*

1.3 Discuss the two sources of loss in quality.

1.4 Describe the relationship between quality and productivity. Do efforts at productivity improvement help or hurt quality? Why? Do efforts at quality improvement help or hurt productivity? Why? Construct examples for both cases.

1.5 List the benefits of process improvement.

1.6 List and briefly describe Dr. Deming's Fourteen Points.

1.7 Describe the Deming cycle and discuss its importance to never-ending improvement.

1.8 List some service applications for never-ending improvement in your organization.

1.9 Discuss the similarity in process improvement efforts in service and manufacturing.

NOTES

1. Howard S. Gitlow, "Definition of Quality," *Proceedings—Case Study Seminar—Dr. Deming's Management Methods: How They Are Being Implemented in the U.S. and Abroad* (Andover, Mass.: G.O.A.L., November 1984) pp. 4–18.
2. H. J. Harrington, "Quality's Footprints in Time," IBM Technical Report, TR 02.1064, September 20, 1983, p. 7.
3. Ibid, p. 8.
4. Ibid, p. 8.
5. Howard S. Gitlow and Shelly J. Gitlow, *The Deming Guide to Quality and Competitive Position* (Englewood Cliffs, N.J.: Prentice Hall, 1987), p. 7.
6. Harrington, "Quality's Footprints in Time," pp. 13–20.
7. Gitlow and Gitlow, *The Deming Guide,* pp. 29–31.
8. Ibid., pp. 9–10
9. W. Edwards Deming, *Out of the Crisis* (Cambridge, Mass.: MIT Center for Advanced Engineering Studies, 1986).
10. Gitlow and Gitlow, *The Deming Guide.*
11. Deming, *Out of the Crisis,* pp. 88–89.
12. W. Scherkenbach, *The Deming Route to Quality and Productivity: Road Maps and Roadblocks* (Washington, D.C.: Ceepress Books, 1986), pp. 35–40.
13. Kaoru Ishikawa, *What Is Total Quality Control? The Japanese Way* (Englewood Cliffs, N.J.: Prentice Hall, 1985), pp. 60–71.
14. Deming, *Out of the Crisis,* pp. 184–85.
15. Deming, Ibid., p. 183.
16. Deming, Ibid., p. 190.
17. Deming, Ibid., pp. 197–98.

Chapter 2

Fundamentals of Statistical Studies

STATISTICS DEFINED

Statistics can be broadly defined as the study of numerical data to better understand the characteristics of a population or process. A skilled statistician extracts a clear picture from numerical information, which aids in decision-making. The goal of a statistical study, therefore, is to provide a basis for action on a population or process.[1]

TYPES OF STATISTICAL STUDIES

Statistical studies can be broken down into two types: enumerative and analytic. *Enumerative studies* are statistical investigations that lead to action on a static population; that is, a group of units that exist in a given time period and/or in a given location.

An example of an enumerative problem is the estimation at a point in time of the number of residents in a certain geographical area and the inventory or yield of wheat, rice, or other grain to determine the amount of food that must be shipped in to feed the residents or the amount that could be shipped out without detriment to health. Dynamic questions—such as why the people are where they are or why they produce a certain crop—are not considered.[2]

Other examples of enumerative studies include determining market share for Product X in households with a certain demographic profile; calculating statistics on births, deaths, income, education, and occupation, by area; estimating the frequency of an illness in a given city; and assaying samples from a barge of coal to determine an appropriate price. All of these examples are time specific and static; there is no reference to the past or the future.

Analytic studies are statistical investigations that lead to action on a dynamic process. They focus on the causes of patterns and variations that take place from year to year, from area to area, from class to class, or from one treatment to another[3]—for example, determining why the production of grain in an area is low and how it can be increased in the future.

Other examples of analytic studies include testing varieties of wheat to determine the optimal type for a particular area's future production; comparing the output of two machine types, over time, to determine if one is more productive; comparing ways of advertising a product or service to increase market share; and measuring the effect of an action, such as change in speed, temperature or ingredients, on an industrial process output.[4] All of these examples focus on the future, not on the present. Management uses the information gathered in the above examples to make dynamic decisions about a process: What type of wheat should be grown in the future to achieve a optimal yield? Should we replace Machine A with Machine B to achieve greater production? Is television advertising more effective than print media for increasing market share? If we change the ingredients in our product, can we improve its quality and lower its cost?

Both types of statistical studies focus on quality-related issues in subsequent chapters.

Enumerative Studies

Basic Concepts. As stated before, enumerative studies are statistical investigations that lead to action on static populations. A *population* (or universe) is the total number of units of interest that exist in a given time period and/or a given location. A population to be studied must be operationally defined by listing all of its units; this list is called a *frame*. The population and frame are assumed to be identical; if they are not, then bias and error can occur in any study results. A frame may differ from the population it is supposed to define for numerous reasons—omissions, errors, double counting. Regardless of the type of error, the difference between the frame and the population is called the *gap*.

A *sample* is a portion of the investigated frame selected as the source of information about that frame. For example, 100 randomly selected accounts receivable drawn from a list, or frame, of 10,000 accounts receivable constitute a sample. There are two basic types of samples: nonrandom and random.

Nonrandom samples are selected on the basis of convenience (*convenience sample*); the opinion of an expert (*judgment sample*); or a quota to ensure proportional representation of certain classes of items, units, people, and so on, in the sample (*quota sample*). All nonrandom samples have the same shortcoming—they lack a frame, thus their results are subject to an unknown degree of bias. Because the sample items are not randomly selected from an operationally defined frame, classes of items in the population may be systematically denied representation in the sample (thereby biasing the results). Suppose, for example, that a convenience sample is used to estimate the average salary in the manufacturing area of a company; work crews organize the list of names, and then every fifth name is selected as part of the sample. As work crews have organized the list, it first lists the foreman, then 4 workers, then another foreman, and so on. The sample will either contain all foremen or all workers, depending on the starting point. Thus the average salary will not reflect the true average because of the method of selecting the sample. Using nonrandom samples is advisable only when the cost of obtaining more accurate information is too high.

Random samples are selected so that every element in the frame has a known probability of selection. There are several types of random samples—*simple, stratified,* and *cluster*—which many basic statistics books discuss in detail. All random samples have the same strengths. They allow generalized statements to be made about the frame from the sample. These generalized statements form the basis for action on the population under study.

A random sample is selected by operationally defining a procedure that utilizes random numbers in selecting sampled items from the frame to eliminate bias and hold uncertainty within known limits. There are several steps:

Step 1: Count the number of elements in the frame. Call this number N.
Step 2: Number the elements in the frame from 1 through N. If N is 25, then the elements in the frame should be numbered from 01 through 25, as illustrated in Figure 2.1. All elements *must* receive an identification number with the same number of digits.

FIGURE 2.1   Identification Frame

Item	Identification Number
A	01
B	02
C	03
D	04
E	05
F	06
.	.
.	.
.	.
Y	25

Step 3: Select a page in Table 9, a table of random numbers. For example, select page 589.

Step 4: On the selected page of random numbers, select a column of numbers, select a starting point in that column, and use as many digits as there are digits in N (two digits in the case of N = 25). For example, beginning with the first column in Table 9 on page 589, selecting the seventh line of that column as the starting point, and using the first two digits of each number in that column, the first random number is 19. Figure 2.2(a) shows a list of 39 random numbers obtained in this way.

Step 5: Determine the necessary sample size. This calculation is discussed in many textbooks. Let n be this number. Assume n = 6 in this example.

Step 6: Select the first six two-digit numbers in the chosen column on the selected page that are between 01 and 25, inclusive. If a number is smaller than 01 (e.g., 00) or larger than 25 (e.g., 31) ignore the number and continue down the column. If an acceptable number appears more than once, ignore every repetition and continue

---

FIGURE 2.2   Selection of a Random Sample

(a)   Random Numbers from Pages 589 of Table 9

19	30	40
09	28	78
31	13	98
67	60	69
61	13	39
04	34	62
05	28	56
73	59	90
54	87	09
42	29	34
27	62	12
49	38	69
29	40	93

(b)   Random Sample Size n = 6 from Frame of Size N = 25

Sample Number	Identification Number	Items
1	19	S
2	09	I
3	04	D
4	05	E
5	13	M
6	12	L

moving down the column until six unique numbers have been selected. If the bottom of the page is reached before six unique random numbers are obtained, go to the top of the page and move down the next two-digit column. Figure 2.2(b) shows the six random numbers selected and the corresponding items comprising the sample.

Step 7: Finally, analyze the information as a basis for action.

Remember that different samples of six each may yield different results. The difference between the result that would be obtained from a complete count and that which is obtained from any sample is called the *sampling error*. Random samples, however, do not have bias; and increasing the sample size can hold the sampling error to any given limits. These are the advantages of random sampling compared with nonrandom sampling. Again, nonrandom samples are rarely used in enumerative studies—even though they are easier to implement—because they are inherently biased.

Conducting an Enumerative Study. The following steps present a guide for conducting an enumerative study. All steps are the same whether the study is a complete count or a sample.[5]

Step 1: Specify the reason(s) you want to conduct the study; for example, to estimate the average number of sick days per employee in the XYZ Company in 1988. If this average is greater than 8.0 days, a preventive health care plan will be instituted. If it is less than or equal to 8.0 days, the current plan will be maintained.

Step 2: Specify the population to be studied. In our example, the population would be all full-time employees in the XYZ Company in 1988; an employee designated full-time at any time during 1988 is considered full-time in 1988.

Step 3: Construct the frame—a list of all full-time employees. Everyone who will use the study's results as a basis for action must agree on the frame.

Step 4: Perform secondary research, such as the examination of prepublished data, to determine how much information is already available about the problem under investigation; for example, check the records kept by the Personnel Department.

Step 5: Determine the type of study to be conducted; for example, mail survey, personal interviews, or chemical analysis of units. In this example, we would survey employee absentee cards for 1988.

Step 6: Make it possible for the respondents to give clear, understandable information and/or for the researcher to elicit clear, understandable information. For example, the method for analyzing absentee cards should be clear and straightforward. Consider the problem of nonresponse—refusal to answer; no one at home; missing items to be studied are all possible causes of nonresponse. How significant is the nonresponse? Will it impair the study results? What can be done to reduce it? Establish a procedure for dealing with and reducing nonresponse problems. In this example, be sure no absentee cards are missing, and make sure the data gatherers understand how to interpret the absentee cards.

Step 7: Establish the sampling plan to be used, determine the amount of allowable error in the results, and calculate the cost of the sampling plan. At this stage, steps 1 and 2 may need revisions because of cost considerations. For example, we may decide to draw a simple random sample of employee absentee cards using random numbers, at a cost of one dollar per card, assuming an allowable error of one quarter day in the estimate.

Step 8: Establish procedures to deal with differences between interviews, testers, inspectors, and so on—for example, assessing the differences in collected data resulting from differences in the abilities of the data gatherers.

Step 9: Prepare unambiguous instructions for the data gatherers covering all phases of data collection. Supervisors may require special additional instructions. Conduct training for all data gatherers and supervisors. Use statistical methods to determine when their training is complete, as discussed in Deming's Point 6 of Chapter 1 on page 21.

Step 10: Establish plans for data handling, including format of tables, headings, and number of classes.

Step 11: Pretest the data-gathering instrument and data-gathering instructions. If pretests show a high refusal rate or generate unsatisfactory quality data, the study may be modified or abandoned. Conduct a "dry run" with the data gatherers to ensure they understand and adhere to established procedures. Revise the data-gathering instrument and instructions based on the information collected during the pretest. Put all aspects of the study's procedure into final form.

Step 12: Conduct the study and the tabulations. The study *must* be carried out according to plans. From the gathered data, calculate the sampling errors of interest; users of the study will then be able to understand the degree of uncertainty present in the study results. In our example, we would calculate the standard error for the average number of days absent per employee.

Step 13: Interpret and publish the results so that decision makers can take appropriate action. For example, if the average number of days absent per employee is greater than 8.0, then establish the preventative health are program.

In conducting an enumerative study, it is important to realize that Step 7 requires random sampling, not nonrandom sampling, so that the information can serve as a basis for action with a known degree of uncertainty. The result of a nonrandom sample in an enumerative study is worth no more than the reputation of the man who signs the report, because the margin of uncertainty in the estimate reported is entirely dependent on his knowledge and judgement, rather than on objective, quantifiable methods.[6]

Analytic Studies

Basic Concepts. Analytic studies are statistical investigations that lead to action on a dynamic process; that is, a process that has a past, present, and will have a future. Process improvement actions based on a process's past behavior are rational only if the process's past behavior was stable, hence predictable to the future. The assumption of stability implies that the process has an identity—therefore, the process will generate an identifiable population of output. This output population is called a *conceptual population*. The characteristics of a conceptual population generated from past outputs can be predicted to a conceptual population generated from future output only for stable processes.

A *conceptual frame* is a listing of all the outputs comprising the conceptual population, based on past data. It is important that randomization (using random numbers) is used in selecting items from the conceptual frame; otherwise, the resulting information may be biased and hence misleading. In essence, if the process under study does not exhibit a conceptual population, or if it is impossible to randomly study the items in the conceptual frame (for example, because items in the conceptual frame are at the bottom of a pile of items), then the information obtained from such a study may be misleading.[7]

Samples in analytic studies are almost always judgment samples. Nevertheless, judgment samples with appropriate randomization are amenable to the methods of *conditional inference,* or statistical inference dependent upon the conditions that existed at the time of the study. As these conditions may never exist again, we cannot generalize the results to other conditions. The ability to generalize requires knowledge of the process being studied and is not a statistical inference in the enumerative sense.

Randomization is critical in analytic studies because it permits the researcher to state, with a known probability of being wrong, that his results are significant or not, where we define in advance what constitutes a significant difference, as determined by an expert in the process. The results may be stated even though the same conditions under which the study was conducted may never be encountered again. The determination of a significant difference under any conditions is an important contribution to knowledge. The conditional inferences, when combined with knowledge of the process being studied, provide a basis for action on the process.

There are numerous situations in which judgment samples can be drawn in analytic studies; Chapters 8 through 10 discuss the procedure by which judgment samples are drawn with appropriate randomization to allow conditional inference.

Conducting an Analytic Study. Subsequent chapters detail the steps required to conduct an analytic study, a major focus of this text.

Distinguishing Between Enumerative and Analytic Studies

A simple rule exists for deciding whether a study should be enumerative or analytic.[8] If a 100 percent sample of the frame answers the question under investigation, the study is enumerative; if not, the study is analytic. A 100 percent sample can be obtained in an enumerative problem because items are drawn from the frame, which is composed of all the elements in the population. A 100 percent sample cannot be obtained in an analytic problem because the conceptual frame contains items existing in the future, which, of course, cannot be sampled.

STATISTICAL STUDIES AND QUALITY

Statistical studies are critical components of quality improvement efforts. It is important to understand the different types of statistical studies when conducting research for the improvement of quality of design, quality of conformance, and quality of performance, so that the information generated by the study can be used rationally as a basis for quality improvement action.

SUMMARY

The goal of a statistical study is to provide a basis for action on a population or process. There are two types of statistical studies: 1) enumerative, which lead to action on a static population, and 2) analytic, which lead to action on a dynamic process.

There are two basic types of sampling plans, random and judgment. Random samples are required to provide useful information from an enumerative study; judgment samples with appropriate randomization serve the same function for analytic studies.

Knowing whether a study should be enumerative or analytic is essential, for each type of study requires different sampling and computational techniques. A simple rule is: if a 100 percent sample of the frame answers the question under investigation, the study is enumerative; if not, the study is analytic. It is important to understand the different types of statistical studies so that the results can be used for quality improvement actions.

EXERCISES

2.1 List the two types of statistical studies. What is the purpose of any statistical survey? Describe each type of statistical study. List three examples of each type of study in quality improvement efforts.

2.2 Define the following terms: *population, frame, gap, nonrandom sample, judgment sample, random sample, conceptual population, conceptual frame*.

2.3 List the steps required to draw a simple random sample.

2.4 List the steps required to perform an enumerative study.

2.5 Discuss why judgment samples cannot be used as a rational basis for action in enumerative studies.

2.6 Discuss why judgment samples can be used as a rational basis for action in analytic studies.

2.7 Specify a rule to distinguish enumerative from analytic studies.

NOTES

1. W. Edwards Deming, *Some Theory of Sampling* (New York, N.Y.: John Wiley and Sons, 1950), p. 247.
2. Ibid., pp. 247–48.
3. Ibid., p. 249.
4. W. Edwards Deming, "On the Use of Judgement-Samples," *Reports of Statistical Applications*, Vol. 23, March 1976, p. 26.
5. Deming, *Some Theory of Sampling*, pp. 4–9.
6. Deming, "On the Use of Judgement-Samples," p. 29.
7. Donald Wheeler and David Chambers, *Understanding Statistical Process Control* (Knoxville, Tenn.: Statistical Process Controls, Inc., 1986), pp. 42–43.
8. Deming, "On the Use of Judgement-Samples," p. 26.

Chapter 3

Documenting and Defining a Process

INTRODUCTION

Analytic studies lead to action on a process. Before an analytic study can be performed, however, the process must be documented and defined. This chapter focuses on the procedures used to define and document a process.

A PROCESS DEFINED

What exactly is a process? We have used the term previously in our discussion of analytic studies, but now let's be more specific about its definition.

A *process* is the transformation of *inputs* (manpower/services, equipment, materials/goods, methods, and environment) into *outputs* (manpower/services, equipment, materials/goods, and methods). The transformation involves the addition or creation of value in one of three aspects: time, place, or form. To illustrate these value aspects:

- Time value: Something is available *when* it is needed; for example, you have food when you are hungry, or material inputs are ready on time.
- Place value: Something is available *where* it is needed; for example, gas is in your tank, not in Saudi Arabia, or woodchips are in a paper mill.
- Form value: Something is available *how* (in the form in which) it is needed; for example, bread is sliced so it can fit in a toaster, or paper has three holes so it can be placed in a binder.

Figure 3.1 illustrates a basic process.

FIGURE 3.1   Basic Process

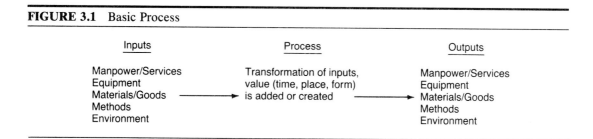

Inputs	Process	Outputs
Manpower/Services	Transformation of inputs,	Manpower/Services
Equipment	value (time, place, form)	Equipment
Materials/Goods	is added or created	Materials/Goods
Methods		Methods
Environment		Environment

Processes exist in all aspects of organizations, and understanding them is very important. Many people mistakenly think of only production processes when *process improvement* is mentioned. However, administration, sales, service, human resources, training, maintenance, paper flows, interdepartmental communication, and vendor relations are all processes that can be examined and improved using statistical methods. It is critical that we approach process improvement holistically—related to every area of the organization.

A firm is a multiplicity of micro subprocesses, all synergistically building to the macro process of that firm. It is important to realize that all processes have customers and suppliers. These customers and suppliers can be internal or external to the organization. A customer can be an end user or the next operation down the line; the customer does not have to be human. A vendor, which also does not have to be human, could be another firm supplying subassemblies or services, or the prior operation up the line.

The Feedback Loop

An important aspect of processes is the *feedback loop,* which relates information about outputs back to the input stage so that an analysis of the transformation process can be made. Figure 3.2 depicts the feedback loop in relation to a basic process.

FIGURE 3.2   Feedback Loop

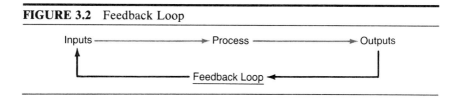

Inputs ——————→ Process ——————→ Outputs

—————— Feedback Loop ←——

The tools and methods discussed herein provide vehicles for relating information about outputs to the input stage. Transmitting this information aids decision making about process improvement. A major purpose of an analytic study is to provide the information (flowing through a feedback loop) needed for process improvement.

Examples of Processes

An example of a well-known process is the hiring process, depicted in Figure 3.3. The inputs, which include the candidate's resume; the information gathered from the interview with the candidate; and other data from references, former employers, and schools attended are transformed into the output, an employee to fill a vacant position. The process of transforming inputs into output includes synthesizing and evaluating the information, making a decision, and hiring the candidate. An important aspect of this process is the feedback loop, which enables the supervisor of the new employee to report back to the personnel decision maker concerning the appropriateness of the employee. The personnel decision maker/supplier and supervisor/customer can use this information to work together to improve the hiring process.

Clerical functions are also processes. Figure 3.4 depicts the process of sending a memo in an organization. First, the manager inputs the information for the memo and the instructions regarding its distribution. The secretary then transforms the input, by typing it and distributing it, into the output, which is communication to the employees. The feedback loop is important because the manager and the employees can work together to improve the communication process.

Figure 3.5 illustrates a generic production process. The inputs (component parts, machines, and operators) are transformed in the process of

FIGURE 3.3   Hiring Process

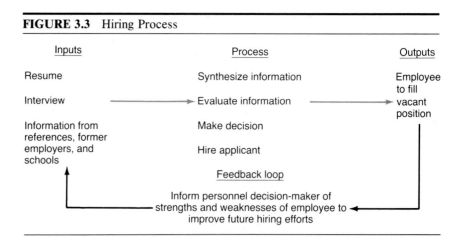

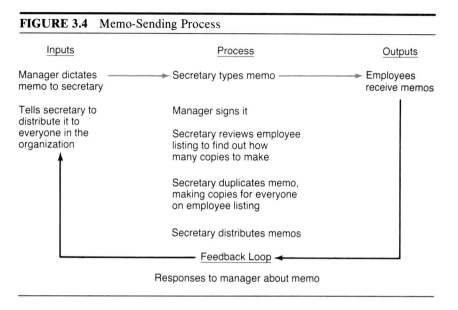

FIGURE 3.4 Memo-Sending Process

making the final product and shipping it to the customer. The output is the customer receiving the product. Again, the feedback loop, which, in this case is the customer's reporting back to the supplier on the product's performance, is extremely important. Communicating this way promotes cooperation on process improvement.

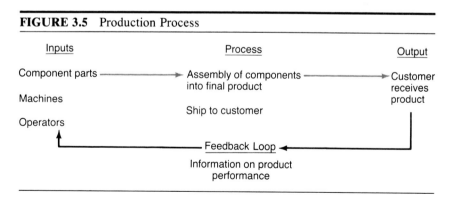

FIGURE 3.5 Production Process

DOCUMENTING A PROCESS

Improving a process is the ultimate goal of process documentation. The following example demonstrates why processes documentation is an important step in process improvement.

In a study of an industrial laundry, an analyst began to diagram the flow of paperwork using a flowchart. While walking the "green" copy of an invoice through each step of its life cycle, he came upon a secretary feverishly transcribing information from the "green" invoice copy into large black looseleaf books. He asked her what she was doing so he could record it on his flowchart. She responded: "I'm recording the numbers from the green papers into the black books." He then asked, "What are the black books?" "I don't know," she said. He continued: "What are the green papers?" She said, "I don't know." He asked, "Who looks at the black books?" She answered, "Nobody." He then asked, "How long have you been doing this job?" She responded, "Seven and one half years." He finally asked, "What percentage of your time do you spend working on the black books?" Her answer: "About 60 percent."

The analyst examined the black books, then asked the secretary how long ago the person who hired and trained her had left the company. She said that the person who hired her had left the company seven years ago. At this point, the consultant realized he had solved a problem the company didn't know it had. On examination he had determined that the black books were sales analysis books. Every day all sales, by item, were entered into the appropriate page in the black book; and at the end of each month, page totals were calculated, yielding monthly sales, by items. Nobody was looking at the books because nobody knew what the secretary was doing. The current manager assumed the secretary was doing something important. After all, she seemed so busy.

This is an example of failure to document a process to ensure that it is logical, complete, and efficient. As an epilogue, the secretary was reassigned to other needed duties, for the sales analysis had been computerized five years earlier.

The Process Documentation Procedure

It is important to ask the following questions when first documenting a process:

1. Who owns the process? Who is responsible for process improvement?
2. What are the boundaries of the process?
3. What is the flow of the process?
4. What are the objectives of the process? How is the success of the process in meeting objectives being measured?
5. Are the measurements being taken on the process valid?

Thinking about the above questions will enable an individual or group to document a process; the following paragraphs discuss each of the questions.

Who Owns the Process? Every process must have an owner—that is, an individual responsible for the state and improvement of the process. As process owners may have responsibilities extending beyond their departments, they must be high enough in the organization to have influence over the resources necessary to implement process improvements.

In some cases, the process owner is the focal point for the functions that operate within his process. Nevertheless, each function is still controlled by the line management within that function. The process owner will require representation from each function that comprises the process. These individuals are assigned by the line managers of the subprocesses, provide functional expertise to the process owner, and are the implementors of changes within their functions. The owner is the "voice" of the process throughout the organization.

What Are the Boundaries of the Process? Boundaries must be set for a process under study to make it easier to establish process ownership and highlight the process's key interfaces with other (customer/vendor) processes. Process interfaces frequently are the source of process problems stemming from confusion about downstream requirements. Constructing operational definitions for critical process characteristics agreed upon by process owners on both sides of a process boundary will go a long way towards eliminating process interface problems. (Operational definitions are discussed in depth later on in this chapter.)

What Is the Flow of the Process? A *flowchart* is a pictorial summary of the flow of the various operations of a process used to document that process. Figure 3.6 is an example of a flowchart for a quality-of-design study.

Figure 3.6 depicts the elements in a type of quality-of-design study called a *production design study*. In the design phase, the design engineers determine detailed specifications for the product and prepare a full-scale prototype of the product. They then test and evaluate the design as either "good" or "bad." If it is bad it returns to the design stage for

FIGURE 3.6 Quality-of-Design Study

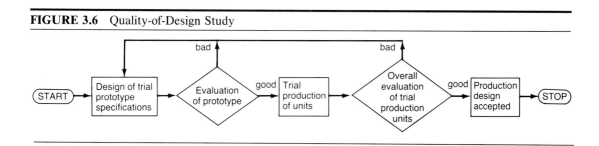

redesign. If it is good, it moves to the next phase, trial production. After this stage, the product is again evaluated and if judged bad, returns to the design stage. If it is good, the production design is accepted and full-scale production begins. During appropriate stages of the process, the results are fed back to the designers. Also, overall evaluations determine how the product compares with that of competitors and whether it surpasses consumers' expectations.

After a flowchart has been constructed, it can provide important information about a process that can help a manager, designer, analyst, or anyone seeking to understand, modify, or improve that process.

Flowchart Symbols. The American National Standards Institute, Inc. (ANSI), has approved a standard set of flowchart symbols, used in documenting a process. The shape of the symbol and the information written within the symbol provide insight into that particular step of the process. Figure 3.7 presents the basic symbols for flowcharting.[2] Using these symbols standardizes process documentation for better communication in process improvement.

FIGURE 3.7   Flowchart Symbols

Basic input/output symbol

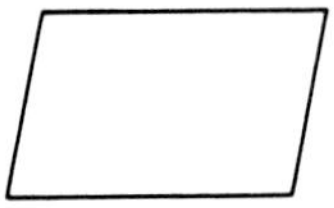

The general form that represents input or output media, operations, or processes is a parallelogram.

Basic processing symbol

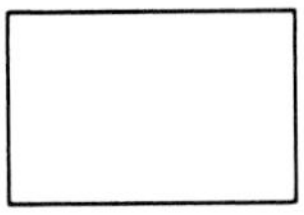

The general symbol used to depict a processing operation is a rectangle.

Decision symbol

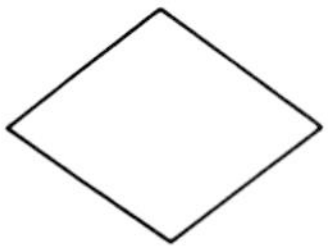

A diamond is the symbol that denotes a decision point in the process. This includes attribute type decisions such as pass-fail, yes-no. It also includes variable type decisions such as which of several categories a process measurement falls into.

Flowline symbol

A line with an arrowhead is the symbol that shows the direction of the stages in a process. The flowline connects the elements of the system.

Start/stop symbol

The general symbol used to indicate the beginning and end of a process is an oval.

Types of Flowcharts. This section describes two types of flowcharts that are used in documenting a process.[3]

Systems Flowchart. A systems flowchart is a pictorial summary of the sequence of operations that make up the process. The flowchart depicted in Figure 3.6, an example of a systems flowchart, represents the phases or stages in the process of a quality-of-design study; it pictorially represents each of the operations, using the standard flowcharting symbols.

Layout Flowchart. A layout flowchart depicts the floor plan of an area, usually including the flow of paperwork or goods and the location of equipment, file cabinets, storage areas, and so on. These flowcharts are especially helpful in improving the layout to utilize the space more efficiently. Figure 3.8 shows a layout flowchart, before and after flowcharting analysis and flow improvement. Figure 3.8(a) shows the existing system, and Figure 3.8(b) illustrates the new proposed system. This flowchart includes the flow of work, along with the floor plan and location of desks and files. Comparing the existing and proposed systems is simple when the flow of the process is documented this way.

FIGURE 3.8 Layout Flow Chart

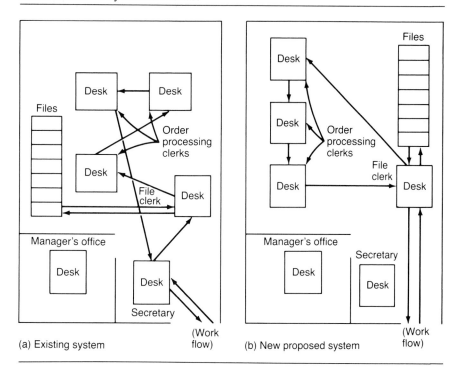

(a) Existing system   (Work flow)   (b) New proposed system   (Work flow)

Advantages of Flowcharts. Flowcharting a process, as opposed to using written or verbal descriptions, has several advantages.

Functions as a Communications Tool. A flowchart provides an easy way to convey ideas among engineers, managers, hourly personnel, vendors, and others in the extended process. It is a concrete, visual way to represent complex systems.

Functions as a Planning Tool. Flowcharts greatly aid process designers. Flowcharts enable them to visualize the elements of new or modified processes and their interactions while still in the planning stages.

Provides an Overview of the System. It is easy to view the critical elements and steps of the process in the context of a flowchart. It removes unnecessary details and breaks down the system so designers and others get a clear, unencumbered look at what they are creating.

Defines Roles. A flowchart demonstrates the roles of personnel, work stations, and system subprocesses. It also identifies the personnel, operations, and locations involved in the process.

Demonstrates Inter-Relationships. Flowcharts show how the elements of a process relate to each other.

Promotes Logical Accuracy. Flowcharts enable the viewer to easily spot errors in logic. Planning becomes easier because designers have to clearly break down all of the elements and operations of the process.

Facilitates Troubleshooting. Flowcharts are an excellent diagnostic tool and help detect problems in the process, failures in the system, or barriers to communication.

Documents a System. A flowchart provides a record of a system that enables anyone to easily examine, understand, or change that system.

A flowchart is a widely utilized tool that can be applied in any type of organization to aid in process documentation and, ultimately, improvement.

Guidelines for Systems Type Flowcharts. Flowcharts are simple to use, provided that certain guidelines are followed, in keeping with standard practices.[4]

1. Draw the flow chart from the top of a page to the bottom and from left to right.
2. Carefully define and clarify the activity being flowcharted.
3. Determine where the activity starts and ends.
4. Describe each step of the activity using single-verb descriptions; for example, ''prepare statement'' or ''design prototype.''

5. Keep each step of the activity in proper sequence.
6. Carefully observe the scope or range of the activity being flowcharted. Do not draw in any branches that leave the activity being charted.
7. Use the standard flowcharting symbols.

Using Flowcharts to Facilitate Process Improvement. After a flowchart has been constructed, it can provide useful information about a process. This information can help a designer, analyst, manager, or anyone seeking to improve the process. Finding opportunities to better the system begins with indicating strong and weak points on the flowchart. For example, the communication between the engineering and research and development (R&D) departments might be a strength; but the engineering department's relationship with the production department might be a weakness. Indicating points beyond the analyst's control also helps in problem solving. For example, if the board of directors sets policy and refuses to change a regulation, nothing can be done. Noting points that actually affect the customer (internal or external) will also aid process improvement. For example, is the marketing department feeding back information on customer satisfaction to the production, engineering, design, and R&D departments? Indicating the strengths and weaknesses in the process, the points out of the analyst's control, and the elements actually affecting customers makes process improvement possible.

What Are the Objectives of the Process? The process owner is responsible for clearly stating process objectives that are consistent with organizational objectives. An example of an organizational objective: Strive to continually provide customers with higher quality products/services at an attractive price that will meet their needs. By adapting this organizational objective to the process objectives a process owner can find meaning and a starting point for stating process objectives. For example, a process owner in Purchasing could translate the above organizational objective into the following subset of objectives:

1. Continuously monitor Purchasing's customers (e.g., maintenance department, administration department) to determine their needs regarding:
   a. Number of days from purchase request to item/service delivery.
   b. Ease of filling out purchasing forms.
   c. Satisfaction with purchased material.
2. Continuously train and develop purchasing department personnel with respect to job requirements; for example, take measurements on the following as a basis for process improvement:
   a. Number of errors per purchase order.
   b. Number of minutes to complete a purchase order.

Whatever the process objectives, all persons involved with that process must understand its objectives and devote their efforts to meeting *those* objectives. Clearly stating the process objectives helps everybody involved with the process work toward the same goals.

Are the Measurements Being Taken on the Process Valid?

Operational Definitions. Management's (process owner's) attempts to improve a documented process must comprise precise definitions of process objectives and process subobjectives, including precise definitions of specifications, products/services, and jobs. Such definitions bring about increased communication and improve process and organizational definitions. They are a prerequisite for understanding among process members. Operational definitions used for collecting data must have the same meaning to everyone so that the organization can use them as a basis for action.

It is useful to illustrate the confusion that can be caused by the absence of operational definitions. The label on a shirt reads "75% cotton." What does this mean? Three quarters cotton, on the average, over this shirt, or three quarters cotton over a month's production? What is three quarters cotton? Three quarters by weight? If so, at what humidity? By what method of chemical analysis? How many analyses? Does 75 percent cotton mean that there must be some cotton in any random cross-section the size of a silver dollar? If so, how many cuts should be tested? How do you select them? What criterion must the average satisfy? And how much variation between cuts is permissible? Obviously, the meaning of "75% cotton" must be stated in operational terms, otherwise confusion results.

As another example, one operation in a production process is a deburring operation; hence, it is reasonable to ask for the definition of a *burr*. The supervisor in charge of the deburring operation was asked for a definition and stated that a *burr* is a bump or protrusion on a surface (a *surface* is a face). Also, he said, "Deburring's five inspectors all have at least 15 years' experience and certainly know a burr when they see one."

A test was conducted to determine if the definition of a burr was consistent among all five inspectors. Ten parts were drawn from the production line and placed into a tray so that each part could be identified by a number; each of the inspectors was shown the tray and asked to determine which parts had burrs. Figure 3.9 records the results.

Although inspectors B and C always agree, they always disagree with inspector E. Inspector A agrees with inspectors B and C 40 percent of the time, with inspector E 60 percent of the time, and with inspector D 50 percent of the time. Inspector D agrees with inspectors, B, C, and E 50 percent of the time. The above does not paint a pretty picture. The absence of an operational definition of a burr creates mayhem. The deburring department's manager (process owner) and inspectors have no concept of their jobs. This creates fear (Deming's point 8) and steals their pride of workmanship (Deming's point 12).

FIGURE 3.9 Identification of Burrs on 10 Parts

	Inspector				
Part Number	*A*	*B*	*C*	*D*	*E*
1	0	1	1	1	0
2	0	1	1	0	0
3	1	1	1	1	0
4	1	1	1	0	0
5	0	1	1	1	0
6	0	1	1	0	0
7	1	1	1	1	0
8	1	1	1	0	0
9	0	1	1	1	0
10	0	1	1	0	0

Legend: 0 = no burr on part
1 = burr on part

The Importance of Operational Definitions.   Major problems can arise when process measurement definitions are inconsistent over time, or when their application and/or interpretation are different over time. Employees may be confused about what constitutes their job. Also, major problems between customers and suppliers may result from the absence of agreed-upon operational definitions. Endless bickering and ill will are inevitable results.

Operational definitions establish a language for process improvement and put communicable meaning into a process, product, service, job, or specification. Specifications such as "defective," "safe," "round," "5 inches long," and so on have no communicable meaning until they are operationally defined. Everyone concerned must agree on an operational definition before process improvement can proceed.

A given operational definition is neither right nor wrong; its importance lies in its acceptance by all extended process members. As conditions change, the operational definition may change to meet new needs.

Establishing Operational Definitions.   An operational definition consists of: 1) a criterion to be applied to an object or to a group, 2) a test of the object or of the group, and 3) a decision as to whether the object or group did or did not meet the criterion. The three components of an operational definition are best understood through some examples.

A firm produces washers. One of the critical quality characteristics is roundness. The following procedure is one possible procedure to use to arrive at an agreed-upon operational definition of roundness.

Step 1: Criterion for roundness.

Buyer:

Use calipers that are in reasonably good order. (You perceive at once the need to question every word.)

Seller:

What is "reasonably good order"? (We settle the question by letting you use your calipers.)

Seller:

But how should I use them?

Buyer:

We'll be satisfied if you just use them in the regular way.

Seller:

At what temperature?

Buyer:

The temperature of this room.

Buyer:

Take 6 measurements of the diameter about 30 degrees apart. Record the results.

Seller:

But what is "about 30 degrees apart"? Don't you mean exactly 30 degrees?

Buyer:

No, there is no such thing as exactly 30 degrees in the physical world. So try for 30 degrees; we'll be satisfied.

Buyer:

If the range between the six diameters does not exceed .007 centimeters, we will declare the washer to be round. (They have determined the criterion for roundness.)

Step 2: Test of roundness.

a. Select a particular washer.
b. Take the six measurements and record the results in centimeters: 3.365, 3.363, 3.368, 3.366, 3.366, and 3.369.
c. The range is 3.369 to 3.363, or a 0.006 difference. They test for conformance by comparing the range of 0.006 with the criterion range of less than or equal to 0.007 (Step 1).

Step 3: Decision on roundness.

Because the washer passed the prescribed test for roundness, they declare it to be round.

If a seller has employees who understand what round means and a buyer who agrees, the problems the company may have had satisfying the customer will disappear.[5]

As another illustration, a salesperson is told that her performance will be judged with respect to the percentage of change in this year's sales over last year's sales. What does this mean? Average percentage change each month? Each week? Each day? For each product? Percentage change between December 31, 1986, and December 31 1987, sales?

How are we measuring sales—gross, net, gross profit, net profit? Is the percentage change in constant or inflated dollars? If it is in constant dollars, is it at last year's prices or this year's prices and under what economic conditions?

A loose definition of percentage change will lead to confusion, frustration, and ill will between management and the sales force—hardly the way to improve productivity. How should management operationally define a percentage change in sales?

Step 1: Criterion for percentage change in sales.

A percentage change in sales is the difference between 1987 (January 1, 1987, to December 31, 1987) sales and 1986 (January 1, 1986, to December 31, 1986) sales divided by 1986 sales:

Percentage change (86, 87) = (S87 − S86)/S86

where

S87 = dollar sales volume for the period January 1, 1987, through December 31, 1987, and

S86 = dollar sales volume for the period January 1, 1986 through December 31, 1986.

S86 is measured in constant dollars, with 1985 as the base year, using June 15, 1985, and June 15, 1986, prices to derive the constant dollar prices and total unit sales less returns (resulting from any cause) as of December 31, 1986, for each product.

$$S86 = \sum_{i=1}^{m} [(P_i85/P_i86)(TS_i86 - R_i86)]$$

where

m = the number of products in the product line,
P_i85 = the price of product i as of June 15, 1985,
P_i86 = the price of product i as of June 15, 1986,
TS_i86 = the total unit sales for product i between January 1, 1986 and December 31, 1986, and
R_i86 = the total of unit returns (for any reason) for product i between January 1, 1986 and December 31, 1986.

R_i86 recognizes that products sold late in 1986 that may be returned will be reflected in R_i87, next year's return for product i. S87 is measured in constant dollars, with 1985 as the base year, using June 15, 1985, and June 15, 1987, prices to derive the constant dollar prices and total unit sales less returns (for any reason) as of December 31, 1987, for each product. (P_i85 remains the same for all products.)

$$S87 = \sum_{i=1}^{m} [(P_i85/P_i87)(TS_i87 - R_i87)]$$

where all items are defined as in S86 with the appropriate shift in time frame. This procedure for computing the percentage change in sales between 1986 and 1987 will be in effect, regardless of economic conditions. Further, management may revise the definition of a percentage change in sales after the 1990 sales evaluation but not before, unless the sales force and sales management agree.

Steps 2 and 3: Test and decision on percentage change in sales.

a. The salesperson and her sales records are the object under study.
b. The sales manager will use all 1986 and 1987 invoices and sales return slips to compute the net number of units sold for each product in 1986 and 1987. The sales manager will record the computations and results.

The prior definition of sales might not suit another manager and sales force; however, if the sales manager adopts it and the sales force understands it, it is an operational definition.

Operational definitions are not trivial; if management doesn't operationally define many critical variables and attributes so that workers as well as customers agree, serious problems will follow. Statistical methods become useless tools in the absence of operational definitions. It is management's responsibility to operationally define relevant quality characteristics to eliminate this problem.

SUMMARY

In this chapter we have discussed documenting and defining a process. A process is the transformation of inputs into outputs that involves the addition of value in time, place, or form. Processes exist in all aspects of all organizations, including administration, sales, training, vendor relations, and production. Two important aspects of processes are that they all have customers and suppliers and that they all need feedback loops for improvement. A process must be documented—broken down into its elements—before it can be improved. Flowcharts provide a means to break down a process and help anyone seeking to improve that process.

Operational definitions bring about increased communication and an ability to move toward process and organizational objectives by providing precise meanings for specifications, products/services, jobs, and so on; they establish a language for process improvement. If operational definitions are not used or are not agreed upon by all concerned, serious problems can occur.

EXERCISES

3.1 What is a process boundary? Why should process boundaries be established?

3.2 What is a flowchart? List and describe the different types of flowcharts.

3.3 Construct a flowchart for a simple process that requires using at least two flowcharting symbols.

3.4 State several examples of process objectives.

3.5 List and explain the three parts of an operational definition. Why are operational definitions important?

3.6 Construct an operational definition for a specification of your choice. Is your operational definition right? What makes your operational definition right and another operational definition wrong?

NOTES

1. E. J. Kane, "IBM's Quality Focus on the Business Process," *Quality Progress,* April 1986, pp. 26–33.
2. G. A. Silver and J. B. Silver, *Introduction to Systems Analysis* (Englewood Cliffs, N.J.: Prentice-Hall, 1976), pp. 142–47.
3. J. M. Fitzgerald, and A. F. Fitzgerald, *Fundamentals of Systems Analysis* (New York: John Wiley & Sons, 1973), pp. 230–37.
4. Fitzgerald and Fitzgerald, *Fundamentals of Systems Analysis,* pp. 227–84.
5. H. Gitlow and P. Hertz, "Product Defects and Productivity," *Harvard Business Review,* September–October 1983, pp. 138–139.

PART II

Foundations for Analytic Studies

This section presents the underlying foundations for analytic studies, which include both the probability and statistics required for managers to think statistically. Thinking statistically is key to developing the ability to manage by data rather than guesswork and opinion.

Chapter 4 outlines the concepts and axioms of probability. Chapter 5 summarizes the basic notions of statistics. Chapter 6 discusses important probability models used in analytic studies; and Chapter 7 analyzes the fundamentals of sampling distributions, which provide the foundation for the construction of control charts in analytic studies.

Chapter 4

Basic Probability

INTRODUCTION

Almost everyone makes statements such as: "It will probably rain tomorrow" or "Chances are that Bert will outperform Ernie at this new task." These are simple, intuitive statements of probability that usually do not require any further explanation. For example, it would be unusual if someone said, "Exactly what probability do you attach to rain tomorrow?" or, "Exactly what probability do you attach to Bert outperforming Ernie at this new task?"

Consider the following statements, however: "The probability is 0.993 that the case hardness depth of a camshaft will be between 3.5 mm and 10.5 mm, provided that the case hardening operation is in statistical control"; or, "The market share for Brand X widgets among males younger than 18 years of age in Florida is nine percentage points." These are more complex statements and may require further explanation. For example, someone might ask, "What do you mean by statistical control?" or, "What do you mean by market share?" Such questions must be answered if these complex probabilities are to be understood and be useful for decision-making purposes.

PROBABILITY DEFINED

The two examples of complex probabilities above highlight the two traditional definitions of probability, the classical definition and the relative frequency definition.

The Classical Definition of Probability

The *classical* approach to probability states that if an experiment has N equally likely and mutually exclusive (that is, the occurrence of one event precludes the occurrence of another event) outcomes, and if n of those outcomes correspond to the occurrence of event A, then the probability of event A is:

$$P(A) = \frac{n}{N} \qquad (4.1)$$

where

   n = the number of experimental outcomes that correspond to the occurrence of event A, and

   N = the total number of experimental outcomes.

For example, tossing a fair die results in one of six equally likely and mutually exclusive outcomes occurring (N = 6). If we are interested in the probability of tossing an even number (e.g., 2 or 4 or 6, n = 3), then the probability of tossing an even number on a die is:

$$P(Even) = \frac{3}{6} = 0.50000$$

The Relative Frequency Definition of Probability

The *relative frequency* approach to probability states that if an experiment is conducted a large number of times, then the probability of event A occurring is:

$$P(A) = \frac{k}{M} \qquad (4.2)$$

where

   M = the maximum number of times that event A could have occurred during these experiments, and

   k = the number of times A occurred during these experiments.

For example, suppose in the die-tossing example that a fair die was tossed one hundred thousand times and that a two or four or six appeared in 50,097 throws. Consequently, the relative frequency probability of tossing an even number on a fair die is:

$$P(Even) = \frac{50,097}{100,000} = 0.50097$$

The classical definition of the probability of an even toss and the relative frequency definition of an even toss differ as a result of the methods used in their respective calculations.

   Analytic studies are conducted to determine process characteristics. Still, process characteristics have a past and present and will have a future; hence, classical probabilities have no frame for computation.

Probabilities concerning process characteristics must be obtained empirically through experimentation and must therefore be relative frequency probabilities.

The preceding camshaft example typifies the relative frequency approach to probability. No frame of events exists from which to make probabilistic statements because we are dealing with a process that has a past and present and will have a future. We must rely on the relative frequency approach to probability to describe the process' behavior.

As another illustration, a newly hired worker is supposed to perform an administrative operation, entering data from sales slips into a computer terminal. Is it possible to predict the percentage of sales slips per hour entered into the computer by the worker that will be in error? Unfortunately not; the best we can do is to train workers properly, then observe them performing their jobs over a long period of time. If a worker's efforts represent a stable system (more on this in Chapters 8 through 10), we can compute the relative frequency of sales slips entered with errors per hour as an estimate of the probability of a sales slip's being entered with errors. This empirical probability can be used to predict the percentage of sales slips per hour entered with errors.

Calculating Classical and Relative Frequency Probabilities

A bin contains 4,000 screws; 3,000 are good, and 1,000 are defective. We will use this bin of screws to compare the two definitions of probability.

Calculating the Classical Probability.   The classical definition of the probability of drawing a defective screw is, from Equation 4.1, 1,000/(1,000 + 3,000) = 0.250.

Calculating the Relative Frequency Probability.   The relative frequency definition of the probability of drawing a defective screw is very different. Suppose we place the screws in the bin, mix them thoroughly, and draw one screw at random. We record the status of the screw (0 if good and 1 if defective) and replace the screw in the bin. We again thoroughly mix the screws, draw another screw, and record the number 0 or 1. Suppose we repeat this process 1,250 times. Further, suppose we put the numbers into groups of 50 and record the number of defective screws in each group, as shown in the first four columns of Figure 4.1.

If we examine more samples of size 50, we will note that the fluctuations in the cumulative fraction defective diminish because the estimate of the cumulative fraction of defective screws is based on a larger sample. The last three columns of Figure 4.1 and the whole of Figure 4.2 illustrate this. From Figure 4.2, we see that after accumulating 23 subgroups the fraction defective of screws appears to have stabilized around 0.208.

FIGURE 4.1 Cumulative Samples of 50 Screws

1	2	3	4	5	6	7
Subgroup Number	Subgroup Size	Number of Defective Screws in Subgroup	Fraction of Defective Screws in Subgroup	Cumulative Number of Defective Screws	Cumulative Number of Screws	Cumulative Fraction of Defective Screws
1	50	10	.20	10	50	.200
2	50	9	.18	19	100	.190
3	50	15	.30	34	150	.227
4	50	7	.14	41	200	.205
5	50	9	.18	50	250	.200
6	50	17	.34	67	300	.223
7	50	13	.26	80	350	.229
8	50	8	.16	88	400	.220
9	50	10	.20	98	450	.218
10	50	7	.14	105	500	.210
11	50	7	.14	112	550	.204
12	50	11	.22	123	600	.205
13	50	9	.18	132	650	.203
14	50	9	.18	141	700	.201
15	50	8	.16	149	750	.199
16	50	9	.18	158	800	.198
17	50	12	.24	170	850	.200
18	50	17	.34	187	900	.208
19	50	8	.16	195	950	.205
20	50	7	.14	202	1000	.202
21	50	17	.34	219	1050	.209
22	50	9	.18	228	1100	.207
23	50	11	.22	239	1150	.208
24	50	11	.22	250	1200	.208
25	50	10	.20	260	1250	.208
		Total: 260				

This number is the relative frequency estimate of the probability of drawing a defective screw. We use the term *stabilized* to be synonymous with the concept of a constant system of variation. In other words, the subgroup fraction defective is within stable and predictable limits. Chapters 8 through 10 will extensively discuss the concept of stability.

Comparing the Probabilities. Figure 4.3 compares the two definitions of probability. We have seen that the classical probability of selecting a defective screw from the bin is 0.250. The relative frequency probability of selecting a defective screw—or the long-run cumulative fraction defective—is 0.208. In the absence of a complete frame, hence in the absence

FIGURE 4.2   Cumulative Fraction of Defective Screws

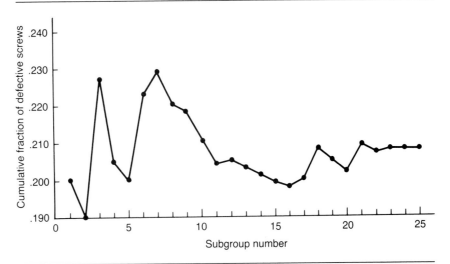

of the classical fraction defective in the bin, the best we can do is to use the most recent cumulative fraction defective.

It is important to consider why the relative frequency approach did not yield a 0.250 probability of selecting a defective screw. The difference between the two probabilities indicates that the mechanical procedure of mixing and selecting the screws under the relative frequency definition

FIGURE 4.3   Comparison of the Two Definitions of the Probability of a Defective Screw

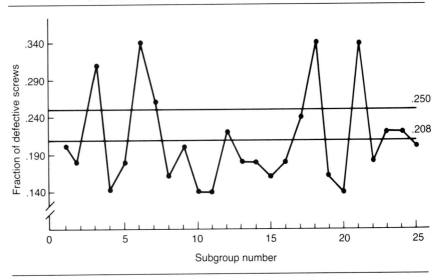

must have been biased against defective screws. A possible explanation of this bias is that defective screws are broken and consequently smaller. As a result, they either fall to the bottom of the bin during mixing or present a smaller target for selection. If the mixing and selecting procedure used under the relative frequency approach were truly random, the two probabilities would be the same on average.

RULES OF CLASSICAL PROBABILITY

Sets

Classical probability is defined with respect to a frame, or a finite set of events. A *set* is a collection of objects, such as people or products, described by listing its members. If, for example, we wanted to describe the set of employees in Arco's cost department, we would write:

A = {Jones, Hopper, Klein, Parker, Smith}

Any portion of a set may itself be a set. Sets that are wholly contained in other sets are known as *subsets*. If A = {Jones, Hopper, Klein, Parker, Smith} and B = {Hopper, Parker}, then B is a subset of A; symbolically, B ⊂ A. More formally, set B is a subset of A if every element in B is also an element of A.

The set containing all of the elements in the population is known as a universal set, and labeled *U*. For example, if U is the set of all employees of the Arco Company, the elements of the universal set would be a list of all employees' names. A set containing no elements at all is known as the empty or null set and is labeled ϕ.

Often we are interested in elements that are members of one set, or another set, or both. For example, we might be looking for all elements that belong to set A, or set D, or both. This is the *union* of sets A and D and is written A ∪ D.

If A = {Jones, Hopper, Klein, Parker, Smith}, and
D = {Jones, Church, Dole}, then
A ∪ D = {Jones, Hopper, Klein, Parker, Smith, Church, Dole}.

Notice that even though Jones was included in both A and D, it was only necessary to list her once in the new set.

At times we are interested in the elements of both of two overlapping sets; if an element is a member of only one set but not the other, it is not a member of both. The elements common to both comprise the *intersection* of the two sets, which is written as A ∩ D.

If A and D are defined as above, then
A ∩ D = {Jones}

because Jones is the only element that appears in both sets.

If the intersection of two sets is the null set, the sets are said to be *mutually exclusive*. For example, if S = {1, 2, 3} and T = {4, 5}, then S ∩ T = φ, and S and T are mutually exclusive sets.

The *Venn diagram* is often helpful in the examination of sets and the populations they describe. In a Venn diagram, the area of a rectangle represents the universe of interest. Portions of this area, denoted by circles, represent subsets of the universe.

Consider all Arco employees to constitute the universal set, U. If we are interested in two subsets of U, the employees in the cost department and the employees who serve on the quality committee, the Venn diagram would be as shown in Figure 4.4(a).

FIGURE 4.4 Venn Diagram

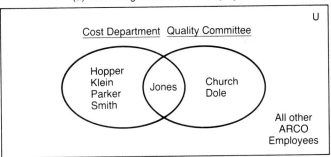

(a) Venn Diagram for ARCO Employees

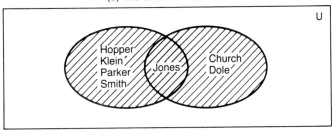

(b) The Union of Two Sets

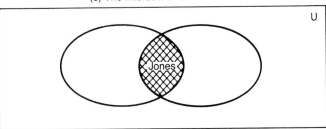

(c) The Intersection of Two Sets

The union of the two sets, consisting of all employees who are in the cost department or who serve on the quality committee, or both, is shown in the hatched area of Figure 4.4(b). The intersection—those employees who are in the cost department *and* serve on the quality committee—is shown in the hatched area of Figure 4.4(c).

Consider a firm employing 100 salesmen, 35 of whom deal exclusively with accounts of $5,000 or less, 26 of whom deal only with accounts of $5,001 or more, 12 of whom deal with accounts represented by East Coast companies only, and the rest of whom do business with any account.

If we let M = {salesmen dealing only in accounts of $5,000 or less}, N = {salesmen dealing only in accounts of $5,001 or more}, and P = {salesmen dealing only with East Coast accounts}, we could draw a Venn diagram to depict our salesforce, as shown in Figure 4.5.

The single-hatched shaded region indicates the salesmen who sell only to accounts of $5,000 or less and on the East Coast, represented by M ∩ P, and the double-hatched shaded region indicates salesmen who sell to accounts of $5,001 or more and on the East Coast, represented by N ∩ P.

If we adopt the notation that A′ (read "A prime") represents elements not belonging to a set A, or the *complement* of A, we can represent salesmen who sell to accounts of $5,000 or less and not on the East Coast by M ∩ P′. It should be obvious that we can now represent any characteristic or combination of characteristics as a portion of our Venn diagram.

Axioms of Classical Probability

Classical probability is defined with respect to a finite set of events. Interest centers on taking action on the finite amount of material (items, people, etc.,) in the frame being studied, at a given point in time.

FIGURE 4.5   Venn Diagram for 100 Salesmen

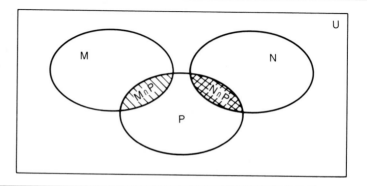

Suppose that the events in a frame can be grouped into subsets called $A_1, A_2, \ldots, A_n$ such that $A_1 \cup A_2 \cup \ldots \cup A_n = U$; that is, the union of the subsets includes all elements of the frame or universe. We assign to each subset, A_i, a nonnegative real number, $P(A_i)$, which can take on values between 0 and 1 inclusive.

$$0 \le P(A_i) \le 1 \qquad (4.3)$$

The nonnegative real number $P(A_i)$ is called the *probability that an event belongs to A_i*. We base the computation of $P(A_i)$ on the classical definition of probability.

The probability that an event belongs to some subset of the frame is 1:

$$P(A_1 \cup \ldots \cup A_n) = 1 \qquad (4.4)$$

The above probability axioms are appropriate for a finite set of events.

To illustrate, consider a one-time purchase of a lot of 1,000 component parts delivered to a factory. The plant manager must ascertain the fraction defective to determine how much to pay for the lot. This is an enumerative problem; action is taken on the lot. Using an operational definition (recall Chapter 3) of what constitutes a good part, the plant manager inspects the entire lot of component parts and divides the parts into two sets: A_1 is made up of defective parts, and A_2 is made up of good parts. The inspection reveals that A_1 has 86 units and A_2 has 914 units. Hence, the classical probability of a defective unit, $P(A_1)$, is, from Equation 4.1,

$$86/(86 + 914) = 0.086$$

Therefore, 8.6% of the incoming lot is defective, and the manager sets the price for the lot based on this information.

RULES OF RELATIVE FREQUENCY PROBABILITY

Relative frequency probability is defined with respect to an infinite set of events. The set is considered infinite because it is generated by a process that has a past, present, and future; it is assumed to continue forever. The probability of selecting any single point in an infinite set of points is zero. Consequently, there is no classical set (frame) from which probability estimates can be made. No frame exists for future events; thus, we must find another way to estimate probabilities.

Axioms of Relative Frequency Probability

Suppose that the events generated by a process can be grouped into subsets called $A_1, A_2, \ldots, A_n$. As in the classical case, we can assign to each subset A_i a nonnegative real number $P(A_i)$ that can take on values between 0 and 1 inclusive, as in Equation 4.3:

$$0 \leq P(A_i) \leq 1$$

The nonnegative real number $P(A_i)$ is called the *probability that an event belongs to A_i*. We base the computation of $P(A_i)$ on the relative frequency definition of probability.

The probability that an event belongs to some subset is 1, as in Equation 4.4:

$$P(A_1 \cup A_2 \cup \ldots \cup A_n) = 1$$

For subsets $A_1, A_2 \ldots, A_n$, which are mutually exclusive (that is, if any element belongs to one of the subsets, it does not belong to any other), the probability of any of the subsets occurring is shown in Equation 4.5:

$$P(A_1 \cup A_2 \cup \ldots \cup A_n) = P(A_1) + P(A_2) + \ldots + P(A_n) \quad (4.5)$$

The above probability axioms are appropriate for an infinite set of events.

To illustrate, a process generates cartons of tuna fish (events) shipped to four different regions (subsets) in the United States:

Region (subset)	Subset Code
Northeast (NE)	A_1
Southeast (SE)	A_2
Middle (M)	A_3
West (W)	A_4

If our concern is solely with past shipments, a frame exists from which we can make classical probability statements. For example, Figure 4.6 depicts a 20-week history of shipments to the regions. The probability that a carton of tuna fish was shipped to the Southeastern United States in the 20-week period is, from Equation 4.1:

$$549/(1081 + 549 + 1594 + 1803) = 549/5027 = 0.109$$

However, if we want to predict the probability that a carton of tuna fish will be shipped to a particular region in the future, we use the cumulative fraction defective for probability estimates, assuming the process is stable. The last column of Figure 4.7 shows that the cumulative fraction defective—from which we estimate the probability that a carton of tuna fish will be shipped to the Southeastern United States—is 0.109.

FIGURE 4.6	Shipments of Cartons of Tuna Fish in a 20-Week Period				
	Cartons Shipped by Region				*Total Cartons*
Week	*NE*	*SE*	*M*	*W*	*Shipped*
1	55	25	85	95	260
2	61	20	79	85	245
3	49	21	76	95	241
4	41	23	79	99	242
5	59	30	82	90	261
6	67	32	83	96	278
7	63	22	82	97	264
8	58	21	74	87	240
9	41	22	77	89	229
10	44	33	75	86	238
11	47	34	74	89	244
12	64	38	76	87	265
13	59	24	82	91	256
14	62	34	88	92	276
15	48	24	86	88	246
16	56	30	79	83	248
17	44	27	85	98	254
18	56	38	84	86	264
19	58	26	70	88	242
20	49	25	78	82	234
Totals	1,081	549	1,594	1,803	5,027

MARGINAL, JOINT, AND CONDITIONAL PROBABILITY

Probabilities may be classified as marginal probabilities, joint probabilities, or conditional probabilities.

Marginal Probability

We define the *marginal probability* that event A will occur as the number of times A occurs divided by the total number of events, denoted by P(A).

In analytic studies, we compute marginal probabilities by determining the number of units in set A for a given time period (say a day); dividing this quantity by the total number of units of interest in the same time period; repeating this process for a long period of time; and determining if the process is stable (Chapters 8 through 10 discuss this). If it is stable, we then estimate the marginal probability by summing all items in set A over all the days in question and dividing this number by the total number of items of interest for all the days (see Equation 4.2). If the marginal prob-

FIGURE 4.7 Shipments of Cartons of Tuna Fish

1	2	3	4	5	6	7	8	9	10
					Total Cartons Shipped	Fraction of Cartons Shipped to SE	Cumulative Number of Cartons Shipped to SE	Cumulative Total Number of Units Shipped	Cumulative Fraction of Units Shipped to SE
Week	NE	SE	M	W					
1	55	25	85	95	260	.096	25	260	.096
2	61	20	79	85	245	.082	45	505	.089
3	49	21	76	95	241	.087	66	746	.088
4	41	23	79	99	242	.095	89	988	.090
5	59	30	82	90	261	.115	119	1249	.095
6	67	32	83	96	278	.115	151	1527	.099
7	63	22	82	97	264	.083	173	1791	.097
8	58	21	74	87	240	.088	194	2031	.096
9	41	22	77	89	229	.096	216	2260	.096
10	44	33	75	86	238	.139	249	2498	.100
11	47	34	74	89	244	.139	283	2742	.103
12	64	38	76	87	265	.143	321	3007	.107
13	59	24	82	91	256	.094	345	3263	.106
14	62	34	88	92	276	.123	379	3539	.107
15	48	24	86	88	246	.098	403	3785	.106
16	56	30	79	83	248	.121	433	4033	.107
17	44	27	85	98	254	.106	460	4287	.107
18	56	38	84	86	264	.144	498	4551	.109
19	58	26	70	88	242	.107	524	4793	.109
20	49	25	78	82	234	.107	549	5027	.109
Total	1,081	549	1,594	1,803	5,027				

ability stream is not stable, we must take steps to achieve process stability. If the marginal probability stream is stable, we can use the cumulative fraction defective to estimate the probability of future occurrence of event A.

Joint Probability

We define the *joint probability* that events A and B will occur together as the number of times A and B occur together divided by the total number of events; this is denoted by P(A ∩ B).

In analytic studies, we view joint events as the simultaneous occurrence of events A and B. In other words, we call the simultaneous occurrence of events A and B "event C"; and event C is treated as a marginal event. For example, an examination of the first 1,000 sales invoices for MCG, Inc., in 1987 yielded information on the size of sales and the industry in which a sale was made, as shown in Figure 4.8.

Assuming the selling process is stable with respect to sale size and industry, then the joint probability of a sale "more than $5,000" and in the "automobile" industry is:

P(More than $5,000 ∩ Automobile industry) = 250/1000 = 0.25

In an analytic study, we treat the above joint event as a marginal event that occurs when a sale is more than $5,000 and in the automobile industry.

Conditional Probability

We define the *conditional probability* of event A, given that event B has already occurred, as the number of times A and B occur together divided by the number of times event B occurs, denoted by P(A|B). In other

---

FIGURE 4.8   The First 1,000 Sales Invoices in 1987, MCG, INC.

Industry Classification	Size of Sale		Total
	More than $5,000	*Less than or Equal to $5,000*	
Automobile	250	50	300
Other	200	500	700
TOTAL	450	550	1000

words, we are setting a condition on the population by looking only at set B:

$$P(A|B) = P(A \cap B)/P(B) \qquad (4.6)$$

In analytic studies, conditional events are viewed as the occurrence of event A in a particular trial given event B has already occurred in the said trial. In other words, the occurrence of A given B in a trial is called *event C,* and event C is treated as a marginal event. For example, a reexamination of the first 1,000 sales invoices for MCG, Inc. presented in Figure 4.8 (assuming stability as before) yields the conditional probability of a sale of more than \$5,000 to a firm in the automobile industry as:

P(More than \$5,000|Automobile industry) = 250/300 = 0.833

In an analytic study, we treat the above conditional event as a marginal event that occurs when a sale of more than \$5,000 is made to a firm, given the firm is in the automobile industry.

PROBABILITY THEOREMS

The Addition Theorems

The addition theorems are important because they allow for the computation of the probability of whether a randomly selected event belongs to one or more of a group of sets.

Theorem 4.1: If two sets are mutually exclusive, the probability of the union, or the probability of belonging to one set, or the other set, or both sets, is the sum of the individual probabilities. Symbolically,

$$P(A_1 \cup A_2) = P(A_1) + P(A_2) \qquad (4.7)$$

Theorem 4.2: If two sets are not mutually exclusive, the probability of the union, or the probability of belonging to one set, or the other set, or both sets, is the sum of the individual probabilities minus the probability of the intersection, or the probability of belonging to both sets. Symbolically,

$$P(A_1 \cup A_2) = P(A_1) + P(A_2) - P(A_1 \cap A_2) \qquad (4.8)$$

where $P(A_1 \cap A_2)$ refers to the joint probability of belonging to both sets. $P(A_1 \cap A_2)$ is subtracted so that the events in the subset $A_1 \cap A_2$ are not double-counted, once in $P(A_1)$ and a second time in $P(A_2)$.

The above theorems can be extended to any number of sets. The extension to three sets is illustrated below:

$$P(A_1 \cup A_2 \cup A_3) = P(A_1) + P(A_2) + P(A_3) \text{ for mutually exclusive sets}$$
$$P(A_1 \cup A_2 \cup A_3) = P(A_1) + P(A_2) + P(A_3)$$
$$- P(A_1 \cap A_2) - P(A_1 \cap A_3) - P(A_2 \cap A_3)$$
$$+ P(A_1 \cap A_2 \cap A_3) \text{ for non-mutually exclusive sets}$$

FIGURE 4.9 Absentee Data for Two-Employee Work Station

		Fred		
		Working (B₁)	*Absent (B₂)*	*Total*
Max	*Working (A₁)*	494	1	495
	Absent (A₂)	3	2	5
	Total	497	3	500

Classical Application of the Addition Theorems. To illustrate the classical interpretation of the addition theorems, consider a work station in a factory that employs two people, Max and Fred, to do the required work. On any given day, one of four possible working conditions may exist. Max may be working, but Fred may be absent; Fred may be working, but Max may be absent; they both may be absent; or they both may be working. Approximately two years of absentee data (500 days) were collected to show the frequency of occurrence for the above events, as shown in Figure 4.9.

From this information and Equation 4.8, we can see that the classical probability of Fred's or Max's being absent is:

$$P(A_2 \cup B_2) = P(A_2) + P(B_2) - P(A_2 \cap B_2)$$
$$= 5/500 + 3/500 - 2/500 = 6/500$$
$$= 0.012$$

It is important to point out that the probability just computed (0.012) is an enumerative statistic, valid for the time period studied. It is useful for future decision-making purposes (in an analytic study), but only if the absentee patterns of Fred and Max are stable over time.

Relative Frequency Application of the Addition Theorem. Let us consider Max's and Fred's absenteeism patterns over time. If we plotted the proportion of days per month each man was absent (assuming an average of 30 days per month) and applied the statistical methods to be discussed in Chapters 8 through 10, we would see that both Max's and Fred's absentee patterns are stable over time. Hence, $P(A_2)$ and $P(B_2)$ can be used to predict Fred's and Max's future absentee behavior. Further, if we plotted the proportion of days per month both men were absent and applied the statistical methods of Chapters 8 through 10, we would see that the joint absentee patterns (the event that both are absent) for Fred and Max are stable over time. Consequently, we can use $P(A_2 \cap B_2)$ to predict Fred's and Max's future joint absentee behavior.

From Equation 4.8, then, we can see that the probability of Fred's or Max's being absent is:

$$P(A_2 \cup B_2) = P(A_2) + P(B_2) - P(A_2 \cap B_2)$$
$$= 5/500 + 3/500 - 2/500 = 6/500$$
$$= 0.012$$

Consequently, we can predict that either Fred or Max (or both) will be absent 1.2 percent of every month on average.

The Multiplication Theorems

The multiplication theorems are important because they indicate whether sets are statistically independent or dependent. Sets are statistically independent if the occurrence of an element from one set does not affect the probability of occurrence of an element from another set. We can state statistical independence symbolically as:

$$P(A_i|B_j) = P(A_i) \tag{4.9}$$

In other words, the occurrence of B_j in no way affects the occurrence of A_i. The above statement of statistical independence leads to the following theorems.

Theorem 4.3: The joint probability of A_i and B_j is equal to the probability of A_i times the conditional probability of B_j given A_i. Symbolically, we can state this as:

$$P(A_i \cap B_j) = P(A_i)\ P(B_j|A_i) \tag{4.10}$$

Theorem 4.4: A_i and B_j are statistically independent events if and only if the joint probability of A_i and B_j is equal to the probability of A_i times the probability of B_j. Symbolically:

$$P(A_i \cap B_j) = P(A_i)\ P(B_j) \tag{4.11}$$

We can extend Theorem 4.4 to more than two sets.

Theorem 4.5: A_i, B_j, and C_k are statistically independent sets if and only if the joint probability of A_i and B_j and C_k is equal to the probability of A_i times the probability of B_j times the probability of C_k. Symbolically, we can state Theorem 4.5 as:

$$P(A_i \cap B_j \cap C_k) = P(A_i)\ P(B_j)\ P(C_k) \tag{4.12}$$

Classical Application of the Multiplication Theorems. Returning to the two-person factory work station, the probability that Max and Fred are absent is, from Figure 4.9,

$$P(A_2 \cap B_2) = 2/500 = 0.004$$

As the probability that Max is absent times the probability that Fred is absent is

$$P(A_2) \cdot P(B_2) = 5/500 \cdot 3/500 = 15/250,000$$
$$= 0.00006$$

we can say that Max's and Fred's absentee status are not independent:

$$P(A_2 \cap B_2) \neq P(A_2) \, P(B_2)$$

Again, the above probability is enumerative. If we want to use it to predict future behavior, in an analytic study, we must determine whether it is a stable probability, as illustrated in the Addition Rule example.

Relative Frequency Application of the Multiplication Theorems. If marginal and joint event patterns are stable over time, then we can predict statistical independence into the future. Symbolically, if P(A), P(B), and P(A $\cap$ B) are found to be stable over time, then we can predict whether or not P(A $\cap$ B) = P(A) $\cdot$ P(B) over time.

SUMMARY

In this chapter, we have presented and discussed the classical and relative frequency definitions of probability. The former allows us to calculate the probability of an event for enumerative studies, while we use the latter in analytic studies.

We introduced the notion of a set as a collection of objects, and we presented the fundamental elements and operations of set theory, using Venn diagrams for illustration. Given these concepts, the axioms of probability can be defined.

We examined the marginal, joint, and conditional probabilities and discussed the Addition Theorems and Multiplication Theorems in light of classical and relative frequency probability concepts.

EXERCISES

4.1 Consider all 500 employees of the XYZ Company as the universal set under consideration. Denote all male employees by the set A; all employees who are college graduates by the set B; and all employees younger than 35 by set C. Describe in words the elements of:
   a. A'
   b. A $\cap$ B
   c. A $\cup$ C
   d. A' $\cap$ (B $\cup$ C)

4.2 A local store has 96 suitcases in stock, 24 of which were manufactured by company M, 60 by company N, and the rest by company P. What is the probability of selecting at random a suitcase which was:
   a. manufactured by company N?
   b. not manufactured by company P?

   c. manufactured by either company M or company P?

   d. manufactured by neither company M nor company N?

4.3 Based on past history, the probability that product A conforms to specifications is 0.8 and the probability that a nonrelated product, product B, does not conform to specifications is 0.3. What is the probability that

   a. both products conform?

   b. product A conforms, but product B does not?

   c. neither product A nor product B conforms?

   d. product A does not conform, but product B does?

   e. Add the probabilities calculated in *a, b, c,* and *d* and comment on the result.

4.4 A radio manufacturer is known to produce, on average, 2 percent defective radios. A random sample of three radios was taken from his warehouse. What is the probability that

   a. all three were defective?

   b. none was defective?

4.5 A company manufactures two models of hand-held calculators, model K and model L. From previous experience, they know that the probability that the firm will be unable to meet demand for model K in October is 0.64 and the probability that they will be unable to meet demand for model L in October is 0.36. Furthermore, the probability of being unable to meet demand for both K and L in October is 0.24.

   a. Draw a Venn diagram of this system.

   b. What is the probability of being unable to meet demand for either model K or model L in October?

4.6 A motorist has noticed that 8 percent of the time, when he fills his tank up with Brand S of gasoline, he experiences engine knock. He does, however, like to save money, so he frequently buys his fuel at a "generic" gas station. This unaffiliated station buys odd lots of gasoline from various suppliers, and about 20 percent of the time it is Brand S. The motorist also knows that on other brands of gas (not Brand S), he will experience engine knock 10 percent of the time.

   After filling up one morning, the motorist notices engine knock.

   a. Define and calculate all marginal and joint probabilities.

   b. Find the probability that the motorist has purchased Brand S given that his engine is knocking.

4.7 You have been newly appointed as the plant manager in a factory that produces socks. On your first day you are presented with the following production history.

Grade	Number of Socks Produced January 1985–December 1987	
Firsts (good)	73,197	(72.21%)
Seconds (sell at lower price)	17,877	(17.64%)
Scrap (junk)	10,286	(10.15%)
Total	101,360	

a. What does the above information tell you about the factory's capability to produce good socks (firsts) in the future?
b. Would more information about the above figures improve their usefulness? What information?
c. Assuming that the production process is stable:
   1. What is the probability of the process turning out a "first" sock? A "second" sock? A "scrap" sock?
   2. What is the probability of the process turning out either a second or junk sock?
   3. Suppose you took two socks randomly from the production process; what is the probability that they both will be seconds? Will be seconds or worse?
   4. Suppose you took five socks randomly from the production process; what is the probability that they all will be seconds? Will be seconds or worse?

4.8 Seven years of departmental accident data for an industrial laundry are recorded below. The data have been analyzed chronologically (by day) and found to exhibit stable patterns of variation with respect to accidents per day.

| Department | Number of Accidents | | | | | |
	Mon.	Tue.	Wed.	Thu.	Fri.	Total
Receiving/Shipping	12	8	9	8	13	50
Washroom	30	19	21	27	33	130
Drying	10	10	11	9	9	49
Ironing & Folding	22	20	21	19	20	102
Total	74	57	62	63	75	331

a. What is the probability that an accident will occur in the Washroom on Monday? In Ironing & Folding on Thursday?

b. What is the probability that an accident will occur on Friday? In Receiving/Shipping?

c. What is the probability that an accident will occur in Drying, given that it is Tuesday? In the Washroom, given it is Wednesday?

d. Classify the types of probabilities you computed in *a*, *b*, and *c*. Which are marginal probabilities? Which are joint probabilities? Which are conditional probabilities?

e. What is the probability that an accident will occur on a Monday or Friday? On a Tuesday, Wednesday, or Thursday?

f. What is the probability that an accident will occur in the Washroom or Ironing & Folding?

g. Are "department an accident occurs in" and "day of the week" statistically independent? Why?

4.9 An accounting clerk is responsible for computing the effective number of hours worked daily (regular time + (1.5 × overtime)) for each of 500 production workers and for recording the number of days absent each week on the time card for each worker. The above two functions are the job characteristics upon which she is judged (computing effective daily hours per worker and recording number of days absent each week per worker).

The accounting supervisor draws weekly random samples of 50 time sheets to monitor and help improve the clerk's performance. The following chart reflects 30 weeks of analysis.

| Recorded Number of Days Absent per Week | Computation of Effective Hours per Week | | Total |
	Correct (no mistakes)	Incorrect (one or more mistakes)	
Correct (recorded properly)	1,235	190	1,425
Incorrect (recorded improperly)	65	10	75
Total	1,300	200	1,500

a. Can the accounting supervisor use the above information to help the clerk improve her performance? Why?

b. What assumptions are required for the supervisor to be able to use the above figures to aid him in his supervisory efforts?

c. Assuming that the accounting clerk's work is in statistical control in relation to both quality characteristics:

1. What is the probability that the effective hours will be computed correctly? The number of days absent will be recorded correctly?
2. What is the probability that a time card will be completed correctly? Incorrectly?
3. What is the probability that the number of days absent will be recorded incorrectly, given that the number of effective hours was computed incorrectly?
4. Classify the types of probability you calculated in *1*, *2*, and *3* above. Which are marginal, joint, or conditional?
5. What is the probability that a time card will have errors in effective hours per week or recorded absences per week?
6. What is the probability that a time card will have errors in effective hours per week and recorded absences per week?
7. Are computation of effective hours per week and recording of absent days per week statistically independent job characteristics? Does improvement in one automatically effect improvement in the other? (This is a loaded question that goes beyond statistics!)

Chapter 5

Basic Statistics

INTRODUCTION

In the last chapter, we looked at the issue of sampling from a population or a process and quantifying the results as probabilities. In this chapter, we continue our examination of the characteristics of data taken from a population or a process. We discuss the types of data that may be encountered in statistical studies and how they are classified; methods for visually displaying data; and the calculation of numerical measures to describe the data.

TYPES OF DATA

Information collected about a product, service, process, person, or machine is called *data*. In the analytic study of a process, data are often used as a basis for action. For example, if we collect data about the weight of a particular grade of paper, we can monitor and understand the production process and take actions that will allow us to make paper with a weight consistently around some desired value. As another example, if we collect data about the number of payroll entry errors by pay period, we can understand when and why errors are made so that we can take appropriate actions to reduce the error rate.

We classify data into two types, attribute data and variables data.

Attribute Data

Attribute data arise from (1) the classification of items, such as products or services, into categories; from (2) counts of the number of items in a given category or the proportion in a given category; and from (3) counts of the number of occurrences per unit.

Classification of Items into Categories. Often data are classified into two categories or grades, representing conformity or nonconformity with some quality characteristic. Whether or not a purchase order is filled out properly; whether or not a bearing in a piece of equipment has excessive vibration; whether or not a jar of instant coffee is broken; whether or not a nine-volt battery is operating; or whether or not an automobile radio is defective are all examples of attributes.

Items may be similarly classified into three or more categories or grades. Classifying the grade of a stick of butter as Grade AA, Grade A, or Grade B; classifying employees by department; or classifying the type of defect in a roll of paper as dished, crushed core, or wrinkled are all examples of attributes.

As an illustration, union grievances are classified as first level, second level, third level, or fourth level; a first-level grievance can be resolved by a foreman, whereas a fourth-level grievance must be handled by a special arbitration board. That is, the higher the level of the union grievance, the more serious the grievance. Figure 5.1 illustrates a list of 11 union grievances and their classifications for 1987.

Counts of the Number of Items in a Given Category, or Proportion in a Given Category. The number of items in a given category also represents attribute data. For example, the second column of Figure 5.2 shows the number of union grievances in each of the four categories discussed in the above example.

FIGURE 5.1 Attribute Data: Classification of 1987 Union Grievances into Four Categories

Grievance Number	Grievance Level
1	1
2	1
3	1
4	2
5	4
6	1
7	1
8	1
9	3
10	1
11	1

FIGURE 5.2   Attribute Data: Number of 1987 Union Grievances per Level

Category	Number of Grievances	Proportion of Grievances
First level	8	8/11 = .73
Second level	1	1/11 = .09
Third level	1	1/11 = .09
Fourth level	1	1/11 = .09

The *proportion* of items in a given category is also an attribute. From Figure 5.2, we see that the proportion of second level grievances can be calculated as:

$$\text{proportion} = \frac{\text{number of second level union grievances}}{\text{number of union grievances}}$$

$$= 1/11 = 0.09$$

The last column of Figure 5.2 shows the proportion of grievances in each category.

Counts of the Number of Occurrences per Unit.   An attribute of interest in some quality studies is the number of defects per unit, where a unit can be a single item, a batch of the item under question, or a unit of time or space. For example, the number of defects per reel of paper; the number of errors per typed page; the number of accidents per month; the number of union grievances per week; the number of blemishes per square yard of fabric; or the number of defective automobile radios in a lot of 1,000 such radios are all examples of attribute data.

   As an illustration, in the production of stainless steel washers, the manufacturing process occasionally results in a defective product. Four types of defects are observed: cracks, dents, scratches, and discoloration. An individual washer may have some, all, or none of these defects. Every hour, 10 washers are examined for defects. Figure 5.3 shows the number of defects for each washer examined in an eight-hour day.

Variables Data

Variables data arise from (1) the measurement of a characteristic of a product, service, or process; and from (2) the computation of a numerical value from two or more measurements of variables data.

Measurement of a Characteristic.   When we make precise measurements of quality characteristics, such as the outside diameter of stainless steel

FIGURE 5.3 Attribute Data: Number of Defects per Washer in Eight Hourly Samples of 10 Stainless Steel Washers

Hour #	Item #	# Defects	Hour #	Item #	# Defects	Hour #	Item #	# Defects
1	1	1		8	0		5	1
	2	0		9	3		6	0
	3	0		10	0		7	0
	4	2	4	1	1		8	0
	5	1		2	1		9	3
	6	0		3	0		10	2
	7	0		4	4	7	1	0
	8	0		5	0		2	1
	9	4		6	1		3	0
	10	0		7	0		4	0
2	1	1		8	0		5	1
	2	0		9	3		6	0
	3	0		10	1		7	0
	4	3	5	1	0		8	2
	5	1		2	2		9	0
	6	0		3	0		10	0
	7	2		4	0	8	1	1
	8	0		5	0		2	0
	9	0		6	3		3	0
	10	0		7	0		4	0
3	1	1		8	0		5	1
	2	0		9	1		6	0
	3	0		10	2		7	0
	4	2	6	1	1		8	0
	5	4		2	1		9	2
	6	0		3	0		10	0
	7	1		4	0			

washers (in centimeters); the weights of boxes of detergent (in ounces); or the lifetimes of incandescent bulbs (in hours), we are recording the values of a *variable*.

As an illustration, in a logging operation, trees are cut at age 20 years. In Figure 5.4, the length, in centimeters, of each such log is recorded, as measured in 19 daily samples of five trees each.

Computation of a Numerical Value from Two or More Measurements of Variables Data. Figure 5.5 shows the recorded values of two variables for each of six trucks: the number of miles driven since the last refueling, and the number of gallons of fuel consumed since the last refueling.

FIGURE 5.4   Variables Data: Length of Logs in a Sample of 95 Trees

Day #	Log #	Length (cm)	Day #	Log #	Length (cm)	Day #	Log #	Length (cm)
1	1	783.2		3	1393.9		5	1719.7
	2	1322.8		4	1968.6	14	1	638.8
	3	1012.5		5	1561.2		2	1834.0
	4	1408.6	8	1	1908.7		3	730.0
	5	1589.6		2	1612.2		4	1338.5
2	1	400.2		3	1585.6		5	1621.4
	2	1456.1		4	1510.4	15	1	1192.1
	3	1126.7		5	898.7		2	1542.6
	4	1563.8	9	1	1087.8		3	1886.1
	5	1362.5		2	938.1		4	1688.2
3	1	1537.8		3	1472.3		5	1069.7
	2	505.4		4	1072.0	16	1	1835.3
	3	1686.7		5	389.7		2	1485.2
	4	1508.0	10	1	1208.6		3	1521.4
	5	1235.8		2	1868.5		4	2104.4
4	1	1827.2		3	1321.7		5	1225.2
	2	2148.3		4	2067.7	17	1	929.4
	3	1175.2		5	835.9		2	1732.1
	4	1657.5	11	1	1830.6		3	1340.2
	5	410.3		2	1227.1		4	1445.5
5	1	1577.7		3	1423.3		5	460.7
	2	827.6		4	1022.2	18	1	1021.8
	3	1353.1		5	1690.0		2	1515.2
	4	1410.2	12	1	625.6		3	753.6
	5	1050.2		2	1131.1		4	1117.3
6	1	1468.7		3	1193.9		5	1490.0
	2	1229.2		4	1252.8	19	1	572.5
	3	1346.3		5	1551.8		2	1436.8
	4	1725.6	13	1	1481.4		3	1356.6
	5	693.1		2	1763.6		4	1112.6
7	1	1639.9		3	1392.2		5	1512.7
	2	1786.3		4	1498.9			

From these measurements, we may calculate another variable, miles per gallon for each truck, by calculating the ratio of the number of miles driven since the last refueling divided by the number of gallons of fuel consumed since the last refueling. This computed ratio, shown in the last column of Figure 5.5, is an illustration of variables data computed from two other measurements of variables data.

As another illustration, the computation of the volume of a rectangular container from the product of the measurements of its length, width, and height results in a new computed variable value.

FIGURE 5.5 Variables Data: Calculation of the Variable Miles per Gallon

Truck #	# Miles Since Refueling	# Gallons Fuel Consumed	Miles per Gallon
1	308	17.3	17.8
2	256	15.3	16.7
3	274	16.5	16.6
4	310	16.9	18.3
5	302	17.1	17.7
6	296	17.3	17.1

In any quality study, we usually want to rearrange and/or manipulate data to gain some insight into the characteristics of the population or process under study. In the remainder of this chapter we discuss ways of interpreting and appropriately describing data, both visually and numerically.

CHARACTERIZING DATA

Enumerative Studies

A complete census of the frame in an enumerative study provides all the information needed to take action on the frame. For example, suppose a lumber supplier must determine how much to charge for a particular large shipment to a lumberyard. The total charge is based on the lengths of the logs in the shipment (the frame). Thus, if we knew the length of each log in the shipment (that is, 100 percent sampling in an enumerative study), we would know exactly how much to charge for the shipment, assuming accuracy in the measurement process.

As discussed in Chapter 2, we often sample rather than perform a complete census; we then use that information to develop visual and/or numerical measures that enable us to take action on the frame. As we are examining only a portion of the frame, the possibility of taking an incorrect action does exist. In our example, suppose the lumber supplier bases the charge on the average of the 95 log lengths shown in Figure 5.4. Two types of error are possible in this procedure: (a) his charge is higher by an amount C than what it would be if he had measured every log (often called the *consumer's risk*); or, (b) his charge is lower by an amount C′ than what it would be if he had measured every log (the *producer's risk*).

As another illustration, consider the sample of stainless steel washers examined for defects in Figure 5.3. If all the washers produced in this eight-hour day make up a single lot to be shipped (the frame), in an enumerative study the manufacturer may wish to determine the number of defects in the entire lot to decide whether or not to ship the lot. In using the sample data as a basis for action, again there are two possible errors: (a) he decides to ship a lot that, in fact, has an unacceptably large number of defects (the consumer's risk); or (b) he decides to reject a lot that, in fact, has an acceptably small number of defects (the producer's risk).

If the information used as the basis for action constitutes a *random sample* from the frame, these errors can be quantified, and valid statistical inferences can be made on the frame in question. We have discussed in Chapter 2 the nature of and procedures for selecting a random sample. In this chapter we discuss how to describe and characterize such data to ultimately make appropriate inferences.

Analytic Studies

A complete census of the frame is impossible in an analytic study, for a frame consists of all past, present, and future observations, and the latter cannot be measured. As we are dealing with an ongoing process, we wish to characterize the data to take action on that process for the future. For example, in the manufacture of washers, suppose we are not interested, as before, in one particular day's output but in the process itself. That is, we wish to determine the necessity for action: whether the process should be left unchanged for the future or modified in some way. A sample selected from the process, such as that in Figure 5.3, may lead to two types of error: (a) decide to retain the process, even though it should be modified; or (b) decide to modify the process, although it is actually unnecessary to do so.

However, unlike the enumerative study, as the frame is unknown (the future cannot be measured) it is not possible to precisely quantify these errors. Any inferences we make in an analytic study are conditional on the environmental state when the sample was selected; information on such a problem can never be complete. If, however, a knowledge of the process and the environment and an analysis of the data indicate that the process is *stable* and predictable (as discussed in Chapters 8 through 10), the visual and numerical characterizations discussed in this chapter can be used to make inferences and take action in the future.

VISUALLY DESCRIBING DATA

Tabular Displays

Frequency Distributions.   A *frequency distribution* shows us, in tabular form, the number of times, or the frequency with which, a given value or group of values occurs. For example, when analyzing variables type data,

FIGURE 5.6 Frequency Distribution of Lengths of Logs in a Sample of 95 Trees

Length (cm)	Class Midpoint	Absolute Frequency	Relative Frequency (percent)
400 but less than 700	550	9	9.5
700 but less than 1,000	850	8	8.4
1,000 but less than 1,300	1,150	20	21.1
1,300 but less than 1,600	1,450	35	36.8
1,600 but less than 1,900	1,750	18	18.9
1,900 but less than 2,200	2,050	5	5.3

we generally group the data into class intervals and count the number of items whose values fall within each interval. Thus, Figure 5.6 shows a frequency distribution for the data of Figure 5.4, where the class intervals are: 400 but less than 700 cm, 700 but less than 1,000 cm, 1,000 but less than 1,300 cm, 1,300 but less than 1,600 cm, 1,600 but less than 1,900 cm, and 1,900 but less than 2,200 cm. The class limits are the endpoints of each interval; in our example, the lower class limits are the lower limits of each interval: 400, 700, 1,000, 1,300, 1,600, and 1,900. The upper class limits are 700, 1,000, 1,300, 1,600, 1,900, and 2,200. The midpoint of each class is the value halfway between the class limits.

Figure 5.6 indicates relative frequencies (the percent of the total number of observations in each class) as well as absolute frequencies (the actual number of observations in each class). Especially in situations where we want to compare two frequency distributions with unequal total numbers, we often use *relative frequencies*—or the class frequencies divided by the total number of observations—expressed as a percent.

For attribute type data, we may construct frequency distributions by counting the number of items possessing a particular attribute. Thus, Figure 5.7 shows a frequency distribution of the number of defects in 80 stainless steel washers.

As a general rule of thumb, we select anywhere from 5 to 20 classes in the construction of a frequency distribution for variables type data; too few or too many classes may fail to reveal patterns of interest. For example, taking the data of Figure 5.4 and constructing a frequency distribution consisting of two classes, 400 but less than 1,300, with 37 observations, and 1,300 but less than 2,200, with 58 observations, offers little insight in comparison with the distribution of Figure 5.6. Where possible, we make the class widths the same for all intervals. This allows us to logically compare frequencies in different classes. For example, if we are measuring the lifetime of 100 special purpose incandescent bulbs and count 3 bulbs in a class interval of 1 to 5 hours and 35 bulbs in a class interval of

		Relative
	Absolute	Frequency
Number of Defects	Frequency	(percent)
0	46	57.50
1	18	22.50
2	8	10.00
3	5	6.25
4	3	3.75

FIGURE 5.7   Frequency Distribution of Number of Defects in a Sample of 80 Stainless Steel Washers

5.1 to 20 hours, we could not easily determine whether the large difference in frequency results from actual variations in bulb lifetime or simply from the huge difference in class widths. Sometimes, however, we use an open-ended interval to include a small number of highly dispersed values at the top and/or bottom of the distribution. Using the same example, if 1 or 2 of the 100 bulbs burned for 90 hours, while all the rest burned for 50 hours or less, it might make sense to specify the top interval to be "more than 50 hours." The number of classes in a frequency distribution for attribute type data is determined by the number of values the attribute can assume; for example, an attribute frequency distribution depicting the sex of employees would have two classes.

Cumulative Frequency Distributions.   For many processes, we are interested in the frequency of items with a value less than some measurement. For example, how many of the 100 incandescent bulbs burned for less than 30 hours, how many burned for less than 40 hours, etc. This information is presented in a cumulative frequency distribution. Figure 5.8(a) shows the cumulative frequencies for the data of Figure 5.4, calculated by adding the successive frequencies in the distribution of Figure 5.6. For example, the number of logs with a length of less than 1,600 cm is the sum of the frequencies for all the class intervals up to and including "1,300 but less than 1,600," or $9 + 8 + 20 + 35 = 72$. Relative cumulative frequency, or the percent of the items with a value less than the upper limit of each class interval, is calculated as the absolute cumulative frequency divided by the sample size. Thus, for example, 75.8 percent of the logs in the sample have lengths of less than 1,600 cm. Figure 5.8(b) shows the cumulative frequency distribution for the number of defects in the 80 washers.

Limitations of Frequency Displays.   It is important to note that the frequency displays we have discussed do not include information on the time-ordering of data. Recall that the diameters of our 80 stainless steel

FIGURE 5.8   Cumulative Frequency Distributions

(a)   Cumulative Frequency Distribution of Lengths of Logs in a Sample of 95 Trees

Log Length (cm) Less Than or Equal to	Cumulative Frequency	Relative Cumulative Frequency (percent)
700	9	9.5
1,000	17	17.9
1,300	37	38.9
1,600	72	75.8
1,900	90	94.7
2,200	95	100.0

(b)   Cumulative Frequency Distribution of Number of Defects in a Sample of 80 Stainless Steel Washers

Washers with Number of Defects Less Than or Equal to	Cumulative Frequency	Relative Cumulative Frequency (percent)
0	46	57.50
1	64	80.00
2	72	90.00
3	77	96.25
4	80	100.00

washers were actually measured by examining 10 washers every hour over a period of eight consecutive hours (see Figure 5.3). The frequency distribution in Figure 5.7, however, does not indicate that the observations were made in any particular sequence. In an analytic study, where we examine the sample to take action on the process, a frequency display would fail to show trends that may be occurring over time. This loss of information is critical.

Consider the production of brass nuts. The process, as a result of equipment wear, is producing nuts with an inside diameter actually increasing over time. Every hour, for 20 consecutive hours, a nut is selected randomly and its inside diameter recorded. The observations are shown in Figure 5.9(a). A frequency distribution is shown in Figure 5.9(b).

If the nuts manufactured during this 20-hour period constitute a population, the frequency distribution indicates all pertinent information about the lot. However, if we are conducting a study of a manufacturing process for the future, it would be essential to detect the trend over time and look for the cause; the frequency distribution in this case gives us an incomplete picture.

FIGURE 5.9   Inside Diameters of Brass Nuts

(a)   Hourly Observations

Hour #	Inside Diameter (in)	Hour #	Inside Diameter (in)	Hour #	Inside Diameter (in)
1	1.00	8	1.11	15	1.25
2	1.01	9	1.12	16	1.28
3	1.03	10	1.14	17	1.32
4	1.05	11	1.16	18	1.36
5	1.07	12	1.18	19	1.41
6	1.08	13	1.20	20	1.46
7	1.09	14	1.22		

(b)   Frequency Distribution

Inside Diameter (in)	Absolute Frequency
1.00 but less than 1.10	7
1.10 but less than 1.20	5
1.20 but less than 1.30	4
1.30 but less than 1.40	2
1.40 but less than 1.50	2

Graphical Displays

Data are often represented in graphical form. Although the exact measurements may not be as precisely identifiable as they are in a table, such displays offer a composite picture of the relationships and/or patterns at a glance.

Frequency distributions of variables data are commonly presented in frequency polygons or histograms. Frequency distributions of attribute data are commonly presented in bar charts. In all these displays, the class intervals are drawn along the horizontal axis, and the absolute or relative frequencies along the vertical axis.

Frequency Polygons.   In a *frequency polygon,* we plot the midpoint of each class interval on the horizontal axis against the frequency of that class on the vertical axis. We then connect these points with a series of line segments. To complete the polygon, we add two more classes to the distribution—one class preceding the smallest class and one class succeeding the largest class. Both of these are assigned a "0" frequency at their midpoints. In Figure 5.10(a) the frequency distribution of the data in Figure 5.6 is presented as a frequency polygon.

FIGURE 5.10   Graphical Displays of Data

(a) Frequency Polygon for Lengths of Logs in a Sample of 95 Trees

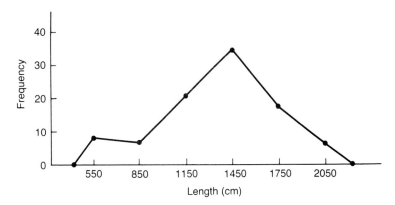

(b) Frequency Polygon for Product Weights of 10 Items

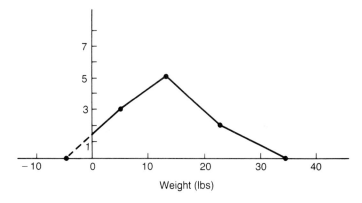

(c) Histogram for Lengths of Logs in a Sample of 95 Trees

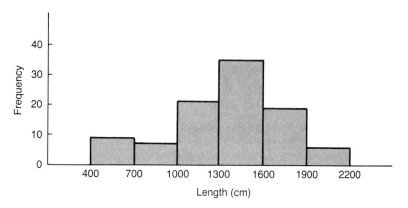

In the particular case where the lower class limit of the smallest class is 0, adding a class preceding the smallest class results in negative values for that class (and its midpoint) on the horizontal axis. To indicate that such values could not really occur, we use a dashed line for the portion of the polygon corresponding to negative observations. Figure 5.10(b) illustrates this procedure for a frequency distribution of 10 product weights, in which three observations are between 0 and 10 pounds, five observations are between 10 and 20 pounds, and two observations are between 20 and 30 pounds.

Histograms. In a *histogram,* vertical bars are drawn, each with a width corresponding to the width of the class interval and a height corresponding to the frequency of that interval. The bars share common sides, with no space between them. Figure 5.10(c) shows a histogram of the frequency distribution of the data in Figure 5.6.

Bar Charts. For attribute data, a *bar chart* is used to graphically display the data. It is constructed in exactly the same way as a histogram, except that instead of using a vertical bar spanning the entire class interval, we use a line or bar centered on each attribute category. Figure 5.11 shows a bar chart for the frequency distribution in Figure 5.7.

Ogive. The graph of a cumulative frequency distribution, called an *ogive* (pronounced ''oh jive''), is constructed by plotting the upper limit of each class interval on the horizontal axis against the cumulative percent for that class on the vertical axis. Figure 5.12 shows the ogive for the cumula-

FIGURE 5.11 Bar Chart for Number of Defects in a Sample of 80 Stainless Steel Washers

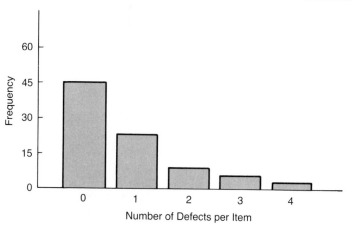

FIGURE 5.12   Ogive for Lengths of Logs in a Sample of 95 Trees

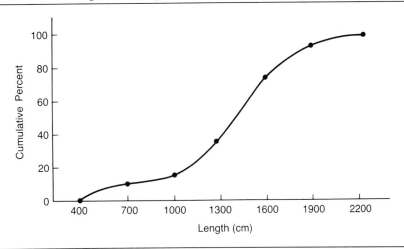

tive frequency distribution of Figure 5.8(a). Note that the lower limit of the first class interval has a cumulative percentage of 0, as there are no observations below that value.

Run Charts: Importance of Time-Ordering in Analytic Studies.   As we indicated in our earlier discussion of the limitations of tabular frequency displays, in analytic studies we want to be able to detect trends or other patterns over time to take action on the process. The graphical displays we have examined do not incorporate the time-ordering of the data, hence are of limited value in such studies. In a *run chart* (also called a *tier chart*), this information is preserved by plotting the observed values on the vertical axis and the time they were observed on the horizontal axis.

For example, suppose we are studying a process in which chocolate rectangles are cut from larger blocks of chocolate and then packaged as six-ounce bars. Every 15 minutes, three chocolate bars are weighed, prior to packaging. Figure 5.13 shows the weights for each bar examined in a seven-hour day.

We can construct a frequency distribution of this data using, as lower class limits, 5.98, 6.00, 6.02, 6.04, and 6.06.

Figure 5.14(a) is a histogram of this distribution indicating, at a glance, that most of the chocolate bars weigh slightly more than the labeled weight of 6 ounces. In an enumerative study, in which we are interested only in the batch of chocolate bars produced during this eight-hour day, such information may be sufficient for taking action on that batch. In an analytic study, however, where we are seeking to take action on the process itself for the future, we will see that the information gained from Figure 5.14(a) is incomplete.

FIGURE 5.13   Weights of Chocolate Bars Examined at 15-Minute Intervals

Time	Observation #	Weight (oz)	Time	Observation #	Weight (oz)
9:15	1	6.01	12:45	1	6.03
	2	5.99		2	6.02
	3	6.02		3	6.03
9:30	1	5.98	1:00	1	6.03
	2	5.99		2	6.00
	3	6.01		3	6.01
9:45	1	6.03	1:15	1	6.04
	2	6.02		2	6.02
	3	6.02		3	6.03
10:00	1	6.02	1:30	1	6.05
	2	6.03		2	6.02
	3	6.02		3	6.04
10:15	1	6.00	1:45	1	6.03
	2	5.99		2	6.04
	3	6.01		3	6.01
10:30	1	5.99	2:00	1	6.02
	2	6.00		2	6.02
	3	6.00		3	6.02
10:45	1	6.02	2:15	1	6.04
	2	6.01		2	6.05
	3	6.00		3	6.03
11:00	1	6.01	2:30	1	6.06
	2	6.03		2	6.03
	3	6.01		3	6.04
11:15	1	6.01	2:45	1	6.05
	2	6.02		2	6.04
	3	6.00		3	6.02
11:30	1	6.00	3:00	1	6.05
	2	6.02		2	6.04
	3	6.01		3	6.03
11:45	1	6.04	3:15	1	6.04
	2	6.02		2	6.06
	3	6.03		3	6.05
12:00	1	6.02	3:30	1	6.05
	2	6.01		2	6.03
	3	6.00		3	6.04
12:15	1	6.03	3:45	1	6.03
	2	6.02		2	6.04
	3	6.04		3	6.03
12:30	1	6.02	4:00	1	6.06
	2	6.02		2	6.06
	3	6.03		3	6.05

FIGURE 5.14 Weights of Chocolate Bars Examined at 15-Minute Intervals

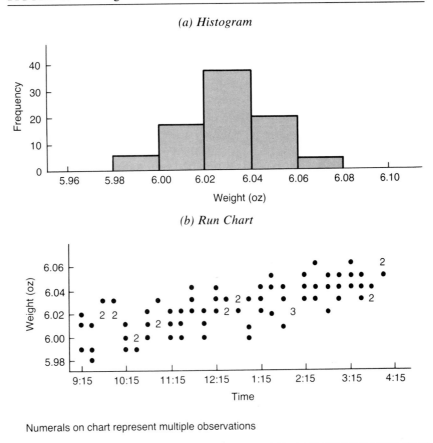

(a) Histogram

(b) Run Chart

Numerals on chart represent multiple observations

Figure 5.14(b) is a run chart that shows the observed weights plotted over time. The graph clearly shows an upward drift in the weights of chocolate bars throughout the day and it indicates the need for action on the process.

NUMERICALLY DESCRIBING DATA

Measures of Central Tendency

While tabular and graphical representations give a broad overview, we often need quantitative measures that summarize important characteristics of a population or a process. One such property is *central tendency,* or the behavior of the middle, or central portion, of the population or process data.

The Mean. In trying to convey the underlying character of variables data by somehow representing the middle of the data, one common numerical representation is the arithmetic average, or *mean*. The mean is simply the sum of the numerical values of the measurement divided by the number of items examined. In an enumerative study, if the items constitute a frame, the average is called the *population mean* and is usually denoted by the Greek letter μ (pronounced "mew"). When the items constitute a sample drawn from a frame, we call the average a *sample mean,* and denote it as $\bar{x}$ ("x bar"). Thus, in an enumerative study, we might make reference to either μ or $\bar{x}$. In an analytic study, the mean of the conceptual population, which includes all past, present, and future values, is also denoted by μ and is estimated from past and present values by $\bar{\bar{x}}$ ("x bar-bar"). When the items constitute a sample from the conceptual population, we similarly call the average a sample mean, and denote it as $\bar{x}$.

To communicate this type of information in compact and precise form we identify each numerical value as an "x" measurement; the sum of all the measurements we call Σx ("the sum of x") where Σ is the capital Greek letter *sigma* and represents, mathematically, "summation."

In an analytic study, the mean of the conceptual population, μ, cannot be calculated from all past, present, and future values. Furthermore, the interpretation of the sample mean in an analytic study is not straightforward. If the process is not acting in a predictable fashion, the sample average computed, say, last week may be dramatically different from the sample average computed next week; consequently, this calculated measure could be grossly misleading. Chapters 8 through 10 explain how we can establish that a process is predictable (or stable), and how we can then interpret the sample mean.

If we have a stable process, the sample mean, $\bar{x}$, can be calculated as

$$\bar{x} = \frac{\Sigma x}{n} \tag{5.1}$$

where n represents the size of the sample, or the number of items included in the determination of the sample mean.

If we calculate sample means for subgroups of size n in the past and present, we can estimate the mean of the conceptual population by the average of these sample means, or:

$$\bar{\bar{x}} = \frac{\Sigma \bar{x}}{\# \text{ subgroups}} \tag{5.2}$$

To estimate the mean of the conceptual population of all chocolate bars, we may calculate sample means, $\bar{x}$, for each subgroup of three bars observed every 15 minutes, as shown in Figure 5.13. These subgroup sample means are shown in Figure 5.15.

FIGURE 5.15 Weights of Chocolate Bars Examined at 15-Minute Intervals: Sample Means of Subgroups of Size 3

Time	$\bar{x}$	Time	$\bar{x}$	Time	$\bar{x}$	Time	$\bar{x}$
9:15	6.01	11:00	6.02	12:45	6.03	2:30	6.04
9:30	5.99	11:15	6.01	1:00	6.01	2:45	6.04
9:45	6.02	11:30	6.01	1:15	6.03	3:00	6.04
10:00	6.02	11:45	6.03	1:30	6.04	3:15	6.05
10:15	6.00	12:00	6.01	1:45	6.03	3:30	6.04
10:30	6.00	12:15	6.03	2:00	6.02	3:45	6.03
10:45	6.01	12:30	6.02	2:15	6.04	4:00	6.06

From these, we may estimate the mean weight of all chocolate bars produced by this process from Equation 5.2:

$$\bar{\bar{x}} = \frac{\Sigma \bar{x}}{\text{\# subgroups}} = \frac{168.68}{28} = 6.02$$

Grouped Data. Sometimes the only numerical data available are grouped data, or data presented as a frequency distribution. As we observed in the previous section, such a display often facilitates the analysis of large sets of numbers. However, without access to the raw data, we cannot calculate the exact value of the mean. For example, if we consider the class intervals in Figure 5.6, in the absence of the individual observations we see it would be impossible to add up the x values, for we do not know them. We make the assumption that the individual observations in a given class interval are distributed uniformly over the interval and have a mean equal to the class midpoint, x_m, so that the mean can be approximated by:

$$\bar{x} = \frac{\Sigma(fx_m)}{n} \tag{5.3}$$

where f is the frequency of a given class interval.

Thus, for the frequency distribution in Figure 5.6, the sample mean could be approximated as:

$$\bar{x} = \frac{\Sigma(fx_m)}{n}$$

$$= \frac{(9)(550) + (8)(850) + (20)(1150) + (35)(1450) + (18)(1750) + (5)(2050)}{95}$$

$$= \frac{127,250}{95} = 1339.5 \text{ cm}$$

This sample mean is not exactly the same as the one obtained from the raw data using Equation 5.1, 1335.2. The latter is the mean value of the 95 log lengths; the value calculated from the frequency distribution, which does not utilize the actual data points, is an approximation of this mean.

If a frequency distribution contains open-ended intervals, we have no midpoint to use for such classes; without additional information from which to calculate a class midpoint, we cannot make the calculation of the sample mean using Equation 5.3.

The Median. Another measure of the central tendency is given by the *median,* or middle value when the data are arranged in numerical order. When there are an even number of observations, the median value is taken to be the arithmetic average of the middle two values.

Symbolically, if n is the number of data points, the median can be calculated as:

$$M_e = [(n + 1)/2]^{th} \text{ item, if n is odd, or}$$

$$= \text{average of } (n/2)^{th} \text{ item and } (n/2 + 1)^{th} \text{ item, if n is even}$$

(5.4)

Given an ogive, or the graph of the cumulative frequencies for a given set of data points, the median may also be approximated by reading the data value corresponding to a cumulative frequency of 50 percent. That is, as the median is the middle value, 50 percent of the data points must have values less than the median.

The central tendency of some data is not adequately represented by an arithmetic average. Consider the measurement of burning times of small birthday candles from a production lot, in minutes. Seven candles are randomly selected from the lot; and the burning times are found to be 3 minutes, 2 minutes, 4 minutes, 6 minutes, 19 minutes, 5 minutes, and 3 minutes. As these data points are a sample from the population of all such candles in this production lot, and we are interested in making some inference on the burning characteristics of this lot, this is an enumerative problem; we may calculate the sample mean from Equation 5.1:

$$\bar{x} = \frac{3 + 2 + 4 + 6 + 19 + 5 + 3}{7} = \frac{42}{7} = 6 \text{ minutes}$$

Note, however, that in six of the seven data points, the burning time was six minutes or less, so that the average value as a measure of central tendency is somewhat misleading. If we look at a frequency polygon for the data points, shown in Figure 5.16, we see that it is asymmetrical, or *skewed* to the right; only a small proportion of the data has high values.

The median burning time is the middle data point in the sequence of burning times written in ranked order (from lowest value to highest value): 2, 3, 3, 4, 5, 6, 19, or 4 minutes. Note that half the data points have

FIGURE 5.16 Frequency Polygon for Burning Time of Small Birthday Candles

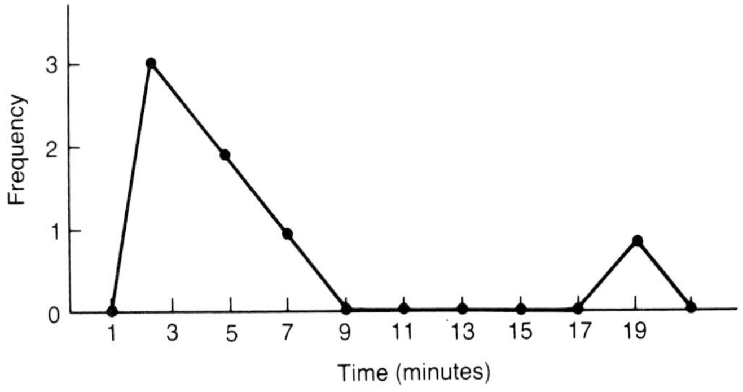

a value lower than the median, and half have a value higher than the median. The median is not as influenced by the magnitude of the extreme items as is the mean. That is, if the longest data point had been 500 minutes instead of 19 minutes, the median would remain 4 minutes, although the mean would increase by a great deal. Hence, the median is particularly appropriate in an enumerative example such as ours, where a single extreme value appears atypical.

The interpretation of the median in an analytic study, however, can be misleading. Where extreme data points are observed in a sample, we have seen that they would not affect the computation of the median. In an enumerative study in which such data points are deemed atypical, the median is an appropriate measure of central tendency. However, in an analytic study, where we are concerned with the process itself, the existence of extreme data points is a critical factor in our analysis. That is, extreme data points may indicate process disturbances and instability, and the need for corrective action on the process. A measure like the median, which is insensitive to extreme data points, must be used with caution.

Grouped Data. For grouped data, where the values of the individual data points are unknown, we first locate the median class, or the class in which the middle item occurs. To illustrate using the frequency distribution in Figure 5.6, the middle item is the $[(95 + 1)/2]^{th}$, or 48th item. From the cumulative frequencies shown in Figure 5.6, we see that the 48th item occurs in the class interval "1,300 but less than 1,600." If we make the assumption that the data points within each class are evenly dispersed, then we can visualize the median class as follows:

$$\overbrace{}^{\text{300 cm}}$$

$$\overbrace{}^{\text{x cm}}$$

Length $1{,}300 \ldots M_e \ldots \ldots \ldots 1{,}600$

$$\overbrace{}^{\text{10 items}}$$

Item Number $38 \ldots 48 \ldots \ldots \ldots 73$

$$\underbrace{}_{\text{35 items}}$$

That is, if the data points are spread evenly throughout the interval, and the median item is 10/35 of the way from item #38 to item #73, then the median occurs 10/35 of the way from 1,300 to 1,600. Hence, the median item would be $1{,}300 + (10/35) \cdot (1{,}600 - 1{,}300) = 1{,}385.7$ cm. Symbolically, we can represent this calculation as:

$$M_e = L + (n_1/n_2)W \qquad (5.5)$$

where L is the lower limit of the median class, n_1 is the number of items in the median class which are below the median, n_2 is the frequency of the median class, and W is the width of the median class.

From the raw data of Figure 5.4, we find that if the 95 data points are arranged in ascending order, the median, or 48th item, is 1,408.6. This is the exact median for the sample; the value computed from the frequency distribution, 1,385.7, is an approximation.

The Mode.   The *mode* of a distribution is the value that occurs most frequently, or the value corresponding to the high point on a frequency polygon or histogram. Like the median, and unlike the mean, it is not affected by extreme data points. A frequency distribution with one such high point is called *unimodal;* distributions with more than one high point of concentration are called *multimodal.*

For the data of Figure 5.3, an examination of the frequency distribution in Figure 5.7 shows that the mode, or modal number of defects, is 0, for that value has the largest frequency. This is a unimodal distribution. The frequency polygon in Figure 5.16, with two high points of concentration, indicates a *bimodal* distribution.

In the sample of the burning times of seven randomly selected birthday candles discussed earlier, we found that $\bar{x} = 6$ minutes and the median was 4 minutes. The mode in this example is the most frequently occurring value, or 3 minutes. As in the case of the median, had the longest observation been 500 minutes instead of 19 minutes, the mode would remain 3 minutes, although the mean would increase.

As with the median, an important characteristic of the mode is that it is not affected by extreme data points. In analytic studies, where extreme data points may reveal a great deal about the process under investigation, the mode must be used with caution.

Grouped Data. For the data of Figure 5.4, we see that because no one data point occurs more than once, it is meaningless to refer to the mode. Nevertheless, when we examine the frequency distribution in Figure 5.6, we see that there is, in fact, a *modal class*. In grouped data with equal class intervals, the modal class is the class interval with the highest frequency. If we make the assumption that the distribution is relatively smooth, the mode can be approximated by:

$$M_o = L_o + [d_1/(d_1 + d_2)]W \qquad (5.6)$$

where L_o is the lower limit of the modal class, d_1 is the difference between the frequency of the modal class and the preceding class interval, d_2 is the difference between the frequency of the modal class and the succeeding class, and W is the width of the modal class.

For the sample of 95 log lengths in the frequency distribution of Figure 5.6, the modal class is the class interval "1,300 but less than 1,600"; $d_1 = 35 - 20 = 15$; $d_2 = 35 - 18 = 17$; the width of the inverval W = 1,600 − 1,300 = 300. Thus, the mode can be approximated by:

$$1,300 + \left[\frac{15}{15 + 17}\right](300) = 1,440.625$$

In general, the mode has some major shortcomings. For grouped data, the computed value given by Equation 5.6 is only a rough approximation. It tends to vary from sample to sample to a much greater degree than the mean and median.

The Proportion. Often data is classified into two nonnumerical attributes, such as broken–not broken, operating–not operating, defective–conforming. The proportion or fraction of the data possessing one of two such attributes is then a meaningful measure of central tendency.

The data of Figure 5.17 represent the classification of 38 radios as defective or nondefective in an analytic study of a production process.

As in the computation of a sample mean in an analytic study, if the process is not a stable one, the proportion defective computed from this particular sample may be misleading because of process variability. Although we will not discuss process predictability and stability until Chapter 8, under the assumption that the process is stable, the *sample proportion,* or *sample fraction defective,* can be calculated as:

$$p = \frac{x}{n} \qquad (5.7)$$

where x is the number of defective items and n is the total number of items in the sample. Thus,

$$p = \frac{8}{38} = 0.21$$

FIGURE 5.17 Defective Units in a Sample of 38 Radios

Item #	Condition	Item #	Condition
1	Nondefective	20	Defective
2	Defective	21	Nondefective
3	Nondefective	22	Nondefective
4	Nondefective	23	Nondefective
5	Nondefective	24	Nondefective
6	Nondefective	25	Nondefective
7	Nondefective	26	Defective
8	Defective	27	Nondefective
9	Nondefective	28	Nondefective
10	Nondefective	29	Nondefective
11	Nondefective	30	Nondefective
12	Nondefective	31	Nondefective
13	Nondefective	32	Nondefective
14	Defective	33	Defective
15	Nondefective	34	Defective
16	Nondefective	35	Nondefective
17	Nondefective	36	Nondefective
18	Nondefective	37	Defective
19	Nondefective	38	Nondefective

Often, in analytic studies of processes that operate over long periods of time, data on defectives are taken on a continuing basis to provide information on the necessity for action on the process. Figure 5.18 shows such data for the manufacture of radios over a seven-day period, where the data of Figure 5.14 were taken on the first day.

FIGURE 5.18 Defective Units in Daily Samples of 38 Radios

Day #	# Radios Inspected	# Defective Units
1	38	8
2	38	6
3	38	0
4	38	2
5	38	5
6	38	8
7	38	3

If the process is a stable one, the average fraction defective can be obtained by treating the data as a single sample, so that

$$p = \frac{\Sigma x}{\Sigma n} \tag{5.8}$$

$$= \frac{32}{266} = 0.12$$

Measures of Variability

All populations and processes have some degree of variability, given appropriate sensitivity of the measuring instrument. It is therefore necessary to be able to quantify not only the central tendency but also the degree of variability in a set of data. To illustrate the need for some numerical measure, consider the three sets of data in Figure 5.19, which represent the output from three different processes. For each process, the mean weight is 8.0 grams. If we limited ourselves to reporting this measure of central tendency only, we would actually fail to adequately characterize the three sets of measurements. While the three groups have the same mean, they differ with respect to how the data are spread around that mean, which we call the variability, or *dispersion*. The outputs from processes A and B weigh between 5 and 10.5 grams, while those from process C weigh between 7.6 and 8.4 grams. Further, the weights from processes B and C are clustered around the mean, while those from process A are more widely dispersed away from the mean. The two commonly used quantitative measures of such variability are the range and the standard deviation.

The Range.   The *range* is the simplest measure of dispersion; for raw data from an enumerative or an analytic study, it is defined as the differ-

FIGURE 5.19   Weights of Ball Bearings from Three Manufacturing Processes

Process A		Process B		Process C	
Item #	*Weight (g)*	*Item #*	*Weight (g)*	*Item #*	*Weight (g)*
1	5.0	1	5.0	1	7.6
2	5.3	2	7.8	2	7.8
3	8.0	3	7.9	3	8.0
4	9.2	4	8.0	4	8.1
5	10.0	5	8.8	5	8.1
6	10.5	6	10.5	6	8.4

ence between the largest data point and the smallest data point in a set of data.

$$R = x_{max} - x_{min} \qquad (5.9)$$

Thus, for process A, the range is $10.5 - 5.0 = 5.5$ g; for process B the range is $10.5 - 5.0 = 5.5$ g; and for process C it is $8.4 - 7.6 = 0.8$ g. The larger the range, the more dispersed the data. In our illustrations, the outputs from processes A and B have more variability than the output from process C.

Grouped Data.   For *grouped data*, the range can be approximated by the difference between the upper limit of the highest interval and the lower limit of the lowest interval. Thus, for the 95 log lengths in Figure 5.6, the range of lengths can be approximated by $2{,}200 - 400$, or $1{,}800$ cm. The actual range in the sample, from the data of Figure 5.4, is $2{,}148.3 - 400.2 = 1{,}748.1$ cm.

One shortcoming of the range as a measure of dispersion is that it is a function only of the extreme values of a set of data. For example, the data of both process A and process B have the same range, although the overall data points of process B are clustered more closely about the mean than those of process A.

The Standard Deviation.   The *standard deviation* as a measure of dispersion takes into account each of the data points and their distances from the mean. The more dispersed the data points, the larger the standard deviation will be; the closer the data points to the mean, the smaller the standard deviation will be. The population standard deviation is computed as the square root of the average squared deviations from the mean and is denoted by the lower case Greek letter σ (sigma).

For analytic studies, where the conceptual population cannot be measured, we estimate its value by the standard deviation of a sample (or subgroup) of n observations, called the *sample standard deviation,* s:

$$s = \sqrt{\frac{\Sigma(x - \bar{x})^2}{n - 1}} \qquad (5.10a)$$

or, the computationally simpler formula

$$s = \sqrt{\frac{\Sigma x^2 - \dfrac{(\Sigma x)^2}{n}}{n - 1}} \qquad (5.10b)$$

The square of the standard deviation is called the *variance,* σ^2. The square of the sample standard deviation is the *sample variance,* s^2.

For the data of Figure 5.19, the calculation of the sample standard deviation is illustrated below for process A:

x	x^2
5.0	25.00
5.3	28.09
8.0	64.00
9.2	84.64
10.0	100.00
10.5	110.25
$\Sigma x = 48.0$	$\Sigma x^2 = 411.98$

Using Equation 5.10b:

$$s = \sqrt{\frac{411.98 - \dfrac{(48)^2}{6}}{5}} = \sqrt{5.596} = 2.37 \text{ grams}$$

Similarly, the sample standard deviation from process B is 1.79 grams, and from process C is 0.28 grams. The standard deviation, then, clearly shows not only that the output from process C is less variable than that from the other two processes but that the overall variability from process B is smaller than that for process A, a distinction that the range did not exhibit.

Grouped Data. For grouped data, where individual values are not known, to approximate the standard deviation we make the assumption that the individual observations in a given class interval all lie at the class midpoint, x_m. Thus, in calculating the standard deviation of the frequency distribution in Figure 5.6, we would assume that all logs in the first class interval had length 550 cm, all those in the second class interval had length 850 cm, and so on. Given this assumption, we can calculate the squared deviation for each class interval once and then multiply it by the number of items in that class, f, to get the sum of the squared deviations for the entire class. This yields:

$$s = \sqrt{\frac{\Sigma f(x_m - \bar{x})^2}{n - 1}} \tag{5.11a}$$

or, the computationally simpler formula

$$s = \sqrt{\frac{\Sigma f x_m^2 - \dfrac{(\Sigma f x_m)^2}{n}}{n - 1}} \tag{5.11b}$$

For the frequency distribution of Figure 5.6 we calculate:

Class Intervals	f	x_m	x_m^2	fx_m	fx_m^2
400 but less than 700	9	550	302,500	4,950	2,722,500
700 but less than 1,000	8	850	722,500	6,800	5,780,000
1,000 but less than 1,300	20	1,150	1,322,500	23,000	26,450,000
1,300 but less than 1,600	35	1,450	2,102,500	50,750	73,587,500
1,600 but less than 1,900	18	1,750	3,062,500	31,500	55,125,000
1,900 but less than 2,200	5	2,050	4,202,500	10,250	21,012,500

$$\Sigma fx_m = 127{,}250 \quad \Sigma fx_m^2 = 184{,}677{,}500$$

Using Equation 5.11b:

$$s = \sqrt{\frac{184{,}677{,}500 - \dfrac{(127{,}250)^2}{95}}{94}} = 389.1 \text{ cm}$$

As the individual values are not used in this computation, it is an approximation of the sample standard deviation based on the 95 actual values in Figure 5.4 and calculated from Equation 5.10a or 5.10b, which is 405.2.

Measures of Shape

In addition to the measures of central tendency and variability that we have discussed, a population or the output of a process can also be characterized by its shape.

Skewness. We have already mentioned one property of shape in our discussion of the median, the *skewness,* or lack of symmetry of a set of data. Many variables are naturally skewed, such as surface areas, volumes, warpage, and incomes. The frequency distributions in Figure 5.20 illustrate different degrees of skewness.

A numerical measure of skewness, *Pearson's coefficient of skewness,* is defined as:

$$\text{Skewness}_P = \frac{3(\bar{x} - M_e)}{s} \tag{5.12}$$

Note that for a symmetric distribution, as in Figure 5.20(a), where the mean, the median, and the mode are the same, the coefficient will be 0; for a *positively skewed* distribution, as in Figure 5.20(b), where the mean is

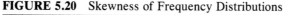

FIGURE 5.20   Skewness of Frequency Distributions

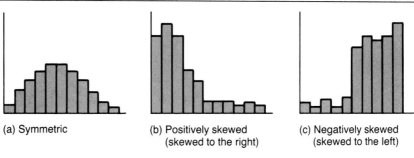

(a) Symmetric   (b) Positively skewed   (c) Negatively skewed
                                 (skewed to the right)   (skewed to the left)

larger than the median, which is larger than the mode, the coefficient will be positive; and for a *negatively skewed* distribution, as in Figure 5.20(c), the mode is larger than the median, which is larger than the mean, and the coefficient will be negative. For our 95 log lengths, an examination of the frequency displays in Figure 5.10(a) and (c) indicates approximate symmetry. We have already found the mean length to be $\bar{x} = 1335.2$ cm. The median is 1408.6 cm, and the standard deviation is 405.2 cm. Using these statistics, we can calculate the Pearson coefficient of skewness to be:

$$\text{Skewness}_P = \frac{3(1335.2 - 1408.6)}{405.2} = -0.543$$

Another way to view skewness is as a measure of the relative sizes of the tails of the distribution. Symmetric distributions, where the two tails are the same, thus have a 0 coefficient of skewness; where the difference between the frequencies in the two tails is great, the magnitude of the coefficient of skewness will be large.

Kurtosis.   Another characteristic of the shape of a frequency distribution is its peakedness, or *kurtosis*. A distribution with a relatively high concentration of data in the middle and at the tails, but low concentration in the shoulders, has a large kurtosis; one that is relatively flat in the middle, with fat shoulders and thin tails, has little kurtosis. In Figure 5.21, we see an illustration of three curves with the same mean and standard deviation, 0 skewness, and different degrees of kurtosis.

A numerical measure of kurtosis is given by[1]

$$\text{Kurtosis} = \frac{\Sigma (x - \bar{x})^4}{ns^4} - 3 \tag{5.13}$$

The bell-shaped frequency distribution (or normal curve), illustrated in Figure 5.21(a), has a kurtosis of 0 and is called *mesokurtic*. A more peaked curve, such as the one in Figure 5.21(b), is called *leptokurtic,* and

FIGURE 5.21 Kurtosis

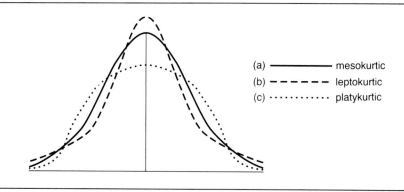

would have a positive numerical kurtosis; a flatter curve, as in Figure 5.21(c), is called *platykurtic,* and would have a negative kurtosis.

Necessary Sample Sizes for Estimating Measures of Shape.   The measures of shape we have discussed depend largely upon the tails of the distribution. As the central portion of most distributions usually contains a sizable fraction of the observed data points, measures of skewness and kurtosis are often based on the relatively small fraction of the data in the tails. Hence, some statisticians suggest sample sizes of at least 100 for estimating skewness, and at least 200 for estimating kurtosis for populations or stable processes.[2] Neither of these statistics can be interpreted for unstable processes.

SUMMARY

In this chapter we have looked at how to describe the characteristics of data taken from a population or process.

We began with a classification of data into two types: attribute data and variables data. Attribute data arise from the classification of items into categories; from counts of the number of items, or proportion, in a given category; and from counts of the number of occurrences per unit. Variables data arise from the measurement of a characteristic and from the computation of a numerical value from two or more variables measurements.

We may characterize data by both visual and numerical means. We have seen that frequency distributions provide a tabular display of numerical data grouped into non-overlapping intervals. Graphical displays offer a composite picture of the relationships and/or patterns in the data at a glance. Frequency polygons, histograms, bar charts, and run charts are useful visual tools for characterizing data.

Numerical measures summarize important characteristics of a population or process in a more quantitative way. We have seen that both grouped and ungrouped data can vary with respect to central tendency, variability, and shape; and we have discussed some important measures of these characteristics. We use the mean, median, and mode to describe the central tendency of variables data; the proportion is a useful measure for attribute data. The range and standard deviation provide useful characterizations of the variability in grouped and ungrouped variables data, and the skewness and kurtosis of a frequency distribution are measures of the shape.

In the next chapter we will explore some mathematical models that can be used to further describe the populations and processes we study.

EXERCISES

5.1 A consumer protection testing agency wants to study the life expectancy of a particular job lot of a new radial tire. Ten tires were randomly selected from the job lot. The mileages these tires achieved before the minimum tread depth was reached (and the tires declared worn out) were: 32,800; 41,700; 35,200; 39,000; 36,200; 35,600; 35,700; 45,200; 42,800; 35,700.
   a. Construct a frequency distribution and cumulative frequency distribution for these data.
   b. Using the frequency distribution above, draw a histogram of the recorded mileages for the 10 new radial tires.
   c. Is a run chart an appropriate display for these data? Explain why or why not, and if so, construct a run chart.
   d. Calculate the mean life expectancy for this sample of 10 tires.
   e. Calculate the median life expectancy for this sample of 10 tires.
   f. Calculate the modal life expectancy for this sample of 10 tires.
   g. Estimate the standard deviation of the life expectancy of all new radial tires from which this sample of 10 tires has been drawn.
   h. Calculate the range for these data.

5.2 A machine shop manager wishes to study the time it takes an assembler to complete a given small subassembly. Measurements, in minutes, are made at 15 consecutive half-hour intervals, and the times to complete the task are 12, 10, 18, 16, 4, 16, 11, 15, 15, 13, 19, 10, 15, 17, 11.
   a. Construct a frequency distribution and cumulative frequency distribution for these data.

b. Using the frequency distribution above, draw a frequency polygon for the times to complete the task. Comment on the skewness of the data.

c. Is a run chart an appropriate display for these data? Explain why or why not, and if so, construct a run chart.

d. Calculate the mean time to complete the task based upon this sample of 15 observations.

e. Calculate the median time to complete the task based upon this sample of 15 observations.

f. Calculate the modal time to complete the task based upon this sample of 15 observations.

g. Estimate the standard deviation of the time to complete the task.

h. Calculate the range for these data.

i. Calculate Pearson's coefficient of skewness for these data.

5.3 A buyer for a large chain of restaurants is about to purchase a large number of chicken breasts. According to the supplier, the breasts weigh an average of one pound each. The buyer selects a random sample of 10 breasts and weighs them, revealing the following weights, in pounds: 1.04, 1.00, 0.94, 1.10, 1.02, 0.90, 0.97, 1.03, 1.05, 0.95.

a. Calculate the mean weight for this sample of 10 chicken breasts.

b. Calculate the median weight for this sample of 10 chicken breasts.

c. Estimate the standard deviation of the weight of chicken breasts from this supplier.

d. Calculate the range of weights of chicken breasts, based upon the sample of 10 chicken breasts.

e. Calculate the kurtosis for these data.

5.4 The ABC Company is planning to analyze the average weekly wage distribution of its 58 employees during fiscal year 1987. The 58 weekly wages are available as raw data corresponding to the alphabetic order of the employees' names:

241	253	312	258	264	265
316	242	257	251	282	305
298	276	284	304	285	307
263	301	262	272	271	265
249	229	253	285	267	250
288	248	276	280	252	258
262	314	241	257	250	275
275	301	283	249	288	275
281	276	289	228	275	
170	289	262	282	260	

a. Calculate the range of the data.
b. Calculate the mean, median, and mode of the data.
c. Construct a frequency distribution for the data.
d. Calculate the mean and median from the frequency distribution above (i.e., from the grouped data).

5.5 The following frequency distribution shows the distance covered (in miles) by a sample of 80 trucks belonging to a long-distance moving company during 1987:

Distance Covered (in miles)	Number of Trucks
30,000 but less than 40,000	2
40,000 but less than 50,000	3
50,000 but less than 60,000	7
60,000 but less than 70,000	12
70,000 but less than 80,000	18
80,000 but less than 90,000	24
90,000 but less than 100,000	14

a. Represent this frequency distribution on a histogram.
b. Represent this frequency distribution on a frequency polygon.
c. Calculate the mean distance covered by the trucks in 1987.
d. Calculate the standard deviation of the distance covered by the trucks in 1987.

5.6 The following frequency distribution shows the number of minutes spent on an elementary task in an assembly line operation performed by 35 employees over a one-month period:

Number of Minutes	Number of Employees
2 but less than 4	3
4 but less than 6	8
6 but less than 8	20
8 but less than 10	4

a. Represent this information on a frequency polygon.
b. Represent this information on a histogram.
c. Calculate the mean number of minutes spent on this task.
d. Calculate the standard deviation of the time to complete this task.

5.7 Ten parts per hour are examined for defects in an assembly line process, with the following results in an eight-hour day:

Time	Number of Parts Inspected	Number of Defective Units
9:00	10	3
10:00	10	4
11:00	10	2
12:00	10	3
1:00	10	5
2:00	10	4
3:00	10	6
4:00	10	5

a. Construct a cumulative frequency distribution for these data.
b. Draw a bar chart for these data.
c. Is a run chart an appropriate display for these data? Explain why or why not, and if so, construct a run chart.
d. Calculate the average fraction defective from these data.

NOTES

1. The kth moment about the mean is $\Sigma(x - \bar{x})^k/n$, where the first moment is always 0; the second moment is the variance; the third moment is an absolute measure of skewness; and the fourth moment is an absolute measure of kurtosis.
2. Donald J. Wheeler and David S. Chambers, *Understanding Statistical Process Control* (Knoxville, Tenn.: Statistical Process Controls, Inc., 1986), p. 57.

Chapter 6

Probability Models

THE NEED FOR PROBABILITY MODELS

We have discussed the notions of obtaining and characterizing sample information from an ongoing process or a population as a basis for decision making. In any such procedure, to make inferences about the process or population we need to construct a mathematical model to describe the behavior of that process or population.

Any process or population governed by the rules of probability presented in Chapter 4 is called a *stochastic* or *probabilistic process* or *population,* and its behavior may be described by a *probability model.* More precisely, the model is a mathematical function that tells us the distribution of all possible values of some *random variable* (representing the characteristic of interest in the study) and the probability of occurrence of each of those values in the process or population.

For example, consider a stable process in which we are interested in the number of major defects per unit in valves. The random variable of interest, X, is the number of defects per valve. In a subgroup of 100 valves, we examine each valve and find that the values of this random variable range between 0 and 3, as shown in the frequency distribution in Figure 6.1(a). The probability distribution tells us the probability, f(x), that the number of defects per unit in the subgroup is x; this is shown in tabular form in Figure 6.1(b).

Naturally, processes and populations that behave differently will have different probability distributions for the random variable of interest. Moreover, as we never have all the data when conducting an analytic study, and rarely have all the data when conducting an enumerative study, we usually will not be able to develop the probability distribution directly from the data but will have to deduce the appropriate probability model, based on our knowledge of the process or population. For analytic studies, probability distributions can provide only approximations of the probabilities of a process's behavior because all processes vary over time,

---

FIGURE 6.1   Defects per Valve in a Subgroup of 100 Valves

---

*(a)   Frequency Distribution of Number of Defects per Valve
in a Subgroup of 100 Valves*

Number of Defects	Number of Valves
0	92
1	3
2	4
3	1

*(b)   Probability Distribution of Number of Defects per Valve
in a Subgroup of 100 Valves*

Number of Defects, X	f(x)
0	0.92
1	0.03
2	0.04
3	0.01

---

even stable processes. However, for analytic studies we will compute all probabilities as if they are exact but will realize that all such computations are only approximations. In this chapter we introduce some important probability distributions and the conditions under which they can serve as mathematical models in quality studies.

TYPES OF RANDOM VARIABLES AND PROBABILITY MODELS

In Chapter 5 we distinguished between attribute data, such as the number of errors in a set of invoices or the number of defects in a batch of TV picture tubes, and variables data, such as the carbon content in a molten metal bath in weight percent, or the tensile strength of steel parts in pounds. Attribute data come from the measurement of a *discrete random variable,* which can take on only specific values within a given interval. Variables data come from the measurement of a *continuous random variable,* which can take on any value within an interval. In a stochastic population or process, if the characteristic being measured is a discrete random variable, its probability distribution is given by a *discrete probability model;* the probability distribution of a continuous random variable is a *continuous probability model.*

Discrete Probability Models

A *probability distribution,* f(x), of a discrete random variable X must satisfy two mathematical conditions:

1. All values of f(x) must be between 0 and 1, inclusive. This, of course, is consistent with our intuitive notion that probabilities cannot be negative (the probability of two defects per valve cannot be −0.6) and can be at most 1 (the probability of three defects per valve cannot be 7.5).
2. The sum of the probabilities for all values of a random variable must be equal to 1. This quantifies the intuitive notion that the random variable must take on one of its values.

We see that the function f(x) in Figure 6.1(b) satisfies the two conditions, and hence is a probability distribution.

The *cumulative distribution function,* F(x), of a discrete random variable X is the probability that X takes on a numerical value of x or less. For example, F(2) is the probability that the number of defects per valve is 2 or less, and is equal to f(0) + f(1) + f(2) = 0.99.

In Chapter 5 we discussed the calculation of the mean and variance as measures for describing data. We use the same quantities to characterize the central tendency and dispersion of a probability distribution.

The *mean,* or *expected value,* of a discrete random variable is denoted by the Greek letter μ, and can be calculated as:

$$\mu = \Sigma \; x \; f(x) \tag{6.1}$$

That is, the expected value is obtained by multiplying the values of the random variable X by the probabilities that those values occur, and summing the products. For the probability distribution in Figure 6.1(b), the expected number of defects per valve in a subgroup of 100 valves is, from Equation 6.1:

$$\mu = \Sigma \; x \; f(x) = (0)(0.92) + (1)(0.03) + (2)(0.04) + (3)(0.01)$$

$$= 0.14 \text{ defects per valve}$$

The *variance* of a discrete probability distribution is denoted by the Greek symbol σ^2, and can be calculated as:

$$\sigma^2 = \Sigma \; (x - \mu)^2 \; f(x) \tag{6.2a}$$

or using the equivalent but sometimes computationally simpler expression:

$$\sigma^2 = \Sigma \; x^2 \; f(x) - \mu^2 \tag{6.2b}$$

Thus, for the probability distribution in Figure 6.1(b), whose mean we found to be 0.14, we calculate the variance to be, from Equation 6.2b:

$$\Sigma \; x^2 f(x) - \mu^2 = (0)^2(.92) + (1)^2(.03) + (2)^2(.04) + (3)^2(.01) - (.14)^2$$

$$= 0.2604$$

We now examine some discrete probability distributions.

The Binomial Distribution.   The binomial probability model describes populations or processes where items have the following properties:

1. Each item has exactly one of two values of the attribute (e.g., the item is either defective or not defective; the item passes inspection or it fails inspection).
2. The probability of each of the two values of the attribute, p and $1 - p$, respectively, are the same for each item (e.g., the probability that any single item is defective is $p = 0.05$; the probability that any single item is not defective is $1 - p = 0.95$).
3. The items are independent of each other (i.e., the probability of the occurrence of an item with one of the values of the attribute is unaffected by the occurrence or nonoccurrence of any other item with that value of the attribute).

If we consider the examination of n such items, the probability that exactly x of them possess one of the attribute values (e.g., the probability that x items are defective) is given by:

$$f(x) = \binom{n}{x} p^x (1 - p)^{n-x} \qquad \text{for } x = 0, 1, 2, \ldots, n \qquad (6.3a)$$

where $\binom{n}{x}$ is called the *binomial coefficient* and is the number of combinations of n things taken x at a time, or,

$$f(x) = \frac{n!}{x!(n - x)!} p^x (1 - p)^{n-x} \qquad \text{for } x = 0, 1, 2, \ldots, n \qquad (6.3b)$$

n! ("n factorial") is the product of all positive integers less than or equal to n. Thus, $5! = (5)(4)(3)(2)(1) = 120$; $2! = (2)(1) = 2$; $1! = 1$. By definition, $0! = 1$. In Table 4, binomial coefficients $\binom{n}{x}$ are given for n between 0 and 20.

To illustrate the use of the binomial distribution, suppose a random sample of 10 items is chosen from a population known to contain 5 percent defective units. Thus, every item is either defective or nondefective. If the proportion defective in the population is known to be 5 percent, then every item has a probability $p = 0.05$ of being defective, and a probability $1 - p = 0.95$ of being nondefective. As the sample is randomly selected, the items are independent of each other. Thus, we have a binomial model; if we want to know the probability that exactly 2 of the 10 chosen items are defective, we use Equation 6.3b, where $n = 10$, $p = 0.05$, and $x = 2$:

$$f(2) = \frac{10!}{2! \, 8!} (0.05)^2 (0.95)^8$$

$$= \frac{(10)(9)(8)(7)(6)(5)(4)(3)(2)(1)}{(2)(1)(8)(7)(6)(5)(4)(3)(2)(1)} (0.05)^2 (0.95)^8$$

$$= 0.0746$$

For sample sizes from n = 2 to n = 25, Table 5 gives the value of f(x) for selected values of p. In our example, for p = 0.05 and n = 10, the table shows f(2) = 0.0746.

In an analytic study of a process, the probability of a defective in the supply is unknown. Recalling our discussion of the proportion in an analytic study in Chapter 5, if we assume the process is stable, we may estimate the probability of a defective in the future by the average proportion in a randomly selected sample or subgroup, $\bar{p}$:

$$\bar{p} = \frac{r}{n} \tag{6.4}$$

where r is the number of defective items in the sample and n is the total number of items in the sample.

Suppose we examine a display of 12 light bulbs from a stable process for cracks in the glass to determine the need for future action on the process. An analytic study of the process yields an estimated fraction defective of $\bar{p} = 0.05$. If these cracks occur independently, the probability that of 12 light bulbs examined exactly 2 will be defective is, from Equation 6.3b:

$$f(2) = \frac{12!}{2!10!} (0.05)^2 (0.95)^{10} = 0.0988$$

The cumulative probability of 2 or fewer cracked bulbs in a sample of 12 from the process is:

$$F(2) = f(0) + f(1) + f(2)$$

$$= \frac{12!}{0!12!} (.05)^0 (.95)^{12} + \frac{12!}{1!11!} (.05)^1 (.95)^{11} + \frac{12!}{2!10!} (.05)^2 (.95)^{10}$$

$$= 0.5404 + 0.3413 + 0.0988 = 0.9805$$

Put another way, in a process where cracked bulbs occur independently with constant probability 0.05, the probability that in a sample of 12 bulbs more than 2 will be cracked is approximately 2 percent.

The Mean and Variance of the Binomial Distribution. The average number of items possessing the attribute value of interest, or the expected value of the binomial distribution, is:

$$\mu = np \tag{6.5}$$

(This result can be derived mathematically from the formula for the expected value of a probability distribution in Equation 6.1, where f(x) is given by Equation 6.3a). For example, the average number of defective items in a sample of 10 items, where the fraction defective is p = 0.3, would be (10)(0.3) = 3. When the process fraction defective can be estimated by $\bar{p}$, the average number of defective items in the sample of n items would be $n\bar{p}$.

Thus, when we examine 12 light bulbs, we would expect to find $(12)(0.05) = 0.6$, or less than 1, cracked bulb, on average, given that the process is stable.

The variance of the number of defectives in a sample of size n can be shown to be:

$$\sigma^2 = np(1 - p) \tag{6.6}$$

Estimating p by $\bar{p}$, we have, for the manufacture of light bulbs, the variance of the number of cracked bulbs in a sample of 12 as $(12)(0.05)(0.95) = 0.57$, assuming a stable process.

The Poisson Distribution.   For events that occur randomly and independently at a constant rate of λ occurrences per unit, the probability of x such occurrences per unit is given by the Poisson probability distribution:

$$f(x) = \frac{\lambda^x e^{-\lambda}}{x!} \qquad \text{for x = 0, 1, 2, . . .} \tag{6.7}$$

where e is the base of the natural logarithm, or 2.71828... Table 6 shows values of $e^{-\lambda}$ for λ between 0.0 and 9.9. Table 7 gives values of f(x) for the Poisson distribution for values of λ between 0.1 and 10.0.

Applications of the Poisson distribution are the probabilities associated with the number of imperfections per square foot in rolls of metal; the number of complaints per day at a retail business; the number of cashews per can in cans of mixed nuts; or the number of telephone calls per hour from a bank of pay phones in an airline terminal. Where the events can be assumed to occur randomly and independently, and the average number of occurrences per unit, λ, is known, Equation 6.7 can be used to calculate the probability of a given number of occurrences per unit.

In an accounting office, bad checks arrive randomly and independently. The process is studied over a period of 12 months; is found to be stable (via methods discussed in Chapters 8 through 10); and the average number of bad checks per month, λ, is estimated to be 3. The probability that five bad checks arrive in a given month is then:

$$f(5) = \frac{(3)^5 e^{-3}}{5!} = 0.1008$$

Alternatively, from Table 7, f(5), the probability that five bad checks arrive in a given month, is 0.1008.

The Poisson Distribution as an Approximation to the Binomial Distribution.   We have seen that the binomial distribution can be used to calculate the probability of x occurrences in n items when the probability of the occurrence remains constant. When n is large and p is small (for example, when large numbers of items are examined and the occurrence of defects is rare), the binomial coefficients in Equation 6.3a become unwieldy (even for hand-held calculators). In such cases, however, the exact binomial probabilities can be approximated quite accurately by the Poisson distribution:

$$f(x) = \frac{(np)^x e^{-np}}{x!} \tag{6.8}$$

where n is the number of items and p is the probability that a single item is defective. The larger n, and the smaller p, the closer these probabilities are to the exact binomial probabilities.

Records show that 0.04 percent of the trucks crossing a particular bridge have brake failures. If, through statistical methods, we determine that the brake failure process is stable over time, we may use such historical data to estimate p as $\bar{p} = 0.0004$.

The probability that among 6,000 trucks crossing the bridge exactly 1 will have a brake failure is then given by the binomial probability distribution. From Equation 6.3b:

$$f(1) = \frac{6000!}{1!5999!} (.0004)^1 (.9996)^{5999} = 0.2177.$$

As the calculation is a cumbersome one, and as n is large and $\bar{p}$ is small, we may use the Poisson distribution to approximate the probability of exactly one brake failure. In Equation 6.8, $n\bar{p} = (6,000)(0.0004) = 2.4$, and

$$f(1) = \frac{(2.4)^1 e^{-2.4}}{1!} = 0.2177.$$

The Mean and Variance of the Poisson Distribution. The average number of occurrences per unit in a Poisson process is $\mu = \lambda$. Thus, in our bad check example, the expected number of bad checks per unit is 3 per month. For the Poisson approximation to the binomial distribution, the mean is given by:

$$\mu = np \tag{6.9}$$

If we observed 6,000 trucks crossing the bridge in our last example, the expected number of brake failures would be 2.4.

The variance of the Poisson distribution is identical to its mean. Thus, $\sigma^2 = \lambda$, or, for the Poisson approximation to the binomial distribution:

$$\sigma^2 = np \tag{6.10}$$

For example, the variance in the number of brake failures would be 2.4.

Continuous Probability Models

A probability distribution f(x) of a continuous random variable, often called a *probability density function,* must satisfy two mathematical conditions analogous to the discrete distributions we have discussed:

1. All values of f(x) must be nonegative.
2. The total area under the graph of the probability function f(x) must be 1.

From these two conditions, the probability that a random variable takes on some value between a and b is given by the area under the graph of f(x) between X = a and X = b; the probability that a continuous random variable takes on a particular individual value is always 0. That is, for continuous random variables, numerical probabilities can be calculated only over intervals and are defined to be 0 at any individual point.

The *cumulative distribution function*, F(x), of a continuous random variable X, is the probability that the random variable X takes on a value of x or less. Thus, F(a) is the probability that the random variable is less than or equal to a, and is the area under the graph of f(x) to the left of X = a.

As an example, consider the function f(x) = 2x where $0 \leq X \leq 1$. Its graph is shown in Figure 6.2(a). We see that all the values of f(x) are nonnegative; the total area under the curve shown in Figure 6.2(b), can be calculated as the area of a triangle with base 1 and height 2, or (1/2)bh = (1/2)(1)(2) = 1. Thus, f(x) is a probability distribution.

The probability that the random variable X in this example takes on a value between, say, 0 and 1/2, is given by the area under the curve between X = 0 and X = 1/2, shown in Figure 6.2(c). This region is a triangle with base 1/2 and height 1, or an area of (1/2)(1/2)(1) = 1/4. The probability that X takes on a value between 0 and 1/2 is equal to 1/4, or, symbolically, F(1/2) = P(0 < X < 1/2) = 1/4. Note that because X is a continuous random variable, P(X = 1/2) = 0.

For most continuous probability distributions, the area of interest is not a simple geometric figure and must be calculated by means of integral calculus. However, we will see that for many distributions, these areas are conveniently tabulated.

As we have discussed earlier, the mean and variance are important characteristics of probability distributions. Although their computation for continuous distributions in general requires a knowledge of integral calculus, we will be able to specify them for some particular distributions of interest. We examine these now.

The Normal Distribution. One of the most important continuous probability models is the normal distribution. A normally distributed random variable has the probability density function:

$$f(x) = \frac{1}{\sigma\sqrt{2\pi}} e^{-1/2 \left(\frac{x - \mu}{\sigma}\right)^2} \quad \text{for } -\infty < X < \infty \qquad (6.11)$$

where μ denotes the mean, σ denotes the standard deviation, π is the

FIGURE 6.2   Continuous Probability Model

(a) A Continuous Probability Function f(x)

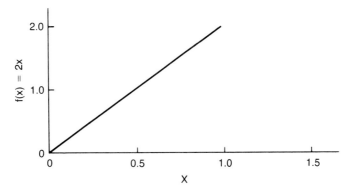

(b) Total Area Under the Curve of f(x)

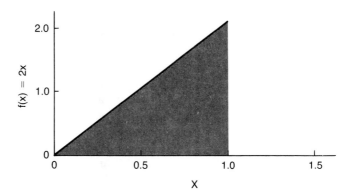

(c) Area Under the Curve of f(x) between X = 0 and X = ½

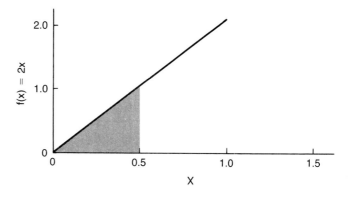

FIGURE 6.3 The Normal Probability Distribution

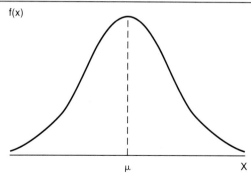

constant 3.14159..., and e is the base of the natural logarithm, 2.71828...
Figure 6.3 illustrates its familiar bell-shaped graph.

Although all normal curves are bell-shaped, different means and standard deviations result in the different appearances shown in Figure 6.4. In Figure 6.4(a), different means cause a shift in the curve along the x-axis. In Figure 6.4(b), the values of the random variable in the curve with $\sigma = 2$ vary less from the mean than those of the random variable with $\sigma = 8$, hence the first has a higher peak and lower tails. Figure 6.4(c) illustrates two normal curves with different means and different standard deviations.

Note that all normal curves are symmetrical; the mean, median, and mode occur at the same value of the random variable X. A normally distributed random variable X can take on any real values, although from Figure 6.4, we see that values of X are always more likely to fall close to the mean, μ, rather than in the tails of the distribution. In fact, while the total area under any normal curve is 1 (as f(x) is a probability distribution), for any values of μ and σ, 68.27 percent of the area falls within one standard deviation of the mean, or between the x-values of $\mu - \sigma$ and $\mu + \sigma$, shown in Figure 6.5(a). Similarly, 95.45 percent of the area under the curve falls within two standard deviations of the mean, or between the x-values of $\mu - 2\sigma$ and $\mu + 2\sigma$, shown in Figure 6.5(b); 99.73 percent of the area falls within three standard deviations of the mean, or between the x-values of $\mu - 3\sigma$ and $\mu + 3\sigma$, shown in Figure 6.5(c). Only 0.27 percent of the area under the curve lies outside this "3-sigma" interval.

Recall that the area under a probability distribution between two values of its random variable corresponds to the probability that the random variable takes on any value in that interval. Thus, for a normally distributed random variable X with mean μ and standard deviation σ, the probability that X takes on values between $\mu - \sigma$ and $\mu + \sigma$ is 0.6827; the probability that X takes on values between $\mu - 2\sigma$ and $\mu + 2\sigma$ is 0.9545,

FIGURE 6.4 The Normal Probability Distribution

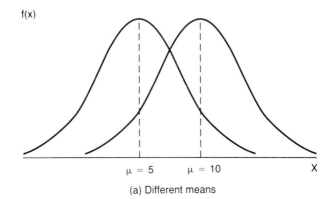

(a) Different means

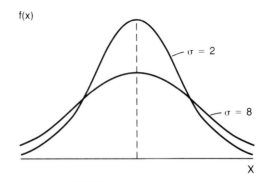

(b) Different standard deviations

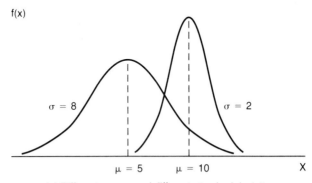

(c) Different means and different standard deviations

FIGURE 6.5   Some Areas Under the Normal Curve

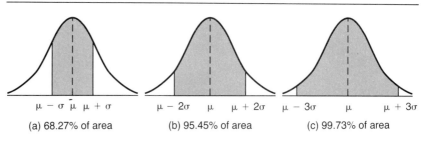

$\mu - \sigma$ μ $\mu + \sigma$	$\mu - 2\sigma$ μ $\mu + 2\sigma$	$\mu - 3\sigma$ μ $\mu + 3\sigma$
(a) 68.27% of area	(b) 95.45% of area	(c) 99.73% of area

and the probability that X takes on values between $\mu - 3\sigma$ and $\mu + 3\sigma$ is 0.9973.

To illustrate, suppose we know that the lifetimes of electric fuses manufactured by a particular stable process are normally distributed with a mean of μ 10,000 hours and a standard deviation of $\sigma = 170$ hours. Then we can make the following statements:

1. The probability that a randomly selected fuse will last more than 10,000 hours is 0.50, and the probability that a randomly selected fuse will last fewer than 10,000 hours is 0.50 (because the normal probability distribution is symmetric about its mean).
2. The probability is 0.6827 that a randomly selected fuse will last between (10,000 − 170 =) 9,830 hours and (10,000 + 170 =) 10,170 hours.
3. The probability is 0.9545 that a randomly selected fuse will last between (10,000 − 2[170] =) 9,660 hours and (10,000 + 2[170] =) 10,340 hours.
4. The probability is 0.9973 that a randomly selected fuse will last between (10,000 − 3[170] =) 9,490 hours and (10,000 + 3[170] =) 10,510 hours.
5. The probability is only 0.0027 that a fuse will last fewer than 9,490 hours or more than 10,510 hours. Because of the symmetry of the normal distribution, we can say that the probability is 0.00135 that a fuse will last fewer than 9,490 hours.

While our example makes certain unrealistic assumptions (namely, that we know that the distribution of our random variable happens to be normal and that we know the process mean μ and process standard deviation σ), we will see that in realistic situations the normal distribution is, nevertheless, an essential one. We will first discuss, in general terms, how we can calculate areas (and therefore probabilities) under the normal curve between any two values of the random variable (not simply between integer multiples of the standard deviation); we then will examine how the normal distribution is used in statistical studies.

Probabilities Under the Normal Curve.   To calculate the probability that an electric fuse lasts between, say, 10,100 hours and 10,200 hours, we would need to calculate the area under a normal curve with mean 10,000 and standard deviation 170 between $x_1 = 10,100$ and $x_2 = 10,200$. This area is the shaded region of Figure 6.6(a).

We can calculate such an area by *standardizing* our random variable X and using a table of areas under a *standard normal curve*. We define a new random variable Z as the number of standard deviations between the value of the random variable X and the mean of the distribution μ:

$$Z = \frac{X - \mu}{\sigma} \tag{6.12}$$

The random variable Z has a normal distribution with a mean of 0 and a standard deviation of 1, as shown in Figure 6.6(b).

Any value on the x-axis will yield exactly one z-value. Thus, $x_1 = 10,100$ hours corresponds to a z-value of:

$$z_1 = \frac{10,100 - 10,000}{170} = 0.59$$

(that is, $x_1 = 10,100$ is 0.59 standard deviations from the mean of 10,000 hours). $x_2 = 10,200$ hours corresponds to a z-value of:

$$z_2 = \frac{10,200 - 10,000}{170} = 1.18$$

(that is, $x_2 = 10,200$ is 1.18 standard deviations from the mean of 10,000 hours). The area under the standard normal curve between $z_1 = 0.59$ and $z_2 = 1.18$, shown in Figure 6.6(c), is identical to the desired area in Figure 6.6(a).

Thus, given a single table that allows us to find the areas between different values of Z under the standard normal curve, we could find the area between any two values of any normally distributed random variable X by standardizing to z-values first.

Table 1 provides the areas between 0 and positive values of Z. To use

FIGURE 6.6 Standardizing Normal Probabilities

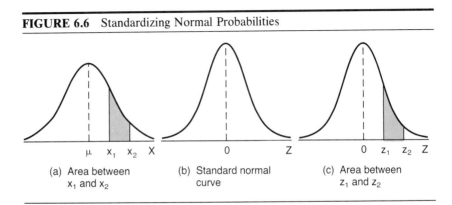

(a) Area between x_1 and x_2

(b) Standard normal curve

(c) Area between z_1 and z_2

the table, values for Z should always be read to two digits after the decimal point. The units and tenths portion are found in the column on the extreme left side of the table. For $z_1 = 0.59$ this is the 0.5 row of the table. The hundredths digit can be found by moving out to the right into the column with the appropriate digit in the top row, or .09 in our case. The value in the body of the table corresponding to $Z = 0.59$ is found at the intersection of the 0.5 row and the .09 column, or .2224. This means that the area between $Z = 0$ and $Z = 0.59$ under the standard normal curve is 0.2224 [see Figure 6.7(a)]. For $z_2 = 1.18$, the value in the table at the intersection of the 1.1 row and the .08 column is .3810. This is the area between $Z = 0$ and $Z = 1.18$ under the standard normal curve [see Figure 6.7(b)].

But in our example, we see from Figure 6.6(c) that the desired area is the difference between the areas in Figure 6.7(b) and Figure 6.7(a); or, the probability that Z is between 0.59 and 1.18 is $0.3810 - 0.2224 = 0.1586$. Symbolically, we can write:

$$P(10,100 < X < 10,200) = P(0.59 < Z < 1.18)$$
$$= P(0 < Z < 1.18) - P(0 < Z < 0.59)$$
$$= 0.3810 - 0.2224$$
$$= 0.1586$$

Thus, 15.86 percent of all electric fuses from this process will have lifetimes of between 10,100 hours and 10,200 hours.

To further illustrate how Table 1 can be used to find any area under a normal curve, consider the areas in Figure 6.8, each under our normal distribution with mean $\mu = 10,000$ hours and standard deviation $\sigma = 170$ hours.

FIGURE 6.7   Calculating Areas under the Standard Normal Curve

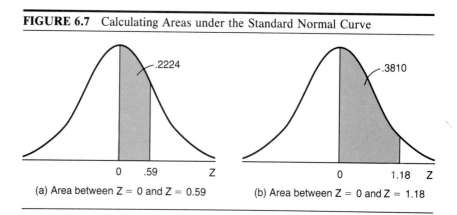

(a) Area between Z = 0 and Z = 0.59   (b) Area between Z = 0 and Z = 1.18

FIGURE 6.8 Areas under the Normal Curve with $\mu = 10,000$ and $\sigma = 170$

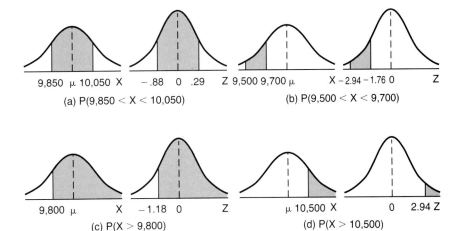

(a) P(9,850 < X < 10,050)

(b) P(9,500 < X < 9,700)

(c) P(X > 9,800)

(d) P(X > 10,500)

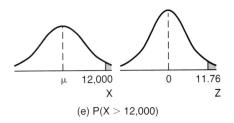

(e) P(X > 12,000)

1. To calculate the probability that X is between 9,850 and 10,050, we standardize the two x-values using Equation 6.12:

$$z_1 = \frac{9,850 - 10,000}{170} = -0.88$$

$$z_2 = \frac{10,050 - 10,000}{170} = 0.29$$

Thus we need to find the area under the standard normal curve between $z_1 = -0.88$ and $z_2 = 0.29$, or $P(9,850 < X < 10,050) = P(-0.88 < Z < 0.29)$. This probability can be written as the area between $Z = -0.88$ and $Z = 0$ plus the area between $Z = 0$ and $Z = 0.29$:

$$P(-0.88 < Z < 0.29) = P(-0.88 < Z < 0) + P(0 < Z < 0.29)$$

But as the standard normal curve is symmetric about its mean of 0, the area under the curve between any negative value of Z and 0 is identical to the value in the table corresponding to the absolute (or positive) value of Z. Thus, $P(-0.88 < Z < 0.29) = 0.3106 + 0.1141 = 0.4247$.

2. $P(9,500 < X < 9,700) = P(-2.94 < Z < -1.76)$. Again, because of symmetry, this can be written as $P(0 < Z < 2.94) - P(0 < Z < 1.76) = 0.4984 - 0.4608 = 0.0376$.

3. $P(X > 9,800) = P(Z > -1.18)$. Recall that the area to the right of the mean of any normal distribution is 0.5000. Thus, $P(Z > -1.18) = P(-1.18 < Z < 0) + 0.5000$, or, because of symmetry, $= P(0 < Z < 1.18) + 0.5000 = 0.3810 + 0.5000 = 0.8810$.

4. $P(X > 10,500) = P(Z > 2.94)$. Again because the area to the right of the mean is 0.5000, $P(Z > 2.94) = 0.5000 - P(0 < Z < 2.94) = 0.5000 - 0.4984 = 0.0016$.

5. $P(X > 12,000) = P(Z > 11.76) = 0.5000 - P(0 < Z < 11.76)$. However, the table only shows values of Z up to 3.09. This is because, as we mentioned, almost all of the area under the normal curve is within three standard deviations of the mean; or, in terms of the standard normal curve, $P(-3 < Z < 3)$ is approximately 1. From the table, we see that $P(0 < Z < 3.09) = 0.4990$. Thus, $P(Z > 3.09) = 0.5000 - 0.4990 = 0.0010$; the probability that Z is larger than 11.76 is even smaller. Thus, $P(Z > 11.76) \cong 0$.

A convenient notation for any z-value of interest is z_α, or the z-value such that the area to the right of z_α is α. From our last example, $z_{.001} = 3.09$ because the area to the right of $Z = 3.09$ is 0.001.

If we wanted to know the z-value that would yield an area of 0.4750 between 0 and z, then the area to the right of z is 0.025 and we are looking for $z_{.025}$. From Table 7, an area of 0.475 corresponds to a z-value of $z_{.025} = 1.96$.

Figure 6.9 shows weights in ounces of 40 bags of flour marked "5 lbs net weight" randomly selected from a stable packaging process. As we cannot examine the entire process, past, present, and future, in an analytic study, we do not know the probability distribution of weights nor do we know its mean or its standard deviation. We will see later in this chapter how we can deduce the functional form of a probability distribution from an examination of a sample from the process. Let us assume for now that the weights of the bags from this process are normally distributed and use some of the techniques discussed in Chapter 5 to estimate the parameters of that distribution.

We can estimate the mean of the distribution by $\bar{x}$, as given by Equation 5.1; we find $\bar{x} = 81.86$ ounces. Similarly, we can estimate the standard deviation from Equation 5.10a or 5.10b; we find $s = 1.67$ ounces. Suppose we wish to find the proportion of bags that weigh five pounds or more. Put another way, if we assume the probability distribution is normal with an estimated mean of $\bar{x} = 81.86$ and an estimated standard deviation of $s = 1.67$, we wish to find the probability that a randomly selected bag from

FIGURE 6.9 Weights of Flour in a Sample of 40 "5 lb" Bags

Bag #	Weight (oz)	Bag #	Weight (oz)	Bag #	Weight (oz)
1	80.0	15	79.5	29	81.5
2	81.5	16	82.5	30	80.5
3	81.0	17	82.0	31	83.0
4	83.0	18	81.0	32	83.0
5	83.5	19	82.5	33	86.0
6	82.0	20	80.5	34	81.0
7	81.5	21	79.5	35	81.0
8	78.5	22	85.0	36	83.0
9	82.5	23	79.5	37	80.0
10	79.5	24	83.0	38	81.0
11	82.0	25	85.0	39	81.5
12	83.5	26	83.0	40	83.5
13	82.0	27	80.0		
14	83.0	28	83.0		

this process weighs 80 ounces or more, P(X > 80), as shown in Figure 6.10(a). As we are using estimated values,

$$z = \frac{x - \bar{x}}{s} \tag{6.13}$$

or, $z = \dfrac{80 - 81.86}{1.67} = -1.11$. Since P(X > 80) = P(Z > -1.11), as shown in Figure 6.10(b), the probability that a randomly selected bag weighs 80 ounces or more is 0.5000 + 0.3665 = 0.8665, or 86.65 percent of the bags weigh five pounds or more.

FIGURE 6.10 Proportion of Bags in Shipment Weighing 80 Ounces or More

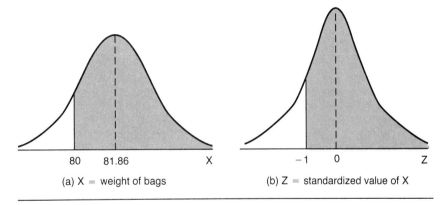

(a) X = weight of bags

(b) Z = standardized value of X

Suppose we wish to determine the weight, x_1, such that 90 percent of the bags weigh x_1 ounces or more; then the probability that a randomly selected bag weighs x_1 ounces or more is 0.90. This means that the area to the right of x_1 under the normal curve under consideration must be 0.90, as shown in Figure 6.11(a). As this required area is larger than 0.5, we know x_1 must lie to the left of the mean. Put another way, we know the area between x_1 and the mean must be $0.90 - 0.50 = 0.40$. But if we standardize x_1 to

$$z_1 = \frac{x_1 - \overline{x}}{s} = \frac{x_1 - 81.86}{1.67}$$

then the area between z_1 and 0 must be 0.4000, as shown in Figure 6.11(b). Note that although z_1 is negative, because of symmetry the area between 0 and the absolute value of z_1 must also be 0.4000. In Table 7, we look in the body of the table for the area closest to 0.4000. We see that an area of 0.3997 corresponds to a z-value of 1.28. Recall, however, that z_1 is actually negative, so we have

$$z_1 = -1.28 = \frac{x_1 - 81.86}{1.67}$$

or $x_1 = 79.72$ ounces.

The computations we have illustrated are based upon a normal probability distribution for the random variable of interest. In Chapter 7, in our discussion of sampling distributions, we will see that even where we cannot assume that a particular random variable has a normal probability distribution, other numerical characteristics of the frame or process are distributed normally nonetheless. In our subsequent discussions of quality control procedures and analysis, it is such normally distributed characteristics that help us to make predictions about a process's future behavior.

FIGURE 6.11   Minimum Weight of 90 Percent of Bags in Shipment of Flour

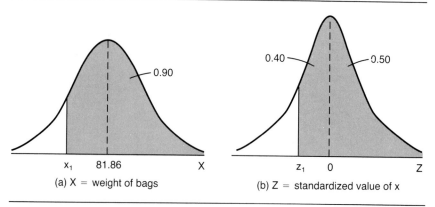

(a) X = weight of bags

(b) Z = standardized value of x

The Normal Approximation to the Binomial Distribution.   We have already seen that over certain ranges of parameter values, probability distribution values that are difficult to evaluate numerically can be approximated by other distributions. For example, we saw that when sample sizes are large and the probability of an occurrence, p, is small, the binomial probability can be approximated by the Poisson distribution.

When n is fairly large and the probability of an occurrence, p, is 0.50, the binomial distribution is a symmetric one whose histogram has the characteristic bell-shaped pattern of the normal distribution. Under these conditions, the cumulative binomial probabilities are very close to those of a normal distribution with mean

$$\mu = np \tag{6.14}$$

and standard deviation

$$\sigma = \sqrt{np(1 - p)} \tag{6.15}$$

As a rule of thumb, the normal approximation to the binomial probability is valid when p is close to 0.50 and both np and $n(1 - p)$ are larger than 5.

Suppose a department store prepares labels for its catalogues and sorts them by zip codes. Samples from this stable process are selected and checked for errors in zip codes, such as reversed digits, missing digits, and incorrect digits. Then the number of labels with errors in a sample of size n has a binomial distribution.

Suppose in a sample of 50 randomly selected labels 22 had errors. Then, from Equation 6.4, $\bar{p} = 22/50 = 0.44$. The probability that if 20 randomly selected labels were examined 4 would have errors is a binomial probability given by Equation 6.3b:

$$f(4) = \frac{20!}{4!16!} (0.44)^4(0.56)^{16} = 0.0170$$

Note that $\bar{p}$, our estimate of p, is close to 0.50, $n\bar{p} = (20)(0.44) = 8.8$ and $n(1 - \bar{p}) = (20)(0.56) = 11.2$. To use the normal approximation, we must first make a *continuity correction*. That is, the binomial probability distribution is a discrete distribution, and f(4) is the probability that exactly four labels have errors. However, for a continuous distribution, recall that the probability of a particular value of X is always 0. To circumvent this conceptual difficulty, we calculate the probability that X is between 3.5 and 4.5; that is, any discrete integer value j from the binomial distribution is represented by the interval $(j - 0.5, j + 0.5)$ under the normal curve.

Thus, we approximate the binomial probability that 4 labels in 20 would have errors by the normal probability $P(3.5 < X < 4.5)$. From Equations 6.14 and 6.15, we estimate the mean of this distribution by $n\bar{p} = (20)(0.44) = 8.8$ and the standard deviation by $\sqrt{n\bar{p}(1 - \bar{p})} = \sqrt{(20)(0.44)(0.56)} = 2.22$.

Standardizing the values $x_1 = 3.5$ and $x_2 = 4.5$:

$$z = \frac{x - 8.8}{2.22}$$

$$P(3.5 < X < 4.5) = P(-2.39 < Z < -1.94) = 0.0178$$

The Student t-Distribution. The *Student t-distribution,* a continuous probability distribution first investigated by W. S. Gosset, who wrote under the pseudonym of "Student," is closely related to the normal distribution. We will see in Chapter 7 that it is often used in studies where sample sizes are small. We introduce the distribution here as an important example of a continuous probability model.

Shown in Figure 6.12, the t-distribution is a symmetrical one that in its standard form has a mean of $\mu = 0$ and a standard deviation of $\sigma = \sqrt{(n - 1)/(n - 3)}$ where n is the sample size in the study. The exact shape of the t-distribution depends on the number of *degrees of freedom,* generally n − 1, where n is the sample size. As the number of degrees of freedom increases, the shape of the t-distribution approaches that of the normal distribution with mean 0 and standard deviation 1. We find areas of interest under the t-distribution from Table 2.

To illustrate, suppose we wish to find the t-value such that 47.5 percent of the area under the curve is between 0 and t. Using the notation we introduced for z-values, we are looking for the t-value such that 2.5 percent of the area under the curve is to its right, or $t_{.025}$. As the t-distribution has different shapes (and hence different areas) for different numbers of degrees of freedom, we must specify the sample size in our study. If n = 10, then the number of degrees of freedom is n − 1 = 9, and $t_{.025}$ is 2.262. If

FIGURE 6.12   The Student t-Distribution

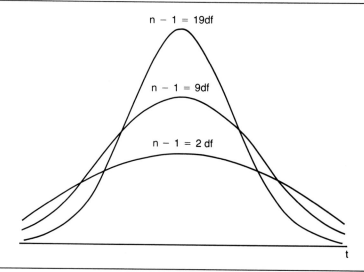

$n = 20$, then there are $n - 1 = 19$ degrees of freedom, and $t_{.025} = 2.093$. As the sample size increases, note that the t-value decreases; when the number of degrees of freedom is 30 or more, we find the t-value in the line corresponding to an infinite number of degrees of freedom, or $t_{.025} = 1.960$. Note that $z_{.025}$ is the value of Z such that the area between 0 and z is 0.4750; from Table 1, the z-value corresponding to an area of 0.4750 is $z = 1.96$. Thus, when the number of degrees of freedom is 30 or more, the t-distribution is approximately the same in shape as the standard normal distribution.

FITTING PROBABILITY MODELS TO FREQUENCY DISTRIBUTIONS

In some cases, as we have seen, the nature of the process under study enables us to determine the appropriate probability model from theory alone. For example, a stable process where each item is defective or nondefective, the probability of a defective item is the same for all items, and the items are independent of each other is a *binomial process*. Given these conditions, the probability of x defective items in a sample of n items is given by the binomial probability function, Equation 6.3a or 6.3b.

In other cases, however, we are not able to identify the appropriate probability function based upon such theoretical considerations alone. In these cases we must use sample information to determine the probability distribution that describes the behavior of the stable process or frame; in general, the larger the sample size, the better our information concerning the probability distribution.

We illustrate below the use of qualitative methods of identification, based on a visual and/or graphical inspection of frequency distributions. We may examine frequency distributions of sample data to determine, without extensive quantitative considerations or calculations, the appropriate probability distributions for analytic studies. We illustrate using a randomly selected sample of 40 containers of fertilizer from a stable process, shown in Figure 6.13(a). A "quick and dirty" test is to examine the frequency distribution of the sample data visually for indications as to the appropriate probability model.

Visual Inspection

Figures 6.13(b) and (c) show a frequency distribution and histogram, respectively, of the sample data in Figure 6.13(a). We may estimate the mean from Equation 5.1 and the standard deviation from Equation 5.10a or 5.10b; we find $\bar{x} = 81.96$ and $s = 1.77$. We see that the histogram possesses some of the characteristics that distinguish a normal probability distribution: it is symmetrical and somewhat bell-shaped; 63 percent of the values fall between $\bar{x} - s$ and $\bar{x} + s$; approximately 98 percent fall

FIGURE 6.13   Weights of Fertilizer in a Sample of 40 Containers

(a) Raw Data

#	Weight (lb)	#	Weight (lb)	#	Weight (lb)	#	Weight (lb)
1	82.0	11	84.0	21	79.5	31	81.0
2	83.0	12	81.0	22	79.5	32	86.0
3	84.5	13	81.5	23	80.5	33	83.0
4	82.0	14	80.0	24	81.0	34	83.0
5	82.5	15	84.0	25	82.5	35	80.5
6	79.5	16	80.0	26	82.0	36	81.5
7	78.5	17	83.0	27	82.5	37	80.0
8	81.5	18	83.0	28	79.5	38	81.0
9	82.0	19	85.0	29	84.0	39	81.5
10	83.5	20	85.0	30	81.0	40	83.5

(b) Frequency Distribution

Weight (lb)	Absolute Frequency	Relative Frequency (percent)	Relative Cumulative Frequency (percent)
78 to less than 79	1	2.5	2.5
79 to less than 80	4	10.0	12.5
80 to less than 81	5	12.5	25.0
81 to less than 82	9	22.5	47.5
82 to less than 83	7	17.5	65.0
83 to less than 84	7	17.5	82.5
84 to less than 85	4	10.0	92.5
85 to less than 86	2	5.0	97.5
86 to less than 87	1	2.5	100.0

(c) Histogram

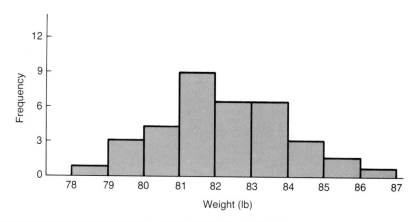

between $\bar{x} - 2s$ and $\bar{x} + 2s$; 100 percent fall between $\bar{x} - 3x$ and $\bar{x} + 3s$, or within a range of six standard deviations. Such an examination often provides a good clue to the probability distribution of the random variable of interest.

Our observation that in a normal distribution approximately 100 percent of the values fall within six standard deviations provides another "quick and dirty" test. That is, if the range of the distribution is approximately six standard deviations, there is some evidence of normality. In our example, the range, from Equation 5.9, is $86.0 - 78.5 = 7.5$, or approximately $7.5/1.77 = 4.24$ standard deviations, an inconclusive result.

Normal Probability Paper

Normal probability paper is so designed that the ogive, or graph of the cumulative frequency distribution described in Chapter 5, will be a straight line if the probability distribution is normal. Thus, if we plot the upper limit of each class versus the relative cumulative frequencies of each class, such a graph provides an indication of whether the frame or process output is normally distributed. As an illustration, consider the ogive in Figure 5.12 based on the relative cumulative frequencies in Figure 5.8(a) of a sample of 95 log lengths. Figure 6.14(a) shows this data on probability paper; we see a clear degree of curvature. In Figure 6.14(b), we have graphed the relative cumulative frequencies for the sample of 40 containers shown in Figure 6.13(b) on probability paper. The linear graph indicates that the container weights are likely to be normally distributed. Note that the relative cumulative frequencies are the cumulative probabilities that the random variable X is less than or equal to the upper class limit.

The methods for estimating (or validating) a probability distribution from sample data that we have studied have been approximate methods based on a visual analysis of histograms and cumulative frequency distributions.

QUANTIFYING VARIATION

It is clear that when we are able to specify the appropriate probability distribution for a random variable in an enumerative or analytic study, the ability to calculate associated probabilities enables us to predict within what ranges our observations are likely to occur.

For example, if we are studying a stable process that can be described by a Poisson distribution, given an estimate of the mean number of defects per unit, λ, we can calculate the probability of x or fewer defects by

FIGURE 6.14 Cumulative Frequency Distributions Graphed on Normal Probability Paper

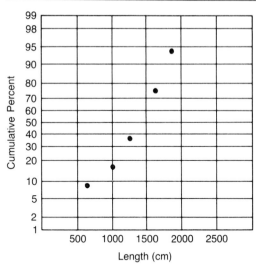

(a) Log length in a sample of 95 trees

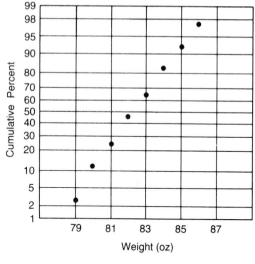

(b) Weights in a sample of 40 containers

summing the probabilities given by Equation 6.7. In our example of bad checks which arrive at an accounting office at an average rate of three per month, we can say that the probability of no more than five bad checks arriving in a given month is f(0) + f(1) + f(2) + f(3) + f(4) + f(5). From Table 7, this would be 0.916.

Another way to look at the variation of a random variable from the mean value is in terms of standard deviations. That is, we might be interested in the probability that the number of bad checks arriving in a given month is within one standard deviation of the mean. We know that the variance of the Poisson distribution is equal to its mean, or the standard deviation is $\sigma = \sqrt{\lambda}$. Thus, the desired probability is the probability that X is between $3 - \sqrt{3} = 1.27$ and $3 + \sqrt{3} = 4.73$, or, as X must be a discrete integer, f(2) + f(3) + f(4) = 0.616.

Such calculations are of particular interest for continuous random variables representing some measurable characteristic of a stable process. Again, where we know the form of the probability distribution, we can generally compute such probabilities; however, even where the form of the distribution is unknown, we are able to define limits on the proportion of the observations (or on the probability that a single observation is) within a given number of standard deviations from the mean.

Variation in Normally Distributed Random Variables

As we have seen in this chapter, given estimates of the mean and standard deviation of a normally distributed random variable, we know that approximately 68 percent of the sample values of that random variable will be within one standard deviation of the mean; approximately 95 percent will be within two standard deviations of the mean, and almost 100 percent will be within three standard deviations of the mean. Furthermore, as a standardized z-value is actually the number of standard deviations between the corresponding value of X and the mean, we can calculate the proportion of values within any number of standard deviations from the mean. Figure 6.15 shows some of these proportions for a normally distributed random variable.

We will see that the more we know about the distribution of the stable process from which our sample data is drawn, the more precise our statements about the variation of the sample values from the mean can be.

Variation in Unimodal Probability Distributions

Recall that the mode of a probability distribution is the value of X that corresponds to the maximum frequency on a histogram. A unimodal probability distribution has one such value. For any distribution that is unimodal, or decreases on both sides of its mode, the *Camp-Meidell inequality* states that the probability that X is within j standard deviations of its mean is *at least* $1 - (1/2.25j^2)$:

$$P(\bar{x} - js < X < \bar{x} + js) \geq 1 - \frac{1}{2.25j^2} \qquad (6.16)$$

FIGURE 6.15 Proportion of Observations within j Standard Deviations of the Mean

j	Interval	Normal Distribution	Unimodal Distribution	Unknown Distribution
1.5	$\bar{x} \mp 1.5s$	0.866	at least 0.802	at least 0.556
2.0	$\bar{x} \mp 2.0s$	0.954	at least 0.889	at least 0.750
2.5	$\bar{x} \mp 2.5s$	0.988	at least 0.929	at least 0.840
3.0	$\bar{x} \mp 3.0s$	0.997	at least 0.951	at least 0.889
3.5	$\bar{x} \mp 3.5s$	~1.000	at least 0.964	at least 0.918
4.0	$\bar{x} \mp 4.0s$	~1.000	at least 0.972	at least 0.938
4.5	$\bar{x} \mp 4.5s$	~1.000	at least 0.978	at least 0.951
5.0	$\bar{x} \mp 5.0s$	~1.000	at least 0.982	at least 0.960

Thus, if we have an unknown probability distribution that is unimodal, the probability that X is within two standard deviations of its mean, $\bar{x}$, is greater than or equal to $1 - [1/(2.25)(2)^2] = 0.889$. Note that if we know, for example, that the distribution is normal, we can make the stronger and more precise statement that the probability is, in fact, 0.954. Figure 6.15 shows the proportion of values within j standard deviations of the mean for various values of j under the assumptions of the Camp-Meidell inequality.

Variation in Unknown Probability Distributions

When nothing is known about the probability distribution of the random variable of interest beyond the estimates $\bar{x}$ and s, information about the variation in values from the mean is even more limited, although not entirely absent. The *Tchebycheff inequality* states that the probability that X is within j standard deviations of its mean is *at least* $1 - 1/j^2$:

$$P(\bar{x} - js < X < \bar{x} + js) \geq 1 - 1/j^2 \qquad (6.17)$$

Suppose we have a random variable about whose probability distribution we have no information. If we can estimate the mean and standard deviation from a sample, we can say that the probability that X is within two standard deviations of the mean is greater than or equal to $1 - 1/(2)^2 = 0.750$. If also know that the distribution is unimodal, we can make the stronger statement that at least 88.9 percent of all possible observations will be within two standard deviations of the mean. When we know the functional form of the probability distribution, we can make precise statements. Thus, if we know that the probability distribution is normal, we can say that 95.4 percent of all possible observations will be within two standard deviations of the mean. Figure 6.15 shows the proportion of values within j standard deviations of the mean as given by the Tchebycheff inequality.

In Figure 6.13, we showed the weights of 40 containers of fertilizer randomly selected from a stable packaging process. Our subsequent examination of the probability distribution using a visual inspection of the histogram and using normal probability paper allowed us to conclude that the conceptual population had a normal distribution. Our estimated mean of $\bar{x} = 81.96$ pounds and estimated standard deviation of $s = 1.77$ pounds allow us to say that the probability of being within j standard deviations of the mean is given by the proportions under the "Normal Distribution" column in Figure 6.15. That is, for example, 99.7 percent of the bags have weights within three standard deviations of the mean, or between $(\bar{x} - 3s =)$ 76.65 and $(\bar{x} + 3s =)$ 87.27 pounds.

If we have no information about the distribution, we can only make the weaker statement that at least 88.9 percent of the containers weigh between 76.65 and 87.27 pounds.

In Figure 5.4, we showed log lengths in a sample of 95 trees cut at age 20 years. If we wish to study log lengths of 20 year old trees over time, assuming that the underlying pattern of growth is stable, we used the frequency distribution of Figure 5.6 to estimate $\bar{x}$ (from Equation 5.3 for grouped data) as 1339.5 cm and to estimate s (from Equation 5.11b for grouped data) as 389.1 cm.

When we graphed the associated cumulative frequencies from Figure 5.8(a) on probability paper in Figure 6.14, we noted that the degree of curvature implied the data were not normally distributed. In the absence of any further information, to calculate the proportion of log lengths within three standard deviations of the mean we use the values under the "Unknown Distribution" column in Figure 6.15. This tells us that at least 88.9 percent of the log lengths of trees cut at age 20 years will be between $(\bar{x} - 3s =)$ 172.2 cm and $(\bar{x} + 3s =)$ 2506.8 cm.

Recall, however, that in our discussion of the mode, we found that the distribution of Figure 5.6 was unimodal, where our rough estimate of the mode was 1440.625. For a unimodal distribution whose mode is close to the value of its mean, the Camp-Meidell inequality holds; using this added information about the distribution of log lengths, the "Unimodal Distribution" column of Figure 6.17 tells us that at least 95.1 percent of the log lengths of trees cut at age 20 years will be between 172.2 and 2506.8 cm.

SUMMARY

We have seen in this chapter how probability distributions may be developed, characterized, and evaluated for validity to serve as models to describe a process or population.

We first discussed the properties of discrete probability models and the calculation of the mean and variance of a discrete random variable. As examples of discrete probability distributions, we introduced the binomial distribution, used to describe the probability that x items out of n items possess a particular attribute of interest; and the Poisson distribution, used to describe the probability of x occurrences per unit when the average number of occurrences per unit is known. We also discussed the use of the Poisson distribution as an approximation to the binomial distribution when n, the number of items, is large and p, the probability that an item possesses a particular attribute, is small.

For a continuous random variable, defined over every value in an interval, we discussed the properties of the probability density function and defined the cumulative distribution function. We introduced the normal distribution, one of the most important probability models in our subsequent work in quality control, and discussed the calculation of probabilities under the normal curve. We also noted that the normal distribution serves as an approximation to the binomial distribution when n is large and p is close to 0.50. We discussed another continuous distribution, the Student t-distribution.

We then discussed how to use sample information to determine the probability distribution that describes the behavior of a stable process or frame. We demonstrated qualitative analysis of the sample data, or a visual inspection of the frequency distribution and the use of normal probability paper.

We considered next the quantification of variation from the mean given information about the probability distribution of the random variable. We introduced the Camp-Meidell and Tchebycheff inequalities.

In the next chapter we will learn how to determine the appropriate probability model for some important statistical measures in quality control.

EXERCISES

6.1 Are the following functions of X probability functions? Explain why or why not.

a.
X	1	3	5
f(x)	1/3	1/3	1/3

b.
X	−1	0	1	2
f(x)	1/3	0	1/2	1/6

c.
X	8	10	12
f(x)	1/2	3/4	1

d.
X	1/2	3	6
f(x)	0.1	0.5	0.4

e.
X	1	5	8
f(x)	−1/3	2/3	2/3

f. $f(x) = 1/4$ for $x = 0, 1, 2, 3, 4$

g. $f(x) = (x − 1)/2$ for $x = 0, 1, 2$

6.2 Find the expected value and variance for the following probability distributions:

a.
X	2	3	4	5	6	7	8
f(x)	0.40	0.25	0.15	0.11	0.07	0.01	0.01

b.
X	2	4	6	8
f(x)	1/8	1/2	1/8	1/4

c.
X	10	15	20	25	30
f(x)	0.20	0.10	0.40	0.25	0.05

6.3 The probabilities of 0, 1, or 2 fires in a small northeastern town in a given month are 0.6, 0.3, and 0.1. What is the expected number of fires in any given month? State any assumptions required for your analysis.

6.4 In a stable manufacturing process, the probabilities of observing 0, 1, 2, 3, or 4 defectives are given by the following distribution:

Number of defectives	0	1	2	3	4
Probability	0.40	0.20	0.10	0.10	0.20

Find the mean and variance for the number of defectives.

6.5 If a random sample of 20 items is chosen from a stable process known to produce 10 percent defective items, what is the probability that three of the items are defective?

6.6 If a random sample of 20 items is chosen from a stable process known to produce five percent defective items, what is the probability that three of the items are defective?

6.7 What is the expected number of defective items in problem 6.5? In problem 6.6?

6.8 In a 100 percent inspection of 50,000 bulbs, 2,000 were discovered to be defective. However, the sorting machine that was supposed to remove the defective bulbs has malfunctioned, so defective bulbs are now intermingled with nondefective bulbs. What is the probability that
  a. a random sample of 10 bulbs withdrawn from the lot will include exactly 3 defective bulbs?
  b. a random sample of 25 bulbs will include at least 2 defective bulbs?
  c. a random sample of 8 bulbs will include at most one defective bulb?

6.9 A bank processes 100 checks per day. The probability of bouncing a check because of a missing signature is 0.05. The check-bouncing process is assumed to be stable.
  a. Find the probability that the bank bounces exactly three checks using the binomial distribution.
  b. Find the probability that the bank bounces exactly three checks using the Poisson approximation to the binomial distribution.
  c. Compare the two results and comment.

6.10 Commuters leave a subway station at an average rate of five per minute.

    a. What is the probability that in a given minute two or fewer people leave the station?

    b. Comment on the validity of using the Poisson distribution in this problem.

6.11 Sales in a men's shop are normally distributed with a mean of $900 per day and a standard deviation of $200 per day. If S is a random variable representing sales, find

    a. the proportion of days when sales will be less than $750.

    b. the proportion of days when sales will be more than $1,400.

    c. the proportion of days when sales will be more than $1,000.

    d. the proportion of days when sales will be less than $400.

    e. the proportion of days when sales will be between $400 and $600.

    f. the proportion of days when sales will be between $400 and $1,400.

6.12 A company packages tuna fish in 7.5-ounce cans. Government regulations require that at least 98 percent of cans labeled 7.5 ounces actually weigh at least 7.5 ounces. If the standard deviation is known to be 0.25 ounces and the process is stable, at what level should the company set the mean fill level to comply with government regulations? Assume that the contents of cans follows a normal distribution.

6.13 The packaging company in problem 6.12 has the opportunity to purchase a new filling machine that is much more accurate. This new machine will fill cans to a desired limit but with a standard deviation of only 0.025 ounces instead of the old machine's value of 0.25 ounces.

    a. Find the new setting for the average fill level per can so that government regulations are met.

    b. If tuna fish costs an average of $0.005 per ounce, and total annual volume is 20 million cans, calculate the expected savings.

6.14 Find the following values:

    a. $t_{.05}$ (5 degrees of freedom)

    b. $t_{.05}$ (6 degrees of freedom)

    c. $t_{.05}$ (29 degrees of freedom)

    d. $z_{.05}$

    e. $t_{.10}$ (10 degrees of freedom)

    f. $t_{.005}$ (10 degrees of freedom)

    g. $z_{.005}$

    h. $-z_{.005}$

    i. $-t_{.005}$ (8 degrees of freedom)

6.15 800 people are employed on an hourly basis, according to the following frequency distribution:

Hourly Wage (dollars)	Number of Employees
$4.60–$4.79	18
4.80– 4.99	135
5.00– 5.19	366
5.20– 5.39	187
5.40– 5.59	80
5.60– 5.79	10
5.80– 5.99	4

a. Construct a histogram for the above data and comment on an appropriate probability model for the hourly wages.
b. Use normal probability paper to determine whether the distribution of hourly wages is approximately normal.
c. Calculate the sample mean and sample standard deviation for the hourly wage data.

6.16 For the data of problem 6.15, what can be said about its variability if
a. nothing is known about its probability distribution.
b. it is known to be unimodal.
c. it is assumed to be normal.

Chapter 7

Sampling Distributions

INTRODUCTION

In the last chapter, we introduced a number of important probability models that may be used to describe output from a stable process or population; and we discussed the problem of deciding on the appropriate model form for a given process or population. Many measurable process or population characteristics are not easily described by these common probability distributions; nevertheless, we will see that the probability distributions of the statistics of these measurable characteristics follow certain known probability models. In analytic studies, these distributions, called *sampling distributions,* of characteristics such as the fraction defective, number of defects, number of defects per unit, the mean, the range, and the standard deviation, enable us to construct control charts to ascertain the stability of a process and the opportunities for corrective action. In enumerative studies, sampling distributions enable us to make inferences from samples to the frame as a basis for action on the population under study.

SAMPLING DISTRIBUTIONS DEFINED

It is most convenient to introduce the notion of a sampling distribution by considering how, in theory, one would arise. Suppose we have a stable process from which we sample for defects every hour. That is, each hour we select a sample of n items, or a *subgroup;* count the number of defective items; and calculate the fraction defective. Clearly, not every hourly fraction defective will be the same. If we construct a frequency distribution of these hourly sample fractions defective, the resulting relative frequencies would tell us how likely we would be to observe a particular sample fraction defective in an hour of inspection. This relative frequency distribution for the fraction defective in all possible hourly inspections of n items is called the sampling distribution of the fraction defective for sample size n.

As another example of a sampling distribution, consider a stable process in which the diameter of bushings is measured in inches. Every 15 minutes four bushings are selected and measured and the average diameter of the subgroup, $\bar{x}$, computed. If we continued this inspection process indefinitely, we would have a collection of all possible $\bar{x}$ values, from which we could construct a relative frequency distribution. This distribution would be called the sampling distribution of the mean for sample size 4.

USING SAMPLING DISTRIBUTIONS IN ANALYTIC STUDIES

We will see in Chapters 8 through 10 that a knowledge of the sampling distribution of a sample characteristic enables us to construct a statistical tool called a *control chart* by means of which we can evaluate the stability (and departures from stability) of the process. For example, for a stable process, the fraction defective in subgroups of n items should vary from subgroup to subgroup; but the amount and nature of the variation is given by the sampling distribution of the fraction defective. On the other hand, when a process is unstable over time, its sample fractions defective would display uncontrolled variation inconsistent with the known sampling distribution. Control charts for such characteristics as fraction defective, based upon the sampling distribution, enable us to ascertain the presence or absense of process stability. Recall that we based all our computations in Chapter 6 of process parameters and estimates in analytic studies on the assumption of process stability. When we know a process is stable, we can predict its behavior in the future; where variation is uncontrolled, such prediction is impossible.

Knowledge of the sampling distributions of some important characteristics is crucial to quality studies. Since a sampling distribution is based on all possible sample values of a particular characteristic, its actual construction requires examining all possible subgroups as described above. In practice, of course, this is either impossible and/or economically infeasible. The functional forms and parameters for sampling distributions are derived, instead, from a theoretical analysis of the behavior of the characteristics being measured. In the remainder of this chapter we discuss the sampling distributions and parameters of some of these characteristics. In particular, as the next few chapters deal with the construction and analysis of control charts, our major objective in this chapter is to lay the groundwork for their use in analytic studies.

ATTRIBUTE SAMPLING DISTRIBUTIONS

Recall from Chapter 5 that measurements can be characterized as either attribute data or variables data. Data which arise from the classification of items into categories or from counts of the number or proportion of items in a given category or from counts of the number of occurrences per unit are called attribute data. In this section, we discuss the sampling distributions of some important attributes: *fraction defective, number of defects, and number of defects per unit.*

The Sampling Distribution of Fraction Defective

Constant Sample Size n. In an analytic study, with an infinite conceptual frame, assume that the fraction defective for a stable process is p. For a randomly selected sample of n items from an infinite frame, we can assume that the items are independent and that the probability of a defective item is p. As there are only two possibilities for each item—defective or nondefective—recall from Chapter 6 that the number of defective items, X, in our examination of n items under these conditions has a *binomial distribution*. That is, Equation 6.3b tells us that the probability of x defectives in n items, where p is the fraction defective for the stable process, is:

$$f(x) = \frac{n!}{x!(n - x)!} p^x(1 - p)^{n-x} \quad \text{for } x = 0, 1, 2, \ldots, n$$

The probability that the fraction defective in a subgroup of size n, X/n, takes on the numerical value x/n is identical to f(x) above:

$$f(x/n) = \frac{n!}{x!(n - x)!} p^x(1 - p)^{n-x} \tag{7.1}$$

To illustrate, suppose we have a stable process whose fraction defective is p = 0.10. We take hourly samples of size 5, count the number of defective items, x, in each, and calculate the fraction defective as x/5. The sampling distribution of the fraction defective is the distribution of the values of x/5. The probability that the fraction defective in a sample of size n is x/n is computed from Equation 7.1 where p = 0.10 and n = 5; Figure 7.1 illustrates the sampling distribution.

Given this probability distribution, we can calculate the average fraction defective from Equation 6.1: $\mu = \Sigma \, x \, f(x)$

$$\mu = (0)(0.5905) + (1/5)(0.3281) + (2/5)(0.0729) + (3/5)(0.0081)$$

$$+ (4/5)(0.0005) + (1)(0.0000)$$

$$= 0.1000$$

FIGURE 7.1 Sampling Distribution of Fraction Defective in an Infinite Frame
(p = 0.10 and n = 5)

Fraction Defective, X/n	*f(x/n)*
0	0.5905
1/5	0.3281
2/5	0.0729
3/5	0.0081
4/5	0.0005
1	0.0000

and, from Equation 6.2a, $\sigma^2 = \Sigma (x - \mu)^2 f(x)$

$$\sigma^2 = (0 - .1)^2(0.5905) + (1/5 - .1)^2(0.3281) + (2/5 - .1)^2(0.0729)$$

$$+ (3/5 - .1)^2(0.0081) + (4/5 - .1)^2(0.0005) + (1 - .1)^2(0.0000)$$

$$= 0.0180$$

and $\sigma = 0.134$. More generally, note that for the sampling distribution of the fraction defective given in Equation 7.1 the mean is

$$\mu = p \qquad (7.2)$$
$$= 0.1$$

and the standard deviation, called the *standard error of the fraction defective,* is

$$\sigma_p = \sqrt{\frac{p(1 - p)}{n}} \qquad (7.3)$$

$$= \sqrt{(.1)(.9)/5} = 0.134$$

That is,

the sampling distribution of the fraction defective in an analytic study with stable process fraction defective p and constant subgroup size n is a binomial distribution with mean fraction defective p and standard error $\sqrt{p(1 - p)/n}$

In reality, we of course do not know the population fraction defective p but estimate it for a stable process, as discussed in Chapter 6. However, our objective in this chapter is not to estimate p but to describe the theoretical sampling distribution of the fraction defective for subsequent use in determining the presence of stability. We will see in Chapter 9 how estimates are used to construct control charts given the theoretical sampling distributions presented here.

Nonconstant Sample Size n. In some situations the subgroups may not all contain the same number of items. Although the sampling distribution remains binomial, the standard error varies with each subgroup:

$$\sigma_{p_i} = \sqrt{\frac{p(1 - p)}{n_i}} \tag{7.4}$$

That is, if n_1 is the number of items in the first subgroup, n_2 is the number of items in the second subgroup, . . . , and, in general, n_i is the number of items in the ith subgroup,

the sampling distribution of the fraction defective in an analytic study with stable process fraction defective p and nonconstant subgroup size n_i is a binomial distribution with mean fraction defective p and standard error $\sqrt{p(1 - p)/n_i}$.

In Chapter 9, we will see how nonconstant subgroup size affects the construction of the control chart.

The Sampling Distribution of Number of Defectives

In studies where the actual number of defectives in each subgroup is recorded, the calculation of the fraction defective in each subgroup requires an additional computation. In addition, the concept of the number of defectives in a sample is often a clearer one than the concept of the fraction defective. For these reasons, we often consider as the random variable of interest the number of defective units rather than the fraction defective. As we will see, the functional forms of these sampling distributions are the same, with means and standard errors adjusted for the change in the definition of the random variable.

In discussing the sampling distribution of fraction defective in an analytic study, we mentioned the case of nonconstant subgroup size n_i, resulting in a standard error for the sampling distribution that varies with subgroup size. In order to consider the sampling distribution of number of defectives for the purpose of constructing control charts, the subgroup size must be constant; if it were not, not only would the standard error vary with each subgroup but the mean as well.

Where the fraction defective for the stable process is p and the constant subgroup size is n, the number of defectives in each subgroup, X, has a binomial distribution, as given by Equation 6.3b:

$$f(x) = \frac{n!}{x!(n - x)!} p^x (1 - p)^{n-x} \quad \text{for } x = 0, 1, 2, \ldots, n \tag{7.5}$$

with mean given by Equation 6.5:

$$\mu = np \tag{7.6}$$

and standard deviation, from Equation 6.6:

$$\sigma = \sqrt{np(1 - p)} \tag{7.7}$$

Thus, if we consider again the stable process whose fraction defective is $p = 0.10$, where hourly samples of size 5 were inspected, the sampling

FIGURE 7.2 Sampling Distribution of Number of Defectives
(p = 0.10 and n = 5)

Number of Defectives, X	f(x)
0	0.5905
1	0.3281
2	0.0729
3	0.0081
4	0.0005
5	0.0000

distribution, f(x/n), is shown in Figure 7.1. For the random variable X defined as the number of defectives, f(x) is calculated in exactly the same way, using Equation 7.5. Figure 7.2 shows the identical probabilities for the random variable X.

The mean number of defectives is, of course, not the same as the mean fraction defective, as the random variable X is not the same as the random variable X/n. From Equation 7.6, the mean number of defectives in each subgroup is np = (5)(0.10) = 0.5; and, from Equation 7.7, the standard deviation, or *standard error of the number of defectives,* is $\sigma = \sqrt{np(1-p)} = \sqrt{(5)(0.1)(0.9)} = 0.671$.

In general, then:

the sampling distribution of the number of defectives in an analytic study with stable process fraction defective p and constant subgroup size n is a binomial distribution with mean number of defectives np and standard error $\sqrt{np(1-p)}$.

The number of defectives is considered as an alternative to the fraction defective for constant subgroup size only for ease of interpretation and/or to save computational time. A knowledge of the sampling distribution of the number of defectives offers no additional information or insight over a knowledge of the sampling distribution of the fraction defective.

The Sampling Distribution of Number of Defects per Unit

An important random variable in many industrial situations is the number of defects in a given "area of opportunity," such as a given item, time period, or geographical area. It is important here to distinguish between defects and defectives. An item is defective, or nonconforming, if it fails to meet the specifications required for customer satisfaction; this means the item will have one or more defects. For example, in Chapter 6 we talked about zip code errors on labels, such as reversed digits. Each such error is a defect. A label is defective if it has one or more such errors. We know from our earlier discussion in this chapter how to determine the sampling distribution of the fraction defective or number of defectives in an examination of such labels. In this section we discuss the sampling distribution of the number of defects per item. It is an especially important random variable in industrial quality control for it allows us to consider the total of many different types of defects in a single, often complex, item.

Consider the manufacture of insulated wire, where we are interested in the number of weak spots in a given length of wire. As long as the inspection unit (length in this example) remains constant, we may sample one or more items at a time, each for a number of defects per (constant) unit.

Constant Subgroup Size.   Suppose that every day for 20 days we record the number of weak spots in 6-foot lengths of wire. Further, suppose that the process average number of weak spots per foot of wire is 2. Then for the unit, or subgroup size, in our example, 6 feet, the average number of defects is 12.

Recall from Chapter 6 that events (such as the occurrence of defects) that occur randomly and independently at a constant mean rate (i.e., a stable process) can be described by the Poisson distribution. If we define our random variable X to be the number of defects per unit, then if the mean number of defects per unit for the process is c, the probability of x defects in a sample consisting of one unit is:

$$f(x) = \frac{c^x \, e^{-c}}{x!} \tag{7.8}$$

In our example, the mean, or average number of defects per 6 feet, is c = 12; the probability of observing x defects in a given daily inspection of 6 feet of insulated wire is:

$$f(x) = \frac{12^x \, e^{-12}}{x!}$$

For a Poisson distribution with mean rate c, the mean number of defects per unit is:

$$\mu = c \tag{7.9}$$

or 12 defects per six feet, and, as the variance of the Poisson distribution is equal to its mean, its standard deviation, or *standard error of the number of defects,* is:

$$\sigma = \sqrt{c} \tag{7.10}$$

or 3.5 defects per 6 feet.

In general, then:

the sampling distribution of the number of defects per subgroup in an analytic study with stable process mean number of defects per subgroup c and constant subgroup size is a Poisson distribution with mean number of defects per subgroup c and standard error $\sqrt{c}$.

Nonconstant Subgroup Size. The Poisson distribution describes the occurrence of defects per unit, or per subgroup, when the unit or subgroup size is constant. If our subgroup size varies from sample to sample, it is necessary to redefine our random variable in terms of some constant unit. That is, if our daily measurements of the number of weak spots per subgroup in insulated wire, X, were taken over different lengths n_i each day, we could calculate the average number of defects per foot from each subgroup as x/n_i. Then the average number of weak spots per foot, x/n_i, has a Poisson distribution with mean number of weak spots per foot for the stable process denoted by u:

$$f(x/n_i) = \frac{u^{x/n_i} e^{-u}}{(x/n_i)!} \tag{7.11}$$

If the mean number of weak spots for the stable process is u = 5 per foot of wire, the probability of observing an average of three weak spots per foot (as X = 3 defects in an inspection of n = 1 foot of wire, or as X = 6 defects in an inspection of n = 2 feet of wire, etc.) is

$$f(3) = \frac{5^3 e^{-5}}{3!} = 0.1404$$

The mean of the probability distribution is given by the process average per unit:

$$\mu = u \tag{7.12}$$

The standard deviation, or *standard error for the average number of defects per unit* is:

$$\sigma_i = \sqrt{u/n_i} \tag{7.13}$$

Note that as in the case of variable sample size for fraction defective, the standard error varies with the subgroup size. In general,

the sampling distribution of the average number of defects per unit in an analytic study with stable process mean number of defects per unit u and nonconstant subgroup size n_i is a Poisson distribution with mean number of defects per unit u and standard error $\sqrt{u/n_i}$.

VARIABLES SAMPLING DISTRIBUTIONS

Recall from Chapter 5 that data representing continuous measurements of quality characteristics made over an interval are characterized as variables data. In this section we discuss the sampling distributions of some important characteristics of variables data: the *mean, range,* and *standard deviation.*

The Sampling Distribution of the Mean

In considering the sampling distribution of the mean, $\bar{x}$, we can make precise statements about the sampling distribution when the distribution of the individual x values is known to be normal; note that this was not a requirement in specifying the sampling distributions of fraction defective, number of defectives, or number of defects per unit. Nevertheless, we will see that even where the distribution of the process is unknown, we can specify the sampling distribution of the mean with almost as high a degree of reliability as when the process is known to be normal.

Consider the measurement of some stable quality characteristic, X. Subgroups of size n are selected, and their averages, $\bar{x}$, are computed for all possible subgroups. The relative frequency distribution of all possible $\bar{x}$ values is the sampling distribution of the mean. If we denote the mean of all possible subgroup averages by $\bar{\bar{x}}$, and the mean of the individual values is μ, then the mean of the sampling distribution is:

$$\mu_{\bar{x}} = \bar{\bar{x}} = \mu \tag{7.14}$$

If the standard deviation of the individual x values for the stable process is σ, then the standard deviation of the sampling distribution, often called the *standard error of the mean,* is:

$$\sigma_{\bar{x}} = \sigma/\sqrt{n} \tag{7.15}$$

Normally Distributed Process.   When the individual x values can be described by a normal distribution, the sampling distribution of the mean is normal as well:

> The sampling distribution of the mean in an analytic study of a normally distributed stable process with mean μ, standard deviation σ, and subgroup size n is a normal distribution with mean μ and standard error $\sigma/\sqrt{n}$.

To illustrate, suppose the distribution of weights of cans of string beans, X, generated by a stable process, is known to be normal with $\mu = 10$ ounces and standard deviation $\sigma = 1.5$ ounces. Subgroups of size 6 are selected and their sample means $\bar{x}$ calculated; these $\bar{x}$ values are themselves normally distributed with mean $\mu_{\bar{x}} = 10$ oz. and standard error $\sigma_{\bar{x}} = 1.5/\sqrt{6} = 0.612$ oz. Note that the standard error of the sampling distribution is smaller than the standard deviation of the weights themselves, although both distributions are normal with the same mean. The variability associated with average values, $\bar{x}$, is always smaller than the variability associated with individual values from the process, as illustrated in Figure 7.3.

FIGURE 7.3 Normally Distributed x and $\bar{x}$

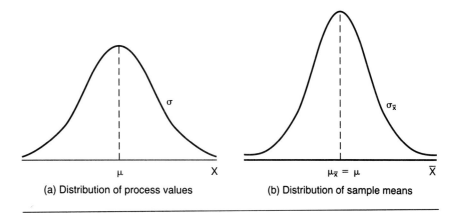

(a) Distribution of process values   (b) Distribution of sample means

Nonnormally Distributed Process. When the distribution of individual values from the stable process is unknown, the mean and standard deviation of the sampling distribution are as given by Equations 7.14 and 7.15. However, we cannot make the statement that the sampling distribution of the mean is normal.

Nevertheless, the *Central Limit Theorem* tells us that if the subgroup size n is large, the sampling distribution of the mean can be approximated by the normal distribution. Although subgroup sizes of 30 or more are generally considered necessary for the sampling distribution to be normal when the process is not, for n as small as 4 the sampling distribution does not deviate substantially from normality. Thus, in general, we can say:

> the sampling distribution of the mean in an analytic study of a stable process with mean μ, standard deviation σ, and subgroup size n is approximately normal with mean μ and standard error $\sigma/\sqrt{n}$.

The Sampling Distribution of the Mean when σ Is Unknown. In describing the sampling distribution of the mean, parameters were defined in terms of the known mean μ and standard deviation σ for the process. Where the x values are known to be normally distributed but the standard deviation must be estimated by s, the sampling distribution of the mean has a t-distribution with $n - 1$ degrees of freedom, where n is the subgroup size. The standardized t-value is given by:

$$t = \frac{\bar{x} - \mu}{s/\sqrt{n}} \qquad (7.16)$$

and its associated probabilities are given in Table 2, as discussed briefly in Chapter 6. Recall, however, that when n is 30 or more, the t-distribution is approximately the same in shape as the normal distribution. Furthermore, even when the individual x values are not too far from normally distributed and σ is unknown, the sampling distribution can be approximated by the t-distribution.

The Sampling Distribution of the Range

Recall that the range of a set of data is the difference between the largest data point and the smallest data point and serves as a measure of dispersion.

If we are interested in the stability of the variability of a process, and sample sizes are small $(2 \leq n \leq 9)$, the range is a valid and easier-to-calculate measure than the standard deviation. The rationale for using the range as a substitute for the standard deviation is as follows. For small sample sizes, the range and standard deviation vary proportionately: if one is large, the probability is high that the other is large, and vice-versa. On the other hand, for larger sample sizes $(n > 10)$, the occurrence of an extreme value will inflate the range but may not appreciably affect the standard deviation. Consequently, in analytic studies using small sample sizes the range is often used as a substitute measure of variability.

Normally Distributed Process.   We saw in Chapter 6 that the more we know about the probability distribution of quality characteristics, the more precise are the inferences we can make about the process or frame. When the characteristic of interest, X, is approximately normally distributed and sample sizes are small $(2 \leq n \leq 9)$, the sampling distribution of the range is known. Although the functional form of this distribution is a complex one, its mean is $\overline{R}$, the average of all subgroup ranges taken from the process, or:

$$\mu_R = \overline{R} = d_2 \, \sigma \qquad (7.17)$$

where σ is the process standard deviation and values of d_2 are tabulated as a function of subgroup size in Table 8. Its standard deviation, or the *standard error of the range*, is σ_R:

$$\sigma_R = d_3 \, \sigma \qquad (7.18)$$

where values of d_3 are also tabulated as a function of subgroup size in Table 8.

The sampling distribution of the range in an analytic study of a stable normally distributed process with mean μ, standard deviation σ, and subgroup size n $(2 \leq n \leq 9)$ has a mean of $d_2\sigma$ and a standard error of $d_3\sigma$.

Nonnormally Distributed Process. The sampling distribution of the range is robust with respect to departures from normality. Thus, the results found above for a normally distributed process can be used with confidence, even when the measurements are not normally distributed.[1]

The Sampling Distribution of the Standard Deviation

Recall that the standard deviation of a set of data is a measure of its dispersion from the mean. The standard deviation takes into account the deviation between each data point and the mean rather than only the largest and smallest values, as does the range.

If we are interested in the stability of the variability of a process, we may wish to describe the behavior of the standard deviations of subgroups of size n. That is, as we discussed in the last section, the range is an appropriate substitute for the standard deviation when the sample size is small. However, when sample sizes are 10 or more, the standard deviation must be used.

Normally Distributed Process. When the distribution of the individual values of the characteristic being measured is normal, or even approximately normal, the sampling distribution of the standard deviation of subgroups of size n has mean:

$$\mu_\sigma = \left(\sqrt{\frac{2}{n-1}} \, \frac{\left(\frac{n-2}{2}\right)!}{\left(\frac{n-3}{2}\right)!} \right) \sigma \qquad (7.19a)$$

This can be more conveniently represented by letting the constant c_4 denote the coefficient of σ in Equation 7.19a:

$$\mu_\sigma = c_4 \sigma \qquad (7.19b)$$

Values of c_4 are tabulated as a function of subgroup size in Table 8. The *standard error of the standard deviation* is:

$$\sigma_\sigma = \sigma \sqrt{1 - c_4^2} \qquad (7.20)$$

The sampling distribution of the statistic $(n-1)s^2/\sigma^2$ is a chi square distribution with $n-1$ degrees of freedom.

The sampling distribution of the standard deviation in an analytic study of a stable normally distributed process with mean μ, standard deviation σ, and subgroup size n has a mean of $c_4\sigma$ and a standard error of $\sigma \sqrt{1 - c_4^2}$. The sampling distribution of the statistic $(n-1)s^2/\sigma^2$ is chi square.

Nonnormally Distributed Process. When the distribution of the individual values from the process is not normal, if the subgroup size n is large (at least 30), the mean of the distribution of subgroup standard deviations is approximately equal to the process standard deviation:

$$\mu_\sigma = \sigma \qquad (7.21)$$

The standard error of the standard deviation is approximately

$$\sigma_\sigma = \frac{\sigma}{\sqrt{2n}} \sqrt{\frac{\gamma_2}{2} + 1} \qquad (7.22)$$

where γ_2 is the kurtosis of the process, as given by Equation 5.13:

$$\gamma_2 = \frac{\Sigma(x - \bar{x})^4}{ns^4} - 3$$

The sampling distribution of the standard deviation in an analytic study of a stable process with mean μ, standard deviation σ, and subgroup size n has a mean of σ and a standard error of $\frac{\sigma}{\sqrt{2n}} \sqrt{\frac{\gamma_2}{2} + 1}$.

SUMMARY

In this chapter we have examined the behavior of subgroup statistics drawn from a process whose probability distribution may or may not be known. We have seen that the sampling distribution of the fraction defective for the case of constant sample size is a binomial distribution; where sample sizes are not constant, the sampling distribution remains binomial, but the standard error varies with each subgroup. The sampling distribution of the number of defectives is also a binomial distribution.

The sampling distribution of the number of defects per unit was found to be a Poisson distribution for both constant and nonconstant subgroup sizes, although again the standard error varies with each subgroup in the latter case.

In an examination of variables sampling distributions, we found that the sampling distribution of the mean, when the process is normal, is a normal distribution. When the underlying process is not normal, but the subgroup size is sufficiently large, we may approximate the sampling distribution by the normal distribution. When the underlying process is normal, but the process standard deviation is unknown, we have seen that subgroup means have a t-distribution.

We found that the sampling distribution of the range also depends upon the underlying process distribution and the sample size; we can characterize the sampling distribution of the standard deviation for constant subgroup size for both normal and nonnormal process distributions, as long as the subgroup size is sufficiently large in the latter case.

In subsequent chapters we will show how the theory of sampling distributions provides the groundwork for an understanding of the behavior of processes and, through the use of control charts, the stabilization and improvement of processes.

EXERCISES

7.1   Random samples of size 10 are drawn from a stable process with fraction defective $p = 0.01$.
   a.   What is the probability that in a single subgroup, exactly 0 units are defective?
   b.   What is the probability that the fraction defective in a single subgroup is 0.00?
   c.   What is the standard error of the fraction defective?
   d.   If the subgroup size is changed to 20, how does the standard error of the fraction defective change?

7.2   For the process of problem 7.1, describe the sampling distribution of the number of defective items.

7.3   Air conditioners are manufactured by a stable process with an average of 2.5 defects per unit.
   a.   What is the probability of observing five defects in a single inspected air conditioner?
   b.   What is the probability of observing five defects in a sample of two air conditioners?
   c.   Compute the standard error of the sampling distribution in part *b.*

7.4   A process has an average value of 40 grams and a standard deviation of 4 grams. Samples of size 4 are taken at periodic intervals and their means are calculated.
   a.   Describe the sampling distribution of the subgroup means, stating any necessary assumptions.
   b.   Assuming the process is stable and normally distributed with the given parameter values, calculate the probability that a single item from the process will weigh 38 grams or less.
   c.   Assuming the process is stable and normally distributed with the given parameter values, calculate the probability that a subgroup of size 4 will have a mean weight of 38 grams or less.

7.5   If the standard deviation of the process in problem 7.4 is unknown but estimated to be 4, describe the sampling distribution of the subgroup means, stating any necessary assumptions.

7.6   Eight subgroups of size 5 are selected from a stable, normally distributed process with a standard deviation of 3.
   a.   What is the mean and standard error of the sampling distribution of the range?
   b.   What is the mean and standard error of the sampling distribution of the standard deviation?

NOTE

1. Donald J. Wheeler and David S. Chambers, *Understanding Statistical Process Control* (Knoxville, Tenn.: Statistical Process Controls, Inc., 1986), p. 94.

PART III

Tools and Methods for Analytic Studies

This section discusses the tools and methods needed to conduct analytic studies. These techniques, when used in conjunction with the PDCA cycle, form a powerful arsenal that can be used to pursue continuous and never-ending improvement of any organization.

Chapter 8 provides an overview of procedures for stabilizing and improving a documented and defined process. The chapter includes a section on the general theory and purpose of statistical control charts, as well as a section on rational subgroups.

Chapter 9 presents a detailed discussion of attribute control charts: p charts (for both constant and variable sample sizes), np charts, c charts, u charts, and single measurement and moving range charts for attribute data. Many examples of attribute control charts are discussed.

Chapter 10 presents a detailed examination of variables control charts: x-bar and R charts, x-bar and s charts, median charts, single measurement and moving range charts for variables data. Many examples of variables control charts are discussed as well.

Chapter 11 presents specific control chart patterns that can be used to detect special sources of variation, and statistical rules that are useful in detecting special sources of variation.

Chapter 12 discusses in detail some other tools and methods that are useful in determining the causes of special and common sources of variation: brainstorming, cause-and-effect diagrams, check sheets, Pareto diagrams, and stratification. These techniques provide information that may lead to plans for action resulting in the resolution of special sources of variation and the reduction of common sources of variation.

Chapter 8

Stabilizing and Improving a Process with Control Charts

INTRODUCTION

A process that has been defined and documented can be stabilized and then improved. In great measure this can be accomplished through the use of statistical control charts, as well as other techniques that will be introduced in Chapter 12.

These tools and methods must be used in an environment that provides a positive atmosphere for process improvement; and top management must sincerely desire real process improvement. W. E. Deming points out that "any attempt to use statistical techniques under conditions that rob the hourly worker of his pride in workmanship will lead to disaster."[1] With this caveat clearly in sight, we may begin to consider the issues of stabilizing and improving a documented and defined process.

THE DEMING CYCLE

The Deming cycle[2] is a method that can aid management in stabilizing a process and pursuing continuous and never-ending process improvement. Recall from Chapter 1 that the Deming cycle is composed of four basic stages: a "planning" stage, a "doing" stage, a "checking" stage, and an "acting" stage. Hence, the Deming cycle is sometimes referred to as the PDCA cycle (*Plan-Do-Check-Act* cycle). A plan is developed (*plan*); the plan is tested in a trial basis (*do*); the effects of the test plan are monitored (*check*); and appropriate corrective actions are taken on the process (*act*). These corrective actions can lead to a new or modified plan, so the PDCA cycle continues forever in an uphill cycle of never-ending improvement. Figure 8.1 illustrates the Deming (PDCA) cycle.

FIGURE 8.1   The Deming (PDCA) Cycle

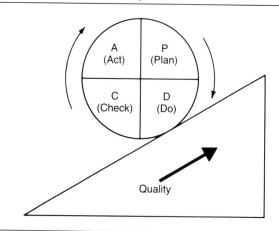

The Deming cycle operates by recognizing that problems (opportunities for improvement) in a process are determined by the difference between customer (internal and/or external) needs and process performance. If the difference is large, customer dissatisfaction may be high, but there is great opportunity for improvement. If the difference is small, the consequent opportunity for improvement is diminished.[3] Nevertheless, it is always economical to continually attempt to decrease the difference between customer needs and process performance, as discussed later in this chapter. We review below the four stages of the Deming cycle.

Stage 1: Plan

The collection of data about process variables is critical when determining a plan of action for what must be accomplished to decrease the difference between customer needs and process performance. In Chapter 1 we discussed the three types of quality that must be understood in the never-ending improvement of the extended process. Data concerning customer needs is collected via *quality-of-performance* studies. These data and more are evaluated and transformed into product/process characteristics (variables) that may be acted upon during *quality-of-design* studies. Data concerning the ability of the process to surpass customer needs are collected via *quality-of-conformance* studies. These data should be collected on product/process characteristics (variables) for which process improvement action can be taken.

A plan must be developed to determine the effect(s) of manipulating process variables upon the difference between process performance and customer needs. The plan must be tested, and the tests are conducted using *quality-of-design, conformance,* and *performance* studies. The tests should be conducted in a laboratory, production setting, office setting, or on a small scale with customers (both internal and external). The results of these tests will lead to a concrete plan for manipulating the process variables to decrease the difference between customer needs and process performance.

Numerical goals should not be used to direct the plan. Deming points out that this will help keep managerial emphasis on those variables that reduce the difference between customer needs and process performance.[4]

It is important that all persons play from the same quality improvement score card; that is, all relevant variables must be operationally defined, as discussed in Chapter 3, and all procedures must be standardized so that everyone conducts a given procedure using the same format. This standardization increases communication and decreases the possibility of error.

Stage 2: Do

The plan established in the first stage is set into motion on a trial basis in the Do stage. This is accomplished through a three part process. First, the organization must educate everyone involved with the trial plan so that they understand the relationship between the manipulated variables and the proposed decrease in the difference between customer needs and process performance (Deming's point 13). Second, the organization must train everyone involved with the trial plan so that they understand how the plan will affect and modify their jobs (Deming's point 6). After the first two steps have been accomplished, the plan can be set into motion on a trial basis as the third part of the Do stage.

Stage 3: Check

The trial plan, which was set into motion in Stage 2 (Do), must be continually monitored (Check) to answer two questions. First, are the manipulated process variables behaving according to the plan and causing a decrease in the difference between customer needs and process performance? And second, are the downstream effects of the plan creating problems or improvements? The results of statistical studies in this Check stage lead to the Act stage.

Stage 4: Act

The purpose of the Act stage is to implement modifications to the plan discovered in the Check stage or to make improvements to the process.

This leads to further narrowing of the difference between customer needs and process performance. Hence, the PDCA cycle continues forever in the never-ending improvement of the extended process.

The Act stage operates by considering whether the manipulated process variables have effectively diminished the difference between customer needs and process performance. If at the Act stage we learn that it has not been effective, the PDCA cycle returns to the Plan stage to search for other process variables that may decrease the difference. However, if the manipulation of process variables has produced the desired results, then the Act stage leads back to the Plan stage to determine the optimal levels at which to set the manipulated process variables. Finally, if at the Act stage we find that the plan is optimal in decreasing the difference between customer needs and process performance, we implement and standardize the process improvements specified by the plan.

The Deming (PDCA) cycle is the basic method used to stabilize a process and continuously work towards improving the extended process.[5] In the next sections we will consider control charts, which are important components of the quality studies required by the Deming cycle.

USING CONTROL CHARTS TO STABILIZE AND IMPROVE A PROCESS

Control charts are statistical tools used to analyze and understand process variables, to determine a process's capability to perform with respect to those variables and to monitor the effect of those variables on the difference between customer (either internal and/or external) needs and process performance. Control charts accomplish this by allowing a manager to understand the sources of variation in a process, and hence to manipulate and control those sources to decrease the difference between customer needs and process performance. This decrease can be managed only if the process under study is stable and capable of improvement.

PROCESS VARIATION

All processes exhibit variation. It is unavoidable, yet, like a "wild beast," must be controlled. For example, one-inch bolts will vary over several thousandths of an inch; the proportion of data entry errors will vary from day to day; a random sample of 100 two-liter bottles of cola will vary in contents from container to container. Despite the omnipresence of process variation, we must always endeavor to control and reduce it.

We can classify process variation as the result of either *common causes* or *special causes*.

Common Causes of Variation

Common causes of variation are inherent in a process. Common variation is comprised of a myriad of small sources that are always present in a process and affect all elements of the process. Examples of possible causes of common variation[6] are:

Procedures not suited to requirements.

Poor product design.

Machines out of order.

Machines not suited to requirements.

Barriers that rob the worker of the right to do a good job and take pride in his or her work.

Poor instruction and/or supervision of workers.

Poor lighting.

Incoming materials not suited to requirements.

Vibration.

Failure to provide the hourly workers with statistical information to help them improve performance and reduce variation.

Humidity too high or too low.

Unpleasant working conditions, such as noise, dirt, extremes of heat or cold, or poor ventilation.

Management should not hold the workers responsible for such problems of the system; the system is management's responsibility. If management is unhappy with the amount of common variation in the system, it must act to remove it. Professionals estimate that common variation causes about 85 percent of the problems in a process, the remaining 15 percent being caused by special variation.

Special Causes of Variation

Variations created by special causes lie outside the system. Frequently their detection, possible avoidance, and removal are the responsibility of the people directly involved with the process. However, sometimes management must try to find these special causes. When found, policy must be set so that if these special causes are undesirable, they will not recur. If, on the other hand, these special causes are desirable, policy must be set so that they do recur.

CONTROL CHARTS AND VARIATION

We will see in this chapter that control charts are used to identify and differentiate between these two different causes of variation. When a process no longer exhibits special variation, but only common variation, it is said to be stable and is capable of being improved.

When only common causes of variation are present in a process, management must take action to reduce the difference between customer needs and process performance by endeavoring to move the centerline of the process closer to a desired level (*nominal*) and/or by reducing the level of common variation. These types of changes will aid in the quest for never-ending improvement.

THE NEED FOR THE CONTINUAL REDUCTION OF VARIATION

During the last two centuries, most "mass production" concerned itself with meeting engineering specifications most of the time; variation was not the central focus. As long as an item or a part served its intended purpose, it was classified as "good" and passed on to its next operation or final use. When excessive variation caused an item to be nonconforming, it was classified as "bad" and downgraded, reworked, discarded, or somehow removed from the mainstream of output. Little if any effort was made to investigate the causes of such variation; it was accepted as a way of life. High output was maintained by overproducing and then sorting into items that met specifications and items that did not. It is still common to find firms increasing output by relaxing engineering specifications to include marginally defective items with good ones.[7]

Deming has written, "It is good management to reduce the variation in any quality characteristic, whether this characteristic be in a state of control or not, and even when few or no defectives are being produced."[8] When variation is reduced, parts will be more nearly alike. Finished products will work better and be more reliable. Customer satisfaction will increase because customers will know what to expect. Process output and capability will be known with greater certainty, and the results of any changes to the process will be more predictable.

Therefore, management must constantly attempt to reduce process variation around desired characteristic specification levels (nominal levels) to achieve the degree of uniformity required to get products to function during their life cycle as promised to the customer.

The belief that there is no loss associated with products that conform to specifications regardless of the size of the deviation from nominal is fallacious.[9] Figure 8.2 shows the traditional view of losses arising from deviations from nominal: losses are zero until the specification limit is reached;

FIGURE 8.2   Traditional View of Losses Arising from Deviations from Nominal

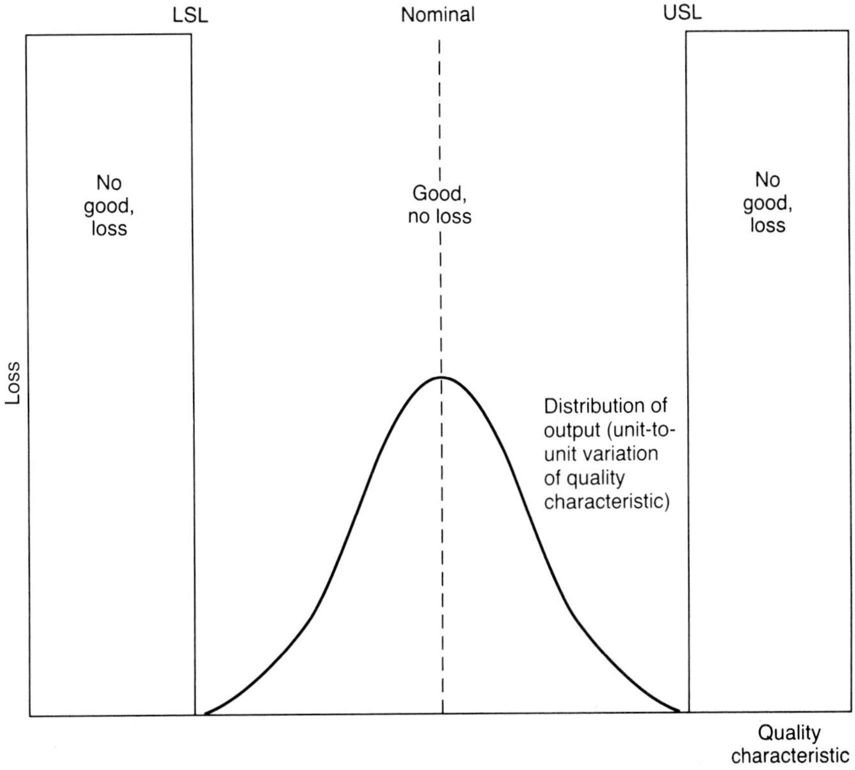

and suddenly they become positive and constant, regardless of the size of the deviation from nominal.

Figure 8.3 shows a more realistic loss function: losses begin to accrue as soon as products deviate from nominal. As can be seen, the view represented in this figure requires the never-ending reduction of process variation around nominal to be maximally cost-efficient and provide the degree of customer satisfaction demanded in today's marketplace.[10]

To illustrate, consider an automobile battery charged by an alternator. The alternator has a voltage regulator that controls the charge to the battery. The alternator-voltage regulator assembly must put out a charge of 13.2 volts to keep the battery's charge at 12 volts.

If the alternator produces a charge of less than 13.2 volts, the electrolyte (acid) in the battery will gradually turn into water, resulting in failure of the battery. The lower the alternator output, the more quickly this will happen. If the alternator output is more than 13.2 volts, excessive heat

FIGURE 8.3   Realistic View of Losses Arising from Deviations from Nominal

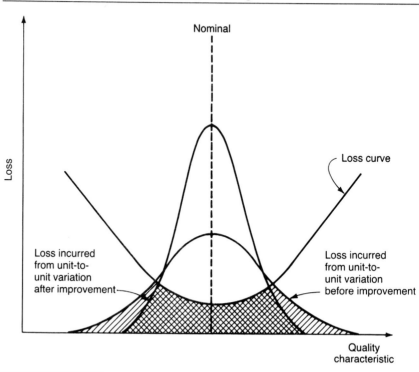

will build up in the battery, causing the battery plates to warp, which will also result in failure of the battery. As the alternator output increases, this effect will occur more quickly.[11]

Any deviation from nominal clearly causes a loss. The greater the deviation, the greater the loss. Consequently, we see the logic in the necessity for the continual reduction of process variation.

Donald J. Wheeler and David S. Chambers discuss another rationale for the continual reduction of process variation.[12] A process can be described as existing in one of four states: chaos, the brink of chaos, the threshold state, and the ideal state.

When a process is in a state of *chaos,* it is producing some nonconforming product and it is not in a state of statistical control (i.e., special causes of variation are present). There is no way to know or predict the percentage of nonconforming product that the process will generate.

A process on the *brink of chaos* produces 100 percent conforming product; however, the process is not stable (i.e., there is variation resulting from special causes). Hence, there is no guarantee that the process will continue to produce 100 percent conforming product indefinitely. As it is unstable, the process may wander and the product's characteristics may change at any time, entering a state of chaos.

The *threshold state* describes a stable process that produces some non-conforming product; process variation results from common causes that are an inherent part of the system. The only way to reduce this variation is to change the process itself.

The *ideal state* describes a stable process producing 100 percent conforming product. The ideal state is not a natural state; forces will always exist to push the process away from the ideal state. Wheeler and Chambers liken this phenomenon to entropy, in that there is similarly a trend towards disorder in the universe. It may help to visualize a process in the ideal state as a perfectly swept lawn. There will always be winds to mar its perfect appearance by depositing leaves, twigs, or other debris. Keeping the lawn perfectly swept is a never-ending challenge. In the same way striving toward an ideal state for a process requires constant attention on the part of management.

Control charts are statistical tools that make possible the distinction between common and special causes of variation. Consequently, control charts permit management to relentlessly pursue the continuous reduction of process variation and strive toward the ideal state for a process.

THE STRUCTURE OF CONTROL CHARTS

All control charts have a common structure. As shown in Figure 8.4, they have a centerline representing the process average and upper and lower control limits that provide information on the process variation.

Control charts are constructed by drawing samples, called subgroups, from some process/product/service characteristic. Control limits are based on the variation that occurs within the sampled subgroups. In this way, variation between the subgroups is intentionally excluded from the

FIGURE 8.4 Structure of a Control Chart

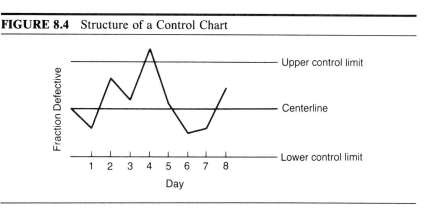

computation of the control limits; the common process variation becomes the variation on which we calculate the control limits. The control limit computations assume that there are no special causes of variation affecting the process. If a special cause of variation is present, the control chart, based solely on common variation, will highlight when and where the special cause occurred. Consequently, the control chart makes possible the distinction between common and special variation and provides management and workers with a basis on which to take corrective action on a process.

The centerline of a control chart is taken to be the estimated mean of the sampling distribution (or the process average); the upper control limit is the mean plus three times the estimated standard error; and the lower control limit is the mean minus three times the estimated standard error.

FIGURE 8.5   Formulation of Control Chart for Data Entry Operation

Raw Data for Construction of Control Chart

Day	Number of Entries Inspected	Number of Defective Entries	Fraction of Defective Entries
1	200	6	.030
2	200	6	.030
3	200	6	.030
4	200	5	.025
5	200	0	.000
6	200	0	.000
7	200	6	.030
8	200	14	.070
9	200	4	.020
10	200	0	.000
11	200	1	.005
12	200	8	.040
13	200	2	.010
14	200	4	.020
15	200	7	.035
16	200	1	.005
17	200	3	.015
18	200	1	.005
19	200	4	.020
20	200	0	.000
21	200	4	.020
22	200	15	.075
23	200	4	.020
24	200	1	.005
Total	4,800	102	

Hence, subgroup means that behave nonrandomly with respect to these control limits will be said to be indications of the presence of special causes of variation.

STABILIZING A PROCESS WITH CONTROL CHARTS

As an example of the use of control charts to detect special variation, consider a data entry operation that makes numerous entries daily.[13] On each of 24 consecutive days, subgroups of 200 entries are inspected. Figure 8.5 illustrates the resulting raw data, and Figure 8.6 shows a plot of the fraction of defective entries as a function of time. The plot of Figure 8.6 seems to indicate that on days 5, 6, 10, and 20 something unusually good happened (zero percent defectives), and on days 8 and 22 something unusually bad happened. A simple control chart will help to determine whether these points were caused by common or special variation.

When the data consist of a series of fractions defective, the appropriate control chart is a p chart. This is a depiction of the fraction of process data that has some attribute of interest—in our example, the fraction defective.

Recall from Chapter 7 that the sampling distribution of the fraction defective has mean and standard error given by Equations 7.2 and 7.3,

FIGURE 8.6 Plot of Fraction of Defective Entries Against Time

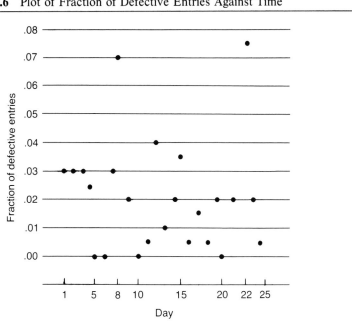

respectively. Thus, the centerline for a p chart is the mean of the sampling distribution, or p, the fraction defective calculated from Equation 5.8:

$$p = \left[\frac{\text{Total number of defectives in all subgroups under investigation}}{\text{Total number of units examined in all subgroups under investigation}} \right] \tag{8.1}$$

Control limits, called *3-sigma limits*, are calculated as p plus and minus three times the standard error. Thus, using the standard error in Equation 7.3, the upper and lower control limits for a p chart are given by:

$$UCL(p) = p + 3\sqrt{\frac{p(1 - p)}{n}} \tag{8.2}$$

$$LCL(p) = p - 3\sqrt{\frac{p(1 - p)}{n}} \tag{8.3}$$

where n is the subgroup size.

We can now use Equations 8.1, 8.2, and 8.3 to find the numerical values for constructing our p chart.

$$p = \frac{102}{4,800} = 0.02125$$

$$\text{centerline}(p) = 0.021$$

$$UCL(p) = 0.02125 + 3\sqrt{\frac{(0.02125)(1 - 0.02125)}{200}}$$

$$= 0.02125 + 3(0.010198)$$

$$= 0.05184$$

$$\text{upper control limit} = 0.052$$

$$LCL(p) = 0.02125 - 3\sqrt{\frac{(0.02125)(1 - 0.02125)}{200}}$$

$$= 0.02125 - 3(0.010198)$$

$$= -0.00934$$

$$\text{lower control limit} = 0.00.$$

Notice that as a negative fraction defective is not possible, the lower control limit is set at 0.00.

Figure 8.7 shows the completed p chart. It is clear that on days 8 and 22 there is some special variation. Notice, however, that the fractions defective on days 5, 6, 10 and 20 are not beyond the control limits. Days with no defectives are not out-of-control; we have merely observed that the process is capable of producing zero defectives some of the time. The

FIGURE 8.7 Control Chart for Fraction Defective

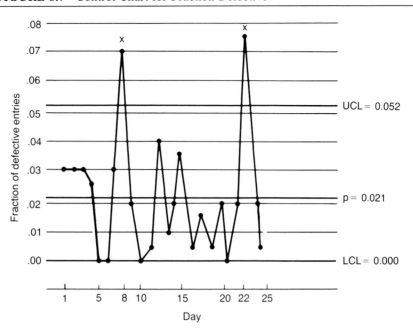

observations on days 8 and 22 indicate that the process exhibits a lack of statistical control.

When a manager or worker determines that the cause of variation is special, he or she should search for and eliminate the causes that may be attributable.to such things as a specific machine, worker, or group of workers, or a new batch of raw materials. After the causes of special variation have been identified and rectified, a stable process that is in statistical control will remain.

In our example, to bring the process under control, management investigated the observations which were out-of-control (days 8 and 22) in an effort to discover and remove the special causes of variation in the process. In this case, management found that on day 8 a new operator had been added to the work force without any training. The logical conclusion was that the new environment probably caused the unusually high number of errors. To ensure that this special cause would not recur, the company added a one-day training program in which data entry operators would be acclimated to the work environment.

A team of managers and workers conducted an investigation of the circumstances occurring on day 22. Their work revealed that on the previous night one of the data entry consoles malfunctioned and was replaced with a standby unit. The standby unit was older and slightly different from the ones currently used in the department. The repairs on the regular console were not expected to be completed until the morning of day 23. To correct this special source of variation, the team recommended purchasing a spare console that would match the existing equipment and disposing of the outdated model presently being used as the backup. Management then implemented the suggestion.

Negative special causes of variation can be eliminated from a process, or positive special causes of variation can be incorporated into a process, by setting and enforcing policy changes. Once this has been done, the process has, in essence, been changed.

The action taken on the process stemming from the investigations of days 8 and 22 should change the process so that the special causes of variation will be eliminated. Consequently, the data from days 8 and 22 may now be deleted. After eliminating the data for the days in which the special causes of variation are found, the control chart statistics are re-computed.

$$p = \frac{73}{4,400} = 0.01659 \cong 0.017$$

$$UCL(p) = 0.01659 + 3 \sqrt{\frac{(0.01659)(1 - 0.01659)}{200}} = 0.04369 \cong 0.044$$

$$LCL(p) = 0.01659 - 3 \sqrt{\frac{(0.01659)(1 - 0.01659)}{200}} = -0.01051 \cong -0.011$$

lower control limit = 0.00.

FIGURE 8.8   Control Chart for Fraction Defective

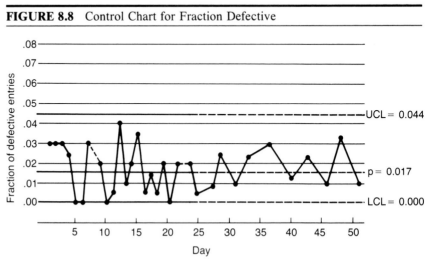

Chart of the process after special causes on days 8 and 22 are found by management and removed.

Figure 8.8 shows the revised control chart. The process appears to be stable and in statistical control. Notice that the revised control chart has somewhat narrower control limits than the original. When special causes have been eliminated, the narrower limits that occur may reveal other points that are now out-of-control. It will then be necessary to again search for special causes. Several such iterations may be required until the process is stable and in statistical control. At least 20 subgroups should remain to determine that the process is indeed stable. If the elimination of subgroups has left fewer than 20 subgroups remaining, additional data should be collected to ensure that the process is in a state of statistical control.

In this example, the control limits were extended into the future (see the dashed lines on the right side of Figure 8.8), and more data were collected and compared to the revised control limits. The process was found to be stable and consequently capable of being improved.

ADVANTAGES OF A STABLE PROCESS

A *stable process* is a process that exhibits only common variation or variation resulting from inherent system limitations. The advantages of achieving a stable process are:

1. Management knows the process capability and can predict performance, costs, and quality levels.
2. Productivity will be at a maximum, and costs will be minimized.
3. Management will be able to measure the effects of changes in the system with greater speed and reliability.
4. If management wants to alter specification limits, it will have the data to back up its decision.

A stable process is a basic requirement for process-improvement efforts.[14]

IMPROVING A PROCESS WITH CONTROL CHARTS

Once a process is stable, it has a known capability. A stable process may, nevertheless, produce an unacceptable number of defects (threshold state) and continue to do so as long as the system, as currently defined, remains the same. Management owns the system and must assume the ultimate responsibility for changing the system to reduce common variation and the difference between customer needs and process performance.

There are two areas for action to reduce the difference between customer needs and process performance. First, action may be taken to change the process average. This might include action to reduce the level

FIGURE 8.9 Control Chart for Fraction Defective

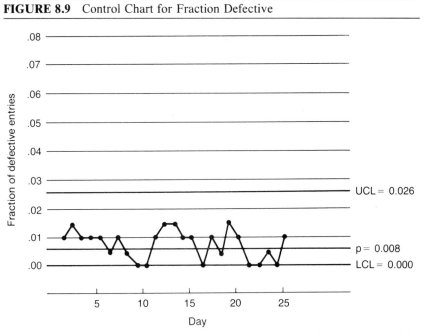

Chart of the process after training and procedure changes have been implemented by management.

of defects or process changes to increase production. Second, management can act to reduce the level of common variation with an eye toward never-ending improvement of the process. Procedures and inputs, such as composition of the work force, training, supervision, materials, tools and machinery, and operational definitions, are the responsibility of management. The workers can only suggest changes; they cannot effect changes to the system.

In our example of the data entry firm, an employee-suggested training program was instituted. The program was aimed at reducing the average fraction of errors and the common variation, which would result in narrower control limits. Figure 8.9 shows the data entry control chart after management instituted the new training program. The average proportion of entries with errors decreased from 0.017 to 0.008, and the process variation decreased as well.

QUALITY CONSCIOUSNESS AND TYPES OF CONTROL CHARTS

Being conscious of quality typically follows a logical pattern. It begins with defect detection, moves to defect prevention, and ends with the realization that the only path to continued prosperity is through never-ending improvement.

Defect Detection

Using a policy of defect detection to achieve quality assumes that defects will be produced. Production and detection of a defective item is more costly than production of a nondefective item. In this first stage of quality consciousness there is no statistical process control, and therefore there are no feedback loops or tools available for correcting the factors that created the defectives in the first place. Once a defect is produced, it is too late to do anything but remove it from the process output.

Defect Prevention: Attribute Control Charts

As the understanding grows that improved quality and its benefits result from process improvement, a quality-conscious organization will move toward defect prevention. The initial entry into this phase generally involves the use of control charts based on attribute data, such as conforming versus nonconforming with respect to some specification.

The most common types of attribute control charts are:

- p chart: used to control the fraction of units with some characteristic (such as the fraction defective).
- np chart: used to control the number of units with some characteristic (such as the number of defectives per batch).
- c chart: used to control the number of events (such as defects) in some fixed area of opportunity (such as a single unit).
- u chart: used to control the number of events (such as defects) in a changeable area of opportunity (such as square yards of paper drawn from an operational paper machine).
- individuals chart: used to control the count when the assumptions for the other attribute charts cannot be met.

Attribute control charts can help move the firm's processes toward a zero percent defective rate. However, attribute control charts do not provide specific information on the cause of the defectives; and as the percent defective approaches zero, larger and larger sample sizes will be needed to detect defective output. For example, if a process is generating an average of one defect in every million units produced, then the average sample size needed to find one defective is one million units. Hence, attribute control charts become ineffective as the proportion of defective output approaches zero. Control charting must continue, but in the face of the limitations of attribute control charts a better means of process improvement and control is required. This will lead management to the next level of quality consciousness—never-ending improvement.

Never-Ending Improvement: Variables Control Charts

Never-ending improvement is the highest level of quality consciousness. Entering this stage necessitates using control charts based on variables data. These types of control charts allow for the never-ending reduction of unit-to-unit variation, even within specification limits. For example, steel rods from a process may all conform to specifications; an attribute control chart would show zero percent defective in every sample. However, by taking actual measurements on rod lengths (variables data), management can collect information that will enable them to consistently strive for the reduction of unit-to-unit variation.

The most common types of variables control charts are:

- x-bar chart: used to control the process average.
- R chart: used to control the process range.
- s chart: used to control the process standard deviation.
- median chart: used as a simple alternative to the combination of an x-bar and R chart.
- individuals chart: used to control subgroups of size one drawn from a process; frequently used when sampling is expensive or only one observation is available per subgroup (e.g., production per month).

Using variables control charts, management may continuously seek to reduce variation, center a process on nominal, and decrease the difference between customer needs and process performance. Chapters 9 and 10 extensively discuss the uses and applications of each of these different types of control charts.

AN EXAMPLE OF AN ATTRIBUTE CONTROL CHART

As another example of a situation in which an attribute control chart is appropriate, consider a new can-forming process used in the orange juice industry. After encountering severe quality problems, management decides to select periodic samples of size 50 from 30 consecutive trial production runs. The data are attribute in nature because each can is either defective or not. Each set of 50 comprises a subgroup that may or may not contain defective units. The number of defectives in each subgroup is thus a discrete random variable that can take on a value from 0 to 50. The fraction defective in each subgroup of 50 cans will be recorded and charted. The proper control chart in this case is the p chart (or, alterna-

FIGURE 8.10   Number of Defectives in 30 Subgroups of Size 50

Production Run	Subgroup Size	Number of Defectives	Subgroup Fraction Defective
1	50	12	0.24
2	50	15	0.30
3	50	8	0.16
4	50	10	0.20
5	50	4	0.08
6	50	7	0.14
7	50	16	0.32
8	50	9	0.18
9	50	14	0.28
10	50	10	0.20
11	50	5	0.10
12	50	6	0.12
13	50	17	0.34
14	50	12	0.24
15	50	22	0.44
16	50	8	0.16
17	50	10	0.20
18	50	5	0.10
19	50	13	0.26
20	50	11	0.22
21	50	20	0.40
22	50	18	0.36
23	50	24	0.48
24	50	15	0.30
25	50	9	0.18
26	50	12	0.24
27	50	7	0.14
28	50	13	0.26
29	50	9	0.18
30	50	6	0.12
	1,500	347	

tively, an np chart, which Chapter 9 discusses in detail). The data derived from the subgroups are shown in Figure 8.10.

Equation 8.1 yields a centerline value of:

$$p = 347/1{,}500 = 0.2313$$

Then from Equations 8.2 and 8.3 the UCL and the LCL values are:

$$\text{UCL}(p) = 0.2313 + 3\sqrt{\frac{(0.2313)(1 - 0.2313)}{50}} = 0.4102 \cong 0.41$$

FIGURE 8.11 Control Chart for Fraction Defective

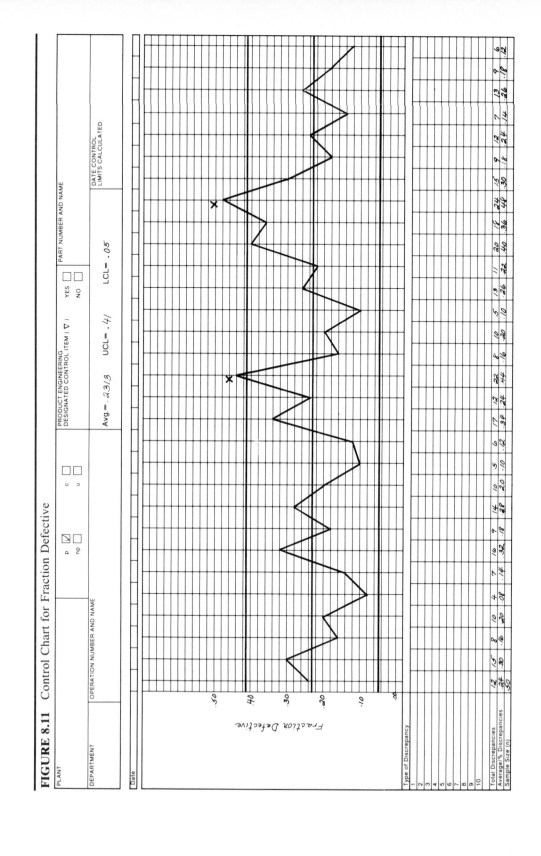

and

$$LCL(p) = 0.2313 - 3 \sqrt{\frac{(0.2313)(1 - 0.2313)}{50}} = 0.0524 \cong 0.05$$

Figure 8.11 illustrates the control chart for this process.

There are two points that indicate a lack of control because they fall above the UCL. There may be other grounds for believing that this process is exhibiting signs of special causes of variation; Chapter 11 discusses more sophisticated methods for determining whether special sources of variation are present in a process.

AN EXAMPLE OF A VARIABLES CONTROL CHART

A large pharmaceutical firm provides vials that are filled to a nominal value (specification) of 52.0 grams. The firm's management has embarked on a program of statistical process control and has decided to use variables control charts for this filling process to detect special causes of variation. Samples of six vials are selected every five minutes during a 105-minute period. Each set of six measurements makes up a subgroup.

As will be discussed more fully in Chapters 9 and 10, the appropriate control chart in this instance is an x-bar and R chart. The purpose of this chart is to see whether the process output is stable with regard to its variability and its average value. The output of each subgroup is summarized by its sample average and range. x-bar is the average for each of the subgroups, as given by Equation 5.1, while R is the range, or the largest data value in each subgroup minus the smallest data value in that subgroup for each subgroup, calculated from Equation 5.9.

For our example, the subgroup ranges are shown in Figure 8.12. They begin with a range at 9:30 of

$$R = 53.10 - 52.22 = 0.88$$

and continue for all 22 subgroups to the last range at 11:15 of

$$R = 52.16 - 51.67 = 0.49$$

The average of the R values is called $\overline{R}$, and it is computed by taking the simple arithmetic average of the R values.

$$\overline{R} = \Sigma R/k \qquad (8.4)$$

where k is the number of subgroups.

Recall from Chapter 7 that the sampling distribution of the range has a mean and standard error that can be written as functions of the process standard deviation σ, as given by Equations 7.17 and 7.18, respectively.

FIGURE 8.12   Vial Weights

Date: 1-10-87   *Nominal Fill: 52.00 gms.*
Obs.   *Measurement (gm)*
#	Time	1	2	3	4	5	6	R	$\bar{x}$
1	9:30	52.22	52.85	52.41	52.55	53.10	52.47	0.88	52.60
2	9:35	52.25	52.14	51.79	52.18	52.26	51.94	0.47	52.09
3	9:40	52.37	52.69	52.26	52.53	52.34	52.81	0.55	52.50
4	9:45	52.46	52.32	52.34	52.08	52.07	52.07	0.39	52.22
5	9:50	52.06	52.35	51.85	52.02	52.30	52.20	0.50	52.13
6	9:55	52.59	51.79	52.20	51.90	51.88	52.83	1.04	52.20
7	10:00	51.82	52.12	52.47	51.82	52.49	52.60	0.78	52.22
8	10:05	52.51	52.80	52.00	52.47	51.91	51.74	1.06	52.24
9	10:10	52.13	52.26	52.00	51.89	52.11	52.27	0.38	52.11
10	10:15	51.18	52.31	51.24	51.59	51.46	51.47	1.13	51.54
11	10:20	51.74	52.23	52.23	51.70	52.12	52.12	0.53	52.02
12	10:25	52.38	52.20	52.06	52.08	52.10	52.01	0.37	52.14
13	10:30	51.68	52.06	51.90	51.78	51.85	51.40	0.66	51.78
14	10:35	51.84	52.15	52.18	52.07	52.22	51.78	0.44	52.04
15	10:40	51.98	52.31	51.71	51.97	52.11	52.10	0.60	52.03
16	10:45	52.32	52.43	53.00	52.26	52.15	52.36	0.85	52.42
17	10:50	51.92	52.67	52.80	52.89	52.56	52.23	0.97	52.51
18	10:55	51.94	51.96	52.73	52.72	51.94	52.99	1.05	52.38
19	11:00	51.39	51.59	52.44	51.94	51.39	51.67	1.05	51.74
20	11:05	51.55	51.77	52.41	52.32	51.22	52.04	1.19	51.89
21	11:10	51.97	51.52	51.48	52.35	51.45	52.19	0.90	51.83
22	11:15	52.15	51.67	51.67	52.16	52.07	51.81	0.49	51.92

Thus, the centerline can be computed as:

$$\bar{R} = \frac{16.28}{22} = 0.740$$

Using a multiple of three standard errors to construct the upper control limit and the lower control limit, we have:

$$UCL(R) = \bar{R} + 3\sigma_R$$

From Equation 7.17, $\bar{R} = d_2\sigma$; from Equation 7.18, $\sigma_R = d_3\sigma$. We can thus replace $\bar{R}$ and σ_R with these functions of the process standard deviation, σ:

$$UCL(R) = d_2\sigma + 3d_3\sigma = d_2\sigma(1 + 3d_3/d_2)$$
$$= \bar{R}(1 + 3d_3/d_2)$$

where d_2 and d_3, the factors introduced in Chapter 7 for stable processes that are normally distributed, are given in Table 8.

Similarly,

$$LCL(R) = \overline{R} - 3\sigma_R$$
$$= d_2\sigma - 3d_3\sigma = d_2\sigma(1 - 3d_3/d_2)$$
$$= \overline{R}(1 - 3d_3/d_2)$$

A more convenient way to represent these control limits is by defining two new constants, D_3 and D_4:

$$D_3 = 1 - 3d_3/d_2$$
$$D_4 = 1 + 3d_3/d_2$$

so that:

$$UCL(R) = D_4\overline{R} \qquad (8.5)$$

and

$$LCL(R) = D_3\overline{R} \qquad (8.6)$$

where D_3 and D_4 are tabulated as a function of subgroup size in Table 8. For our example:

$$UCL(R) = 2.00(0.740) = 1.48$$

and

$$LCL(R) = 0(0.740) = 0$$

Figure 8.13(a) illustrates the resulting average range and upper and lower control limits for the subgroup ranges. The R chart is then examined for signs of special variation. None of the points on the R chart is outside of the control limits, and there are no other signals indicating a lack of control. Thus, there are no indications of special sources of variation on the R chart. More will be said later in this chapter and in Chapter 11 about indications of special causes of variation.

After analyzing the R chart, we construct the x-bar chart. This control chart depicts variations in the averages of the subgroups. To find the average for each subgroup, we add the data points for each subgroup and divide by the number of entries in the subgroup, as given by Equation 5.1. For the pharmaceutical company, the average of the 9:30 subgroup is:

$$\frac{52.22 + 52.85 + 52.41 + 52.55 + 53.10 + 52.47}{6} = 52.60$$

This calculation is repeated for each of the subgroups. The x-bar results for this example can be found in the last column of Figure 8.12.

Recall from Chapter 7 that the sampling distribution of the mean has mean and standard error given by Equations 7.14 and 7.15, respectively, where σ is the process standard deviation. Thus, the centerline of an

FIGURE 8.13 x-Bar and R Charts for Filling Operation

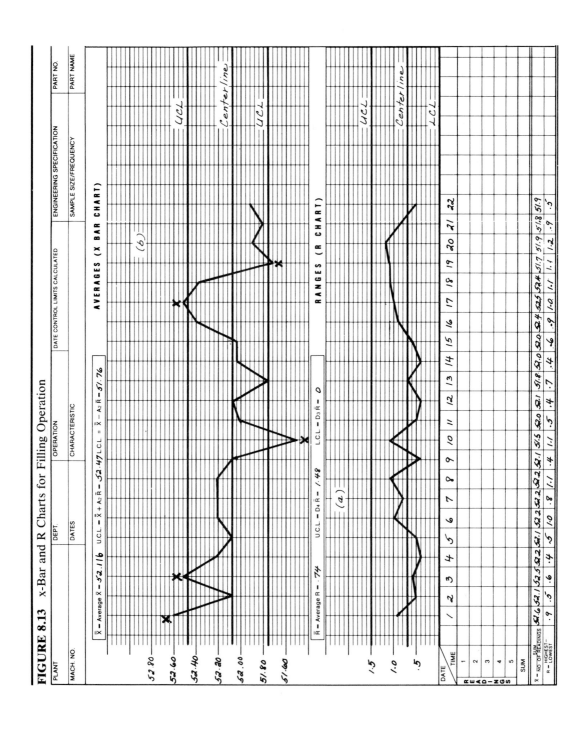

x-bar control chart is found by taking the average of the subgroup averages, $\bar{\bar{x}}$, calculated from Equation 5.2. In our example, the average of the 22 subgroup averages is:

$$\bar{\bar{x}} = \frac{1146.55}{22} = 52.116$$

Using a multiple of three standard errors to construct the control limits, we have:

$$UCL(\bar{x}) = \bar{\bar{x}} + 3(\sigma/\sqrt{n})$$

From Equation 7.17, $\bar{R} = d_2\sigma$, so that

$$UCL(\bar{x}) = \bar{\bar{x}} + 3[(\bar{R}/d_2)/\sqrt{n}]$$

or, more conveniently,

$$UCL(\bar{x}) = \bar{\bar{x}} + A_2\bar{R} \qquad (8.7)$$

and

$$LCL(\bar{x}) = \bar{\bar{x}} - 3(\sigma/\sqrt{n})$$
$$= \bar{\bar{x}} - 3[(\bar{R}/d_2)/\sqrt{n}]$$

or,

$$LCL(\bar{x}) = \bar{\bar{x}} - A_2\bar{R} \qquad (8.8)$$

where $A_2 = 3/(d_2\sqrt{n})$ is tabulated as a function of subgroup size in Table 8.

For the pharmaceutical company, the upper and lower control limits can now be computed as:

$$UCL(\bar{x}) = 52.116 + (0.48)(0.740) = 52.47$$

and

$$LCL(\bar{x}) = 52.116 - (0.48)(0.740) = 51.76$$

Figure 8.13(b) illustrates the x-bar chart. Notice that a total of five points on the x-bar chart are outside of the control limits and therefore indicate a lack of control. Further investigation is warranted to determine the source(s) of these special variations.

It is important to realize that an x-bar chart cannot be meaningfully analyzed if its corresponding R chart is not in statistical control. The reason for this is that x-bar chart control limits are calculated from $\bar{R}$ (i.e. $\bar{\bar{x}} \pm A_2\bar{R}$), and if the range is not stable, no calculations based on it will be stable.

TWO POSSIBLE MISTAKES IN USING CONTROL CHARTS

There are two different types of mistakes that the user of a control chart may make: overadjustment and underadjustment. Proper use of control charts will minimize the economic consequences of making either of these types of errors.

Overadjustment

The *overadjustment* error occurs when the user reacts to swings in the process data that are merely the result of common variation, such as adjusting a process downward if its past output is above average or adjusting a process upward if its past output is below average. When a process is overadjusted, it resembles a car being oversteered, veering back and forth across the highway. In general, processes should not be adjusted on the basis of time-to-time observations but on the basis of information provided by a statistical control chart.

The effects of overadjustment can be seen by examining a frequently used demonstration device known as a Quincunx board, shown in Figure 8.14. The Quincunx board is a rectangular box with an upper chamber containing a large number of beads. A horizontal sliding bar feeds one or more beads at a time into a triangular hopper that then allows the beads to fall at a specified lateral point directly above 10 rows of pegs. Each time a bead hits a peg, it will bounce right or left so that its position after falling through the 10 rows of pegs is a result of 10 random events. It does not seem unreasonable to expect that most beads would tend to fall almost directly beneath the point at which they were released. However, some beads may tend to wander a bit and end up to the right or left of their release point.

Notice that in Figure 8.14 the beads will fall into a series of slots after passing through the rows of pegs. The slots have been numbered from 1 to 10 for purposes of illustration. Note also that the opening of the triangular hopper is set to release the beads directly above the number 5 slot. Provided that the hopper is not moved, the process output (i.e., slot position of the beads) will be stable.

The slots themselves are divided into two portions: a short upper and a longer lower one. The short portion is used to observe subgroups, while the longer portion is used for the accumulation of subgroups. Figures 8.14 through 8.18 viewed as a sequence show the use of the Quincunx board.

In Figure 8.15 we can see the result of allowing a subgroup of the beads to pass through the triangular hopper and fall into the upper portion of the slots. The bottom of the hopper is centered directly over slot number 5, and beads have already fallen in slots 2 through 7. Figure 8.16 shows not only this subgroup but the accumulation of several earlier subgroups. If

FIGURE 8.14   A Quincunx Board

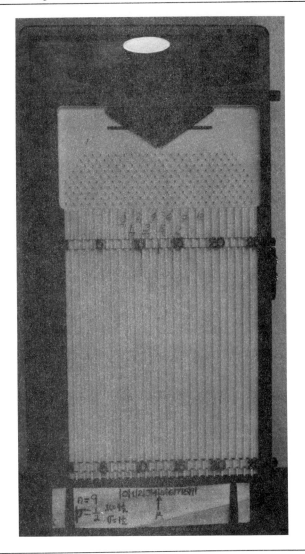

the position of the hopper is not changed, the accumulated results will follow an approximately normal distribution. The variation from the mean is simply common variation.

In Figure 8.16 the latest subgroup is shown as merged with the earlier ones. The process average is 4.5, and the range is 6.

Now suppose instead of having the good sense to leave this process alone, we were to adjust it after each bead dropped. This is *adjusting for*

FIGURE 8.15   Quincunx Bead Falling

Beads being released from the hopper of a Quincunx board. Notice that most beads end up below the release point.

common variation. After each bead passes through the pegs, we count the number of categories above or below 5 and adjust the hopper that many categories in the opposite direction in an attempt to compensate. This scheme will result in increased process variation. Figure 8.17 shows the hopper being adjusted in this manner.

The first subgroup produced this way can be seen in Figure 8.18. The range for this one subgroup alone is 10, as compared to the value of 6 observed for the collected subgroups when the hopper was left fixed. Additional subgroups will, in all likelihood, only increase the variability further. This is the penalty for adjusting on the basis of common variation.

Underadjustment

Underadjustment, or lack of attention, results when a process is out-of-control and no effort is made to provide the necessary regulation. The

FIGURE 8.16 Quincunx Subgroups

Beads in a Quincunx are observed as a subgroup in the upper set of slots in the picture on the right and then added to the aggregated observations in the picture on the left.

process swings up and down in response to one or more special causes of variation, which may have compounding effects.

Avoiding both of these mistakes all of the time is an impossible task. That is, never adjusting the process—so that we never make the mistake of overadjusting—could result in severe underadjustment. On the other hand, if we made very frequent adjustments to avoid the problem of underadjustment, we would probably be overadjusting. Control charts provide an economical means to minimize the loss that results from these two errors. Consequently, control charts provide management with information on when to take action on a process and when to leave it alone.

FIGURE 8.17   Repositioned Quincunx Hopper

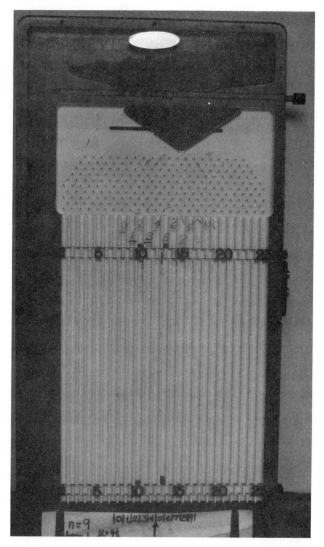

Quincunx hopper is moved to the right to compensate for beads falling to the left.

TWO USES OF CONTROL CHARTS

As we have seen, control charts fall into two broad categories: attribute and variables. In both cases, a particular quality characteristic is mea-

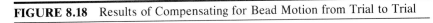

FIGURE 8.18   Results of Compensating for Bead Motion from Trial to Trial

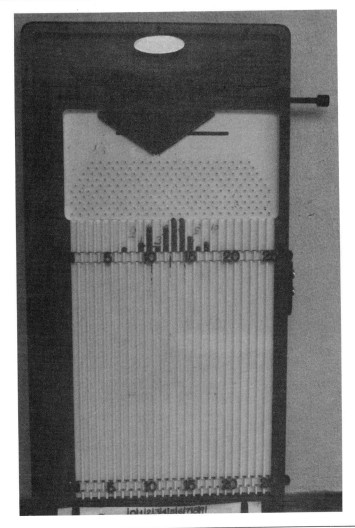

sured and then examined. That examination can be used to either evaluate the history of the process or the present state of the process.

Evaluating the Past

The retrospective examination of the process's completed output using a control chart answers the question of whether the process has been in statistical control. A lack of control, or the presence of special causes of variation, is indicated when one or more of the control chart points is beyond the control limits or is otherwise in violation of one of the several rules introduced later in this chapter. Chapter 11 more fully discusses patterns indicating a lack of control. When no special causes of variation are present, the characteristic measured is said to be in *statistical control,* or stable.

Evaluating the Present

The other use of control charts is to maintain a state of statistical control during a process's operation. Control charts can be used to generate a signal in a real time sense. The signal might, for example, call attention to tool wear or changes in humidity that might require intervention in the process. In this sense control charts are useful in maintaining an existing state of process stability.

SOME OUT-OF-CONTROL EVIDENCE

Chapter 11 will fully discuss out-of-control patterns, but for now let us look at five simple rules to illustrate how specific patterns indicate a lack of control. As mentioned earlier, a process exhibits a lack of statistical control if a point falls beyond either of the control limits. However, it is possible for all points to be within the control limits while there are other factors that indicate a lack of control in the process. Stable processes always exhibit random patterns of variation. Accordingly, most data points will tend to cluster about the mean value, with an approximately equal number of points falling above and below the mean. A few of the values will lie close to the control limits. Rarely, points fall beyond a control limit. Also, there will seldom be prolonged runs upward or downward for a number of subgroups.

In order to examine patterns indicating a lack of control, the area between the control limits is divided into six bands, each band one standard error wide. As shown in Figure 8.19, the bands within one standard error of the centerline are called the *C zones;* the bands between one and two standard errors from the centerline are called the *B zones;* and the outermost bands, which lie between two and three standard errors from the mean, are the *A zones.*

Five simple rules based on these bands are commonly applied to determine if a process is exhibiting a lack of statistical control. Any out-of-control points found are marked with an X directly on the control chart.

Rule 1. *A process exhibits a lack of control if any single value falls outside of the control limits.*

As we have already seen, this is the first criterion, the most obvious one, and the one most often applied. Figure 8.7 exhibits points (indicated by an X) that are out-of-control by virtue of this rule.

FIGURE 8.19 A, B and C Zones for a Control Chart

```
- - - - - - - - - - - - - - - - - - - - - - - - - - -  UCL
              Zone A

              Zone B

              Zone C
                                                       Centerline
              Zone C

              Zone B

              Zone A
- - - - - - - - - - - - - - - - - - - - - - - - - - -  LCL
```

Rule 2. *A process exhibits a lack of control if any two out of three consecutive points fall in one of the A zones or beyond on the same side of the centerline.*

This means that if any two of three consecutive points are in the A zone or beyond, the process exhibits a lack of control when the second A zone point occurs. The two points must be in zone A or beyond on the same side of the centerline; the third point can be anywhere. Figure 8.20 illustrates a process that is exhibiting a lack of control by virtue of Rule 2 at two points in time: at observation 6 and at observation 21.

When applying Rule 2, or any of the other indicators of a lack of control, it is always best to look for patterns demonstrating evidence of a

FIGURE 8.20 Lack of Control by Virtue of Rule 2

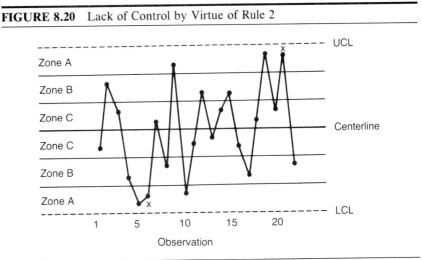

lack of control by looking backwards along the control chart. This makes any patterns or trends more obvious and makes it easier to find the beginning of a pattern or trend.

Consider points 21, 20 and 19. Point 21 is in the upper zone A; point 20 in zone C; and point 19 in the upper zone A again. Point 21 is thus the second of two out of three consecutive points in zone A, and so it is marked as demonstrating a lack of control. Next, notice that observation number 6 is in zone A, preceded by observation number 5 (which was also in zone A) and then observation number 4 (which was in zone B). This makes observation number 6 the second of two out of three consecutive points in the A zone. Therefore, observation number 6 is marked with an X as evidence of a lack of control. The analysis began by looking from right to left on the control chart, and so more recent indications of a lack of control are the first to be found.

Rule 3. *A process exhibits a lack of control if four out of five consecutive points fall in one of the B zones or beyond on the same side of the centerline.*

This means that if any of the four out of five consecutive points are in either one of the B zones or beyond on the same side of the centerline while the fifth is not, the fourth point in the B zone or beyond is deemed to be providing evidence of a lack of control. It should be marked with an X. Figure 8.21 illustrates several possible patterns by which a process may be out-of-control via Rule 3. Observation number 15 is in zone B; number 14 is in zone B; number 13 is also in zone B; number 12 is in zone A (beyond zone B), while number 11 is in zone C. This means that observation number 15 is an indication of a lack of control. Notice that observation 16 lies in zone B also. It, too, should therefore be considered out-of-

FIGURE 8.21   Lack of Control by Virtue of Rule 3

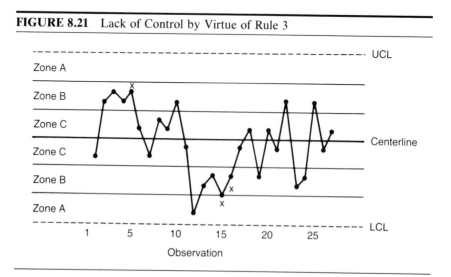

FIGURE 8.22   Lack of Control by Virtue of Rule 4

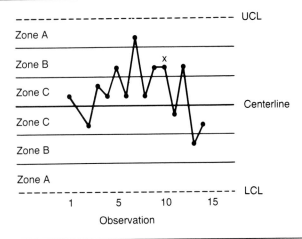

control because it is the fourth of a different set of five consecutive points (points 12, 13, 14, 15, and 16) that lie in zone B or beyond.

Points 5, 4, 3 and 2 all lie in zone B. They constitute four points in a row in this zone (point 1 can be considered the fifth point because it is in zone C); so the fourth point in zone B, point number 5, is marked with an X.

Rule 4. *A process exhibits a lack of control if eight or more consecutive points lie on one side of the centerline.*

The eighth and subsequent points are said to provide evidence of a lack of control by virtue of this rule. Figure 8.22 shows a process that is exhibiting a lack of control by virtue of this rule.

FIGURE 8.23   Lack of Control by Virtue of Rule 5

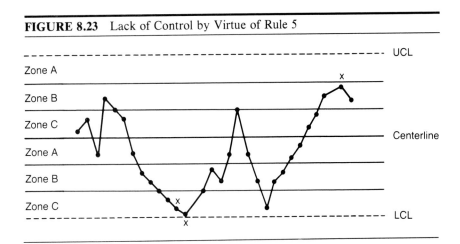

Notice that observation number 10 is the eighth of a string of points on one side of the centerline and can therefore be considered evidence of lack of control.

Rule 5. *A process exhibits a lack of control if eight or more consecutive points move upward in value or if eight or more consecutive points move downward in value.*

The eighth and subsequent points that continue moving up (or down) are said to provide evidence of a lack of control. Figure 8.23 shows a process exhibiting a lack of control by virtue of this rule.

Chapter 11 presents a more detailed discussion of out-of-control rules.

COLLECTING DATA: RATIONAL SUBGROUPING

Proper organization of the data to be control-charted is critical if the control chart is to be helpful in process improvement. We must be certain that we are asking the right questions. In other words, the data must be organized in such a way as to permit examination of variation productively and in a manner that will reveal special sources of variation. The organization of the data defines the question the control chart is addressing to achieve process improvement. Each subgroup should be selected from a small area so that relatively homogeneous conditions exist within each subgroup; this is called *rational subgrouping*. An excellent example can be found in *Understanding Statistical Process Control* by Wheeler and Chambers.[15]

Consider the case of a manufacturer of industrial paints. One-gallon cans are filled four at a time, each one by a separate filling head. The department manager is interested in learning if the weight of the product is stable and within specification and has decided to use statistical process control as an aid.

The supervisor is asked to take five successive cans from each of the four filling heads every hour. The gross weight (in kilograms) of each can is recorded for 20 measurements each hour. The supervisor continues observations for eight consecutive hours, yielding 160 individual observations (as shown in Figure 8.24).

The way in which these observations are arranged may reveal variation from one of three sources: variations over *time,* variations between *measurements,* or variations between *filling heads.* Variation over time, hour to hour in this case, is represented by the differences in the groups of 20 cans; variation between measurements is represented by the difference between the five cans selected at each hour regardless of filling head; and variation between filling heads is represented by the differences between the results of the filling heads for each of the five cans selected per head, per hour.

FIGURE 8.24   Fill Data for 160 Paint Cans

			Time: 8 A.M.							*Time: 9 A.M.*			
			Measurement							*Measurement*			
		1	*2*	*3*	*4*	*5*			*1*	*2*	*3*	*4*	*5*
H	1	6.09	6.10	6.09	6.09	6.09	H	1	6.13	6.13	6.14	6.13	6.11
E	2	6.09	6.09	6.10	6.09	6.09	E	2	6.12	6.12	6.11	6.13	6.10
A	3	6.10	6.11	6.12	6.11	6.11	A	3	6.11	6.13	6.13	6.14	6.14
D	4	6.16	6.16	6.17	6.17	6.17	D	4	6.20	6.20	6.20	6.17	6.17

			Time: 10 A.M.							*Time: 11 A.M.*			
			Measurement							*Measurement*			
		1	*2*	*3*	*4*	*5*			*1*	*2*	*3*	*4*	*5*
H	1	6.14	6.13	6.12	6.13	6.13	H	1	6.10	6.10	6.08	6.15	6.11
E	2	6.11	6.10	6.11	6.10	6.14	E	2	6.08	6.12	6.10	6.11	6.10
A	3	6.13	6.13	6.11	6.13	6.14	A	3	6.13	6.13	6.15	6.15	6.09
D	4	6.16	6.16	6.19	6.19	6.21	D	4	6.20	6.18	6.21	6.21	6.20

			Time: 12 Noon							*Time: 1 P.M.*			
			Measurement							*Measurement*			
		1	*2*	*3*	*4*	*5*			*1*	*2*	*3*	*4*	*5*
H	1	6.14	6.15	6.14	6.13	6.16	H	1	6.12	6.09	6.11	6.10	6.10
E	2	6.10	6.12	6.15	6.13	6.13	E	2	6.16	6.13	6.13	6.09	6.10
A	3	6.13	6.13	6.14	6.14	6.13	A	3	6.16	6.11	6.11	6.13	6.10
D	4	6.17	6.18	6.18	6.18	6.17	D	4	6.19	6.21	6.21	6.19	6.16

			Time: 2 P.M.							*Time: 3 P.M.*			
			Measurement							*Measurement*			
		1	*2*	*3*	*4*	*5*			*1*	*2*	*3*	*4*	*5*
H	1	6.07	6.07	6.08	6.07	6.08	H	1	6.11	6.12	6.13	6.13	6.13
E	2	6.07	6.08	6.07	6.07	6.09	E	2	6.10	6.11	6.13	6.10	6.13
A	3	6.09	6.09	6.09	6.09	6.10	A	3	6.13	6.16	6.14	6.13	6.13
D	4	6.15	6.15	6.16	6.16	6.14	D	4	6.18	6.19	6.20	6.19	6.21

The manager must decide on the proper arrangement of these data; the way in which they are arranged will dictate the variation that might be revealed. Let us consider an arrangement of the data for each possible source of variation.

Arrangement One

If the basic subgroup consists of the four head readings for a given measurement and hour, as shown in Figure 8.25, then the variation within the

FIGURE 8.25   Arrangement One

Time: 8:00 A.M.
Measurement

	1	2	3	4	5
H 1	6.09	6.10	6.09	6.09	6.09
E 2	6.09	6.09	6.10	6.09	6.09
A 3	6.10	6.11	6.12	6.11	6.11
D 4	6.16	6.16	6.17	6.17	6.17

subgroups will be the variation from head to head. That is, our estimate of the process standard error will be based on the measurements taken across all four filling heads. The process is being studied on a measurement-to-measurement and hour-to-hour basis, which means that the process variation is allocated as follows:

Source of Variation	*Allocation*
Hour-to-hour	Between subgroups
Measurement-to-measurement	Between subgroups
Head-to-head	Within subgroups

Arrangement number one asks the following questions:

1. Is there a systematic difference in hour-to-hour observations?
2. Is there a systematic difference in measurement-to-measurement observations?
3. Are the head-to-head differences consistent?

Consequently, the first arrangement of the data looks for hour-to-hour and measurement-to-measurement special causes of variation.

For the eight hours, there will be a total of 40 subgroups. The average and range for each subgroup has been computed and appears below in Figure 8.26. The average of the averages, $\bar{\bar{x}}$, has been calculated to be 6.13. The average of the range values, $\bar{R}$, is 0.08. When the data is analyzed via this arrangement, it shows some evidence of a lack of control, as shown in Figure 8.27. The evidence is the long string of points that are above the centerline on the x-bar portion of the control chart (Rule 4). This, in all likelihood, results from some special source of variation that is unclear at this time.

FIGURE 8.26   Arrangement One: 40 Subgroups

		Time: 8 A.M. Measurement								*Time: 9 A.M.* Measurement				
		1	*2*	*3*	*4*	*5*				*1*	*2*	*3*	*4*	*5*
H	1	6.09	6.10	6.09	6.09	6.09	H	1		6.13	6.13	6.14	6.13	6.11
E	2	6.09	6.09	6.10	6.09	6.09	E	2		6.12	6.12	6.11	6.13	6.10
A	3	6.10	6.11	6.12	6.11	6.11	A	3		6.11	6.13	6.13	6.14	6.14
D	4	6.16	6.16	6.17	6.17	6.17	D	4		6.20	6.20	6.20	6.17	6.17
$\bar{x}$		6.11	6.12	6.12	6.12	6.12	$\bar{x}$			6.14	6.15	6.15	6.14	6.13
R		0.07	0.07	0.08	0.08	0.08	R			0.09	0.08	0.09	0.04	0.07

		Time: 10 A.M. Measurement								*Time: 11 A.M.* Measurement				
		1	*2*	*3*	*4*	*5*				*1*	*2*	*3*	*4*	*5*
H	1	6.14	6.13	6.12	6.13	6.13	H	1		6.10	6.10	6.08	6.15	6.11
E	2	6.11	6.10	6.11	6.10	6.14	E	2		6.08	6.12	6.10	6.11	6.10
A	3	6.13	6.13	6.11	6.13	6.14	A	3		6.13	6.13	6.15	6.15	6.09
D	4	6.16	6.16	6.19	6.19	6.21	D	4		6.20	6.18	6.21	6.21	6.20
$\bar{x}$		6.14	6.13	6.13	6.14	6.16	$\bar{x}$			6.13	6.13	6.14	6.16	6.13
R		0.05	0.06	0.08	0.09	0.08	R			0.12	0.08	0.13	0.10	0.11

		Time: 12 Noon Measurement								*Time: 1 P.M.* Measurement				
		1	*2*	*3*	*4*	*5*				*1*	*2*	*3*	*4*	*5*
H	1	6.14	6.15	6.14	6.13	6.16	H	1		6.12	6.09	6.11	6.10	6.10
E	2	6.10	6.12	6.15	6.13	6.13	E	2		6.16	6.13	6.13	6.09	6.10
A	3	6.13	6.13	6.14	6.14	6.13	A	3		6.16	6.11	6.11	6.13	6.10
D	4	6.17	6.18	6.18	6.18	6.17	D	4		6.19	6.21	6.21	6.19	6.16
$\bar{x}$		6.14	6.15	6.15	6.15	6.15	$\bar{x}$			6.16	6.14	6.14	6.13	6.12
R		0.07	0.06	0.04	0.05	0.04	R			0.07	0.12	0.10	0.10	0.06

		Time: 2 P.M. Measurement								*Time: 3 P.M.* Measurement				
		1	*2*	*3*	*4*	*5*				*1*	*2*	*3*	*4*	*5*
H	1	6.07	6.07	6.08	6.07	6.08	H	1		6.11	6.12	6.13	6.13	6.13
E	2	6.07	6.08	6.07	6.07	6.09	E	2		6.10	6.11	6.13	6.10	6.13
A	3	6.09	6.09	6.09	6.09	6.10	A	3		6.13	6.16	6.14	6.13	6.13
D	4	6.15	6.15	6.16	6.16	6.14	D	4		6.18	6.19	6.20	6.19	6.21
$\bar{x}$		6.10	6.10	6.10	6.10	6.10	$\bar{x}$			6.13	6.15	6.15	6.14	6.15
R		0.08	0.08	0.09	0.09	0.06	R			0.08	0.08	0.07	0.09	0.08

FIGURE 8.27 Control Chart for Arrangement One

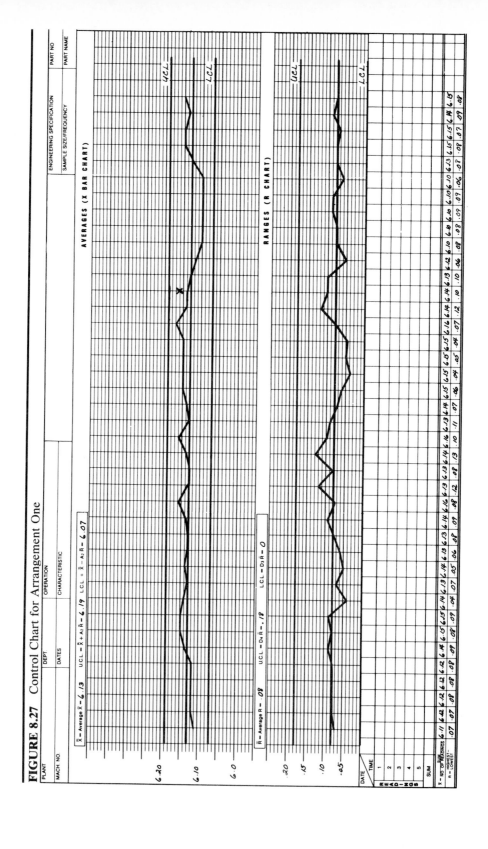

Arrangement Two

If the basic subgroup consists of five measurements for a given head and hour, as shown in Figure 8.28, then the variation from measurement to measurement is the basis for our estimate of the standard error. Our estimate of the process standard error is based on the observations taken across all five measurements for each one of the heads for each hour. The process is being studied on a filling head-to-filling head and hour-to-hour basis. The process variation is allocated as follows:

Source of Variation	*Allocation*
Hour-to-hour	Between subgroups
Measurement-to-measurement	Within subgroups
Head-to-head	Between subgroups

FIGURE 8.28   Arrangement Two

Time: 8 A.M.
Measurement

		1	*2*	*3*	*4*	*5*	$\bar{x}$	*R*
H	1	6.09	6.10	6.09	6.09	6.09	6.09	0.01
E	2	6.09	6.09	6.10	6.09	6.09	6.09	0.01
A	3	6.10	6.11	6.12	6.11	6.11	6.11	0.02
D	4	6.16	6.16	6.17	6.17	6.17	6.17	0.01

Time: 9 A.M.
Measurement

		1	*2*	*3*	*4*	*5*	$\bar{x}$	*R*
H	1	6.13	6.13	6.14	6.13	6.11	6.13	0.03
E	2	6.12	6.12	6.11	6.13	6.10	6.12	0.03
A	3	6.11	6.13	6.13	6.14	6.14	6.13	0.03
D	4	6.20	6.20	6.20	6.17	6.17	6.19	0.03

Time: 10 A.M.
Measurement

		1	*2*	*3*	*4*	*5*	$\bar{x}$	*R*
H	1	6.14	6.13	6.12	6.13	6.13	6.13	0.02
E	2	6.11	6.10	6.11	6.10	6.14	6.11	0.04
A	3	6.13	6.13	6.11	6.13	6.14	6.13	0.03
D	4	6.16	6.16	6.19	6.19	6.21	6.18	0.05

FIGURE 8.28 *Concluded*

Time: 11 A.M.
Measurement

		1	2	3	4	5	$\bar{x}$	R
H	1	6.10	6.10	6.08	6.15	6.11	6.11	0.07
E	2	6.08	6.12	6.10	6.11	6.10	6.10	0.04
A	3	6.13	6.13	6.15	6.15	6.09	6.13	0.06
D	4	6.20	6.18	6.21	6.21	6.20	6.20	0.03

Time: 12 Noon
Measurement

		1	2	3	4	5	$\bar{x}$	R
H	1	6.14	6.15	6.14	6.13	6.16	6.14	0.03
E	2	6.10	6.12	6.15	6.13	6.13	6.13	0.05
A	3	6.13	6.13	6.14	6.14	6.13	6.13	0.01
D	4	6.17	6.18	6.18	6.18	6.17	6.18	0.01

Time: 1 P.M.
Measurement

		1	2	3	4	5	$\bar{x}$	R
H	1	6.12	6.09	6.11	6.10	6.10	6.10	0.03
E	2	6.16	6.13	6.13	6.09	6.10	6.12	0.07
A	3	6.16	6.11	6.11	6.13	6.10	6.12	0.06
D	4	6.19	6.21	6.21	6.19	6.16	6.19	0.05

Time: 2 P.M.
Measurement

		1	2	3	4	5	$\bar{x}$	R
H	1	6.07	6.07	6.08	6.07	6.08	6.07	0.01
E	2	6.07	6.08	6.07	6.07	6.09	6.08	0.02
A	3	6.09	6.09	6.09	6.09	6.10	6.09	0.01
D	4	6.15	6.15	6.16	6.16	6.14	6.15	0.02

Time: 3 P.M.
Measurement

		1	2	3	4	5	$\bar{x}$	R
H	1	6.11	6.12	6.13	6.13	6.13	6.12	0.02
E	2	6.10	6.11	6.13	6.10	6.13	6.11	0.03
A	3	6.13	6.16	6.14	6.13	6.13	6.14	0.03
D	4	6.18	6.19	6.20	6.19	6.21	6.19	0.03

The second arrangement has 32 subgroups each with five measurements (we still have the same 160 measurements). These five are the measurements taken each hour on each of the four filling heads. The questions now being asked of the data are:

1. Is there a systematic difference in hour-to-hour observations?
2. Is there a systematic difference in head-to-head observations?
3. Are measurement-to-measurement differences consistent?

Consequently, the second arrangement of the data looks for hour-to-hour and head-to-head special sources of variation.

The centerline, $\bar{\bar{x}}$, remains 6.13, while the average range, $\bar{R}$, is now computed as 0.03. A control chart illustrating this is shown in Figure 8.29.

Notice that the process can now be seen as being wildly out-of-control, with many points beyond the control limits. Grouping the measurements by fill head reduced the within group variation so that the average range was lowered. This subsequently tightened the control limits and revealed the points out-of-control. Undoubtedly special sources of variation are present, and the control chart indicates where we should begin our investigation; that is, careful examination reveals that many of the out-of-control points on the x-bar portion correspond to the number 4 fill head. Its fill values are consistently above the upper control limit. Obviously, fill head number 4 is putting more product on average into the containers than the other three fill heads.

The reason this was not revealed by the first arrangement of the data and the first control chart is that the first chart was not aimed at finding differences between the fill heads; it was aimed at examining measurement-to-measurement differences and hour-to-hour differences. The second arrangement was grouped to reveal differences between the fill heads and differences from hour to hour.

Arrangement Three

Most revealing at this point would be a third arrangement of the data that keeps separate control charts for each of the fill heads. As each filling head has been separated with its own control chart, there is no longer any variation between the filling heads on our control chart. There are, in fact, now four distinct control charts, none of which can detect filling head-to-filling head variation. Computationally, this third arrangement of the data is merely a slight reorganization of the data found in Figure 8.28; it is shown in Figure 8.30. Using this arrangement, the process variation is allocated as follows:

Source of Variation for a Head	*Allocation*
Hour-to-hour	Between subgroups
Measurement-to-measurement	Within subgroups

Note that head-to-head variation is no longer within the control chart and is only visible by comparing the different control charts.

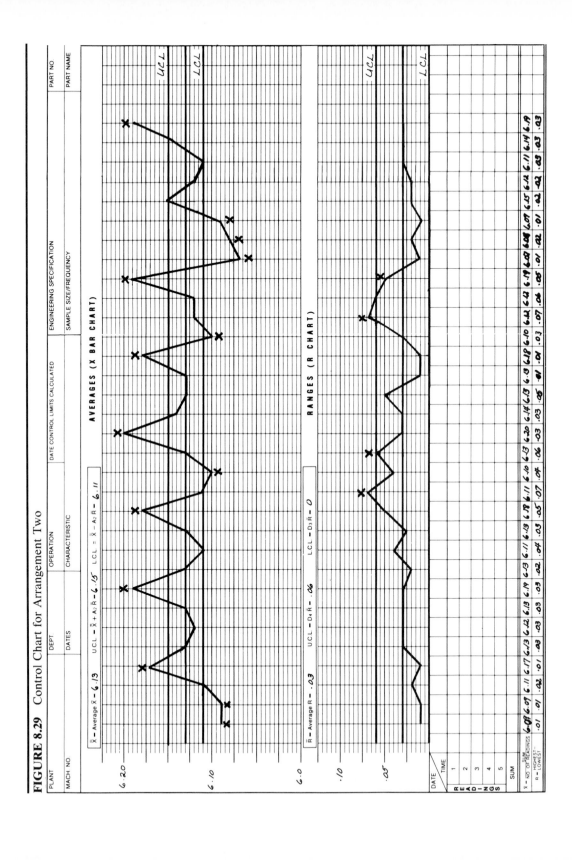

FIGURE 8.29 Control Chart for Arrangement Two

FIGURE 8.30 Arrangement Three

Filling Head 1
Measurement

Time	1	2	3	4	5	$\bar{x}$	R
8 A.M.	6.09	6.10	6.09	6.09	6.09	6.09	0.01
9 A.M.	6.13	6.13	6.14	6.13	6.11	6.13	0.03
10 A.M.	6.14	6.13	6.12	6.13	6.13	6.13	0.02
11 A.M.	6.10	6.10	6.08	6.15	6.11	6.11	0.07
12 NOON	6.14	6.15	6.14	6.13	6.16	6.14	0.03
1 P.M.	6.12	6.09	6.11	6.10	6.10	6.10	0.03
2 P.M.	6.07	6.07	6.08	6.07	6.08	6.07	0.01
3 P.M.	6.11	6.12	6.13	6.13	6.13	6.12	0.02

Filling Head 2
Measurement

Time	1	2	3	4	5	$\bar{x}$	R
8 A.M.	6.09	6.09	6.10	6.09	6.09	6.09	0.01
9 A.M.	6.12	6.12	6.11	6.13	6.10	6.12	0.03
10 A.M.	6.11	6.10	6.11	6.10	6.14	6.11	0.04
11 A.M.	6.08	6.12	6.10	6.11	6.10	6.10	0.04
12 NOON	6.10	6.12	6.15	6.13	6.13	6.13	0.05
1 P.M.	6.16	6.13	6.13	6.09	6.10	6.12	0.07
2 P.M.	6.07	6.08	6.07	6.07	6.09	6.08	0.02
3 P.M.	6.10	6.11	6.13	6.10	6.13	6.11	0.03

Filling Head 3
Measurement

Time	1	2	3	4	5	$\bar{x}$	R
8 A.M.	6.10	6.11	6.12	6.11	6.11	6.11	0.02
9 A.M.	6.11	6.13	6.13	6.14	6.14	6.13	0.03
10 A.M.	6.13	6.13	6.11	6.13	6.14	6.13	0.03
11 A.M.	6.13	6.13	6.15	6.15	6.09	6.13	0.06
12 NOON	6.13	6.13	6.14	6.14	6.13	6.13	0.01
1 P.M.	6.16	6.11	6.11	6.13	6.10	6.12	0.06
2 P.M.	6.09	6.09	6.09	6.09	6.10	6.09	0.01
3 P.M.	6.13	6.16	6.14	6.13	6.13	6.14	0.03

Filling Head 4
Measurement

Time	1	2	3	4	5	$\bar{x}$	R
8 A.M.	6.16	6.16	6.17	6.17	6.17	6.17	0.01
9 A.M.	6.20	6.20	6.20	6.17	6.17	6.19	0.03
10 A.M.	6.16	6.16	6.19	6.19	6.21	6.18	0.05
11 A.M.	6.20	6.18	6.21	6.21	6.20	6.20	0.03
12 NOON	6.17	6.18	6.18	6.18	6.17	6.18	0.01
1 P.M.	6.19	6.21	6.21	6.19	6.16	6.19	0.05
2 P.M.	6.15	6.15	6.16	6.16	6.14	6.15	0.02
3 P.M.	6.18	6.19	6.20	6.19	6.21	6.19	0.03

The third arrangement asks the following questions for each of the filling heads:

1. Is there a systematic difference in hour-to-hour observations?
2. Are measurement-to-measurement differences consistent?

Consequently, the third arrangement of the data is set up to look for hour-to-hour special sources of variation for each head.

Arranging the data in this way permits the construction of individual sets of x-bar and R charts for each filling head. When the four head control charts are drawn on the same scale as in Figure 8.31, the charts reveal things that may have been obscured earlier: head 4 is significantly different from heads 1, 2, and 3. This information allows management to take appropriate action on the process (fix head 4).

Thus, proper subgrouping of the data to be control-charted is critical to process improvement efforts. Knowledge of the process under study is often the best guide to rational subgrouping of data.

SUMMARY

This chapter has reviewed the Deming cycle and has demonstrated the use of control charts to stabilize and improve a process. The basic thrust of the Deming cycle is to decrease the difference between customer needs and process performance. Control charts are important vehicles for accomplishing this.

We took a first look at both p charts and x-bar and R charts (which will be discussed in greater detail in the next two chapters). We presented five rules for determining whether a process is exhibiting a lack of control; we will present a complete set of these rules in Chapter 11, along with an explanation of their function and use.

We next saw that rational subgrouping is important to ensure that our control charts will focus on the proper sources of variation. We saw that when subgroups are not rationally created, control charts may fail to reveal a lack of control and may not produce all of the useful information they are capable of producing.

In the next chapter we will discuss control charts for attributes used to stabilize and improve a process. These charts include p charts, np charts, c charts, u charts and individuals charts. In Chapter 10 we will focus on the specifics of variables control charts: x-bar and R charts, x-bar and s charts, median charts, and single measurement and moving range charts.

FIGURE 8.31 Control Charts for Arrangement Three

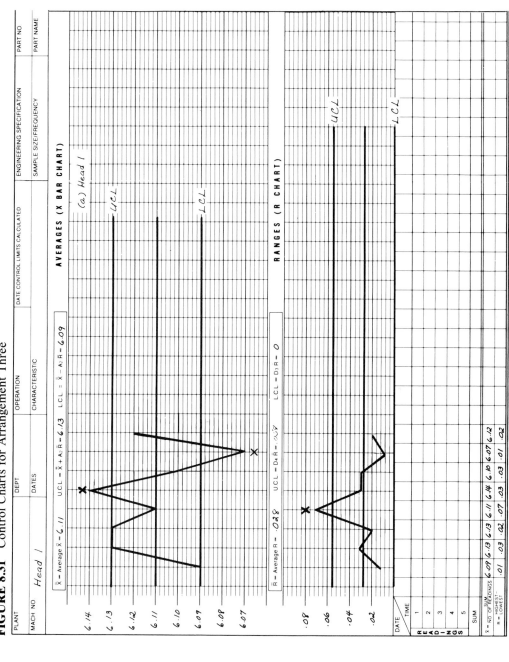

FIGURE 8.31 *Continued*

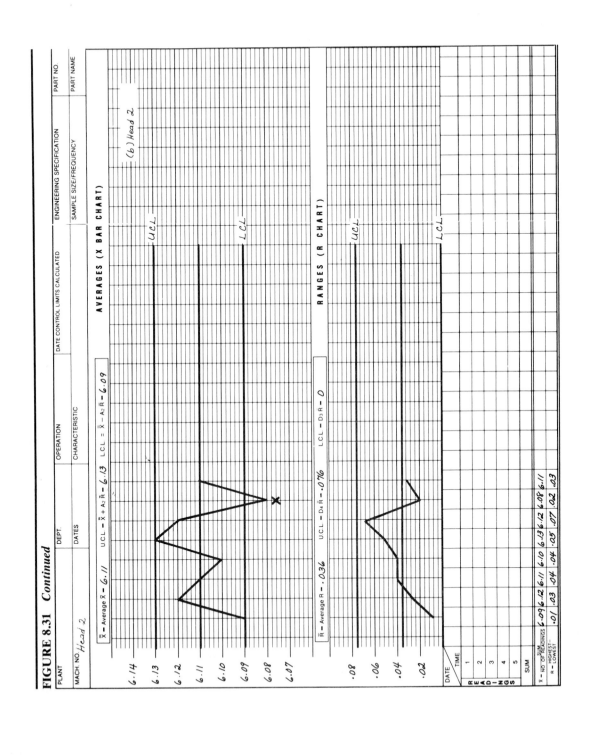

(b) Head 2

FIGURE 8.31 *Continued*

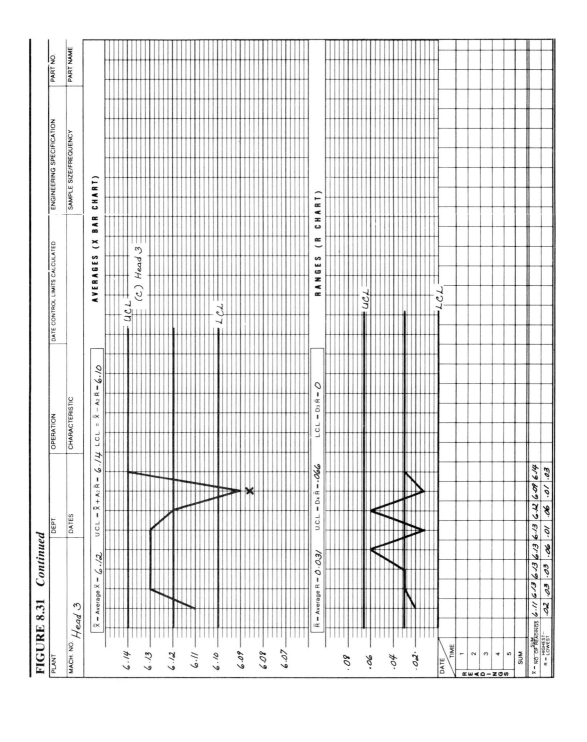

FIGURE 8.31 *Concluded*

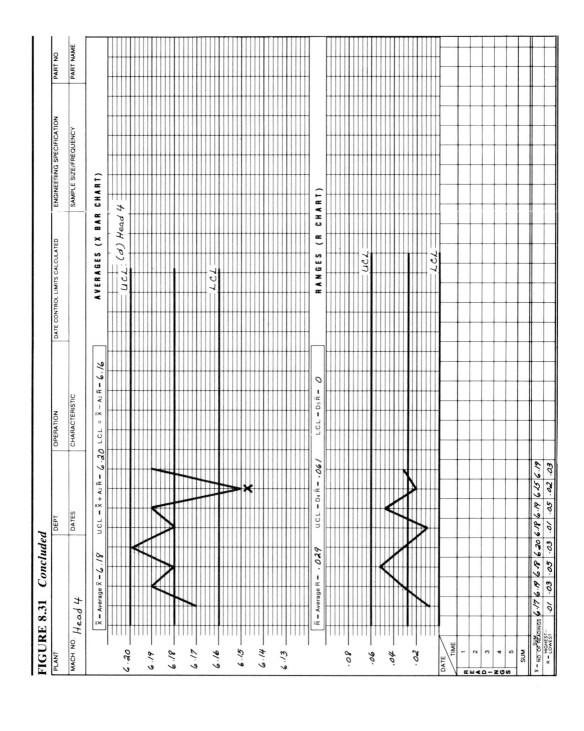

EXERCISES

The control charts in problem 8.1 through 8.6 have been divided into zones. In each problem, identify any points that are indications of a lack of control, and explain why you think the point indicates a lack of control.

8.1

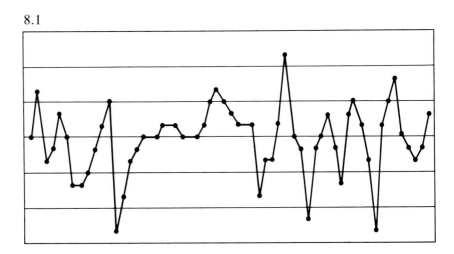

8.2

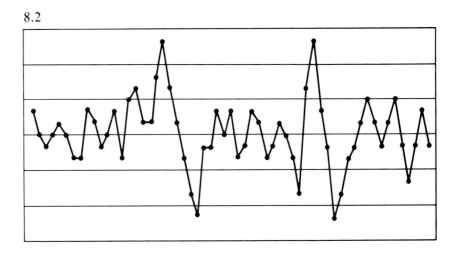

8.3

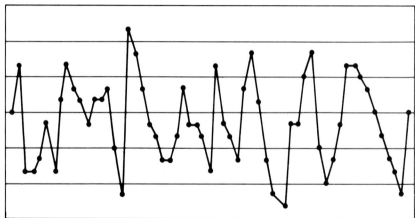

8.4

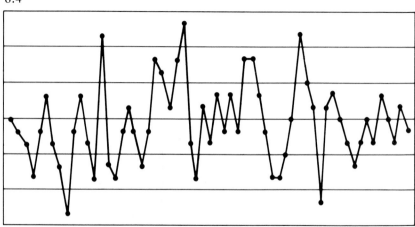

8.5

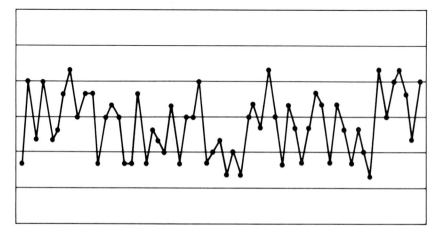

8.6

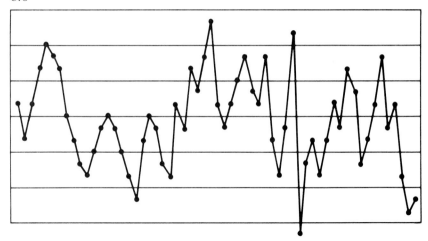

8.7 In the production of steel sheeting, five measurements are made once an hour at each of five indicator positions. The results of those 25 measurements for six consecutive hours are:

Time: 9:00 A.M.
Measurement

		1	2	3	4	5
I	1	20.4	20.1	20.1	19.5	19.8
n	2	19.7	20.0	20.0	19.6	19.5
d	3	20.1	19.7	20.4	19.8	19.6
i	4	21.3	20.8	21.2	20.9	20.8
c.	5	19.6	20.1	19.9	20.0	19.9

Time: 10:00 A.M.
Measurement

		1	2	3	4	5
I	1	20.2	19.9	19.7	20.3	20.3
n	2	20.1	19.7	20.0	19.7	19.9
d	3	19.5	20.0	19.9	20.1	19.5
i	4	20.9	20.9	21.4	21.1	21.1
c.	5	20.4	20.1	20.1	19.8	20.4

Time: 11:00 A.M.
Measurement

		1	2	3	4	5
I	1	19.8	19.9	19.5	19.6	20.0
n	2	19.9	19.6	19.8	20.4	20.2
d	3	19.6	20.3	19.6	20.4	19.7
i	4	20.6	21.2	20.6	21.3	20.7
c.	5	20.4	19.5	20.1	20.3	20.3

Time: 12:00 Noon
Measurement

		1	2	3	4	5
I	1	20.1	20.4	20.1	20.3	19.8
n	2	19.9	20.2	19.9	19.6	19.5
d	3	20.1	19.9	19.9	20.2	20.0
i	4	20.8	21.3	21.4	20.6	21.3
c.	5	20.1	20.3	19.8	20.1	19.5

Time: 1:00 P.M.
Measurement

		1	2	3	4	5
I	1	20.2	19.6	20.1	20.4	20.0
n	2	20.0	20.2	19.7	19.8	19.7
d	3	20.1	19.5	19.6	19.5	20.3
i	4	21.0	21.2	20.8	20.8	20.5
c.	5	20.2	19.8	19.5	20.0	20.2

Time: 2:00 P.M.
Measurement

		1	2	3	4	5
I	1	20.4	19.9	20.1	19.7	19.8
n	2	20.3	20.2	20.1	19.7	19.9
d	3	19.8	19.6	20.0	20.3	19.9
i	4	20.8	20.8	20.6	21.2	20.6
c.	5	19.6	20.2	19.5	19.7	19.6

a. Arrange the data to answer the following questions:
1. Are there systematic differences in the hour-to-hour averages?
2. Are there systematic differences in the measurement-to-measurement averages?
3. Are the indicator-to-indicator differences consistent?
b. Use the results of part *a.* to construct an $\bar{x}$ and R chart, and identify any indications of a lack of statistical control.
c. Arrange the data to answer the following questions:
1. Are there systematic differences in the hour-to-hour averages?
2. Are there systematic differences in the indicator-to-indicator averages?
3. Are the measurement-to-measurement differences consistent?
d. Use the results of part *c.* to construct an $\bar{x}$ and R chart, and identify any indications of a lack of statistical control.

    e. Use the results of part *c.* to allocate the three sources of variation: hour to hour, measurement to measurement, and indicator to indicator.

    f. Arrange the data to answer the following questions:

       1. Is there a systematic difference in hour-to-hour observations?

       2. Are measurement-to-measurement differences consistent?

    g. Construct the appropriate $\bar{x}$ and R charts with reference to your answer to part *f.*

NOTES

1. W. Edwards Deming, *Quality, Productivity and Competitive Position* (Cambridge, Mass: Massachusetts Institute of Technology Center for Advanced Engineering Study, 1982), p. 116.
2. W. Edwards Deming, *Out of the Crisis* (Cambridge, Mass: Massachusetts Institute of Technology Center for Advanced Engineering Study, 1986), pp. 88–89.
3. W. Scherkenbach, *The Deming Route to Quality and Productivity: Road Maps and Roadblocks* (Washington, D.C.: Ceepress Books, 1986), pp. 35–40.
4. Howard Gitlow and Shelly Gitlow, *The Deming Guide to Quality and Productivity* (Englewood Cliffs, N.J.: Prentice-Hall, 1986), pp. 154–70.
5. Kaoru Ishikawa, *What is Total Quality Control? The Japanese Way* (Englewood Cliffs, N.J.: Prentice-Hall, 1985), pp. 60–71.
6. Deming, *Quality, Productivity and Competitive Position,* p. 123.
7. Donald J. Wheeler and David S. Chambers, *Understanding Statistical Process Control* (Knoxville, Tenn.: Statistical Process Controls, Inc., 1986), pp. 2–6.
8. W. Edwards Deming, "On Some Statistical Aids Toward Economic Production," *Interfaces* 5, August 1975, pp. 1–15.
9. Genichi Taguchi and Yu-In Wu, *Off-Line Quality Control* (Nagoya, Japan: Central Japan Quality Control Association, 1980), pp. 7–9.
10. Ibid.
11. Howard S. Gitlow, "Definition of Quality, *Proceedings—Case Study Seminar—Dr. Deming's Management Methods: How They are Being Implemented in the U.S. and Abroad* (Andover, Mass: G.O.A.L., Nov 6, 1984), pp. 4–18.
12. Wheeler and Chambers, *Understanding Statistical Process Control,* pp. 2–6.
13. Howard S. Gitlow and Paul Hertz, "Product Defects and Productivity" *Harvard Business Review,* Sept./Oct. 1983, pp. 131–41.
14. Deming, *Quality, Productivity and Competitive Position,* p. 130.
15. Wheeler and Chambers, *Understanding Statistical Process Control,* pp. 111–20.

Chapter 9

Attribute Control Charts

INTRODUCTION

Attribute data are data based on counts, or the number of times one observes a particular event. The events may be the number of nonconforming items, the fraction of nonconforming items, the number of defects, or any other distinct occurrence that is operationally defined. Attribute data may include such classifications as defective or conforming, go or no-go, acceptable or not acceptable, or number of defects per unit.

In Chapter 8 we saw that the first step on the ladder of quality consciousness consists of sorting conforming from nonconforming product; shipping conforming items; and discarding, reworking, or downgrading nonconforming ones. In the first step, efforts focus on defect detection and on trying to inspect quality in by removing defective items. This stage is characterized by dependence on mass inspection to achieve quality rather than statistical process control. Even today, many firms consider this quality control.

Attribute control charts represent the second step on the ladder of quality consciousness. The second step occurs when organizations move toward defect prevention in the drive toward zero percent defective. At the second step statistical process control is initiated using various attribute control charts. However, information that output failed to meet a given specification does not answer the question of *why* the specification was not met. Total (100 percent) conformance to specifications does not provide a mechanism for never-ending process improvement. As we shall see in Chapter 10, the third step on the ladder leads to continuous and never-ending improvement through the use of variables control charts.

TYPES OF ATTRIBUTE CONTROL CHARTS

There are three basic classifications of attribute control charts: binomial count charts, area of opportunity charts, and individuals charts.

Binomial Count Charts

Binomial count charts deal with either the fraction of items or the number of items in a series of subgroups that have a particular characteristic. The *p chart* is used to control the fraction of items with the characteristic. Subgroup sizes in a p chart may remain constant or may vary. A p chart might be used to control defective versus conforming, go versus no-go, or acceptable versus not acceptable. The *np chart* serves the same function as the p chart except that it is used to control the number rather than the fraction of items with the characteristic and is used only with constant subgroup sizes.

Area of Opportunity Charts

Area of opportunity charts deal with the number of times a particular characteristic appears in some given area of opportunity. An area of opportunity can be a radio, a crate of radios, a roll of paper, a section of a roll of paper, a time period, a geographical region, a stretch of highway, or any delineated observable region in which one or more events may be observed. There are two types of area of opportunity control charts: c charts and u charts.

c Charts. *c charts* are used to control the number of times a particular characteristic appears in a constant area of opportunity. A *constant area of opportunity* is one in which each subgroup used in constructing the control chart provides the same area or number of places in which the characteristic of interest may occur. For example, defects per air conditioner, or accidents per work week in a factory, or deaths per week in a city all provide an approximately constant area of opportunity for the characteristic of interest to occur. The area of opportunity is the subgroup, whether it be the air conditioner, the factory work week, or the week in the city.

u Charts. *u charts* serve the same basic function as c charts; however, they are used when the area of opportunity changes from subgroup to subgroup. For example, we may examine varying square footage of paper selected from rolls for blemishes, or carloads of lumber for damage when the contents of the rail cars varies.

Individuals Charts

Individuals charts are used for attribute data when the assumptions for one of the other charts based on count data cannot be met.

FIGURE 9.1 A Typical Attribute Control Chart Form

Type of Discrepancy																																																		
1.																																																		
2.																																																		
3.																																																		
4.																																																		
5.																																																		
6.																																																		
7.																																																		
8.																																																		
9.																																																		
10.																																																		
Total Discrepancies																																																		
Average/% Discrepancies																																																		
Sample Size (n)																																																		

PLANT

DEPARTMENT   OPERATION NUMBER AND NAME

p ☐   c ☐
np ☐   u ☐

PRODUCT ENGINEERING
DESIGNATED CONTROL ITEM (▽)   YES ☐   PART NUMBER AND NAME
                                  NO ☐

Avg. =   UCL =   LCL =   DATE CONTROL
                                LIMITS CALCULATED:

Date

ATTRIBUTE CONTROL CHART PAPER

Standard forms exist for the construction of attribute control charts. Although there may be some slight individualizing from firm to firm, there are almost always certain standard areas provided on the forms. Figure 9.1 shows an example of an attribute control chart form.

In the upper left corner the plant/factory/office location is entered, and then just to the right the type of control chart is noted. The next two boxes require information on product engineering designation and the part name and number. Other identifying entries include the department and the operation number and name. The next two boxes provide space to enter the process average, UCL and LCL, and the date on which they were calculated.

At the very bottom of the page are spaces for entering the total number of discrepancies (or defects); the percent (or fraction) of discrepancies; and the sample (subgroup) size, n. The ten cells directly above these are for listing the type of discrepancy, usually by code number because of space constraints on the form.

The large open area on the left is for calibration and identification of the control chart's scale. The scale should be created to accommodate all of the observed and anticipated data entries. The control limits should fall well within the created scale so that there is room left for any points beyond the control limits to be entered on the graph.

Notice that the larger, upper portion of the cells (the ones on which the control chart will actually be drawn) is offset by exactly one half cell width from those below. This is to avoid any confusion as to which vertical bar corresponds to which data entry.

Just above this larger area is a single row of boxes for noting the date, time, or other identifying information for each observation.

BINOMIAL COUNT CHARTS

Defect prevention, the second step in the journey toward quality consciousness, relies on the use of attribute control charts to help to begin to reduce the difference between customer specifications and process performance. When the data are in the form of binomial counts, either a p chart or np chart is used.

Conditions for Use

The structure of the area of opportunity in which the characteristic of interest can occur dictates whether a chart is to be based on a binomial count. As we learned in Chapter 6, for n distinct units, such as the n units in a subgroup:

1. Each unit must be classifiable as either possessing or not possessing the characteristic of interest. For example, each unit in a subgroup might be classified as either defective or nondefective, or conforming or nonconforming. The number of units possessing the characteristic of interest is called the *count,* X.
2. The probability that a unit possesses the characteristic of interest is stable from unit to unit.
3. Within a given area of opportunity, the probability that a given unit possesses the characteristic of interest is independent of whether or not any other unit possesses the characteristic.

Data satisfying these conditions will follow a binomial distribution, and we may use one of the associated control charts. However, we must exercise caution to avoid using a p chart or np chart inappropriately.

When Not to Use p Charts or np Charts

Occasionally data based on measurements (variables data) are downgraded into data in terms of conformance or nonconformance (attribute data). This should be avoided because the data based on measurements can provide more information than the data based on conformance or nonconformance.

It also is important that the denominator in the fraction being charted is the proper area of opportunity. If it is not, then the data are not truly a proportion but a ratio. For example, the fraction of defectives found on the second shift will be a useful proportion only if it is computed by dividing the number of defectives found on the second shift by the proper area of opportunity, the number of units produced on the second shift. If a ratio is created using the number of defectives found on the second shift divided by the number of items shipped by the second shift, there is no way of knowing that the items shipped during the second shift were all produced on the second shift. Some of the items shipped on the second shift may have been produced during the first shift and therefore may be an inappropriate area of opportunity.

Last, it is important to exercise caution to ensure that the control chart is being created for a single process. Control charting output from combined different processes will result in irrational subgroups or subgroups that will not identify process problems. Little if anything can be learned from such charts, and the net effect may be a masking of special causes of variation.

Constructing Binomial Count Charts

A version of the Deming cycle, discussed in Chapter 8, may be used to construct and use a p chart or np chart.

I. Plan
   A. The purpose of the chart must be determined.
   B. The characteristic to be charted must be selected and operationally defined.
   C. The manner, size, and frequency of subgroup selection must be established.
   D. The type of chart (i.e., p chart or np chart) must be established.
   E. If subgroup sizes are to vary, it must be decided whether new control limits will be computed for each subgroup or whether one of the approximate methods (to be discussed later in this chapter) will be used.
   F. Forms for recording and constructing the control chart must be established.

II. Do
   A. Data must be recorded.
   B. The fraction of items with the characteristic of interest must be calculated for each of the subgroups.
   C. The average value must be calculated.
   D. The control limits and zone boundaries must be calculated and plotted onto the control chart.
   E. The data points must be entered on the control chart.

III. Check
   A. The control chart must be examined for indications of a lack of control.
   B. All aspects of the control chart must be reviewed periodically and appropriate changes made when required.

IV. Act
   A. Actions must be undertaken to bring the process under control by eliminating any special causes of variation.
   B. Actions must be undertaken to reduce the causes of common variation for the purpose of never-ending improvement of the process.
   C. Specifications must be reviewed in relation to the capability of the process.
   D. The purpose of the control chart must be reconsidered by returning to step I.

The Plan Stage. The first step in the Plan stage is to determine the purpose of the chart. A p chart or np chart may be created to:

1. Search for special causes of variation indicating a lack of statistical control.
2. Help determine and monitor the fraction or number of items with some characteristic of interest, such as the fraction or number of output that is nonconforming or defective.
3. Precede variables control charts in the pursuit of never-ending improvement.

The second step in the Plan stage is to select and operationally define the characteristic for control charting. Very often a single item possesses several characteristics, any of which may cause the item to be considered defective or nonconforming. Most often a single chart will be kept for the entire item, but frequently separate charts will be kept for individual characteristics. It is usually efficient to concentrate initial efforts on control charts for the characteristics that cause problems for the customer and are within control of the process owner studying the problem. Some of the techniques that will be discussed in Chapter 12, such as brainstorming, may be useful in selecting the characteristics to be charted.

The third step in the Plan stage is to determine the manner, size, and frequency for the selection of subgroups. As we discussed in Chapter 8, rational subgroups should be selected to minimize the within-subgroup variation. Frequently, subgroups are selected in the order of production. The decisions concerning the method of selection and the factors to be isolated may require many hours of careful planning by those individuals with knowledge of, and experience with, the process. Early efforts may need revision as a result of unexpected factors that may be revealed while developing the control chart. This may often lead to the creation of several charts where only one was initially contemplated, but this may be of use in isolating special causes of variation and reducing common variation in the areas charted.

The size of subgroups will be discussed in a later section of this chapter.

The frequency with which the subgroups are selected is generally specific to each situation and depends upon factors such as the rate of production, elapsed time, and shift duration. The frequency should be logical in terms of shifts, time periods, or any other rational grouping. The shorter the intervals between subgroups, the more quickly information may be fed back for possible action. Cost will naturally be a factor; but after process stability has been established, frequency of subgroup selection can often be decreased and efforts focused elsewhere.

The fourth step in the Plan stage is to decide whether to use a p chart or np chart. There is no substantive difference between these two charts. The information portrayed is essentially the same; only the form is different. The p chart displays the *fraction* with the characteristic of interest, while the np chart displays the *number* of items with that characteristic of interest. From a technical standpoint, they may be used interchangeably.

Nevertheless, as the np chart permits the data to be entered as whole numbers (rather than as the ratio of the number of nonconforming items to the subgroup size), the np chart may be preferable. Still, as will be discussed later on, if the subgroup size varies from subgroup to subgroup, a p chart is typically used.

If subgroup sizes will vary, it must be decided whether the control limits and zone boundaries will be recalculated for each subgroup. As we saw in Chapter 7 and will explore further in this chapter, the standard error for subgroups from a binomial process will vary inversely with the subgroup size. Varying subgroup size implies varying control limits and zone boundaries; however, we will explore several techniques that we may employ to overcome this complication.

The final step in the Plan stage is to select the control chart form. Standard forms are available from the American Society for Quality Control for attribute control charts.[1] Many firms have developed their own forms, such as the one shown in Figure 9.1.

Occasionally, supplemental forms (checksheets) are used merely to collect data; Figure 9.2 illustrates the configuration for such a form.

Forms such as this can be used to collect data, which are then transferred to a control chart. This technique may be especially convenient if the control chart is to be drawn at another time or with the aid of a computer—or if the work environment is not suitable for drawing the chart.

FIGURE 9.2   Sample Data Collection Form for p Charts and np Charts

DATA SHEET FOR p CHART OR np CHART

Department: _____

Part Name: _____   Part Number: _____

Date	Time	Inspected by	Number Inspected	No. of Defectives	Fraction Defective	Comments

The Do Stage. The Do stage begins with the recording of the required data for each subgroup on either the data collection sheet or directly on the control chart paper. Any abnormalities or unusual occurrences should be recorded in the space provided for comments. This may help provide clues to special causes of variation should a lack of control be found.

If the chart is a p chart, the fraction of items with the characteristic of interest must be calculated for each subgroup. After data for each subgroup have been collected (using no fewer than 20 subgroups), the average value for p is calculated using Equation 8.1. This value provides a centerline for the control chart and is the basis for the calculation of the standard error used to determine the control limits and zone boundaries.

Next, the control limits and zone boundaries are computed—using the equations introduced in Chapter 8 and discussed later in this chapter—and are then drawn onto the control chart.

Last in the Do stage, the p values (or np values for the np chart) are plotted onto the control chart.

It is usually desirable to complete the control chart promptly and display it for those individuals working with the process. It is not unusual for such a display to have some beneficial results, especially if those involved have been educated about the purpose and meaning of control charts.

The Check Stage. Using the rules introduced in Chapter 8 and discussed in detail in Chapter 11, we then examine the control chart for indications of a lack of control. The data should be examined from right to left (looking backward in time) and any indications appropriately noted.

Periodically the centerline, control limits, and zone boundaries should be reviewed. The timing of the review, of course, depends on the process and its history. Typically p charts and np charts are kept for long periods of time. Any change in any aspect of the process is cause to consider a review of the chart parameters.

The Act Stage. Indications of special sources of variation may be revealed during the Check stage. If any special variation is found, steps must be initiated to guarantee that these sources of special variation are removed from the process if they are bad or incorporated into the process if they are good. It is not uncommon for supervisors or foremen to already be aware of problem areas; the control chart helps to discover the cause and reinforce arguments for improvement. Furthermore, control charts help focus attention on areas needing immediate help.

If it appears there is a lack of control on the desirable side of the chart, it is a good practice to examine the inspection procedures; faulty inspection procedures may be to blame. Or perhaps a special cause is responsible for points on the desirable side of the chart that should be formally incorporated into the process—that is, improvements may have spontaneously occurred in the process that, once discovered, should be incorporated.

Although a control chart may reveal no indications of special causes of variation, the overall level of the fraction or number of items with the characteristic of interest may not be at a satisfactory level (threshold state). Other tools and techniques, such as cause-and-effect diagrams and Pareto analysis (to be discussed in Chapter 12), may be used in an attempt to reduce the high fraction of nonconforming items as the Deming cycle rolls as a wheel up the hill of never-ending process improvement.

In the drive towards never-ending improvement, *no* level of defects is low enough; nevertheless, as the proportion of defects shrinks as a result of efforts at process improvement, subgroups will often contain no defects at all. This will make the use of p charts or np charts difficult if not impossible because of the large subgroup sizes needed to detect even a single defective item. This leads to the use of variables control charts, to be discussed in Chapter 10.

Last in the Act stage is the reconsideration of the purpose of the control chart. We return to the beginning of the Plan stage, where the cycle begins again.

The p Chart for Constant Subgroup Sizes

A p chart with constant subgroup size is a relatively straightforward control chart. Constant subgroup size implies that the same number of items is sampled and then classified for each subgroup on the chart. We use a discrete, countable characteristic of output to construct a p chart; for example, the fraction of customers who pay their bills in fewer than 30 days, the fraction of correspondence sent electronically, or the fraction of an airline's flights that arrive within 15 minutes of their scheduled arrival time.

The Sampling Distribution. Categorization of data into two classes suggests a binomial process and probability distribution. That is, for a stable process, every item has approximately the same probability of being in one of the two categories. We say *approximately* because even stable processes exhibit variation. If the fraction of nonconforming items is p for an ongoing stable process, then from our discussions in Chapters 6 and 7, the probability that a subgroup of size n will contain a given fraction, x/n, of these nonconforming items is given by the binomial distribution in Equation 7.1. If we examine a series of subgroups of constant size n, the mean fraction of non-conforming items, from Equation 7.2, is p, and the associated standard error, from Equation 7.3, is:

$$\sqrt{\frac{p(1 - p)}{n}}$$

In Chapter 8 we saw one example of this type of control chart; the centerline for the p chart is established at p, the overall fraction of output that was nonconforming, as given by Equation 8.1. The upper control limit and the lower control limit are found by adding and subtracting three times the standard error from the centerline value using Equations 8.2 and 8.3.

To reiterate:

$$\text{centerline}(p) = p = \left[\frac{\text{Total number of defectives in all subgroups under investigation}}{\text{Total number of units examined in all subgroups under investigation}} \right] \quad (9.1)$$

$$\text{UCL}(p) = p + 3 \sqrt{\frac{p(1 - p)}{n}} \quad (9.2)$$

$$\text{LCL}(p) = p - 3 \sqrt{\frac{p(1 - p)}{n}} \quad (9.3)$$

Construction of a p Chart: An Example.   As an illustration, consider the case of an importer of decorative ceramic tiles. Some tiles are cracked or broken before or during transit, rendering them useless scrap. The fraction of cracked or broken tiles is naturally of concern to the firm. Each day a sample of 100 tiles is drawn from the total of all tiles received from each tile vendor. Figure 9.3 presents the sample results for 30 days of incoming shipments for a particular vendor.

The average fraction of cracked or broken tiles can be calculated from this data using Equation 9.1. This is the centerline for the control chart.

$$\text{centerline}(p) = p = 183/3000 = 0.061$$

The upper and lower control limits can then be computed using Equations 9.2 and 9.3.

$$\text{UCL}(p) = 0.061 + 3 \sqrt{\frac{(0.061)(1 - 0.061)}{100}} = 0.133$$

$$\text{LCL}(p) = 0.061 - 3 \sqrt{\frac{(0.061)(1 - 0.061)}{100}} = -0.011$$

Recall from our discussion in Chapter 8 that a negative lower control limit in a p chart is meaningless. Instead, we use a value of 0 for the lower control limit. Figure 9.4 illustrates this control chart.

For a stable process the probability that any subgroup fraction will be outside the 3-sigma limits is small. Also, if the process is stable, the probability is small that the data will demonstrate any other indications of the presence of special causes of variation by virtue of the other rules discussed in Chapters 8 or 11. However, if the process is not in a state of statistical control, the control chart provides an economical basis upon which to search for and identify indications of this lack of control.

FIGURE 9.3 Daily Cracked Tiles

Day	Sample Size	Number Cracked or Broken	Fraction
1	100	14	0.14
2	100	2	0.02
3	100	11	0.11
4	100	4	0.04
5	100	9	0.09
6	100	7	0.07
7	100	4	0.04
8	100	6	0.06
9	100	3	0.03
10	100	2	0.02
11	100	3	0.03
12	100	8	0.08
13	100	4	0.04
14	100	15	0.15
15	100	5	0.05
16	100	3	0.03
17	100	8	0.08
18	100	4	0.04
19	100	2	0.02
20	100	5	0.05
21	100	5	0.05
22	100	7	0.07
23	100	9	0.09
24	100	1	0.01
25	100	3	0.03
26	100	12	0.12
27	100	9	0.09
28	100	3	0.03
29	100	6	0.06
30	100	9	0.09
Totals	3,000	183	

For this p chart—or, in fact, for any of the attribute control charts—the exact probabilities that a stable process will generate points indicating a lack of control are generally difficult to calculate because even a stable process exhibits variation in its mean, dispersion, and shape. Nevertheless, the exact value of these probabilities is not too important for ordinary applications; the fact that they are small is important. Therefore, if a point does lie beyond the upper or lower control limits, we shall infer that it indicates a lack of control. Additionally, for p charts, the four other rules for out-of-control points described in Chapter 8 can all safely be assumed to be applicable (for a more complete discussion, see Chap-

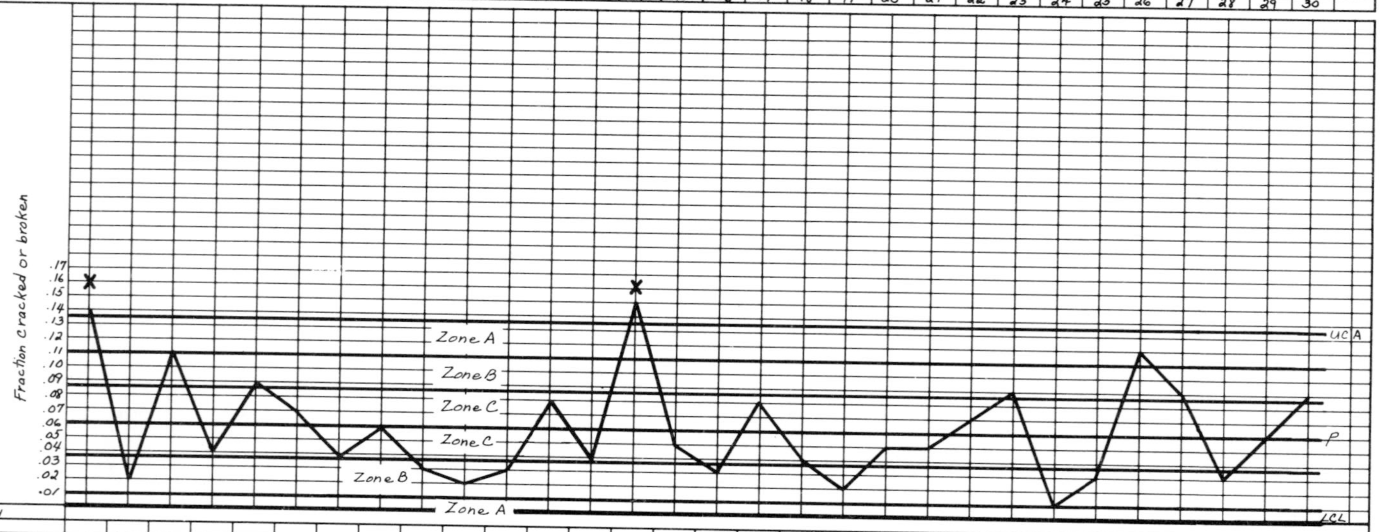

FIGURE 9.4 p Chart for Fraction of Cracked or Broken Tiles

PLANT	p [X] np [ ] c [ ] u [ ]	PRODUCT ENGINEERING DESIGNATED CONTROL ITEM (∇) YES [ ] NO [ ]	PART NUMBER AND NAME
DEPARTMENT	OPERATION NUMBER AND NAME	Avg.= .061 UCL= .133 LCL= 0	DATE CONTROL LIMITS CALCULATED:

Type of Discrepancy
1.
2.
3.
4.
5.
6.
7.
8.
9.
10.

Date	1	2	3	4	5	6	7	8	9	10	11	12	13	14	15
Total Discrepancies	14	2	11	4	9	7	4	6	3	2	3	8	4	15	5
Average/% Discrepancies	.14	.02	.11	.04	.09	.07	.04	.06	.03	.02	.03	.08	.04	.15	.05
Sample Size (n)	100	100	100	100	100	100	100	100	100	100	100	100	100	100	100

Date	16	17	18	19	20	21	22	23	24	25	26	27	28	29	30
Total Discrepancies	3	8	4	2	5	5	7	9	1	3	12	9	3	6	9
Average/% Discrepancies	.03	.08	.04	.02	.05	.05	.07	.09	.01	.03	.12	.09	.03	.06	.09
Sample Size (n)	100	100	100	100	100	100	100	100	100	100	100	100	100	100	100

ter 11). In order to apply all of these rules, we need to compute the boundaries for the A, B, and C zones.

Recall that the boundaries between zones B and C are one standard error on either side of the centerline. Here they are found by adding and subtracting the quantity $\sqrt{p(1-p)/n}$ from the centerline, p.

$$\sqrt{\frac{p(1-p)}{n}} = \sqrt{\frac{(0.061)(1-0.061)}{100}} = 0.024$$

so that

$$\begin{array}{c}\text{boundary between} \\ \text{upper zones B and C}\end{array} = p + \sqrt{\frac{p(1-p)}{n}} \qquad (9.4)$$

In our example this value is $0.061 + 0.024 = 0.085$ and

$$\begin{array}{c}\text{boundary between} \\ \text{lower zones B and C}\end{array} = p - \sqrt{\frac{p(1-p)}{n}} \qquad (9.5)$$

In our example this value is $0.061 - 0.024 = 0.037$.

These boundaries appear in Figure 9.4.

We find the upper and lower boundaries between zones A and B by adding and subtracting, respectively, two standard errors from the centerline, p.

$$\begin{array}{c}\text{boundary between} \\ \text{upper zones A and B}\end{array} = p + 2\sqrt{\frac{p(1-p)}{n}} \qquad (9.6)$$

and

$$\begin{array}{c}\text{boundary between} \\ \text{lower zones A and B}\end{array} = p - 2\sqrt{\frac{p(1-p)}{n}} \qquad (9.7)$$

Using these in our example,

$$0.061 + 2\sqrt{\frac{(0.061)(0.939)}{100}} = 0.109$$

and

$$0.061 - 2\sqrt{\frac{(0.061)(0.939)}{100}} = 0.013$$

Figure 9.4 also shows these boundaries. An examination of the control chart reveals a process that lacks control. On day 1, the mean fraction value is above the upper control limit. The sample mean for day 14 is also above the upper control limit, another indication of lack of control. None of the other rules presented in Chapter 8 seems to be violated. That is, there are no instances when two out of three consecutive points lie in zone A on one side of the centerline; there are no instances when four out of five consecutive points lie in zone B or beyond on one side of the centerline; there are no instances when eight consecutive points move upward or downward; nor are there eight consecutive points on one side of the centerline.

Nevertheless, the incoming flow of ceramic tiles needs further examination. The special causes of these erratic shifts in the fraction of cracked or broken tiles should be eliminated so that expectations for usable portions can be stabilized. Only after this is done can improvements be made in the process.

Further study reveals that on both day 1 and day 14 the regular delivery truck operator was absent because of illness. Another employee loaded and drove the delivery truck on those days. That individual had never been instructed in the proper care and handling of the product, which needs special handling and treatment. To solve this problem and eliminate this special cause of variation, management created and implemented a training program using the experience of the regular driver for three other employees. Any one of these three employees can now properly fill in and perform satisfactorily. Thus, the system has been changed to eliminate this cause of special variation.

After the process has been changed so that special causes of variation have been removed, the out-of-control points are removed from the data. The points are removed from the control chart, and the graph merely skips over them.

Removing these points also changes the process average and standard error. Therefore, the centerline, control limits, and zone boundaries must be recalculated.

The new centerline and control limits are:

$$p = 154/2800 = 0.055$$

$$\text{UCL(p)} = 0.055 + 3\sqrt{\frac{(0.055)(0.945)}{100}} = 0.123$$

$$\text{LCL(p)} = 0.055 - 3\sqrt{\frac{(0.055)(0.945)}{100}} = -0.013 \text{ (use 0.000)}$$

The new upper and lower boundaries between zones B and C are calculated using Equations 9.4 and 9.5:

$$\text{boundary between upper zones B and C} = 0.055 + \sqrt{\frac{(0.055)(0.945)}{100}} = 0.078$$

$$\text{boundary between lower zones B and C} = 0.055 - \sqrt{\frac{(0.055)(0.945)}{100}} = 0.032$$

The new upper and lower boundaries between zones A and B are calculated using Equations 9.6 and 9.7:

$$\text{boundary between upper zones A and B} = 0.055 + 2\sqrt{\frac{(0.055)(0.945)}{100}} = 0.101$$

$$\text{boundary between lower zones A and B} = 0.055 - 2\sqrt{\frac{(0.055)(0.945)}{100}} = 0.009$$

The entire control chart is redrawn, as shown in Figure 9.5. Notice that the data points for day 1 and day 14 do not appear; the chart has been drawn without these points. Figure 9.5 is the revised control chart. None of the five rules discussed in Chapter 8 are violated, so there does not appear to be a lack of control. The process now appears to be stable and in a state of statistical control. Management may now look for ways to reduce the overall process average of the number of cracked or broken tiles to raise the usable number of tiles per shipment and effectively increase the process output.

Iterative Re-evaluations. It is possible—and not altogether uncommon—that by changing the process, removing points that were out of control, and recomputing the control limits and zone boundaries, points that initially exhibited only common variation, will now indicate a lack of control. If and when this happens, the system must again be reevaluated to eliminate the newly revealed special causes of variation.

This may once again uncover even more indications of a lack of control, which also must be removed from the system. The analysis of the process will continue to iterate in this manner until there no longer appears to be a lack of control. Keep in mind that in the course of these iterations, some of the data will be discarded. Hence the data base will shrink, and the control chart will be based on fewer and fewer subgroups. Furthermore, as changes are made, the process may no longer resemble the original process.

We must also keep in mind that if control limits are recalculated too frequently (as might be the temptation with automatic data processing available with many computer control routines), it becomes possible to mistake common variation for special variation. This effect parallels the oversteering many new drivers experience when first learning to drive a car. Knowledge and experience are the best guides here.

At some point a decision must be made to stop analyzing the original data and collect new data. There is no clear rule for the point at which this should be done. Only knowledge and experience can dictate when to stop analyzing previous data and begin collecting and analyzing new data.

Subgroup Size. When constructing a p chart, the subgroup size, or the number of sample items to be observed at each inspection to determine the fraction conforming or non-conforming, is much larger than that required for variables control charts. This is because the sample size must be large enough so that some nonconforming items are likely to be included in the subgroup. If, for example, a process produces 1.0 percent defectives, sample subgroups of size 10 will only occasionally contain a nonconforming item. As a general rule of thumb, control charts based on binomial count data should have sample sizes large enough so that the average count per subgroup is at least 2.00. This allows the A, B, and C

FIGURE 9.5 Revised p Chart for Fraction of Cracked or Broken Tiles

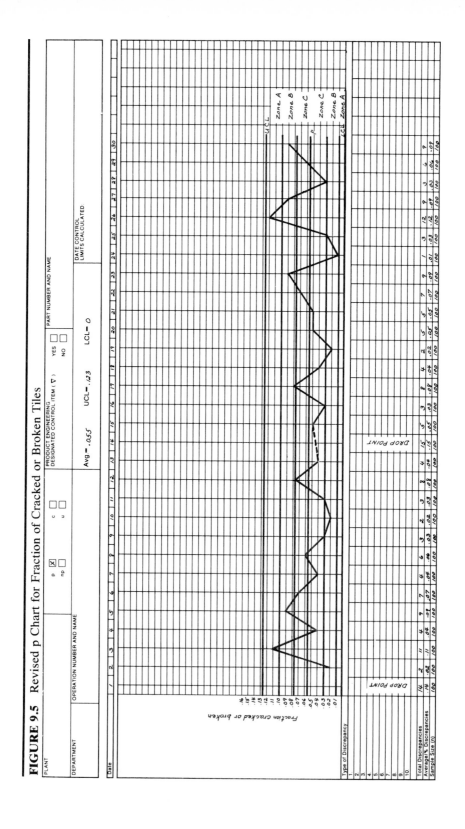

zones to be wide enough to provide a reasonable working region into which data points may fall for analysis. This is true for both the p chart and the np chart (which will be discussed later in this chapter).

Consider, for example, a process producing 5 percent of its output with a characteristic of interest. Subgroups of size 20 would be able to show these items only in counts of 0, 1, 2, and so on, yielding fractions in increments of 5 percent. The centerline would be at 0.050, the lower control limit would be at 0.000, and the upper control limit would be at

$$0.05 + 3 \sqrt{\frac{(0.05)(1 - 0.05)}{20}} = 0.196$$

Only counts of 0, 1, 2, or 3 would fall within the control limits. The examination of patterns such as runs up or down would not be practical; finding eight points moving upward or downward would almost always be redundant because the beginning or end of the run would be beyond the upper control limit and would indicate a lack of control for that reason. Clearly, we would not be able to learn too much from a p chart, based on a subgroup size of 20 items, with its centerline at 0.05 or an np chart with its centerline at 1.0. Samples taken from a process producing nonconforming items at a rate of 1 percent would require samples of 200 to have an average count of 2.00. Even with samples of size 200, samples would provide counts of 0, 1, 2, and so on for subgroup fractions in increments of 0.005. With a centerline at 0.01, the p chart would not be very detailed and might not provide satisfactory indications of a lack of control.

Average subgroup counts of fewer than 2.00 present certain theoretical problems, such as asymmetry of the sampling distribution, which can become extreme, especially if the average count per subgroup falls below 1.00. Hence, subgroup sizes must be made large enough so that the average number of events is at least 2.00.

Ideally, the subgroup sizes should remain the same for all subgroups, but occasionally circumstances require variations in subgroup size. Whether the subgroup size for a p chart varies or remains constant, the larger the subgroup size, the narrower the control limits will be. This is because the subgroup size, n, appears in the denominator of the expression for the standard error; the larger the value for n, the narrower the width of the control limits and zones A, B, and C around the process average will be.

Number of Subgroups. Every process goes through physical cycles, such as shifts and ordering sequences. It is important that control chart calculations are based on a sufficient number of subgroups to encompass all of the cycles of a process to include all possible sources of variation. As a rule of thumb, the number of subgroups should be at least 25.

It is possible to construct control charts for rational subgroups that do not represent chronological events. For example, a p chart for the fraction defective produced by a battery of 100 machines performing the same task (such as spot welding) might be kept on a single control chart. In these situations, the number of subgroups must encompass all machines to encompass all possible sources of variation. Additionally, the rules concerning indications of a lack of control by virtue of trends in the data—such as two out of three consecutive points in zone A or four out of five consecutive points in zone B or beyond—should be ignored.

Another Example. An injection molding process provides a bracket to be used on aircraft passenger seats. Daily samples of 500 brackets are selected from the production output and examined carefully for cracks, splits, or other imperfections that will render them defective. Figure 9.6 lists the results.

FIGURE 9.6 Defective Aircraft Seat Brackets

Date	Sample Size	No. of Defectives	Fraction Defective
July 5	500	12	0.024
6	500	9	0.018
7	500	8	0.016
8	500	10	0.020
9	500	17	0.034
12	500	33	0.066
13	500	15	0.030
14	500	46	0.092
15	500	22	0.044
16	500	13	0.026
19	500	9	0.018
20	500	15	0.030
21	500	4	0.008
22	500	37	0.074
23	500	20	0.040
26	500	15	0.030
27	500	14	0.028
28	500	18	0.036
29	500	45	0.090
30	500	25	0.050
Aug. 2	500	27	0.054
3	500	33	0.066
4	500	17	0.034
5	500	28	0.056
6	500	12	0.024
Totals	12,500	504	

$$p = 504/12500 = 0.040$$

$$LCL(p) = 0.040 - 3 \sqrt{\frac{(0.040)(0.960)}{500}} = 0.014$$

$$UCL(p) = 0.040 + 3 \sqrt{\frac{(0.040)(0.960)}{500}} = 0.066$$

$$\begin{array}{l}\text{boundary between} \\ \text{lower zones A and B}\end{array} = 0.040 - 2 \sqrt{\frac{(0.040)(0.960)}{500}} = 0.022$$

$$\begin{array}{l}\text{boundary between} \\ \text{lower zones B and C}\end{array} = 0.040 - \sqrt{\frac{(0.040)(0.960)}{500}} = 0.031$$

$$\begin{array}{l}\text{boundary between} \\ \text{upper zones B and C}\end{array} = 0.040 + \sqrt{\frac{(0.040)(0.960)}{500}} = 0.049$$

$$\begin{array}{l}\text{boundary between} \\ \text{upper zones A and B}\end{array} = 0.040 + 2 \sqrt{\frac{(0.040)(0.960)}{500}} = 0.058$$

Figure 9.7 illustrates the p chart for these data. Many points indicate that this process was not in a state of statistical control. The operator running the molding process initiates a study of the molding process that reveals that the mold is poorly designed, thus consistent parts cannot be formed. The operator suggests a redesign of the mold that may eliminate most of the special causes of variation and reduce the average fraction defective.

The p Chart for Variable Subgroup Sizes

Sometimes subgroups of observations vary in size. This makes the construction of a p chart somewhat more difficult, but occasionally circumstances make this situation unavoidable. Common among these is when data initially collected for some purpose other than the creation of a control chart are later used to construct a control chart.

The Sampling Distribution.　The standard error of the sampling distribution of the fraction defective varies inversely with the sample size. Changing the sample size naturally changes this value. The formula for the standard error of the fraction defective is $\sqrt{p(1 - p)/n}$. As the sample size increases, the standard error decreases and vice versa. Because control limits and zone boundaries are calculated based on this value, as the standard error changes so will the control limits and the zone boundaries.

Three Alternatives for Coping with Variable Control Limits.　There are three different ways around the problem of variable control limits: the first is to compute the average subgroup size and to use this value for the

FIGURE 9.7 p Chart for Fraction of Defective Aircraft Seat Brackets

PLANT

PART NUMBER AND NAME

DEPARTMENT

OPERATION NUMBER AND NAME

PRODUCT ENGINEERING
DESIGNATED CONTROL ITEM (▽)

p ☒ c ☐
np ☐ u ☐

YES ☐
NO ☐

Avg. = .040   UCL = .066   LCL = .014

DATE CONTROL
LIMITS CALCULATED:
1/27/86

Fraction non-conforming

.11
.10
.09
.08
.07
.06
.05
.04
.03
.02
.01

UCL
Zone A
Zone B
Zone C
Zone C
Zone B
Zone A
LCL

Date	7/5	6	7	8	9	12	13	14	15	16	19	20	21	22	23	26	27	28	29	30	8/2	3	4	5	6
Type of Discrepancy																									
1																									
2																									
3																									
4																									
5																									
6																									
7																									
8																									
9																									
10																									
Total Discrepancies	12	9	8	10	17	33	15	46	22	13	9	15	4	37	20	15	14	18	45	27	33	17	28	12	
Average/% Discrepancies	.024	.018	.016	.020	.034	.066	.030	.092	.044	.026	.018	.030	.008	.074	.040	.030	.028	.036	.090	.054	.066	.034	.056	.024	
Sample Size (n)	500	500	500	500	500	500	500	500	500	500	500	500	500	500	500	500	500	500	500	500	500	500	500	500	

control chart; the second is to compute new control limits and class boundaries for every subgroup based on that subgroup's size; and the third is to compute both a wide and a narrow set of control limits based upon the largest and smallest possible values for n.

When the first method, an average subgroup size, is used, and a point falls in the outer portion of zone A or indicates a lack of control by falling outside the control limits, the true value for the subgroup size should be used to calculate the standard error so that accurate values for the control limits and the zone boundaries may be obtained. The point may then be properly evaluated regarding lack of control. If the point actually indicates a lack of control, appropriate steps must then be taken to find the sources of the special variation and to change the system so that they are rectified.

Additionally, when this method is used, the average value for n should be periodically recalculated to ensure it hasn't drifted too far from the value calculated initially. The first method works best when the subgroup sizes have not been too different in the recent past and will not be too different in the immediate future. The terms *recent* and *immediate* must be carefully considered and defined to make them relevant to the particular application. Many statisticians feel that ± 25 percent in subgroup size is permissible. There is no good substitute for knowledge and experience.

The second method is the most accurate but is also the most tedious and time-consuming. Also, p charts created with variable control limits and zone boundaries may be difficult to read and interpret, especially for a relatively inexperienced user, and therefore undesirable.

The third method utilizes an inner and an outer set of control limits. Of the two sets of control limits, the outer set of control limits is based on the smallest anticipated value for n. The set is wider as a small value of n in the denominator results in a large standard error. Points that fall outside these limits—assuming they are based on subgroup sizes greater than those used for the calculation of the outer limits—clearly indicate a lack of statistical control. The inner set of control limits is based on the largest subgroup's value of n. Points that fall within these limits (and do not otherwise indicate a lack of control) represent points under control. Exact values for control limits will need to be calculated for those points falling between the inner and outer control limits. The use of zone boundaries with this method is extremely cumbersome, and they generally are not used unless a pattern (or some other indication of a lack of control) exists.

Using Average Subgroup Size: An Example. Consider the case of a small manufacturer of low tension electric insulators. The insulators are sold to wholesalers who subsequently sell them to electrical contractors. Each day during a one-month period the manufacturer inspects the production of a given shift; the number inspected varies somewhat. Based on carefully laid out operational definitions, some of the production is deemed nonconforming and is downgraded. Figure 9.8 illustrates the results for 25 weekdays beginning on September 2.

As the subgroup sizes do not vary by more than 25 percent, the centerline and control limits can be calculated using an average value for n:

$$centerline(p) = p = 594/9769 = 0.061$$

$$average\ value\ of\ n = 9769/25 = 390.76$$

$$UCL(p) = 0.061 + 3 \sqrt{\frac{(0.061)(0.939)}{390.76}} = 0.097$$

$$LCL(p) = 0.061 - 3 \sqrt{\frac{(0.061)(0.939)}{390.76}} = 0.025$$

FIGURE 9.8   Nonconforming Electric Insulators

Date	Number Inspected	Number Nonconforming	Fraction Nonconforming
9/2	350	22	0.063
3	420	27	0.064
4	405	20	0.049
5	390	12	0.031
6	410	23	0.056
9/9	384	23	0.060
10	392	25	0.064
11	415	26	0.063
12	364	24	0.066
13	377	29	0.077
9/16	409	12	0.029
17	376	36	0.096
18	399	23	0.058
19	355	21	0.059
20	410	26	0.063
9/23	414	21	0.051
24	366	24	0.066
25	377	22	0.058
26	404	24	0.059
27	387	26	0.067
9/30	402	27	0.067
10/1	358	30	0.084
2	411	28	0.068
3	404	17	0.042
4	390	26	0.067
Totals	9,769	594	

FIGURE 9.9 p Chart for Fraction of Nonconforming Electric Insulators: Average Subgroup Size Used

PLANT

PART NUMBER AND NAME

DEPARTMENT

OPERATION NUMBER AND NAME

PRODUCT ENGINEERING
DESIGNATED CONTROL ITEM (▽) YES ☐ NO ☐

DATE CONTROL
LIMITS CALCULATED:

p ☒   c ☐
np ☐ u ☐

Avg. = .061   UCL = .097   LCL = .025

Zone A
Zone B
Zone C
Zone C
Zone B
Zone A

Fraction non-conforming (.01 – .10)

Date	9/2	3	4	5	6	9	10	11	12	13	16	17	18	19	20	23	24	25	26	27	30	10/1	2	3	4
Total Discrepancies	22	27	20	12	23	23	25	24	26	29	12	36	23	21	26	21	24	22	24	26	27	30	28	17	26
Average/% Discrepancies	.063	.064	.049	.031	.056	.060	.064	.066	.063	.077	.029	.096	.058	.059	.063	.051	.066	.058	.059	.067	.067	.074	.068	.042	.067
Sample Size (n)	350	420	405	390	410	364	362	364	415	377	409	376	399	353	410	414	366	377	404	387	402	358	411	404	390

Type of Discrepancy
1
2
3
4
5
6
7
8
9
10

Equations 9.4, 9.5, 9.6, and 9.7 may be used to compute the zone boundaries:

$$\text{boundary between lower zones A and B} = 0.061 - 2\sqrt{\frac{(0.061)(0.939)}{390.76}} = 0.037$$

$$\text{boundary between lower zones B and C} = 0.061 - \sqrt{\frac{(0.061)(0.939)}{390.76}} = 0.049$$

$$\text{boundary between upper zones A and B} = 0.061 + 2\sqrt{\frac{(0.061)(0.939)}{390.76}} = 0.085$$

$$\text{boundary between upper zones B and C} = 0.061 + \sqrt{\frac{(0.061)(0.939)}{390.76}} = 0.073$$

Figure 9.9 illustrates the control chart generated from these data. The process output does not indicate a lack of control. At this point, efforts should focus on reducing the fraction of nonconforming production and the variation in that fraction. As the fraction nonconforming is decreased, larger subgroups are needed so that the average count remains at least 2.00. Efforts must be made to look for measurable variables to use for a variables control chart to continue the pursuit of never-ending process improvement.

Using Varying Control Limits: An Example. When sample sizes do vary by more than ±25 percent we may either compute two sets of control limits or calculate new zone boundaries and control limits for each subgroup. Although the calculations required for new zone boundaries for

FIGURE 9.10 Number of Vehicles Using Exact Change

Day	n	Number with Exact Change	Day	n	Number with Exact Change
1	465	180	11	406	186
2	123	38	12	415	149
3	309	142	13	379	90
4	83	20	14	341	148
5	116	35	15	258	107
6	306	108	16	270	84
7	333	190	17	480	185
8	265	106	18	350	184
9	354	94	19	433	210
10	256	116	20	479	197
			Totals	6,421	2,569

each subgroup are more tedious, the technique is more sensitive, thus providing quicker feedback when special causes of variation are present.

Consider, for example, the case of an automatic toll barrier with two types of toll collection mechanisms, automatic and manned; the automatic lanes require exact change while the manned lanes do not. The fraction of vehicles arriving with exact change is examined using a control chart for a series of rush hour intervals on consecutive weekdays. As the number of vehicles passing through the toll varies by more than 25 percent, the control limits change day-to-day. One-hour periods (7:30 AM to 8:30 AM) for 20 consecutive weekdays yield the data in Figure 9.10.

Using these data, p, the centerline, can be calculated from Equation 9.1 as:

$$\text{centerline}(p) = p = 2569/6421 = 0.400$$

We can also calculate each UCL, LCL, and zone boundaries using Equations 9.2, 9.3, 9.4, 9.5, 9.6, and 9.7.

For example, for the first data point:

$$\text{UCL}(p) = 0.40 + 3\sqrt{\frac{(0.40)(1-0.40)}{465}} = 0.468$$

$$\text{LCL}(p) = 0.40 - 3\sqrt{\frac{(0.40)(1-0.40)}{465}} = 0.332$$

$$\begin{array}{l}\text{boundary between} \\ \text{lower zones A and B}\end{array} = 0.40 - 2\sqrt{\frac{(0.40)(1-0.40)}{465}} = 0.355$$

$$\begin{array}{l}\text{boundary between} \\ \text{lower zones B and C}\end{array} = 0.40 - \sqrt{\frac{(0.40)(1-0.40)}{465}} = 0.377$$

$$\begin{array}{l}\text{boundary between} \\ \text{upper zones A and B}\end{array} = 0.40 + 2\sqrt{\frac{(0.40)(1-0.40)}{465}} = 0.445$$

$$\begin{array}{l}\text{boundary between} \\ \text{upper zones B and C}\end{array} = 0.40 + \sqrt{\frac{(0.40)(1-0.40)}{465}} = 0.423$$

Since for each of the subgroups the size of the subgroup varies, these control limits and zone boundaries are only valid for the first observation, where n = 465. Each subgroup will have its own standard error, hence its own control limits and zone boundaries. Figure 9.11 shows the results of calculating these in the same manner as for the first point.

We use these values to draw the control limits and zone boundaries in Figure 9.12. The process indicates many instances of a lack of control. Fully 25 percent of the subgroup proportions are out of control, and the data seems to be behaving in an extremely erratic pattern. Days 19, 18, 13, 9, and 7 are all beyond the control limits. Day 5 also indicates a lack of control because it is the second of three consecutive points falling in zone C or beyond on the same side of the centerline.

FIGURE 9.11   Control Limits and Zone Boundaries for Vehicles with Exact Change

Subgroup Number	n	Fraction Defective	UCL	LCL	Upper Zone C	Lower Zone C	Upper Zone B	Lower Zone B
1	465	0.387	0.468	0.332	0.423	0.377	0.445	0.355
2	123	0.309	0.533	0.267	0.444	0.356	0.488	0.312
3	309	0.460	0.484	0.316	0.428	0.372	0.456	0.344
4	83	0.241	0.561	0.239	0.454	0.346	0.508	0.292
5	116	0.302	0.536	0.264	0.445	0.355	0.491	0.309
6	306	0.353	0.484	0.316	0.428	0.372	0.456	0.344
7	333	0.571	0.481	0.319	0.427	0.373	0.454	0.346
8	265	0.400	0.490	0.310	0.430	0.370	0.460	0.340
9	354	0.266	0.478	0.322	0.426	0.374	0.452	0.348
10	256	0.453	0.492	0.308	0.431	0.369	0.461	0.339
11	406	0.458	0.473	0.327	0.424	0.376	0.449	0.351
12	415	0.359	0.472	0.328	0.424	0.376	0.448	0.352
13	379	0.237	0.475	0.325	0.425	0.375	0.450	0.350
14	341	0.434	0.480	0.320	0.427	0.373	0.453	0.347
15	258	0.415	0.491	0.309	0.430	0.370	0.461	0.339
16	270	0.311	0.489	0.311	0.430	0.370	0.460	0.340
17	480	0.385	0.467	0.333	0.422	0.378	0.445	0.355
18	350	0.526	0.479	0.321	0.426	0.374	0.452	0.348
19	433	0.485	0.471	0.329	0.424	0.376	0.447	0.353
20	479	0.411	0.467	0.333	0.422	0.378	0.445	0.355

Careful study is warranted to determine the cause or causes of this special variation. Removing the special cause(s) of variation may require some fundamental changes in the way this system operates. Nevertheless, it is necessary to eliminate the special sources of variation before attempting to improve the process.

Changing the Process.   Management decides it would be advantageous to remove the erratic patterns in the above process. This would enable them to better serve the public by having adequate toll lanes of either the automatic or manned type available during rush hours. As a result of brainstorming, management institutes the sale of tokens that can be used in the exact change lanes. These are sold at a discount (to encourage their purchase) to motorists at the manned lanes. Because the process has now been changed, a new set of observations is made. After a period of two months, to allow for transient effects to die down, the same sample selection method is again employed. The results for those subgroups appear in Figure 9.13, and their corresponding control limits and zone boundaries are shown in Figure 9.14.

FIGURE 9.12 p Chart for Fraction of Vehicles with Exact Change: Subgroup Sizes Vary

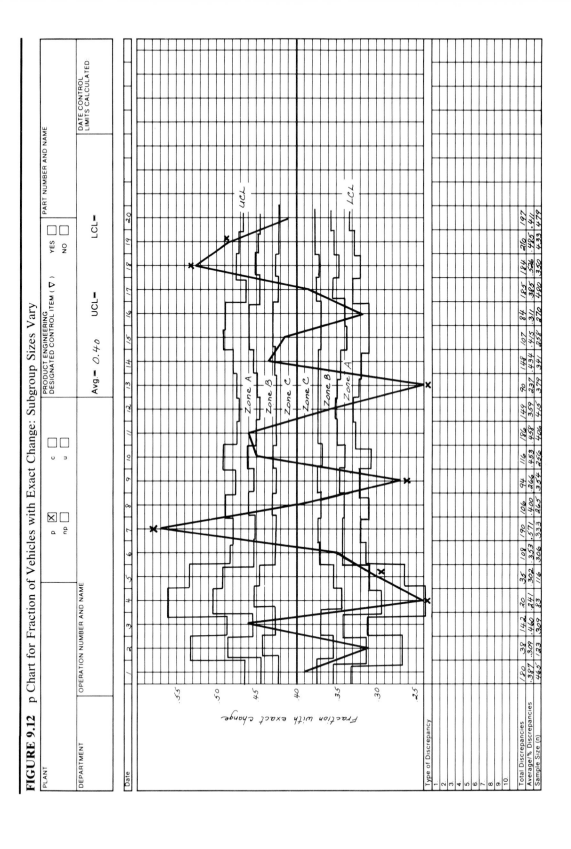

FIGURE 9.13 Number of Vehicles Using Exact Change or Tokens

Day	n	Number with Exact Change	Day	n	Number with Exact Change
1	421	171	11	401	199
2	466	197	12	384	165
3	389	192	13	428	213
4	254	107	14	352	149
5	186	89	15	444	193
6	456	189	16	357	158
7	411	211	17	283	147
8	322	139	18	424	207
9	287	136	19	337	143
10	262	131	20	326	157
			Totals	7,190	3,293

FIGURE 9.14 Control Limits and Zone Boundaries for Vehicles with Exact Change

Subgroup Number	n	Fraction Defective	UCL	LCL	Upper Zone C	Lower Zone C	Upper Zone B	Lower Zone B
1	421	0.406	0.531	0.385	0.482	0.434	0.507	0.409
2	466	0.423	0.527	0.389	0.481	0.435	0.504	0.412
3	389	0.494	0.534	0.382	0.483	0.433	0.509	0.407
4	254	0.421	0.552	0.364	0.489	0.427	0.521	0.395
5	186	0.479	0.568	0.348	0.495	0.421	0.531	0.385
6	456	0.415	0.528	0.388	0.481	0.435	0.505	0.411
7	411	0.513	0.532	0.384	0.483	0.433	0.507	0.409
8	322	0.432	0.541	0.375	0.486	0.430	0.514	0.402
9	287	0.474	0.546	0.370	0.487	0.429	0.517	0.399
10	262	0.500	0.550	0.366	0.489	0.427	0.520	0.396
11	401	0.496	0.533	0.383	0.483	0.433	0.508	0.408
12	384	0.430	0.534	0.382	0.483	0.433	0.509	0.407
13	428	0.498	0.530	0.386	0.482	0.434	0.506	0.410
14	352	0.423	0.538	0.378	0.485	0.431	0.511	0.405
15	444	0.435	0.529	0.387	0.482	0.434	0.505	0.411
16	357	0.443	0.537	0.379	0.484	0.432	0.511	0.405
17	283	0.519	0.547	0.369	0.488	0.428	0.517	0.399
18	424	0.488	0.531	0.385	0.482	0.434	0.506	0.410
19	337	0.424	0.539	0.377	0.485	0.431	0.512	0.404
20	326	0.482	0.541	0.375	0.486	0.430	0.513	0.403

FIGURE 9.15 p Chart for Revised Process for Fraction of Vehicles with Exact Change

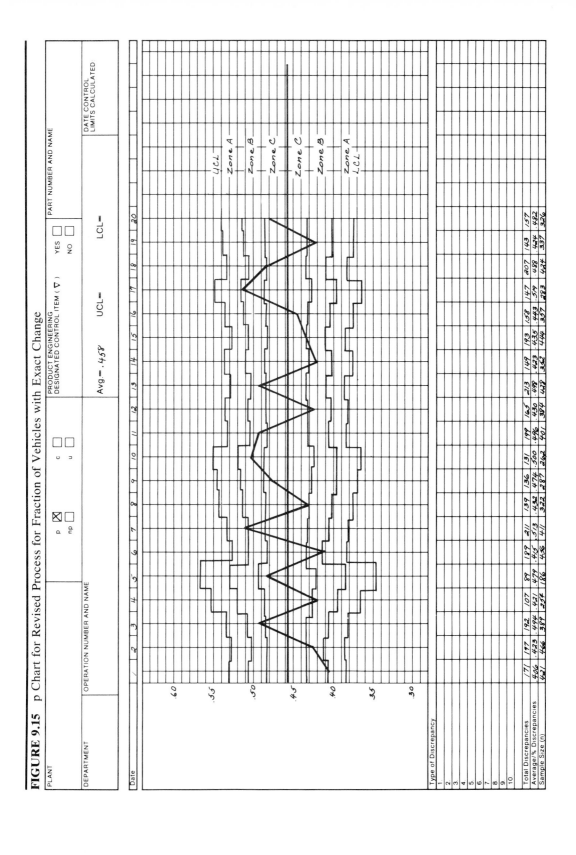

Using Equation 9.1 the centerline is calculated as:

$$p = 3,293/7,190 = 0.458$$

Figure 9.15 illustrates the control chart for the revised process. The process now appears stable, with no indications of a lack of control. Furthermore, the process is now better. Not only is the proportion of motorists using the exact change lanes stable and predictable, but the proportion of those motorists has risen from .400 to .458, which results in a smoother flow of traffic at the toll barrier.

Using Two Sets of Control Limits: An Example. Because the process now appears to be operating in a stable manner, management decides to continue the control chart with less frequent sampling. Based upon the early analysis, the process is deemed sufficiently stable to use the easier to read third method. Instead of computing exact control limits for each of the subgroups, we compute two sets of control limits, a narrow set and a wide set. Zone A, B, and C boundaries are not generally included because they would make the control chart confusing. Management calculates the exact control limits and zone boundaries for any suspicious-looking points or patterns, and any patterns which seem to suggest a lack of control are investigated further. Care must be exercised because the zone boundaries do not appear; patterns indicating a lack of control may be overlooked. For this reason, it is best to change to a sampling plan utilizing fixed subgroup sizes, so that one set of control limits and zone boundaries can be used.

Subgroups at the toll are selected less frequently; Figure 9.16 illustrates the results for 20 nonconsecutive days.

$$p = 3,347/7,405 = 0.452$$

Minimum and maximum values for n are needed to construct the two

---

FIGURE 9.16   Vehicles Using Exact Change or Tokens

Day	n	Number with Exact Change	Fraction	Day	n	Number with Exact Change	Fraction
1	387	182	0.470	11	422	182	0.431
2	404	190	0.470	12	356	160	0.449
3	342	160	0.468	13	431	182	0.422
4	394	177	0.449	14	280	118	0.421
5	411	181	0.440	15	291	139	0.478
6	344	164	0.477	16	248	121	0.488
7	387	177	0.457	17	345	148	0.429
8	390	175	0.449	18	408	199	0.488
9	312	136	0.436	19	388	161	0.415
10	433	205	0.473	20	432	190	0.440
				Totals	7,405	3,347	

sets of control limits. The minimum and maximum values for n should be based upon the smallest and largest subgroup sizes that can reasonably be anticipated to occur. In this case, a decision is made to use 75 as the minimum and 500 as the maximum. As these are slightly beyond the range of past experience, they are likely to provide conservative control limits. The smaller subgroup size will yield the wider set of control limits, and the larger will yield the narrow set of limits.

Recall that points falling outside of the outer control limits are considered to be sure signs of a lack of control; points inside the inner control limits are not considered further (except if they form part of some other nonrandom pattern); and for those points falling between the two sets of control limits, exact control limits are calculated using the actual subgroup size to determine whether they are true indications of a lack of control.

The outer set of control limits are computed as:

$$\text{LCL(p)} = 0.452 - 3 \sqrt{\frac{(0.452)(1 - 0.452)}{75}} = 0.280$$

$$\text{UCL(p)} = 0.452 + 3 \sqrt{\frac{(0.452)(1 - 0.452)}{75}} = 0.624$$

and the inner set of control limits are:

$$\text{LCL(p)} = 0.452 - 3 \sqrt{\frac{(0.452)(1 - 0.452)}{500}} = 0.385$$

$$\text{UCL(p)} = 0.452 + 3 \sqrt{\frac{(0.452)(1 - 0.452)}{500}} = 0.519$$

The control chart is shown in Figure 9.17. No points are beyond the inner control limits. If there were, exact values for the control limits at those points would have to be calculated using the subgroup size actually observed and Equations 9.2 and 9.3. Then it could be determined whether those points truly indicated the presence of special causes of variation. Neither are there any other indications of a lack of control, such as runs up or down or too many consecutive points on one side of the centerline. If there were any signs whatsoever of special causes of variation, the procedure would be same as it would be in the other situations we have discussed in this chapter. That is, we would attempt to identify and correct the special sources of variation using tools such as those to be presented in Chapter 12. After the process has been stabilized, we would recalculate the centerline and control limits and continue our efforts at process improvement.

The np Chart

Binomial count data can sometimes be more easily understood if the data appear as counts rather than fractions. This is especially true when using

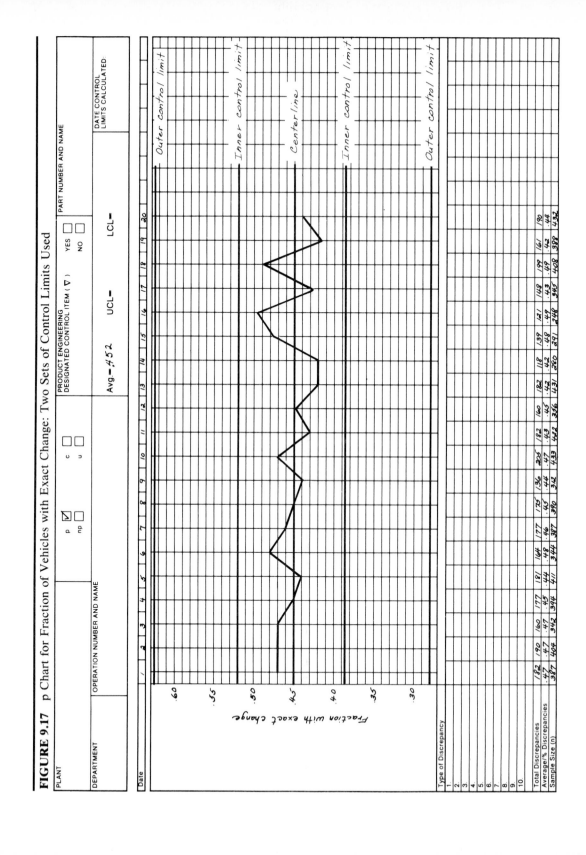

FIGURE 9.17 p Chart for Fraction of Vehicles with Exact Change: Two Sets of Control Limits Used

attribute control charts to introduce control charting and encountering reluctance by some members of the affected community to deal with fractions rather than whole numbers such as the number of defects.

The quantity np is the number of units in the subgroup with some particular characteristic, such as the number of nonconforming units. Traditionally, np charts are used only when subgroup sizes are constant. As the information used is the same as for p charts with constant subgroup sizes, these two charts are interchangeable.

The Sampling Distribution. Just as in the p chart, the categorization of data into two classes suggests a binomial probability distribution. Here, too, for a stable process, every item must have approximately the same probability of being in one of the two categories. Also, if the fraction of nonconforming items is p, then (from our discussions in Chapters 6 and 7) for an ongoing stable process the probability that a subgroup of size n will contain a given number, X, of these, is approximated by the binomial distribution in Equation 6.3a. In a series of subgroups of constant size n, the mean or expected number of nonconforming items is approximated by Equation 7.6 as np and the associated standard error is approximated by Equation 7.7 as $\sqrt{np(1 - p)}$. These results enable us to construct the np chart.

The Construction of the np Chart. Data collected for an np chart will be a series of integers, each representing the number of nonconforming items in its subgroup. Computations for the centerline, the control limits, and the zone boundaries are quite similar to those of the p chart with constant sample sizes.

The centerline is the overall average number of nonconforming items found in each subgroup of the data. For the ceramic tile importer discussed earlier in this chapter (the data appear in Figure 9.3), there are a total of 183 cracked or broken tiles in the 30 subgroups examined; this represents an average count of 183/30 = 6.1 tiles per day. In general, finding the average count is easier than computing np directly as:

$$np = (100) \left[\frac{183}{3000} \right] = 6.100$$

Using Equation 7.7, the standard error for the np chart is:

$$\sqrt{np(1 - p)} = \sqrt{(100)(0.061)(1 - 0.061)} = 2.393$$

The upper and lower control limits are found by adding and subtracting three times the standard error from the centerline respectively.

$$UCL(np) = np + 3 \sqrt{np(1 - p)} \tag{9.8}$$

$$LCL(np) = np - 3 \sqrt{np(1 - p)} \tag{9.9}$$

For the tile importer this yields values of:

$$UCL(np) = (100)(0.061) + 3\sqrt{(100)(0.061)(1-0.061)} = 13.280$$

$$LCL(np) = (100)(0.061) - 3\sqrt{(100)(0.061)(1-0.061)} = -1.080$$

As the LCL value is negative (-1.08), and, just as with the p chart, a negative value is meaningless, a value of 0 is used instead.

As with the p chart, the upper and lower boundaries between zones B and C are found by adding and subtracting one standard error from the centerline, np:

$$\text{boundary between upper zones B and C} = np + \sqrt{np(1-p)} \qquad (9.10)$$

and

$$\text{boundary between lower zones B and C} = np - \sqrt{np(1-p)} \qquad (9.11)$$

The upper boundary between zones B and C for this example is given by:

$$6.1 + \sqrt{(100)(0.061)(1-0.061)} = 8.493$$

and the lower boundary between zones B and C is given by:

$$6.1 - \sqrt{(100)(0.061)(1-0.061)} = 3.707$$

Upper and lower boundaries between zones A and B can be found by adding and subtracting two standard errors from the centerline, np:

$$\text{boundary between upper zones A and B} = np + 2\sqrt{np(1-p)} \qquad (9.12)$$

and

$$\text{boundary between lower zones A and B} = np - 2\sqrt{np(1-p)} \qquad (9.13)$$

The results for this example are:

$$6.1 + 2\sqrt{(100)(0.061)(1-0.061)} = 10.887$$

$$6.1 - 2\sqrt{(100)(0.061)(1-0.061)} = 1.313$$

Figure 9.18 illustrates the np chart for this process. A comparison of Figures 9.4 and 9.18 reveals that these two control charts are mathematically equivalent and present the same information in a slightly different light. The only reason that one is preferred to the other is the form in which the data are presented, or the way in which the user prefers to visualize the control chart. Subsequent actions to stabilize the process are identical to those for the p chart.

FIGURE 9.18 np Chart for Fraction of Cracked or Broken Tiles

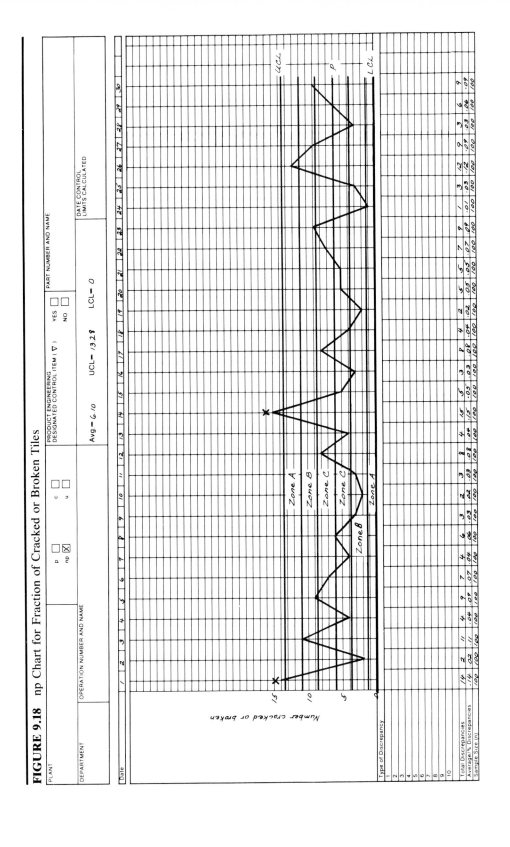

AREA OF OPPORTUNITY CHARTS

A *defective* item is a nonconforming unit. It must be discarded, reworked, scrapped, or downgraded. It is unusable for its intended purpose in its present form. A *defect,* on the other hand, is an imperfection of some type that does not necessarily render the entire product unusable, yet is undesirable. One or more defects may not make an entire good or service defective. For example, we would not scrap or discard a computer, a washing machine, or an air conditioner because of a small scratch in the paint.

An assembled piece of machinery such as an automobile, a dishwasher, or an air conditioner may have one or more defects that may not render the entire unit defective but may cause it to be downgraded or may necessitate its being reworked. Additionally, any product produced in sheets or rolls, such as paper, fabric, or plastic, may have several defects in a sheet or roll and still need not be scrapped as totally defective. In fact, unfortunately there are many instances where more than one factory defect is the norm rather than the exception. This has created situations where products may not even be downgraded as a result of having several flaws. Naturally, in the quest for improvement, we prefer no defects in our product, and control charting is one of the tools to help achieve this end.

When there are multiple opportunities for defects or imperfections in a given unit (such as a large sheet of fabric), we call each such unit an *area of opportunity;* each area of opportunity is a subgroup. When areas of opportunity are discrete units and a single defect will make the entire unit defective, a p chart or np chart is appropriate. However, when areas of opportunity are continuous, or very nearly so, and more than one defect may occur in a given area of opportunity, a c chart or u chart should be used. The c chart is used when the areas of opportunity are of constant size, while the u chart is used when the areas of opportunity are not of constant size.

Conditions for Use

Area of opportunity charts have wide applicability. If we are counting defects, the enamel on an appliance represents a continuous area of opportunity; a roll of cloth or plastic film is a continuous area of opportunity. If we are measuring the number of accidents recorded per month, a month represents a continuous area of opportunity. Measurements of the number of errors per hour in data entry or the number of typographical errors made per page have areas of opportunity (an hour or a page) which present enough opportunities for multiple defects to be considered nearly continuous. Imperfections in a complex piece of machinery, such as a computer, have areas of opportunity that are not strictly continuous; but the large number of individual pieces involved make the areas of opportunity very nearly so, and they are often taken to be continuous.

The Sampling Distributions for c Charts and u Charts

Both the c chart and the u chart are concerned with counts of the number of occurrences of an event over a continuous (or virtually continuous) area of opportunity. Those counts will be whole numbers such as 0, 1, 2, 3, ... The principle behind both types of charts is the same; the primary difference between the two charts is whether the size of the areas of opportunity remains constant from subgroup to subgroup.

In Chapter 6 we explored the probability distribution governing processes such as these. We learned that the probability of observing x events in some large string of opportunities for the event to occur may often be described by a Poisson probability model. Using the Poisson model requires that defects can be described as discrete events; that defects occur randomly within some well-defined area of opportunity; that defects are relatively rare events; and that defects are independent of one another.

Exact conformance to these Poisson model conditions is not always easy to verify. Usually it is not too difficult to tell whether the events are discrete and whether there is some well defined area of opportunity. Whether the events are relatively rare is somewhat subjective and requires some knowledge and experience; the issue of independence is generally revealed by the control chart. That is, if events are not random and independent, they will tend to form identifiable patterns of imperfections that were introduced in Chapter 8 and that will be discussed further in Chapter 11. This will produce indications of a lack of control on the control chart.

Recall from Chapter 7 that a Poisson process with a mean of λ would have the square root of λ as its standard error. We create control limits for such a process by adding and subtracting three standard errors from the process average.

c Charts

Areas of opportunity that are constant in size are easier to handle than those that vary, in much the same way as constant subgroup sizes in a p chart are easier to handle than those that vary. Constant areas of opportunity might be such things as one unit of a particular model of a television set, one printed circuit board, one purchase order, one aircraft canopy, five square feet of paper board, or five linear feet of wire. When all conditions for an area of opportunity chart are met, and when the subgroup sizes remain constant, a c chart is used.

The number of events in an area of opportunity is denoted by c, the count for each area of opportunity. The string of successive c values, taken over time, is used to construct the control chart.

The centerline for the chart is the average number of events observed and is calculated as:

$$\text{centerline(c)} = \bar{c} = \frac{\text{Total number of events observed}}{\text{Number of areas of opportunity}} \quad (9.15)$$

As the standard error is the square root of the mean, its value is $\sqrt{\bar{c}}$. Upper and lower control limits can be found by adding and subtracting three times the standard error from the centerline, $\bar{c}$. Thus,

$$\text{UCL(c)} = \bar{c} + 3\sqrt{\bar{c}} \quad (9.16)$$

$$\text{LCL(c)} = \bar{c} - 3\sqrt{\bar{c}} \quad (9.17)$$

Counts, Control Limits, and Zones. As we have already seen with the p charts and np charts, when a process is in a state of control, only very rarely will points fall beyond the control limits. Therefore, when a point does fall outside the control limits, we shall consider it an indication of a lack of control and take appropriate action. When the lower control limit is calculated to be negative, we will use 0 as the lower control limit because, just as with the p chart and np chart, negative numbers of events (such as −3 defects on a radio) are impossible.

Consider a firm that has decided to use a c chart to help keep track of the number of telephone requests received daily for information on a given product. Each day represents an area of opportunity. Over a 30-day period, 1,206 requests are received, an average of 40.2 per day. This value is $\bar{c}$, the centerline, and the upper and lower control limits can be found using Equations 9.16 and 9.17:

$$\text{UCL(c)} = 40.2 + 3\sqrt{40.2} = 59.2$$

$$\text{LCL(c)} = 40.2 - 3\sqrt{40.2} = 21.2$$

Actual counts occurring in an area of opportunity will always be whole numbers. Thus a count of 59 will be within the control limits, while a count of 60 would be beyond the UCL. The A, B, and C zone boundaries are created at one and two standard errors from the centerline, respectively. The zone boundaries are:

$$\text{boundary between lower zones B and C} = 40.2 - \sqrt{40.2} = 33.9$$

$$\text{boundary between lower zones A and B} = 40.2 - 2\sqrt{40.2} = 27.5$$

$$\text{boundary between upper zones A and B} = 40.2 + 2\sqrt{40.2} = 52.9$$

$$\text{boundary between upper zones B and C} = 40.2 + \sqrt{40.2} = 46.5$$

Because the actual counts are whole numbers, the observations would fall into zones as follows:

Zone	Counts
Upper A	53, 54, 55, 56, 57, 58, 59
Upper B	47, 48, 49, 50, 51, 52
Upper C	41, 42, 43, 44, 45, 46
Lower C	34, 35, 36, 37, 38, 39, 40
Lower B	28, 29, 30, 31, 32, 33
Lower A	22, 23, 24, 25, 26, 27

The zones each contain a reasonable number of whole numbers and are close enough in size to be workable. But consider the problem that would have been encountered if the process average had been $\bar{c} = 2.4$. Here we would get:

$$\text{UCL}(c) = 2.4 + 3\sqrt{2.4} = 7.0$$

$$\text{LCL}(c) = 2.4 - 3\sqrt{2.4} = -2.2 \text{ (use 0.0)}$$

$$\text{boundary between lower zones B and C} = 2.4 - \sqrt{2.4} = 0.9$$

$$\text{boundary between lower zones A and B} = 2.4 - 2\sqrt{2.4} = -0.7 \text{ (use 0.0)}$$

$$\text{boundary between upper zones A and B} = 2.4 + 2\sqrt{2.4} = 5.5$$

$$\text{boundary between upper zones B and C} = 2.4 + \sqrt{2.4} = 3.9$$

As before, because the counts are whole numbers, the observations will fall into zones as follows:

Zone	Counts
Upper A	6, 7
Upper B	4, 5
Upper C	3
Lower C	1, 2
Lower B	0
Lower A	0

These zones are so small that they are practically meaningless. The upper zone C, for example, only has one possible count, 3. The rules set out in Chapter 8 and discussed more fully in Chapter 11 are based on a normal probability of 0.34 that a stable process will generate a point in the upper zone C. However, from Equation 6.8, a Poisson process with a mean of 2.4, as in this example, will generate a count of 3.0 (the single value in zone C) with a probability of only 0.209. The difference between these two probabilities (the normal and the Poisson) is too great in this case to permit the use of the zones. For this reason, when the average count is small, we generally do not make use of the zones in seeking indications of a lack of control. Rather, we focus on points beyond the control limits, runs of points above or below the centerline, and runs upward or downward in the data as indicators of a lack of stability. The exact value of the centerline below which the use of A, B, and C zones becomes impractical requires knowledge of and experience with the particular process involved. As a rule of thumb, the zone boundaries should not be used for c charts with average counts of less than 20.0; when $\bar{c} \geq$ 20, the Poisson distribution can be approximated by the normal distribution. Once again, there is no substitute for knowledge and experience; but as the observable count shrinks, the use of variables control charts must be instituted for continued process improvement.

Furthermore, keep in mind that when the average count is small, larger and larger areas of opportunity will be needed to detect imperfections. This will occur as a natural consequence of improved quality through the use of control charts. When the areas of opportunity needed to find imperfections grow unacceptably large, attributes control charts will have to be abandoned in favor of variables control charts. This usually solves the

FIGURE 9.19 Number of Blemishes Found in 25 Reels of Paper

Reel	Number of Blemishes	Reel	Number of Blemishes
1	4	14	9
2	5	15	1
3	5	16	1
4	10	17	6
5	6	18	10
6	4	19	3
7	5	20	7
8	6	21	4
9	3	22	8
10	6	23	7
11	6	24	9
12	7	25	7
13	11	Total	150

problem of small values for $\bar{c}$ and is another step on the ladder of quality consciousness.

An Example. Let us consider the output of a paper mill as an example. The product appears at the end of a web and is rolled onto a spool called a *reel*. Every reel is examined for blemishes, which are imperfections. Each reel is an area of opportunity. The results of these inspections produce the data shown in Figure 9.19.

The assumptions necessary for using the c chart are well met here, as the reels are large enough to be considered continuous areas of opportunity; imperfections are discrete events and seem to be independent of one another, and they are relatively rare. Even if these conditions are not precisely met, the Poisson process is fairly robust, or insensitive to small departures from the assumptions, so we may still safely utilize the c chart.

In this example the average number of imperfections per reel is:

$$\bar{c} = 150/25 = 6.00$$

and the standard error is $\sqrt{6.00} = 2.45$.

Equations 9.16 and 9.17 yield upper and lower control limits:

$$UCL(c) = 6.00 + 3(2.45) = 13.35$$

$$LCL(c) = 6.00 - 3(2.45) = -1.35 \quad (\text{use } 0.00)$$

Figure 9.20 is a control chart using these values.

Probability Control Limits. Even though the control chart in Figure 9.20 is useful, its theoretical basis can be put on firmer ground by using probability control limits rather than 3-sigma limits. These are control limits based on the percentiles of the Poisson distribution. The Poisson probability distribution is not symmetrical—it is skewed. This skewness is more pronounced when the average is 20 or less. Because of this, using control limits set at three times the standard error from the mean will tend to generate indications of a lack of control where none really exists, especially beyond the lower control limit. These false alarms tend to be more troublesome when the process average is 7.0 or less; but an adjustment based on the Poisson distribution itself is not too difficult to make and should be made when the process average is 20 or less. This will tend to keep the number of false alarms generated at a tolerable level in most instances. False alarms, in and of themselves, can destabilize a stable process by demoralizing employees who may begin to feel that many of their efforts do not result in process improvements.

Statistical theory permits us to use values from the Poisson distribution to establish the control limits. The object is to establish control limits that will indicate a lack of control when that condition exists yet will not generate too many indications of special causes of variation when there

FIGURE 9.20 c Chart for Number of Blemishes on Reels of Paper

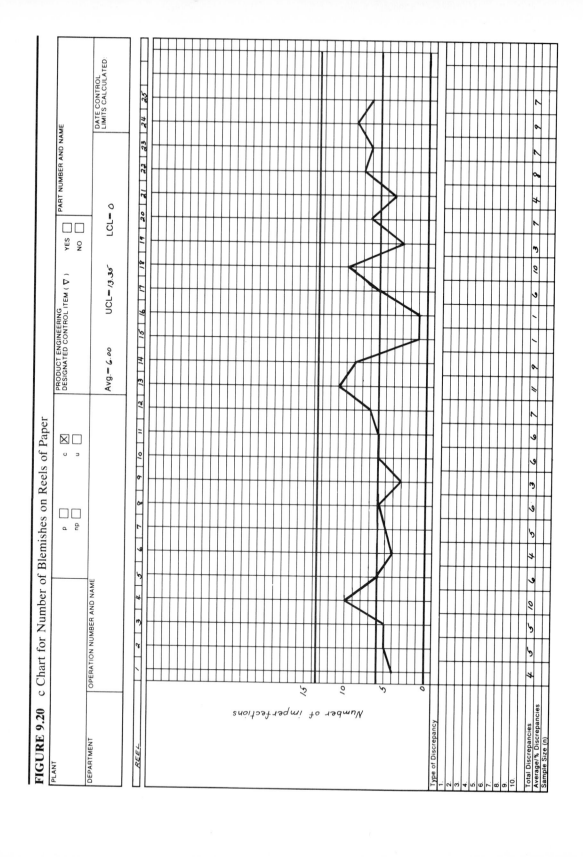

are none. When the process average is 20 or less, the A, B, and C zones are generally not used; thus the c chart will lack some of the sources for indications of a lack of control which use of the zones provides. Without the zones, there will be less opportunity for a process that is really out of control to demonstrate indications of that lack of control. To compensate for this, the probability control limits will be constructed so that a controlled, stable process will generate points outside the control limits one percent of the time. Usually, stable processes, controlled at 3-sigma limits, generate points beyond the control limits about three times in a thousand (0.3 percent). This can be raised to about one percent (or any other value) to compensate for the loss of indications from the use of the A, B, and C zones.

These probability control limits split the one percent probability so that the probability of a controlled process's generating an observation below the lower control limit will be about 0.005 and that above the upper control limit will be about 0.005. In terms of the cumulative distribution, this latter value is a probability of 0.995 of generating an observation falling below the upper control limit. This states that approximately 99 percent of the data generated by a stable process will fall between the

FIGURE 9.21   c Chart Control Limits for 0.005 and 0.995 Probabilities

Process Average	UCL	LCL	Process Average	UCL	LCL
0 to 0.10	0	1.5	9.28 to 9.64	2.5	18.5
0.11 to 0.33	0	2.5	9.65 to 10.35	2.5	19.5
0.34 to 0.67	0	3.5	10.36 to 10.97	2.5	20.5
0.68 to 1.07	0	4.5	10.98 to 11.06	3.5	20.5
1.08 to 1.53	0	5.5	11.07 to 11.79	3.5	21.5
1.54 to 2.03	0	6.5	11.80 to 12.52	3.5	22.5
2.04 to 2.57	0	7.5	12.53 to 12.59	3.5	23.5
2.58 to 3.13	0	8.5	12.60 to 13.25	4.5	23.5
3.14 to 3.71	0	9.5	13.26 to 13.99	4.5	24.5
3.72 to 4.32	0	10.5	14.00 to 14.14	4.5	25.5
4.33 to 4.94	0	11.5	14.15 to 14.74	5.5	25.5
4.95 to 5.29	0	12.5	14.75 to 15.49	5.5	26.5
5.30 to 5.58	0.5	12.5	15.50 to 15.65	5.5	27.5
5.59 to 6.23	0.5	13.5	15.66 to 16.24	6.5	27.5
6.24 to 6.89	0.5	14.5	16.25 to 17.00	6.5	28.5
6.90 to 7.43	0.5	15.5	17.01 to 17.13	6.5	29.5
7.44 to 7.56	1.5	15.5	17.14 to 17.76	7.5	29.5
7.57 to 8.25	1.5	16.5	17.77 to 18.53	7.5	30.5
8.26 to 8.94	1.5	17.5	18.54 to 18.57	7.5	31.5
8.95 to 9.27	1.5	18.5	18.58 to 19.36	8.5	31.5
			19.37 to 20.00	8.5	32.5

control limits as long as the process remains in a state of statistical control. Figure 9.21 gives values for the lower and upper control limits that should be used to construct a c chart when the process average value is 20 or less.

When average count values are greater than 20, the 0.005 and 0.995 control limits are approximated using:

$$\text{LCL(c)} = \bar{c} - 2.47 \sqrt{\bar{c}} \qquad (9.18)$$

and

$$\text{UCL(c)} = \bar{c} + 2.85 \sqrt{\bar{c}} \qquad (9.19)$$

In the example of the reels of paper, the centerline is 6.00. Therefore, for this application of the c chart, the control limits should properly have come from Figure 9.21. As 6.00 is in the 5.59 to 6.23 range, the values for the lower and upper control limits respectively are 0.5 and 13.5. These values have been used to draw the c chart that appears in Figure 9.22.

Notice that these control limits are almost the same as those created using Equations 9.16 (13.35) and 9.17 (0.00). In general, because the number of events is a whole number, both of these control charts may show the same indications of a lack of control. In this particular case the resulting control charts are similar, but a count of 0 will be an indication of a lack of control using the probability limits and not an indication of a lack of control using the 3-sigma limits.

It is not too unusual to find that the control limits resulting from computations using Equations 9.16 and 9.17 and those resulting from Figure 9.21 are similar, and some users merely ignore these table values. The danger in ignoring the table values is that when average counts are small, the 3-sigma limits may generate a high number of false indications of a lack of control. This can lead to overadjustment of a process, which in and of itself may cause the process to become out of control or, as we have said, may lead to some frustration on the part of those employees trying to search for special causes of variation where none exist.

Other Probability Control Limits

Often probability limits that are closer to three standard errors on either side of the centerline are desired to minimize the likelihood of *false alarms,* or observations that fall outside the control limits even though the process is stable. That is, when the distribution of the process output is normal, control limits that are three standard errors from the centerline indicate that the probability of a controlled process's generating an observation outside of the control limits will be 0.0015 above the upper control limit and 0.0015 below the lower control limit. In terms of the cumulative distribution, the lower control limit is such that the probability of an

FIGURE 9.22 c Chart Using 0.005 and 0.995 Probabilities for Number of Blemishes on Reels of Paper

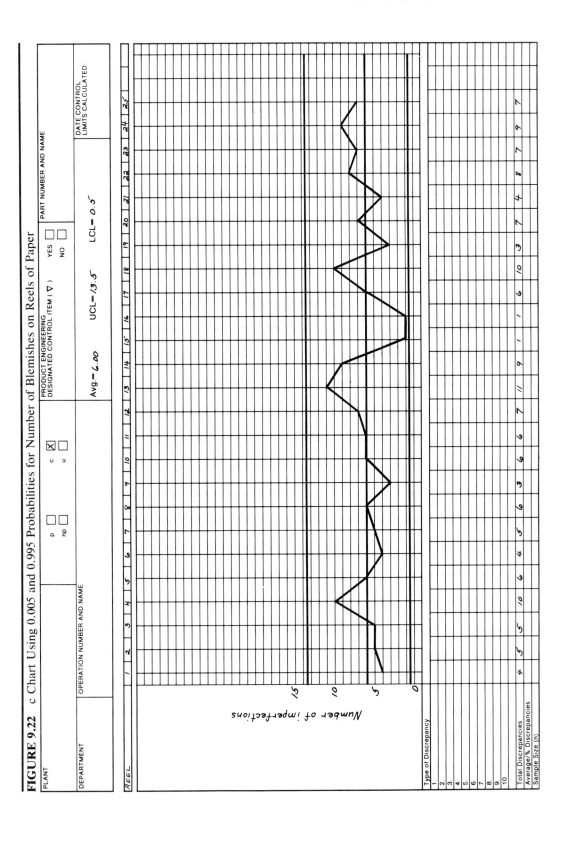

observation's falling below it is 0.0015, and the upper control limit is such that the probability of an observation's falling below it is 0.9985. Because these 0.0015 and 0.9985 limits yield a smaller probability of a controlled process generating an observation beyond the control limits than the 0.005 and 0.995 probability limits, this approach will create fewer false alarms, or observations falling beyond the control limits when the process is actually operating under control. This desirable result must be balanced against the fact that the wider 0.0015 and 0.9985 limits will also mean that it will be slightly less obvious that a process has strayed out of control when it actually has done so.

When the 0.0015 and 0.9985 control limits are used to construct the control chart—and the average count is 18.80 or less—the factors in Figure 9.23 should be used. When the average count, $\bar{c}$, exceeds 18.8 (and the 0.0015 and 0.9985 limits are desired), it is possible to obtain a reasonably good approximation of the control limits by using a normal approximation for the Poisson distribution. This means that the control limits can be obtained by adding and subtracting 3.00 times the square root of $\bar{c}$ from the centerline value. Further, when $\bar{c}$ exceeds 18.8, the A, B, and C zones can be used to help detect out-of-control subgroups.

FIGURE 9.23   c Chart Control Limits for 0.0015 and 0.9985 Probabilities

Process Average	UCL	LCL	Process Average	UCL	LCL
0 to 0.05	0	1.5	8.62 to 8.78	0.5	19.5
0.06 to 0.21	0	2.5	8.79 to 9.27	1.5	19.5
0.22 to 0.47	0	3.5	9.28 to 9.95	1.5	20.5
0.48 to 0.81	0	4.5	9.96 to 10.63	1.5	21.5
0.82 to 1.19	0	5.5	10.64 to 10.74	1.5	22.5
1.20 to 1.63	0	6.5	10.75 to 11.32	2.5	22.5
1.64 to 2.10	0	7.5	11.33 to 12.01	2.5	23.5
2.11 to 2.60	0	8.5	12.02 to 12.54	2.5	24.5
2.61 to 3.12	0	9.5	12.55 to 12.71	3.5	24.5
3.13 to 3.67	0	10.5	12.72 to 13.42	3.5	25.5
3.68 to 4.24	0	11.5	13.43 to 14.13	3.5	26.5
4.25 to 4.82	0	12.5	14.14 to 14.25	3.5	27.5
4.83 to 5.42	0	13.5	14.26 to 14.85	4.5	27.5
5.43 to 6.04	0	14.5	14.86 to 15.57	4.5	28.5
6.05 to 6.50	0	15.5	15.58 to 15.88	4.5	29.5
6.51 to 6.66	0.5	15.5	15.89 to 16.30	5.5	29.5
6.67 to 7.30	0.5	16.5	16.31 to 17.03	5.5	30.5
7.31 to 7.95	0.5	17.5	17.04 to 17.47	5.5	31.5
7.96 to 8.61	0.5	18.5	17.48 to 18.80	6.5	31.5

Figures 9.21 and 9.23 are just two possible choices for the positioning of the control limits based upon the probability that a process in a state of control will generate a signal indicating a lack of control. Other values are, of course, possible; but most often c charts are drawn by adding and subtracting three times the square root of the centerline for the control limits.

A note of caution when dealing with c charts: those charged with determining the number of imperfections must be clear and consistent in the definition of an imperfection. Operational definitions, as discussed in Chapter 3, are extremely important, and the individuals identifying imperfections must be properly trained so that they understand the nature of the process. If the individuals identifying imperfections are not properly trained, some identified imperfections may not actually be imperfections, while some actual imperfections may go undetected—hence, the independence of observations of the occurrence of imperfections may suffer. This, in turn, may result in violations of the underlying assumptions used for the c chart, leading to either the generation of many false alarms or a process that is operating out of control and remaining undetected.

Stabilizing a Process: An Example. A firm that produces industrial washing machines inspects completed units for defects. Figure 9.24 lists counts for the number of defects found on 24 machines.

The centerline for the control chart is:

$$\bar{c} = 1100/24 = 45.8$$

FIGURE 9.24 Defects Found on 24 Machines

Machine Number	Count	Machine Number	Count
1	62	13	51
2	60	14	75
3	36	15	49
4	39	16	52
5	36	17	62
6	47	18	43
7	33	19	70
8	32	20	18
9	74	21	44
10	71	22	20
11	43	23	18
12	39	24	26
		Total	1,100

FIGURE 9.25 Control Chart for 24 Washing Machines

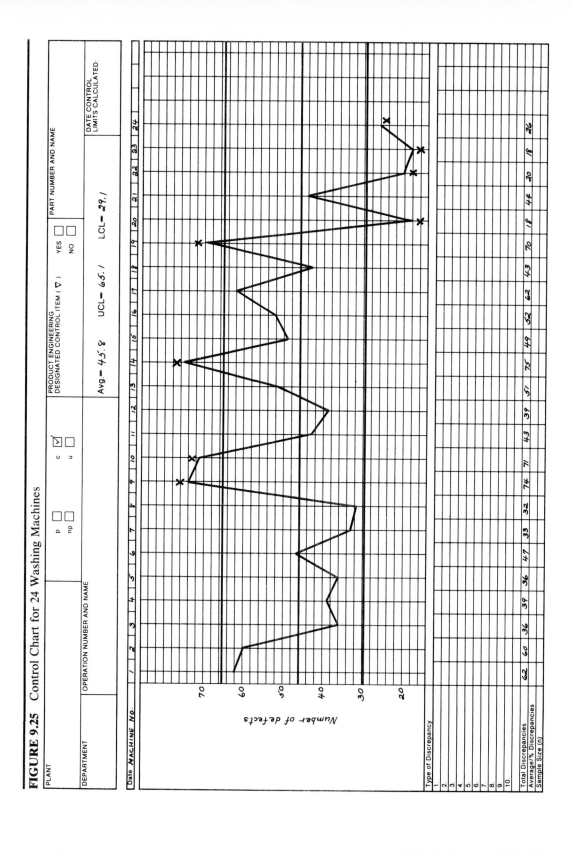

and the 0.005 and 0.995 control limits can be found using Equations 9.18 and 9.19:

$$LCL(c) = 45.8 - 2.47 \sqrt{45.8} = 29.1$$

and

$$UCL(c) = 45.8 + 2.85 \sqrt{45.8} = 65.1$$

Hence, counts of 66 or more and 29 or fewer are one of the indications of a lack of control. Figure 9.25 is a control chart for this process.

The process is not in control. Special causes of variation are present. The local operators responsible for the final inspection act so that the special causes of variation for points 9, 10, 14, 19, 20, 22, and 23 are identified and the appropriate corrections are made. No special cause could be found for machine 24. The data for points affected by known special causes that have been eliminated are deleted from the data set (point 24 could not be eliminated), and the centerline and the control limits are recomputed:

$$\bar{c} = 754/17 = 44.4$$

$$UCL(c) = 44.4 + 2.85 \sqrt{44.4} = 63.4$$

$$LCL(c) = 44.4 - 2.47 \sqrt{44.4} = 27.9$$

The new limits are so close to the old limits that the old limits are used for the next 24 machines produced; Figure 9.26 presents the data.

The first five data points for these next 24 machines are well below the lower control limit. Investigation by the local operators reveals that a substitute for the regular inspector counted the defects on those five

---

FIGURE 9.26   Defects Found on Next 24 Machines

Machine Number	Number of Defects	Machine Number	Number of Defects
25	21	37	46
26	18	38	31
27	7	39	42
28	12	40	44
29	18	41	26
30	32	42	37
31	32	43	26
32	37	44	29
33	39	45	31
34	39	46	34
35	34	47	36
36	39	48	40

FIGURE 9.27 Control Chart for Next 24 Washing Machines

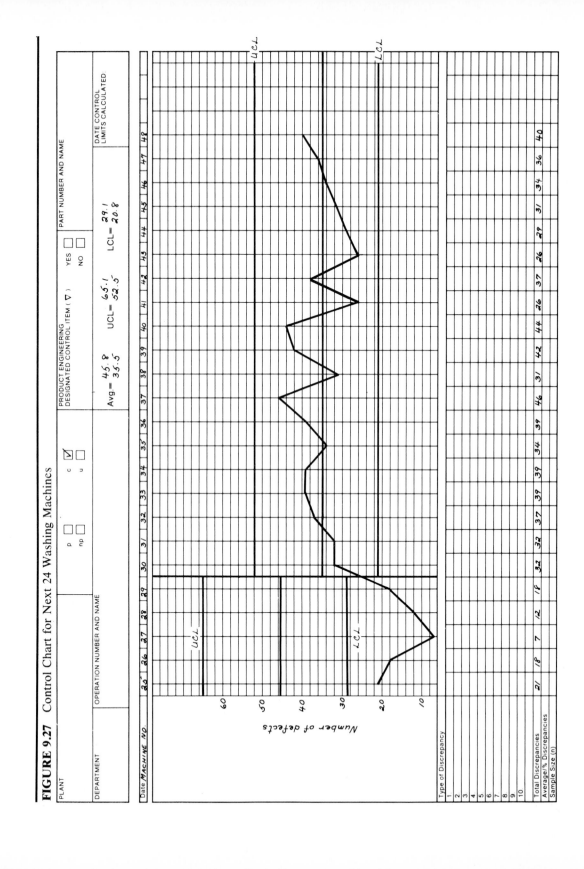

machines. The substitute was not properly trained and did not identify all of the defects correctly. The operators informed management, and management made appropriate changes in policy so that this situation would not recur. These points can now be eliminated from the data. Beginning with machine number 30, all counts are below the process average. Local operators decided that the process has been changed, and so a revised control chart is constructed beginning with point number 30.

$$\bar{c} = 674/19 = 35.5$$

$$LCL(c) = 35.5 - 2.47 \sqrt{35.5} = 20.8$$

$$UCL(c) = 35.5 + 2.85 \sqrt{35.5} = 52.5$$

Figure 9.27 illustrates the revised control chart. The process, as it stands, now appears to be in a state of control.

Another Example.   Consider the case of a mill with a constant work force of 450 employees that posts a sign at the employee entrance reading "SAFETY IS BETTER THAN COMPENSATION." Informal conversations with employees reveal they consider the sign a reminder to be careful. Management has *not* simultaneously made the work environment safer: there still are cluttered aisles, and spills and leaks of liquids on the floor are not attended to hastily. The workers know this but have long ago stopped their fruitless efforts at getting management to allocate the resources necessary to create a safer workplace.

A control chart of the number of accidents per month is constructed in an effort to examine the problem. Figure 9.28 illustrates the data for the past 26 months.

FIGURE 9.28	Accidents per Month		
Month	*Number of Accidents*	*Month*	*Number of Accidents*
Jan.	3	Feb.	2
Feb.	2	Mar.	0
Mar.	0	Apr.	0
Apr.	2	May	3
May	1	June	2
June	1	July	0
July	1	Aug.	1
Aug.	0	Sep.	0
Sep.	0	Oct.	1
Oct.	1	Nov.	0
Nov.	1	Dec.	0
Dec.	3	Jan.	1
Jan.	0	Feb.	1
		Total	26

As the number of man-hours per month remains constant, the area of opportunity is considered constant month-to-month. The centerline for the c chart is:

$$\bar{c} = 26/26 = 1.0$$

From Figure 9.22, the 0.005 and 0.995 probability control limits are:

$$LCL(c) = 0 \text{ and } UCL(c) = 4.5$$

Figure 9.29 displays the control chart for the past 26 months. As there are no indications of any special variation, we can conclude that the process is stable and in a state of statistical control. Whether or not the company knows it, it is in the business of producing accidents at the stable rate of one per month. It will continue to do so until some effort is made to change the underlying process. If no change in the process is made, accidents will continue to be produced at this rate. Consequently, the sign reading "SAFETY IS BETTER THAN COMPENSATION" is unfair. The employees are not empowered to make system changes that would lower the average number of accidents per month; the sign unjustly and subtly shifts the burden of responsibility for safety from management to the employees.

u Charts

In some applications the areas of opportunity vary in size. Generally, the construction and interpretation of control charts are easier when the area of opportunity remains constant, but from time to time changes in that area may be unavoidable. For example, samples taken from a roll of paper may need to be manually torn from rolls, so that the areas sampled—the areas of opportunity—will vary; continuous welds in heat exchangers will have varying areas of opportunity depending on the total number and lengths of the welds present in different units; and the number of typing errors in a document will have areas of opportunity that will vary with the lengths of the documents. When the areas vary, the control chart used is a u chart.

The u chart is similar to the c chart in that it is a control chart for the count of the number of events, such as the number of nonconformities over a given area of opportunity. The fundamental difference lies in the fact that during construction of a c chart, the area of opportunity remains constant from observation to observation, while this is not a requirement for the u chart. Instead, the u chart considers the number of events (such as blemishes or other defects) as a fraction of the total size of the area of opportunity in which these events were possible, thus circumventing the problem of having different areas of opportunity for different observations.

FIGURE 9.29 Control Chart for Accidents per Month

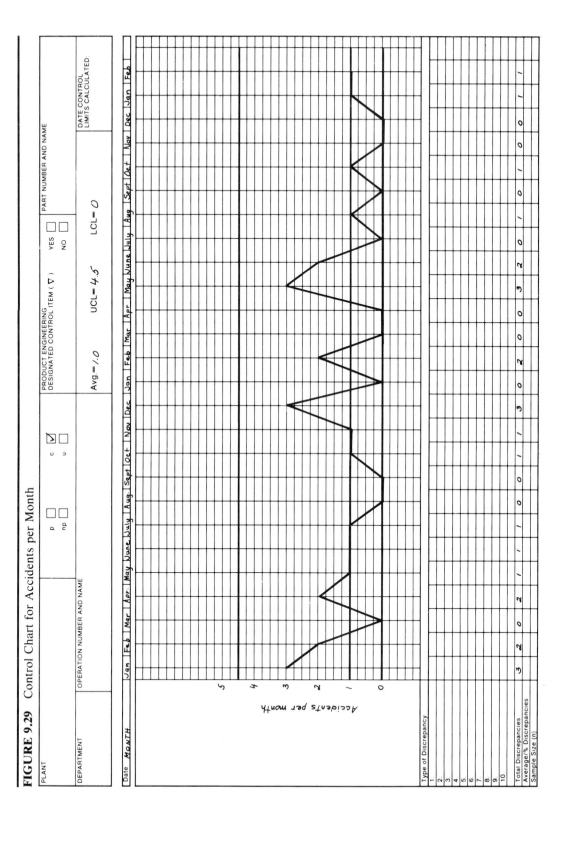

The characteristic used for the control chart, u, is the ratio of the number of events to the area of opportunity in which the events may occur. For observation i, we will call the number of events, such as imperfections, the observed c_i, and the area of opportunity, the a_i. Thus, u_i is the ratio

$$u_i = c_i/a_i \qquad (9.20)$$

for each point.

FIGURE 9.30   Defects in Rolls of Plastic

Insp. Lot (i)	Sq. Ft. of Plastic	Area of Opportunity (in 100 sq. ft.) a_i	Number of Defects in Lot c_i	Defects Per 100 Sq. Ft. u_i
1	200	2.00	5	2.50
2	250	2.50	7	2.80
3	100	1.00	3	3.00
4	90	0.90	2	2.22
5	120	1.20	4	3.33
6	80	0.80	1	1.25
7	200	2.00	10	5.00
8	220	2.20	5	2.27
9	140	1.40	4	2.86
10	80	0.80	2	2.50
11	170	1.70	1	0.59
12	90	0.90	2	2.22
13	200	2.00	5	2.50
14	250	2.50	12	4.80
15	230	2.30	4	1.74
16	180	1.80	4	2.22
17	80	0.80	1	1.25
18	100	1.00	2	2.00
19	140	1.40	3	2.14
20	120	1.20	4	3.33
21	250	2.50	2	0.80
22	130	1.30	3	2.31
23	220	2.20	1	0.45
24	200	2.00	5	2.50
25	100	1.00	2	2.00
26	160	1.60	4	2.50
27	250	2.50	12	4.80
28	80	0.80	1	1.25
29	150	1.50	5	3.33
30	210	2.10	4	1.90
Totals	4,790	$\Sigma a_i = 47.90$	$\Sigma c_i = 120$	

The average of all the u_i values, $\bar{u}$, provides a centerline for the control chart:

$$\text{centerline(u)} = \bar{u} = \Sigma c_i / \Sigma a_i \qquad (9.21)$$

Control limits are usually placed at three standard errors on either side of the centerline for each individual subgroup. As we pointed out earlier in this chapter, the appropriate distribution for both the c chart and the u chart is the Poisson distribution. In the case of the u chart, the Poisson distribution provides a reasonable and workable theoretical framework. However, as the area of opportunity varies from subgroup to subgroup, the calculation for the standard error will also vary from subgroup to subgroup. Hence, the standard error for each subgroup i is calculated as the square root of the average u value divided by the subgroup's area of opportunity, and the control limits for each observation become:

$$\text{LCL(u)} = \bar{u} - 3\sqrt{\bar{u}/a_i} \qquad (9.22)$$

$$\text{UCL(u)} = \bar{u} + 3\sqrt{\bar{u}/a_i} \qquad (9.23)$$

When the lower control limit is negative, a value of 0.0 is used instead.

An Example. Consider the case of the manufacture of a certain grade of plastic. The plastic is produced in rolls, samples of which are taken five times daily. Because of the nature of the process, the square footage of each sample varies from inspection lot to inspection lot. Hence, the u chart should be used here. Figure 9.30 shows the data on the number of defects, c_i, for the past 30 inspection lots. Defects are recorded as number of events per 100 square feet and are calculated from Equation 9.20.

Using Equation 9.21 we find the centerline to be:

average number of defects/100 sq. ft. $= \bar{u} = 120/47.90 = 2.5$

The control limits are different for each of the subgroups and must be computed individually for each subgroup using Equations 9.22 and 9.23. Figure 9.31 lists the resulting values.

Figure 9.32 illustrates control chart for this process. No points indicate a lack of control, so there is no reason to believe that any special variation is present. If sources of special variation were detected, we would proceed as we did with the c chart—that is, we would identify the source or sources of the special variation, eliminate them from the system if detrimental, or incorporate them into the system if beneficial; drop the data points from the data set; and reconstruct and reanalyze the control chart.

Two Alternatives for Coping with Variable Control Limits. Variable control limits can be confusing and can lead to difficulties with calculation and record-keeping. Additionally, as the control limits are based on the

---

FIGURE 9.31   Control Limits for Defects in Rolls of Plastic

Inspection Lot i	Number of Inspection Units a_i	LCL	UCL
1	2.0	0	5.9
2	2.5	0	5.5
3	1.0	0	7.3
4	0.9	0	7.5
5	1.2	0	6.8
6	0.8	0	7.8
7	2.0	0	5.9
8	2.2	0	5.7
9	1.4	0	6.5
10	0.8	0	7.8
11	1.7	0	6.2
12	0.9	0	7.5
13	2.0	0	5.9
14	2.5	0	5.5
15	2.3	0	5.6
16	1.8	0	6.1
17	0.8	0	7.8
18	1.0	0	7.3
19	1.4	0	6.5
20	1.2	0	6.8
21	2.5	0	5.5
22	1.3	0	6.7
23	2.2	0	5.7
24	2.0	0	5.9
25	1.0	0	7.3
26	1.6	0	6.3
27	2.5	0	5.5
28	0.8	0	7.8
29	1.5	0	6.4
30	2.1	0	5.8

---

area of opportunity, future control limits cannot be projected from past control limits even for a stable process. Therefore, a given data point may not be identifiable as indicating a lack of control until some time after it occurs, when control limits can be calculated. This can lead to costly delays in corrective actions; there are two common remedies for this problem.

One possible solution is to calculate approximate control limits based upon an average value for the area of opportunity. If the areas of opportu-

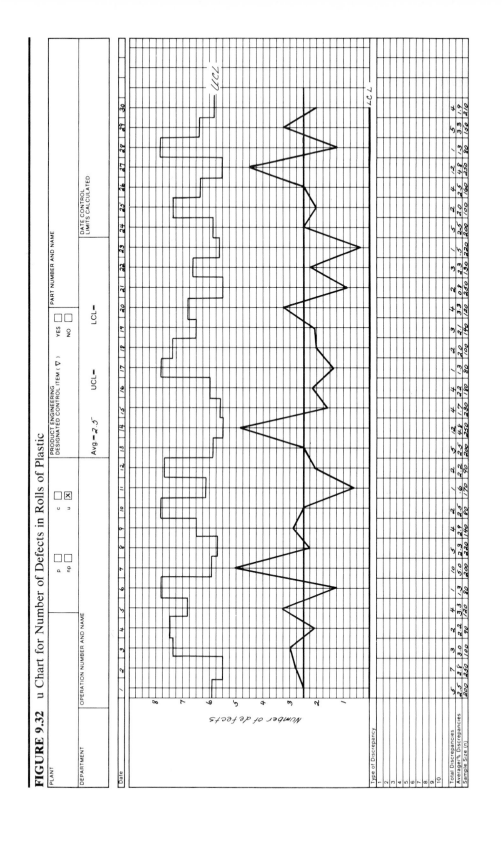

FIGURE 9.32 u Chart for Number of Defects in Rolls of Plastic

nity do not vary by more than about ±25 percent, the control limits can be based on

$$\text{average area of opportunity} = \bar{a} = \Sigma a_i / k \qquad (9.24)$$

where k is the number of subgroups or areas of opportunity.

Then the values for the upper and lower control limits can be approximated using:

$$\text{UCL}(u_{\text{approx.}}) = \bar{u} + 3 \sqrt{\bar{u}/\bar{a}} \qquad (9.25)$$

and

$$\text{LCL}(u_{\text{approx.}}) = \bar{u} - 3 \sqrt{\bar{u}/\bar{a}} \qquad (9.26)$$

Points well within these control limits may be considered to indicate only common variation (assuming there are no other indications of a lack of control, as discussed in Chapters 8 and 11). Exact values for the UCL and LCL should be calculated for those points falling close to the control limits to determine whether any special causes of variation are indicated.

In our example of the rolls of plastic, the average area of opportunity and the approximate control limits are found using Equations 9.24, 9.25, and 9.26:

$$\bar{a} = \text{average area of opportunity} = 47.90/30 = 1.6 \text{ hundred sq. ft.}$$

$$\text{UCL}(u_{\text{approx.}}) = 2.5 + 3 \sqrt{2.5/1.6} = 6.25$$

$$\text{LCL}(u_{\text{approx.}}) = 2.5 - 3 \sqrt{2.5/1.6} = -1.25 \text{ (use 0.00)}$$

Figure 9.33 presents a control chart using these values and the data for the plastic rolls given in Figure 9.30. None of the entries falls close to the approximate control limits, nor are there any other indications of a lack of control. Further investigation of special variation does not appear warranted. One possible drawback when using approximate control limits is that the selection of those points needing calculation of exact UCL and LCL values is somewhat arbitrary; this is because the term "close to the approximate control limits" requires a subjective judgment.

To overcome this problem we can use a second remedy for the variable control limits of the u chart. This alternative involves calculating two sets of control limits: an outer set, using for $\bar{a}$ in Equations 9.22 and 9.23 the smallest value for the area of opportunity that will be encountered; and an inner set, using for $\bar{a}$ the largest possible value for the area of opportunity encountered in the data. Any data point outside the outer control limits clearly indicates a lack of control. Any data point lying within the inner set of control limits clearly does not indicate any special causes of variation (assuming that there are no other runs or nonrandom patterns indicating a lack of control). Those data points falling between the two sets of control limits require the calculation of exact UCL and LCL values.

FIGURE 9.33   u Chart Using Average Area of Opportunity

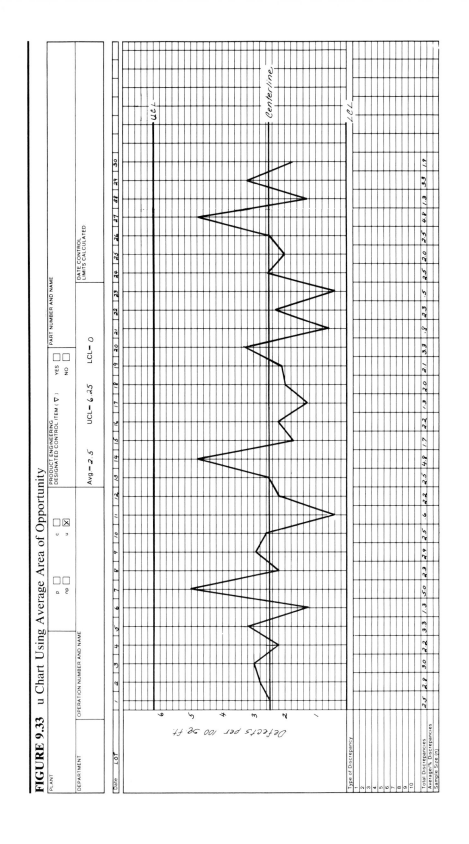

Variable Control Limits: An Example.   A manufacturer of chemical process equipment uses an automatic welding machine for long continuous welds. The welds are x-rayed, and the x-rays are examined for imperfections in the welds. The lengths of the welds are measured in inches and vary with the particular unit or portion of the unit being assembled; and so the appropriate attribute chart for counts of the number of imperfections is a u chart. The firm has traditionally kept track of imperfections in terms of the rate per 100 linear inches of weld. Figure 9.34 shows data from 25 welds as well as the computed rate per 100 inches.

The average rate is computed using Equation 9.21:

$$\bar{u} = 26.76/25 = 1.07 \text{ imperfections per 100 inches}$$

The shortest weld length produced with this equipment is 100 inches, and the longest is 400 inches (areas of opportunity of 1.00 and 4.00 hundred inches, respectively). These values can be used to construct outer and inner control limits, respectively:

$$\text{UCL}(u_{\text{outer}}) = 1.07 + 3 \sqrt{1.07/1.00} = 4.17$$

$$\text{LCL}(u_{\text{outer}}) = 1.07 - 3 \sqrt{1.07/1.00} = -2.03 \qquad \text{(use 0.00)}$$

$$\text{UCL}(u_{\text{inner}}) = 1.07 + 3 \sqrt{1.07/4.00} = 2.62$$

$$\text{LCL}(u_{\text{inner}}) = 1.07 - 3 \sqrt{1.07/4.00} = -.48 \qquad \text{(use 0.00)}$$

Figure 9.35 illustrates a control chart using these values. Units 16 and 17 are between the inner and outer sets of control limits; hence, exact values for the upper control limits are required. For unit 16, the 254 weld inches correspond to an a_i of 2.54 hundred inches; consequently, from Equation 9.23, the UCL value is:

$$\text{UCL}(u) = 1.07 + 3 \sqrt{1.07/2.54} = 3.02$$

The rate of imperfections for unit 16, 2.76, does not indicate the presence of any special causes of variation. For unit 17, the 144 weld inches correspond to an a_i of 1.44 hundred inches; consequently, the UCL value is:

$$\text{UCL}(u) = 1.07 + 3 \sqrt{1.07/1.44} = 3.66$$

The rate of imperfections for unit 17, 4.17, indicates the presence of a special cause of variation.

A careful study of the circumstances surrounding unit 17 reveals that the incorrect welding wire was used in both that unit and unit 16. A change in procedure is instituted to prevent this type of error from recurring—that is, the system has now been changed to eliminate this source of special variation.

As the system has been changed, the data for units 16 and 17 can now be dropped. After this has been done, the process average becomes:

$$\bar{u} = 19.83/23 = 0.86$$

FIGURE 9.34   Imperfections in Welds

Unit No.	Weld Length	Imperfections	Number of Imperfections per 100 Inches (Rate) u
1	187	2	1.07
2	302	1	0.33
3	302	0	0.00
4	172	2	1.16
5	240	5	2.08
6	144	1	0.69
7	120	1	0.83
8	320	2	0.63
9	264	2	0.76
10	180	1	0.56
11	208	1	0.48
12	234	5	2.14
13	180	1	0.56
14	288	3	1.04
15	108	1	0.93
16	254	7	2.76
17	144	6	4.17
18	180	2	1.11
19	288	2	0.69
20	360	3	0.83
21	220	5	2.27
22	156	0	0.00
23	348	1	0.29
24	288	2	0.69
25	144	1	0.69

Total   26.76

The outer and inner sets of control limits can now be re-computed as:

$$\text{UCL}(u_{outer}) = 0.86 + 3 \sqrt{0.86/1.00} = 3.64$$

$$\text{LCL}(u_{outer}) = 0.86 - 3 \sqrt{0.86/1.00} = -1.92 \quad \text{(use 0.00)}$$

$$\text{UCL}(u_{inner}) = 0.86 + 3 \sqrt{0.86/4.00} = 2.25$$

$$\text{LCL}(u_{inner}) = 0.86 - 3 \sqrt{0.86/4.00} = -.53 \quad \text{(use 0.00)}$$

Figure 9.36 shows the revised control chart. Point 21 is beyond the inner upper control limits, so the exact value for the upper control limit must be calculated using Equation 9.23:

$$\text{UCL} = 0.86 + 3 \sqrt{0.86/2.20} = 2.74$$

Hence, the u value 2.27 at point 21 does not indicate the presence of any

FIGURE 9.35 Control Chart for Rate of Weld Imperfections

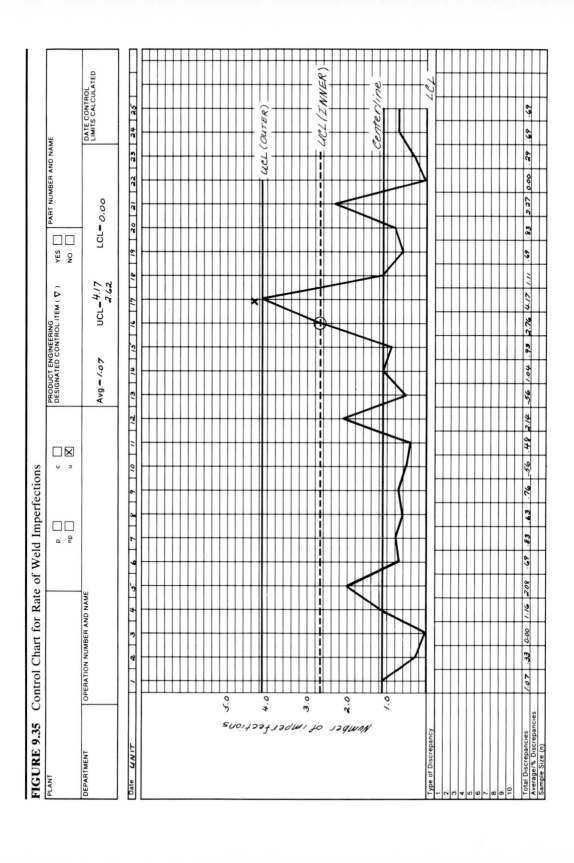

PLANT

PRODUCT ENGINEERING
DESIGNATED CONTROL ITEM (∇)

PART NUMBER AND NAME

DEPARTMENT

OPERATION NUMBER AND NAME

p ☐   c ☐
np ☐   u ☒

YES ☐
NO ☐

DATE CONTROL
LIMITS CALCULATED

Avg = 1.07   UCL = 4.17
                      2.62

LCL = 0.00

Date UNIT

	1	2	3	4	5	6	7	8	9	10	11	12	13	14	15	16	17	18	19	20	21	22	23	24	25

Number of imperfections

5.0
4.0
3.0
2.0
1.0

UCL (OUTER)

UCL (INNER)

Centerline

LCL

Type of Discrepancy

	1	2	3	4	5	6	7	8	9	10	11	12	13	14	15	16	17	18	19	20	21	22	23	24	25
1																									
2																									
3																									
4																									
5																									
6																									
7																									
8																									
9																									
10																									

Total Discrepancies	1.07	.33	0.00	1.16	2.08	.69	.83	.63	.76	.56	.48	2.14	.56	1.04	.93	2.76	4.17	1.11	.69	.83	2.27	0.00	.29	.69	.67
Average/% Discrepancies																									
Sample Size (n)																									

FIGURE 9.36 Revised Control Chart for Rate of Weld Imperfections

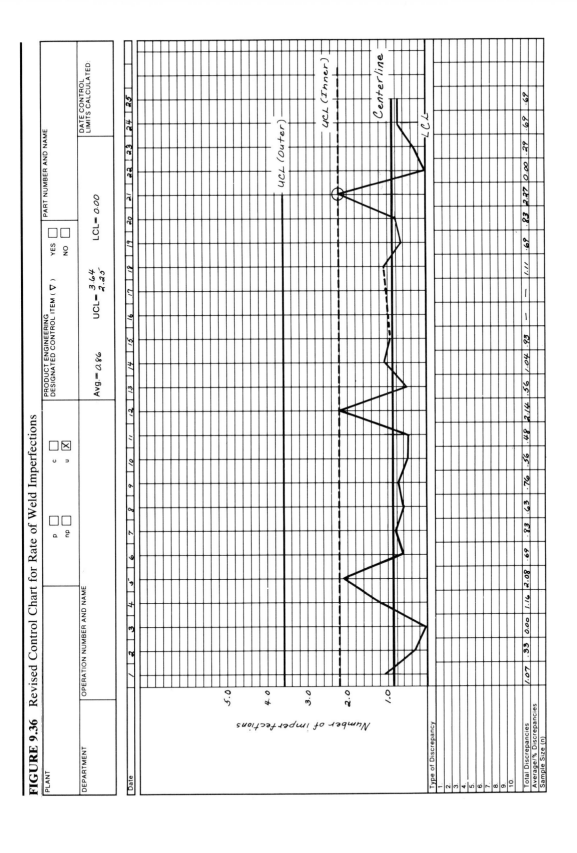

special source of variation, and the process now appears stable and in a state of statistical control.

INDIVIDUALS CHARTS

When attribute data does not lend itself to either a p chart, np chart, c chart, or u chart, individuals charts may be appropriate. The assumptions needed for the other attribute control charts are based on specific, commonly encountered probability distributions. When attribute data cannot safely be assumed to follow a binomial distribution or a Poisson distribution, an individuals chart consisting of a single measurement and a moving range portion is generally used. This approach may be taken as long as the process does not display a gross departure from normality. The individuals charts can be used for either attribute or variables data; we will discuss them further in Chapter 10 in our analysis of variables control charts.

As the Deming (PDCA) cycle progresses, attribute control charts give way to variables control charts. This follows for many reasons, the most obvious being that as the count of the number of defectives or defects shrinks, the subgroup sizes needed to detect them will grow increasingly large. These large subgroups are more costly, and the logical way to continue to improve the process is to determine and measure the appropriate variables instead of the number of defective units or defects. Then variables control charts can be employed for continued process improvement. As we shall see in Chapter 10, variables control charts generally have two parts: one charting the process variability and one charting the process average. Individuals charts introduce this notion by using the variability between the successive individual measurements as an indicator of process variability rather than depending on the underlying assumptions of the distributional form of the process.

Conditions for Use

Single measurement data that do not stray too far from a normal pattern are likely candidates for individuals charting. Caution must be exercised, however, as individuals charts are not as sensitive to changes in the process as some of the variables charts discussed in Chapter 10. It is sometimes worthwhile, therefore, to attempt to form rational subgroups of observations and use a chart such as the x-bar and R chart.

If the data do not strictly follow a Poisson distribution, neither c charts nor u charts are appropriate. Nevertheless, the standard error of the average count may still be approximated using the square root of the average count itself because of the rather robust nature of the Poisson distribution. As a rule of thumb, when the single measurements are whole numbers, the individuals chart can be safely used if the average count is at least 2.00.[2] This is because an upper control limit placed at three standard errors above the centerline will include counts of six or fewer beneath the upper control limit. That is, $2 + 3\sqrt{2} = 6.24$, so the fact that the data are discrete in nature will only slightly affect the structure of the control chart. The detrimental effects of the data being discrete begin only when the standard deviation is smaller than the smallest unit of measurement.[3] Hence, individuals charts can be used whenever attribute data, with an average count of at least 2.00, do not stray too far from normality; when the standard deviation of the data is larger than the smallest unit of measurement; and when the data fail to fulfill the requirements for the other attribute charts.

The Sampling Distribution

Each subgroup in an individuals chart is made up of a single number. For attribute data, these numbers are integers (counts of the number of defects or defectives), provided that the area of opportunity remains constant from observation to observation. If the area of opportunity varies from observation to observation, then, as we will see later on in this section, each data point is divided by the area of opportunity and is represented as a ratio. In either case, subgroups consist of single measurements, so there is no variation within the subgroups themselves. Hence, the estimate of the process variation must be derived from the variation between the successive observations and then employed to estimate the standard error for use in calculating control limits.

The control chart has two parts: one for the process mean and one for the process variability. These are analogous to the x-bar and R charts introduced in Chapter 8. As with the variables charts, it is first necessary to ascertain stability with respect to the variability. In Chapter 7 we saw that the distribution of subgroup ranges has an average, given by Equation 7.17, of $\bar{R} = d_2\sigma$, where σ is the process standard deviation for a normally distributed process. Hence, σ can be estimated using $\bar{R}/d_2$ if the process is distributed normally. In Chapter 8 we saw that control limits for the x-bar and R charts could then be formed using this estimate of the process standard deviation. We follow an analogous procedure for the two portions of the individuals chart. For the range portion of the individuals chart, as there is no variation within the subgroups, we estimate the process standard deviation by calculating a series of moving ranges as the absolute value of the difference between each measurement and the one immediately preceding it:

$$R_i = |x_i - x_{i-1}| \tag{9.27}$$

This produces a sequence of $k - 1$ moving range values, where k is the number of single measurements. We cannot calculate a moving range for the first observation because no point immediately precedes it. We treat each moving range as if it were the range of a subgroup of size 2. From these moving ranges, the average, $\overline{R}$, the centerline for the moving range portion of the control chart, is calculated.

$$\text{centerline(moving range)} = \overline{R} = \Sigma R_i/(k - 1) \qquad (9.28)$$

The process standard deviation, σ, is estimated as $\overline{R}/d_2$, where d_2 for a subgroup of size 2 is, from Table 8, 1.128. Upper and lower control limits for the moving range chart can be calculated as they are for the R chart, from Equations 8.5 and 8.6. However, as the subgroup size for calculating the subgroup range, R, is 2, from Table 8 the value for D_3 will always be 0.00 and for D_4 the value will always be 3.267, so that

$$\text{UCL(moving range)} = D_4\overline{R} = 3.267\ \overline{R} \qquad (9.29)$$

$$\text{LCL(moving range)} = 0.00 \qquad (9.30)$$

After the control chart for the moving ranges has been established, if the chart does not indicate a lack of statistical control, the chart for the single measurements may be constructed. Under the stated assumption that the single measurements, x, are not too far from a normal distribution, the centerline would be the average of the single measurements, $\overline{x}$, and the control limits would be formed as:

$$\text{UCL(x)} = \overline{x} + 3\sigma$$

and

$$\text{LCL(x)} = \overline{x} - 3\sigma$$

Recall that we estimate the process standard deviation by $\overline{R}/d_2$, so that the control limits for the single measurements become:

$$\text{UCL(x)} = \overline{x} + 3\overline{R}/d_2$$

and

$$\text{LCL(x)} = \overline{x} - 3\overline{R}/d_2$$

and as the subgroup size is 2, from Table 8 $d_2 = 1.128$, so that $3/1.128 = 2.66$, and these equations become:

$$\text{UCL(x)} = \overline{x} + (2.66)\overline{R} \qquad (9.31)$$

$$\text{LCL(x)} = \overline{x} - (2.66)\overline{R} \qquad (9.32)$$

Constructing an Individuals Chart: An Example

In the manufacture of a certain variety of blue printing ink, each batch of the product is repeatedly passed over a three-roller mill until a desired fineness and consistency is achieved. The number of passes over the mill

FIGURE 9.37 Number of Mill Passes Required

Batch Number	Number of Mill Passes	Moving Range	Batch Number	Number of Mill Passes	Moving Range
1	3	—	14	2	2
2	4	1	15	2	0
3	2	2	16	3	1
4	3	1	17	2	1
5	2	1	18	1	1
6	2	0	19	3	2
7	3	1	20	2	1
8	2	1	21	2	0
9	1	1	22	3	1
10	3	2	23	2	1
11	2	1	24	4	2
12	3	1	25	2	2
13	4	1	Total	62	27

varies and is recorded. Figure 9.37 lists the results and moving range values for 25 batches.

Notice that by necessity there will be one less moving range value than there are original measurements because a moving range value cannot be calculated for the first data point.

Using Equation 9.28 the average of the 24 moving ranges yield a centerline value of

$$\overline{R} = 27/24 = 1.13$$

The control limits for the moving range portion can be calculated using Equations 9.29 and 9.30:

$$UCL(\text{moving range}) = D_4\overline{R} = (3.267)(1.13) = 3.69$$

$$LCL(\text{moving range}) = 0.00$$

The control chart is shown in Figure 9.38(a). There are no points beyond the control limits, nor are there any other indications of a lack of control; so the single measurements portion may be constructed using Equations 9.31 and 9.32.

The centerline is the average of the 25 passes over the mill:

$$\overline{x} = 62/25 = 2.48$$

$$UCL(x) = 2.48 + (2.66)(1.13) = 5.49$$

$$LCL(x) = 2.48 - (2.66)(1.13) = -0.53 \qquad (\text{use } 0.00)$$

FIGURE 9.38 Individuals Chart for Number of Mill Passes Required

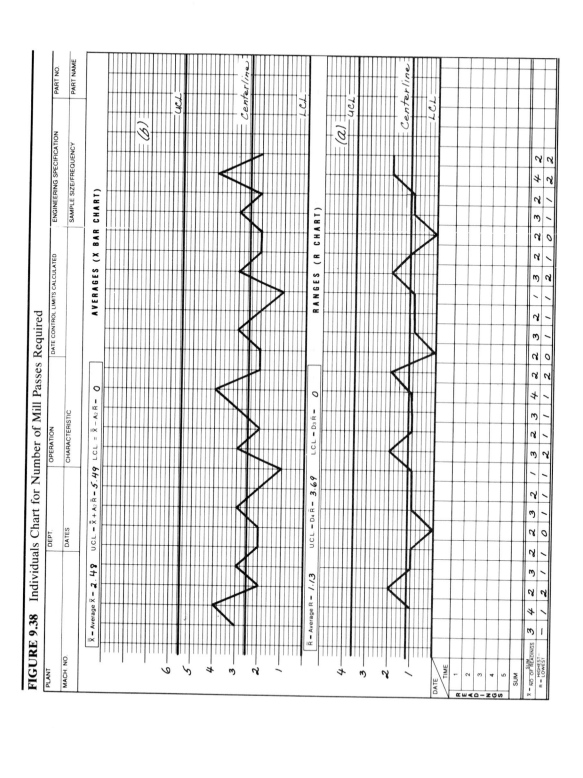

The control chart is shown in Figure 9.38(b); here again there are no points beyond the control limits, nor are there any other patterns indicating a lack of control.

Because we are charting single measurements rather than subgroup means, it is necessary to establish that these measurements are approximately normally distributed. Given a sufficiently large number of observations we may examine the distribution of the data using the methods of Chapter 6. As a rule of thumb, it is usually best to have at least 100 single observations whenever possible as a basis for constructing these charts. Additionally, circumstances may arise wherein the area of opportunity from subgroup to subgroup varies. As was mentioned earlier, when this is the case, the counts in the data should be expressed as a fraction or ratio of the number of events observed to the area of opportunity.

LIMITATIONS OF ATTRIBUTE CONTROL CHARTS

As we have mentioned, as processes improve and defects or defectives become rarer, the number of units that must be examined to find one or more of these events increases. If we consider a p chart where the average fraction of nonconforming items is 0.005, then on average we would need to examine 200 units to have an average count of just 1.00. In the extreme, in order to maintain a reasonable average count as the area of opportunity grows, 100 percent inspection becomes the rule. This implies inspecting all of the items and sorting those that conform to some specification from those that do not. Not only is this inspection costly, but it is equivalent to accepting the fact that the process is producing a constant fraction of its output as defective and will continue to do so. Hence, attribute control charts are limited in terms of the level of process improvement they make possible. Additional process improvement is possible with variables control charts, to be discussed in Chapter 10.

Another disadvantage of using attribute control charts is that if special variation from several different sources is present, it is difficult to identify and isolate the special sources individually. It is possible for one or more of these special sources to mask another, resulting in a process that appears to be stable but is really operating under the influence of several special sources of variation. As the number of special sources of variation increases, their tendency to mask one another can grow, resulting in further difficulties in the future.

On the other hand, variables control charts use numerical measurements, which make them more revealing and powerful than attribute control charts. Attribute data fails to reveal by how much a unit is beyond an upper or a lower specification limit; it only indicates whether a given unit conforms. Therefore the attribute data will not provide as clear a direction for process improvement as will variables data.

SUMMARY

Attribute control charts can be broadly classified into three groups: p charts and np charts, based on binomially distributed counts; u charts and c charts, generally based on a Poisson distribution; and individuals charts, an alternate technique when attribute data are neither binomial nor Poisson in nature.

The p chart can help stabilize a process by providing indications of a lack of statistical control in some characteristic measured as a proportion of output. Subgroup sizes may be constant or may vary from subgroup to subgroup.

The np chart is mathematically equivalent to a p chart; however, the number rather than the proportion of items with the characteristic of interest is charted. Subgroup sizes are generally held constant for each subgroup for the np chart.

A c chart is used when a given single unit of output may have multiple events, such as the number of defects in an appliance or in a roll of paper. The c chart helps stabilize the number of events when the area of opportunity in which the events may occur remains the same for each unit of output from the system.

The u chart is used when counts of the number of events are to be control charted and the area of opportunity varies from unit to unit.

Individuals charts can be used for attributes data when the assumptions needed for any of the other attributes control charts are inappropriate. Their use may herald a change to a variables control chart for the process.

While attribute control charts help identify special process variation, they are only a milestone on the road to never-ending improvement; to continue on the journey, variables control charts must be instituted and used.

EXERCISES

9.1 A manufacturer of wood screws periodically examines screw heads for burrs. Subgroups of 300 screws are selected and examined using a carefully designed procedure. For the observations following:

Obs. No.	Freq.	Proportion with Burrs	Obs. No.	Freq.	Proportion with Burrs
1	19	0.063	11	7	0.023
2	16	0.053	12	13	0.043
3	11	0.037	13	17	0.057
4	6	0.020	14	29	0.097
5	22	0.073	15	1	0.003
6	2	0.007	16	2	0.007
7	4	0.013	17	19	0.063
8	7	0.023	18	28	0.093
9	5	0.017	19	24	0.080
10	27	0.090	20	23	0.077

a. Find the centerline and standard error.
b. Find the control limits and zone boundaries.
c. Identify any indications of a lack of statistical control and state the reason you believe that a lack of control is indicated.

9.2 A bank is studying the proportion of transactions made using an automated teller machine (ATM). The following represents a series of days and the number of transactions using the ATM:

Day	Total Number of Transactions	Number of ATM Transactions
1	320	87
2	356	92
3	280	75
4	325	109
5	344	69
6	410	99
7	385	120
8	324	86
9	367	111
10	312	90
11	276	86
12	342	106
13	387	65
14	312	91
15	390	131
16	354	78
17	322	89
18	353	98
19	317	81

Day	Total Number of Transactions	Number of ATM Transactions
20	374	58
21	409	104
22	366	81
23	298	87
24	311	72
25	339	84

a. Find the appropriate centerline for the control chart.
b. Determine the control limits and zone boundaries.
c. Are there any indications of a lack of control? What are the indications, and why do they indicate a lack of control?

9.3 Samples of 90 retainer rings are examined for the fraction nonconforming. The results for 30 consecutive days are:

Day	Fraction Nonconforming	Day	Fraction Nonconforming	Day	Fraction Nonconforming
1	0.12	6	0.06	11	0.15
2	0.09	7	0.17	12	0.14
3	0.03	8	0.14	13	0.13
4	0.08	9	0.15	14	0.02
5	0.14	10	0.17	15	0.09

Day	Fraction Nonconforming	Day	Fraction Nonconforming	Day	Fraction Nonconforming
16	0.15	21	0.12	26	0.03
17	0.17	22	0.12	27	0.15
18	0.07	23	0.09	28	0.13
19	0.03	24	0.08	29	0.16
20	0.12	25	0.02	30	0.13

Does the process appear to be in a state of statistical control with respect to the fraction nonconforming?

9.4 A given model of a large radar dish represents an area of opportunity in which nonconformities may occur. Results for 25 such assemblies are:

Assembly Number	Number of Nonconformities	Assembly Number	Number of Nonconformities
1	25	14	75
2	60	15	24
3	28	16	50
4	65	17	70
5	91	18	56
6	56	19	21
7	40	20	88
8	54	21	34
9	90	22	82
10	44	23	53
11	62	24	102
12	81	25	64
13	70		

a. Determine the centerline and control limits for the c chart.

b. Are there any indications of a lack of control? What are the indications, and why do they indicate a lack of control?

9.5 A manufacturer of a particular grade of copper tubing experiences flaws in 500 foot coils. The counts of the number of flaws are:

Coil	Count	Coil	Count	Coil	Count
1	7	11	5	21	13
2	2	12	1	22	8
3	6	13	7	23	0
4	4	14	3	24	11
5	5	15	0	25	19
6	15	16	18	26	0
7	4	17	11	27	0
8	0	18	5	28	1
9	0	19	7	29	6
10	2	20	0	30	0

a. Determine the centerline, control limits, and zone boundaries for the c chart.

b. Are there any indications of a lack of control? What are the indications, and why do they indicate a lack of control?

9.6 A large publisher counts the number of keyboard errors that make their way into finished books. The number of errors and the number of pages in the last 26 publications are:

Book Number	Number of Errors	Number of Pages	Book Number	Number of Errors	Number of Pages
1	49	202	14	48	612
2	63	232	15	50	432
3	57	332	16	41	538
4	33	429	17	45	383
5	54	512	18	51	302
6	37	347	19	49	285
7	38	401	20	38	591
8	45	412	21	70	310
9	65	481	22	55	547
10	62	770	23	63	469
11	40	577	24	33	652
12	21	734	25	14	343
13	35	455	26	44	401

a. Determine the centerline and control limits for the u chart.

b. Are there any indications of a lack of control? What are the indications, and why do they indicate a lack of control?

9.7 A large mail order house receives order forms from customers by mail. Each incoming order form contains an integral number of ordered items.

Order Number	Number of Items	Order Number	Number of Items	Order Number	Number of Items
1	1	11	2	21	1
2	5	12	6	22	7
3	4	13	3	23	8
4	7	14	1	24	4
5	5	15	6	25	1
6	1	16	1	26	1
7	1	17	5	27	4
8	7	18	6	28	6
9	1	19	1	29	1
10	1	20	2	30	2

a. Determine the centerline and control limits for each portion of the individuals chart.

b. Are there any indications of a lack of control? What are the indications, and why do they indicate a lack of control?

NOTES

1. Write American Society for Quality Control Inc., 310 West Wisconsin Avenue, Milwaukee, Wi. 53203.

2. Donald J. Wheeler and David S. Chambers, *Understanding Statistical Process Control* (Knoxville, Tenn.: Statistical Process Controls, Inc., 1986), p. 241.

3. Ibid., p. 241.

Chapter 10

Variables Control Charts

INTRODUCTION

Variables data consist of measurements, such as weight, length, width, height, time, temperature, or electrical resistance. Variables data contain more information than attribute data, which merely classify the output of a process as conforming or nonconforming, or else count the number of imperfections. Furthermore, because variables control charts deal with measurements themselves, they do not mask valuable information and therefore are more powerful than attribute charts. They use all information contained in the data; this alone makes variables charts preferable when a choice is possible.

There are four principal types of variables control charts: the x-bar and R chart, the x-bar and s chart, the median chart, and the individuals chart. All are used in the never-ending spiral of process improvement.

Quality consciousness must increase for a firm to continue to reduce the difference between customer needs and process performance. Early efforts to control the quality of output often find firms segregating nonconforming production from conforming production. Nonconforming production is then reworked, discarded, downgraded, or otherwise removed from the mainstream of the process output. Each item so removed incurs a greater overall cost to the firm than the production of a conforming item. This is because special attention is required that is, more often than not, labor intensive and costly. Furthermore, the firm cannot measure the harm done to its reputation and self-esteem by even occasionally shipping a defective item.

As a greater awareness of the need to improve quality grows, organizations will use attribute control charts to control and stabilize factors such as the proportion of defectives or the number of defects. However, as that proportion becomes smaller as a result of these efforts, organizations seeking continued process improvement will eventually begin to use variables measurement and variables control charts. As discussed in Chapter 9, a controlled process producing a relatively low fraction of defects or defective units will require relatively large subgroups to detect those defects. The only way to overcome the need for larger and larger subgroups is to continue upward on the spiral of quality consciousness through the use of variables control charts.

VARIABLES CHARTS AND THE PDCA CYCLE

As with the attributes control charts, the PDCA cycle provides both an important guideline for proceeding with variables control charts and the mechanism for improved quality through continued process improvement.

Plan

The object or purpose of the control chart must be carefully delineated to effectively use the control chart as a vehicle to reduce the difference between customer needs and process performance. A plan must be established that clearly shows what will be control charted, why it will be control charted, where it will be control charted, when it will be control charted, who will do the control charting, and how it will be control charted.

Consequently, the firm must decide which variables to measure. These decisions require the cooperation and input of all those directly or indirectly involved with the process, such as operators, foremen, supervisors, and engineers. Some of the techniques that Chapter 12 discusses, such as brainstorming, cause-and-effect diagrams, check sheets, or Pareto diagrams, may be useful in selecting the process variables that will decrease the difference between customer needs and process performance. Often some feature of the process that has been a source of trouble (resulting in extra cost in scrap or rework) and has failed to yield to corrective efforts is a good place to start. Starting *far upstream,* or near the beginning of a process, will often produce the most dramatic results and may offer the greatest opportunity to alter the factors that are at the root of downstream special causes of variation.

Subgroup selection is crucial to the proper usage of the control chart. Subgroups should be chosen rationally, as discussed in Chapter 8. The choice should minimize the variation within the subgroups. This will allow isolation of special variation between the subgroups while capturing the inherent process variation within the subgroups. The frequency with which subgroups are selected will help minimize the amount of variation between the subgroups by isolating batches, shifts, production runs, machines, or people. This enables the special variation to be captured, isolated, observed, and analyzed.

Decisions concerning rational subgroups often require the combined knowledge of those directly involved with the process and an experienced statistician. Rational subgroup selection may require a trial-and-error solution, may produce several false starts, and may very well require a great deal of patience.

The method by which the measurements are to be made must be studied carefully. Operational definitions, as discussed in Chapter 3, must be constructed and communicated to those who will be involved with the data collection. At this point any forms to be used for data collection should be selected or designed and the responsibility for construction of the charts assigned.

Do

Data collection and the calculation of control chart statistics constitute the Do stage for constructing variables control charts. It is usually best to collect at least 20 subgroups before beginning to construct a control chart. On rare occasions fewer than 20 subgroups may be used, but a control chart should almost never be attempted with fewer than 10 subgroups.

As explained in Chapter 8, variables charts are made up of two parts: one charts the process variability, and one charts the process location. For instance, in the x-bar and R chart, the R values, or subgroup ranges, are used to track the variability. The x-bar values, or subgroup averages, are used for the process location. As the control limits of the portion of the control chart measuring location are based on the average variability ($\overline{R}$ in the x-bar and R chart), the variability-measuring portion of the control chart must be constructed and evaluated first. Only if there is stability in the variability portion can the location portion of the control chart be constructed. For each of the variables control charts, the estimate of the process standard deviation is based on the average value of the measure of variability used for the subgroups. If the process is not stable, its variability will not be predictable, which will result in unreliable estimates of the process standard deviation. If these estimates are used to construct control limits, those limits will also be unreliable and will fail to reveal the presence of special sources of variation when they exist.

After the initial data set has been collected, the centerline, control limits, and zone boundaries (if applicable) should be computed for both portions of the control chart. These should be entered onto the control charts along with the collected set of data points. First, the variability portion of the control chart should be examined for indications of special variation. If there are special causes of variation, they must be studied and the process stabilized before the location portion of the control chart is analyzed. This means skipping ahead to the Check stage of the PDCA cycle before completing the Do stage.

If there are no indications of a lack of control in the variability portion of the control chart, the location portion of the chart can be analyzed in the Check stage of the PDCA cycle. This completes the Do stage.

Check

Indications of a lack of control, such as patterns of the type introduced in Chapter 8 and discussed in detail in Chapter 11, are studied in the Check stage of the PDCA cycle. Indications of special sources of variation may be found in the chart dealing with variation, the chart dealing with location, or in both. Whether the variation is common process variation or variation resulting from special causes, once we have found and identified them we proceed to the Act stage to set policy to formalize process improvements resulting from the analysis of the control chart.

All aspects of the control chart must be reviewed periodically and changes made where appropriate. If the process itself has been changed in some way, the control limits should certainly be recomputed and the analysis of the process begun anew.

Act

If the variation found in the Check stage results only from common causes, then efforts to reduce that variation must focus on changes in the process itself. When indications of special causes of variation are present, the cause or causes of that special variation should be removed if the variation is detrimental or incorporated into the process if the variation is beneficial.

The focus of the Act stage is on formalizing policy that results directly from the prior study of the causes of process variation. This will lead to a reduction in the difference between customer needs and process performance.

Last in the Act stage, the purpose of the control chart must be reconsidered by returning to the Plan stage. This will help to maintain a focus on improvement and point out those areas that can be most beneficial in reducing the difference between customer needs and process performance.

SUBGROUP SIZE AND FREQUENCY

The selection of an appropriate control chart depends, in part, on the subgroup size. In turn, many factors affect the decision of ideal subgroup size. Large subgroups are, of course, more expensive than small ones; nevertheless, large subgroups lead to tighter control limits so long as the subgroups are selected rationally to minimize the within-group variation.

Traditionally, subgroup sizes of four or five have been used ever since they were first suggested by W. A. Shewhart.[1] Subgroups should be large enough to detect points or patterns indicating a lack of control when a lack of control exists. This requires statistical expertise and an understanding of the process under study. Subgroups of two or three are used when the cost of sampling is relatively high. Individuals charts may be used when only one measurement is available or appropriate as a subgroup. Subgroups of 6 to 14 are occasionally used when we want the control chart to be very sensitive to changes in the process average. When subgroup sizes exceed 10, we will see that x-bar and s charts are generally used instead of x-bar and R charts.

The frequency with which subgroups are selected—or how often they are selected—depends upon the particular application. If quick action is required, the frequency should be greater. A stable process merely being monitored will require less frequent subgroup selection than one being studied or being brought into a state of statistical control for the first time. There are no hard and fast rules for determining the frequency of subgroup selection, and decisions are generally made on the basis of the costs and benefits to be gained.

x-BAR AND R CHARTS

As the name implies, the x-bar and R chart uses the subgroup range, R, to chart the process variability, and the subgroup average, $\bar{x}$, to chart the process location. Chapter 8 provided an initial look at x-bar and R charts. There we saw that the periodic selection of small subgroups of process output could be very useful in process stabilization and improvement.

Stable processes yield subgroups that will behave according to predictable laws of probability, enabling us to construct an x-bar and R chart. The parameters of the probability distributions of the two characteristics, $\bar{x}$ and R, can be estimated by relatively simple procedures. As we discussed in Chapter 7, these estimates are related. The estimates of the standard errors of the distributions of both R and $\bar{x}$ are based on the average subgroup range, $\bar{R}$. This not only simplifies the estimation procedure but has a direct impact on the way in which the control charts must be constructed and analyzed.

The Range Portion

The subgroup ranges are used as a measure of dispersion. When the range for each subgroup is calculated, the result is a sequence of values used to construct both the range portion and the x-bar portion of the control chart.

The probability distribution of the process output has a mean and standard deviation that are estimated using the average of the subgroup averages, $\bar{\bar{x}}$, and the average value of the subgroup ranges, $\bar{R}$. Recall from Equation 7.17 that for normally distributed processes:

$$\bar{R} = d_2\,\sigma$$

so that the process standard deviation, σ, can be estimated as $\sigma = \bar{R}/d_2$, where values for d_2 are a function of subgroup size and can be found in Table 8.

It would not be desirable to use all of the data points together, rather than as subgroups, to estimate the process standard deviation because that would include the variation both within the subgroups and between the subgroups in our measure of dispersion. Hence, our efforts would fail to isolate the between subgroup variation, and the control chart would be useless.

The sampling distribution of $\bar{R}$, which was discussed in Chapter 7, has a standard error given by Equation 7.18 as:

$$\sigma_R = d_3\sigma$$

where the value of d_3 depends on subgroup size, and can be found in Table 8. By substitution, the standard error of $\bar{R}$ can be written as:

$$\sigma_R = d_3(\bar{R}/d_2) \tag{10.1}$$

In Chapter 8, these results were used to develop Equations 8.4, 8.5, and 8.6 for the centerline, and upper and lower control limits for the range portion of the control chart. The relationships are:

$$\text{centerline(R)} = \bar{R} = \Sigma\,R/k \tag{10.2}$$

$$\text{UCL(R)} = D_4\,\bar{R} \tag{10.3}$$

$$\text{LCL(R)} = D_3\,\bar{R} \tag{10.4}$$

where k is the number of subgroups and where $D_3 = [1 - 3(d_3/d_2)]$ and $D_4 = [1 + 3(d_3/d_2)]$ are given in Table 8.

In the development of the upper and lower control limits, three times the estimated standard error of the subgroup ranges, σ_R, is added to and subtracted from the average range value, $\bar{R}$, yielding the factors D_4 and D_3. Recall that the zone boundaries for zones A, B, and C, discussed in Chapter 8, are positioned at one and two standard errors on either side of the control chart centerline. Using the estimated standard error of $\overline{R}$ from Equation 10.1, the zone boundaries are given by:

$$\begin{aligned}\text{boundary between} \quad &= \overline{R} - 2d_3\,(\overline{R}/d_2) \\ \text{lower zones A and B} &= \overline{R}(1 - 2d_3/d_2)\end{aligned} \qquad (10.5)$$

$$\begin{aligned}\text{boundary between} \quad &= \overline{R} - d_3\,(\overline{R}/d_2) \\ \text{lower zones B and C} &= \overline{R}(1 - d_3/d_2)\end{aligned} \qquad (10.6)$$

When the result of Equations 10.5 or 10.6 is a negative number, 0.00 is used instead, as negative ranges are meaningless. Also,

$$\begin{aligned}\text{boundary between} \quad &= \overline{R} + d_3\,(\overline{R}/d_2) \\ \text{upper zones B and C} &= \overline{R}(1 + d_3/d_2)\end{aligned} \qquad (10.7)$$

$$\begin{aligned}\text{boundary between} \quad &= \overline{R} + 2d_3\,(\overline{R}/d_2) \\ \text{upper zones A and B} &= \overline{R}(1 + 2d_3/d_2)\end{aligned} \qquad (10.8)$$

Assuming that the range portion of the control chart is stable, the x-bar portion may be built.

The x-Bar Portion

The string of subgroup averages computed from the data is used to construct the x-bar portion of the control chart. As discussed in Chapter 7, the sampling distribution of the mean has mean and standard error given by Equations 7.14 and 7.15, respectively, where σ is the process standard deviation. Thus, the centerline of the x-bar control chart is found using Equation 5.2 by taking the average of the subgroup averages, $\overline{\overline{x}}$:

$$\text{centerline}(\overline{x}) = \overline{\overline{x}} = \Sigma\,\overline{x}/k \qquad (10.9)$$

Using a multiple of three standard errors to construct the control limits, we have:

$$\text{UCL}(\overline{x}) = \overline{\overline{x}} + 3\sigma/\sqrt{n}$$

From Equation 7.17, $\overline{R} = d_2\sigma$, so that $\sigma = \overline{R}/d_2$, and

$$\text{UCL}(\overline{x}) = \overline{\overline{x}} + 3\overline{R}/(d_2\sqrt{n})$$

or, more conveniently:

$$\text{UCL}(\overline{x}) = \overline{\overline{x}} + A_2\overline{R} \qquad (10.10)$$

$$\text{LCL}(\overline{x}) = \overline{\overline{x}} - A_2\overline{R} \qquad (10.11)$$

where $A_2 = 3/(d_2\sqrt{n})$ is tabulated as a function of subgroup size in Table 8.

The product $A_2\overline{R}$ represents three times the standard error of the distribution of the subgroup means. This is useful in forming the A, B, and C zones, used in this chart as well in helping to detect patterns indicating a lack of statistical control. The zone boundaries are placed on both sides of the centerline at a distance of one and two times the standard error, respectively.

$$\text{boundary between lower zones A and B} = \bar{\bar{x}} - (2/3)A_2\bar{R} \qquad (10.12)$$

$$\text{boundary between lower zones B and C} = \bar{\bar{x}} - (1/3)A_2\bar{R} \qquad (10.13)$$

$$\text{boundary between upper zones B and C} = \bar{\bar{x}} + (1/3)A_2\bar{R} \qquad (10.14)$$

$$\text{boundary between upper zones A and B} = \bar{\bar{x}} + (2/3)A_2\bar{R} \qquad (10.15)$$

x-Bar and R Charts: An Example

Consider the case of a manufacturer of circuit boards for personal computers. Various components are to be mounted on each board and the boards eventually slipped into slots in a chassis. The overall length of the boards is very important to assure a proper fit, and this dimension has been targeted as an important item to be stabilized. The boards are cut from large sheets of material by a single rotary cutter continuously fed from a hopper. At the request of the firm's customer, a decision has been made to create a control chart for the length of the circuit boards produced by the process.

After input from many of the individuals involved with the process, it is decided to select five units every hour from the production output. Each group of five items represents a subgroup. This manner of subgroup selection is most likely to isolate the variation over time between the subgroups and, therefore, capture only common process variation within the subgroups.

The boards are measured using an operationally defined method; Figure 10.1 lists the resulting lengths.

The average and the range for each subgroup of five elements in Figure 10.1 have been computed and are shown in the last two columns on the right. This arrangement of the data will be used to determine whether special sources of variation between the subgroups are evident in measurement-to-measurement changes over time.

The construction of the control chart begins with the range portion. The centerline is found by taking the average of the subgroup ranges using

FIGURE 10.1   Cut Circuit Board Lengths

Time	Sample Number	1	2	3	4	5	Average $\bar{x}$	Range R
9 A.M.	1	5.030	5.002	5.019	4.992	5.008	5.010	0.038
10	2	4.995	4.992	5.001	5.011	5.004	5.001	0.019
11	3	4.988	5.024	5.021	5.005	5.002	5.008	0.036
12	4	5.002	4.996	4.993	5.015	5.009	5.003	0.022
1 P.M.	5	4.992	5.007	5.015	4.989	5.014	5.003	0.026
2	6	5.009	4.994	4.997	4.985	4.993	4.996	0.024
3	7	4.995	5.006	4.994	5.000	5.005	5.000	0.012
4	8	4.985	5.003	4.993	5.015	4.988	4.997	0.030
5	9	5.008	4.995	5.009	5.009	5.005	5.005	0.014
6	10	4.998	5.000	4.990	5.007	4.995	4.998	0.017
7	11	4.994	4.998	4.994	4.995	4.990	4.994	0.008
8	12	5.004	5.000	5.007	5.000	4.996	5.001	0.011
9	13	4.983	5.002	4.998	4.997	5.012	4.998	0.029
10	14	5.006	4.967	4.994	5.000	4.984	4.990	0.039
11	15	5.012	5.014	4.998	4.999	5.007	5.006	0.016
12	16	5.000	4.984	5.005	4.998	4.996	4.997	0.021
1 A.M.	17	4.994	5.012	4.986	5.005	5.007	5.001	0.026
2	18	5.006	5.010	5.018	5.003	5.000	5.007	0.018
3	19	4.984	5.002	5.003	5.005	4.997	4.998	0.021
4	20	5.000	5.010	5.013	5.020	5.003	5.009	0.020
5	21	4.988	5.001	5.009	5.005	4.996	5.000	0.021
6	22	5.004	4.999	4.990	5.006	5.009	5.002	0.019
7	23	5.010	4.989	4.990	5.009	5.014	5.002	0.025
8	24	5.015	5.008	4.993	5.000	5.010	5.005	0.022
9	25	4.982	4.984	4.995	5.017	5.013	4.998	0.035
						Totals	125.029	0.569

Equation 10.2:

$$\bar{R} = 0.569/25 = 0.023$$

Not only does this value form the centerline for the R portion of the control chart; it forms the basis for estimating the standard error, hence the control limits and zone boundaries as well. Equations 10.3 and 10.4 yield the control limits:

$$UCL(R) = (2.114)(0.023) = 0.049$$

and

$$LCL(R) = (0)(0.023) = 0.000$$

Equations 10.5, 10.6, 10.7, and 10.8 yield values for the zone boundaries:

$$\begin{aligned}\text{boundary between lower zones A and B} &= 0.023[1 - 2(0.864)/2.326] = 0.006\end{aligned}$$

$$\begin{aligned}\text{boundary between lower zones B and C} &= 0.023(1 - 0.864/2.326) = 0.014\end{aligned}$$

$$\begin{aligned}\text{boundary between upper zones B and C} &= 0.023(1 + 0.864/2.326) = 0.032\end{aligned}$$

$$\begin{aligned}\text{boundary between upper zones A and B} &= 0.023[1 + 2(0.864)/2.326] = 0.040\end{aligned}$$

Figure 10.2(a) displays the control chart. There are no indications of any special sources of variation, and the process appears stable with regard to its range.

After the stability of the range has been established, the x-bar portion of the control chart may be constructed. If the range is stable, $\overline{R}$ can be used as a basis for the estimate of the standard error for the x-bar portion of the chart. Equation 10.9 yields a centerline value of

$$\overline{\overline{x}} = 125.029/25 = 5.001$$

The upper and lower control limits can be found using Equations 10.10 and 10.11:

$$\text{UCL}(\overline{x}) = 5.001 + (0.577)(0.023) = 5.014$$

$$\text{LCL}(\overline{x}) = 5.001 - (0.577)(0.023) = 4.988$$

The zone boundaries are found using Equations 10.12, 10.13, 10.14, and 10.15:

$$\begin{aligned}\text{boundary between lower zones A and B} &= 5.001 - (2/3)(0.577)(0.023) = 4.992\end{aligned}$$

$$\begin{aligned}\text{boundary between lower zones B and C} &= 5.001 - (1/3)(0.577)(0.023) = 4.997\end{aligned}$$

$$\begin{aligned}\text{boundary between upper zones B and C} &= 5.001 + (1/3)(0.577)(0.023) = 5.005\end{aligned}$$

$$\begin{aligned}\text{boundary between upper zones A and B} &= 5.001 + (2/3)(0.577)(0.023) = 5.010\end{aligned}$$

These values are shown in Figure 10.2(b). There are no indications of any special sources of variation present between the subgroups with respect to time in the x-bar portion, so we can conclude that this process is stable and its output is predictable, assuming continued stability.

One important note of caution is appropriate here. All of the above estimates and factors are based upon the assumption that the process

FIGURE 10.2 x-Bar and R Chart for Cut Circuit Board Lengths

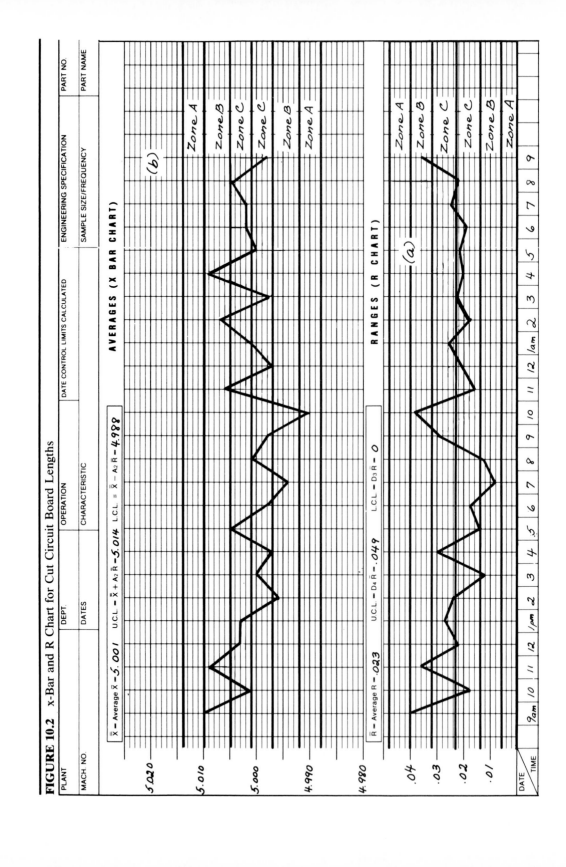

under study is in a state of statistical control. If the process is not in control, these estimates are not reliable; the special causes of variation must be eliminated and the process brought into statistical control before proceeding.

x-Bar and R Charts: Another Example

A manufacturer of high-end audio components buys metal tuning knobs to be used in the assembly of its products. The knobs are produced automatically by a subcontractor using a single machine that is supposed to produce them with a constant diameter. Nevertheless, because of persistent final assembly problems with the knobs, management has decided to examine this process output by requesting that the subcontractor keep an x-

FIGURE 10.3   Tuning Knob Diameters

Time	Sample Number	1	2	3	4	Average $\bar{x}$	Range R
8:30 A.M.	1	836	846	840	839	840.25	10
9:00	2	842	836	839	837	838.50	6
9:30	3	839	841	839	844	840.75	5
10:00	4	840	836	837	839	838.00	4
10:30	5	838	844	838	842	840.50	6
11:00	6	838	842	837	843	840.00	6
11:30	7	842	839	840	842	840.75	3
12:00	8	840	842	844	836	840.50	8
12:30 P.M.	9	842	841	837	837	839.25	5
1:00	10	846	846	846	845	845.75	1
1:30	11	849	846	848	844	846.75	5
2:00	12	845	844	848	846	845.75	4
2:30	13	847	845	846	846	846.00	2
3:00	14	839	840	841	838	839.50	3
3:30	15	840	839	839	840	839.50	1
4:00	16	842	839	841	837	839.75	5
4:30	17	841	845	839	839	841.00	6
5:00	18	841	841	836	843	840.25	7
5:30	19	845	842	837	840	841.00	8
6:00	20	839	841	842	840	840.50	3
6:30	21	840	840	842	836	839.50	6
7:00	22	844	845	841	843	843.25	4
7:30	23	848	843	844	836	842.75	12
8:00	24	840	844	841	845	842.50	5
8:30	25	843	845	846	842	844.00	4
					Totals	21,036.25	129

bar and R chart for knob diameter. Beginning at 8:30 A.M. on a Tuesday, four knobs are selected every half hour. The diameter of each is carefully measured using an operationally defined technique. The average and range for each subgroup are computed; the data, along with these statistics, are shown in Figure 10.3.

The data are arranged this way to attempt to determine whether the differences in the subgroups result from special causes over time. The measurement-to-measurement differences here are arranged to trap special variations over time between the subgroups and confine the common process variation within the subgroups.

Using the subgroup range values and Equation 10.2, the average range can be computed as:

$$\overline{R} = 129/25 = 5.16$$

From this, the control limits can be calculated using Equations 10.3 and 10.4:

$$UCL(R) = (2.282)(5.16) = 11.78$$

$$LCL(R) = 0(5.16) = 0.00$$

The zone boundaries are computed using Equations 10.5, 10.6, 10.7, and 10.8:

$$\text{boundary between lower zones A and B} = 5.16[1 - 2(0.880)/2.059] = 0.75$$

$$\text{boundary between lower zones B and C} = 5.16(1 - 0.880/2.059) = 2.95$$

$$\text{boundary between upper zones B and C} = 5.16(1 + 0.880/2.059) = 7.37$$

$$\text{boundary between upper zones A and B} = 5.16[1 + 2(0.880)/2.059] = 9.57$$

Figure 10.4 illustrates the control chart. The range at subgroup number 23 is beyond the upper control limit. Furthermore, subgroup number 16 is the eighth consecutive point below the centerline and therefore indicates a lack of control by virtue of the fourth rule presented in Chapter 8. Hence, there are two indications of a lack of control.

An investigation reveals that at 7:25 P.M. a water pipe had burst in the lunchroom. The episode was not serious but caused water to leak from the lunchroom onto the floor beneath the machinery involved in the process. This disruption seems to have caused the lack of control observed at subgroup 23. The operators believe this to be a special cause of variation that should not recur once the plumbing has been repaired. The initial study does not reveal any special source of variation for the indication of a lack of control at subgroup 16.

FIGURE 10.4 Initial Range Chart for Tuning Knob Diameters

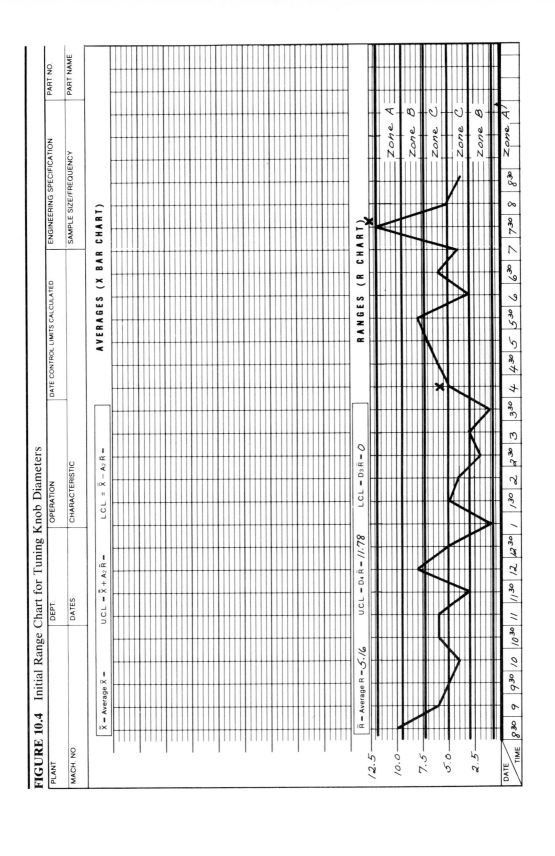

The data for subgroup 23 are then removed from the data set. The repair of the plumbing has permanently removed the conditions leading to this observation. The data for subgroup 16 are left in place, as no special cause of variation can be isolated and removed that would explain its presence. The average value for the subgroup ranges is then recomputed, using Equation 10.2, to reflect the deletion of subgroup 23:

$$\overline{R} = 117/24 = 4.88$$

The revised control limits and zone boundaries are computed using Equations 10.3 through 10.8:

$$UCL(R) = (2.282)(4.88) = 11.14$$

$$LCL(R) = (0)(4.88) = 0.00$$

$$\text{boundary between lower zones A and B} = 4.88[1 - 2(0.880)/2.059] = 0.71$$

$$\text{boundary between lower zones B and C} = 4.88(1 - 0.880/2.059) = 2.79$$

$$\text{boundary between upper zones B and C} = 4.88(1 + 0.880/2.059) = 6.97$$

$$\text{boundary between upper zones A and B} = 4.88[1 + 2(0.880)/2.059] = 9.05$$

The revised R chart appears in Figure 10.5(a); and because the centerline has been shifted, subgroup number 16 is no longer the eighth consecutive point below the centerline. There are no other indications of a lack of control in the data. It is possible that changing the centerline, control limits, and zone boundaries may uncover other indications of a lack of control; but such is not the case here. Hence, the x-bar portion of the chart may be constructed using the average range from the now stable R portion of the chart.

The centerline of the control chart is the average of the subgroup averages, where $\overline{x}$ for subgroup 23 has been subtracted from the total in Figure 10.3. Equation 10.9 yields:

$$\overline{\overline{x}} = 20{,}193.50/24 = 841.40$$

The control limits and the zone boundaries can now be computed using Equations 10.10, 10.11, 10.12, 10.13, 10.14, and 10.15:

$$UCL(\overline{x}) = 841.40 + (0.729)(4.88) = 844.96$$

$$LCL(\overline{x}) = 841.40 - (0.729)(4.88) = 837.84$$

$$\text{boundary between lower zones A and B} = 841.40 - (2/3)(0.729)(4.88) = 839.03$$

$$\text{boundary between lower zones B and C} = 841.40 - (1/3)(0.729)(4.88) = 840.21$$

$$\text{boundary between upper zones B and C} = 841.40 + (1/3)(0.729)(4.88) = 842.59$$

$$\text{boundary between upper zones A and B} = 841.40 + (2/3)(0.729)(4.88) = 843.77$$

The x-bar portion of the control chart has been completed using these values and appears in Figure 10.5(b). Remember from Chapter 8 that a search for indications of a lack of control should always be made from right to left on the control chart; that is, the search should be made by looking backward in time from the present. There are several indications of a lack of control on the control chart of Figure 10.5(b). Four points are beyond the upper control limit. These occurred sequentially from 1:00 P.M. to 2:30 P.M. Also, at 10:00 A.M. (fourth subgroup), the average is 838.00, the second of three consecutive points in the lower zone A. This indicates a lack of control by virtue of Rule 2 of Chapter 8.

There is very little doubt that there is at least one source of special variation acting on this process. Finding that source requires some further investigation in this case. The investigation leads to the discovery that at 12:05 P.M. (just after selection of subgroup number 10), a keyway wedge had cracked and needed to be replaced on the machine. The mechanic who normally makes this repair was out to lunch, so the machine operator made the repair. This individual had not been properly trained for the repair; for this reason, the wedge was not properly aligned in the keyway, and the subsequent points were out of control. Both the operator and the mechanic agree that the need for this repair was not unusual. To correct this problem it is decided to train the machine operator and provide the appropriate tools for making this repair in the mechanic's absence. Furthermore, the maintenance and engineering staffs agree to search for a replacement part for the wedge that will not be so prone to cracking.

No special source of variation can be found for the indication of a lack of control at 10:00 A.M. Indications of special variation will occasionally be found where the special cause will not be identifiable. Failure to identify a special source of variation when one is indicated is no reason to stop analysis; false alarms will happen from time to time.

After the sources of special variation are found and the underlying causes permanently removed from the system, the subgroups generated as a result of those sources are dropped from the data. This changes all of

FIGURE 10.5 x-Bar and R Chart for Tuning Knob Diameters

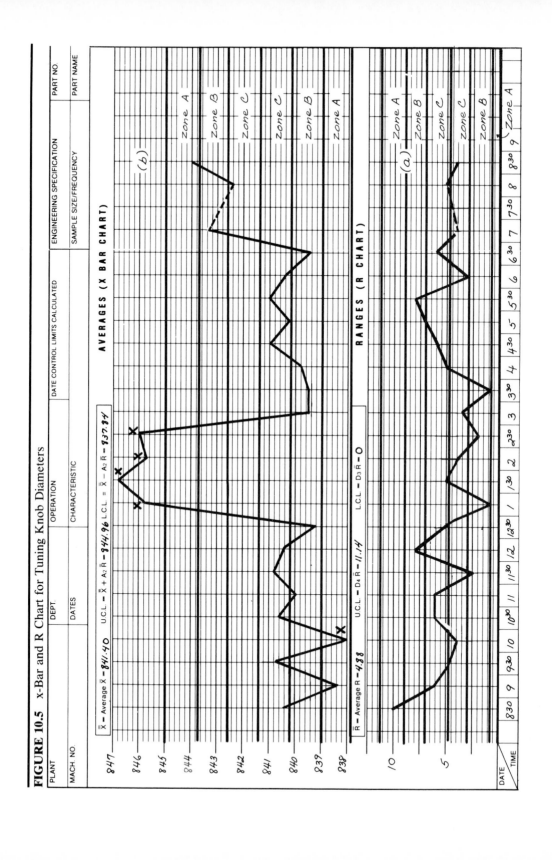

the calculated values for centerlines, control limits, and zone boundaries of the control chart.

Accordingly, for the control knobs, we delete subgroups 10, 11, 12, and 13. We must recompute the values for the relevant statistics.

$$\bar{\bar{x}} = 16,809.25/20 = 840.46$$

$$\bar{R} = 105/20 = 5.25$$

Using these, the control limits and zone boundaries for the revised R chart become:

$$UCL(R) = (2.282)(5.25) = 11.98$$

$$LCL(R) = 0(5.25) = 0.00$$

$$\text{boundary between lower zones A and B} = 5.25[1 - 2(0.880)/2.059] = 0.76$$

$$\text{boundary between lower zones B and C} = 5.25(1 - 0.880/2.059) = 3.01$$

$$\text{boundary between upper zones B and C} = 5.25(1 + 0.880/2.059) = 7.49$$

$$\text{boundary between upper zones A and B} = 5.25[1 + 2(0.880)/2.059] = 9.74$$

The control chart for the range portion is shown in Figure 10.6(a). There are no indications of a lack of control, so the x-bar chart may be constructed. The control limits and zone boundaries are:

$$UCL(\bar{x}) = 840.46 + (0.729)(5.25) = 844.29$$

$$LCL(\bar{x}) = 840.46 - (0.729)(5.25) = 836.63$$

$$\text{boundary between lower zones A and B} = 840.46 - (2/3)(0.729)(5.25) = 837.91$$

$$\text{boundary between lower zones B and C} = 840.46 - (1/3)(0.729)(5.25) = 839.18$$

$$\text{boundary between upper zones B and C} = 840.46 + (1/3)(0.729)(5.25) = 841.74$$

$$\text{boundary between upper zones A and B} = 840.46 + (2/3)(0.729)(5.25) = 843.01$$

The x-bar portion of the control chart is shown in Figure 10.6(b). The last data point, number 25, now indicates a lack of control. That is, if we ignore the missing entry at data point number 23, data point number 25 is

FIGURE 10.6 Revised x-Bar and R Chart for Tuning Knob Diameters

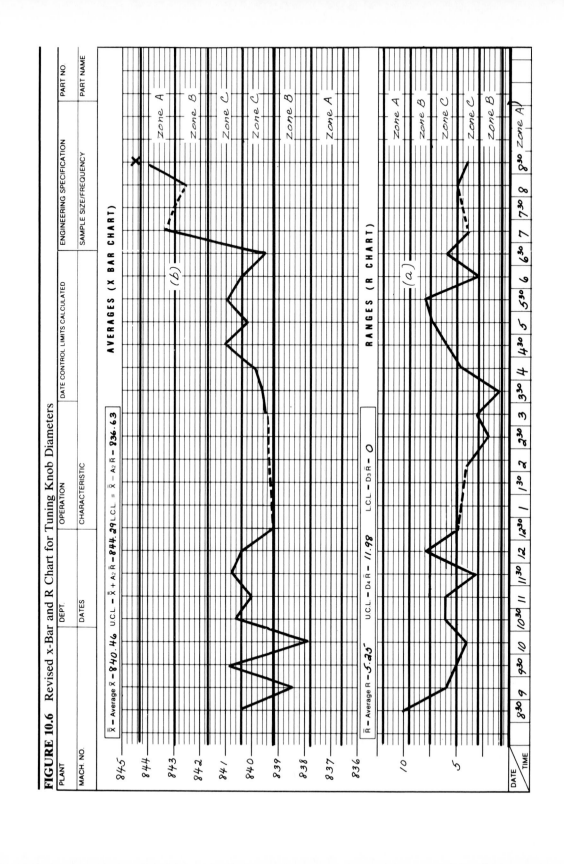

the second of three consecutive points in zone A or beyond, an indication of a lack of control by virtue of the second rule of Chapter 8.

At times a conservative approach is warranted—that is, detection of patterns indicating a lack of control will usually be followed by a somewhat costly search for the special cause(s) of variation. A process that has just been altered to eliminate one or more special sources of variation may be allowed to run for a while longer to determine whether any special sources of variation are really present or whether the indication of a lack of control is only a temporary effect that can be attributed to the removal of some of the data. If special sources of variation are present, the data taken from the process output will soon demonstrate evidence of that variation.

This process can be permitted to run as if it were statistically controlled. However, it must be watched closely to ensure that the special sources of variation have been removed and are no longer affecting the process.

x-BAR AND s CHARTS

x-bar and s charts are quite similar to x-bar and R charts. That is, they both provide the same sort of information—but the x-bar and s charts are used when subgroups consist of 10 or more observations.

In Chapter 5 we saw that the standard deviation of the process output, σ, could be estimated using s, which is computed from Equations 5.10a or 5.10b. s provides an estimate that generally has a smaller standard error than R. The benefits of having a smaller standard error must be weighed against the costs of larger subgroup sizes and more complex calculations. The x-bar and s charts are usually used with larger subgroup sizes, and larger subgroups are not always desirable.

One reason for this is that s may be viewed as a less robust estimator of the population standard deviation than R, for R is more sensitive to shifts in population shape than s. For subgroup sizes of fewer than 10, the range provides a reasonable statistic with which to estimate the standard error. Also, the range is easier to calculate than s, which gives it an advantage in many situations. So when subgroup sizes are small, the range is used as an estimator for σ.

When subgroup sizes are larger than 10, s is almost always used because as subgroup size increases, s becomes a much more statistically efficient estimator for σ. When the subgroup size is increased, the likelihood of encountering an extreme value increases, so that s, which is less affected than R by extreme values in the data, becomes a better estimator for σ.

Historically, subgroup ranges have been preferred because they are easier to compute than subgroup standard deviations. As the use of electronic calculation has grown, the need to avoid tedious computations has decreased, and the reluctance to use x-bar and s charts has decreased.

The s Portion

The construction of the x-bar and s chart parallels that of the x-bar and R chart in that it begins with an examination of the portion of the chart concerned with the variability of the process. The standard deviation, s, must be calculated for each of the subgroups. The value for s is the basis for an estimate of the process standard deviation, from which a set of factors for the control limits is developed.

Equation 5.10b is used to compute s for each subgroup:

$$s = \sqrt{\frac{\Sigma x^2 - (\Sigma x)^2/n}{(n - 1)}}$$

where n is the number of observations in each subgroup. The sequence of s values is then averaged, yielding $\bar{s}$, which is the centerline for the s chart.

$$\text{centerline(s)} = \bar{s} = \Sigma \, s/k \qquad (10.16)$$

where k is, once again, the number of subgroups. Additionally, $\bar{s}$ is used to form an estimate of the process standard deviation, σ:

$$\sigma = \bar{s}/c_4 \qquad (10.17)$$

where c_4 is a factor that depends on the subgroup size, assumes that the process characteristic is normally distributed, and is given in Table 8.

Control limits for the s chart are constructed by adding and subtracting three times the standard error of $\bar{s}$ from the centerline of the control chart. In Chapter 7 the sampling distribution of the standard deviation of the output of a stable, normally distributed process was shown to have a standard error given by Equation 7.20:

$$\sigma_\sigma = \sigma\sqrt{1 - c_4^2}$$

Hence, the upper control limit for the s chart is:

$$\text{UCL(s)} = \bar{s} + 3\,\sigma\sqrt{1 - c_4^2}$$

Using $\bar{s}/c_4$ to estimate σ yields:

$$\text{UCL(s)} = \bar{s} + 3\,\bar{s}\sqrt{1 - c_4^2}/c_4$$
$$= \bar{s}[1 + 3\sqrt{1 - c_4^2}/c_4]$$

This can be simplified by letting $B_4 = 1 + 3\sqrt{1 - c_4^2}/c_4$ so that

$$\text{UCL(s)} = B_4\bar{s} \qquad (10.18)$$

Similarly, letting the factor $B_3 = 1 - 3\sqrt{1 - c_4^2}/c_4$, the lower control limit for the s chart can be found using:

$$LCL(s) = B_3\bar{s} \qquad (10.19)$$

Values for B_3 and B_4 depend on subgroup size, assume that the process characteristic is stable and normally distributed, and can be found in Table 8.

Boundaries for zones A, B, and C for the s chart are placed at the usual multiples of one and two times the standard error on either side of the centerline. As negative values are meaningless, the zone boundaries cease to exist below the 0.00 line on the control chart. To find the zone boundaries, we divide the difference between the upper control limit and the centerline by 3. This value provides an estimate for the standard error of $\bar{s}$. Adding and subtracting this value from the centerline yields the upper and lower boundaries between zones B and C, respectively, while adding and subtracting two times this value yields the upper and lower boundaries between zones A and B, respectively. These boundaries can be expressed as:

$$\text{boundary between lower zones A and B} = \bar{s} - (2/3)\bar{s}(B_4 - 1) \qquad (10.20)$$

$$\text{boundary between lower zones B and C} = \bar{s} - (1/3)\bar{s}(B_4 - 1) \qquad (10.21)$$

$$\text{boundary between upper zones B and C} = \bar{s} + (1/3)\bar{s}(B_4 - 1) \qquad (10.22)$$

$$\text{boundary between upper zones A and B} = \bar{s} + (2/3)\bar{s}(B_4 - 1) \qquad (10.23)$$

If the s portion of the control chart is found to be stable, the x-bar portion may be constructed. However, if the s portion indicates of a lack of statistical control, then the x-bar portion cannot be safely evaluated until any special sources of variation have been removed and the process stabilized. As we shall see below, this is because the estimate of the standard error of the x-bar portion is based on the average value of s. Just as with the x-bar and R chart, if the variability is not in control, then estimates of the standard error are unreliable, which will result in unreliable control limits for $\bar{x}$.

The x-Bar Portion

The centerline for the x-bar chart is the average of the subgroup averages, $\bar{\bar{x}}$, and can be found using Equation 10.9. The control limits are found by adding and subtracting three times the standard error of $\bar{x}$ from the centerline.

$$\bar{\bar{x}} \pm 3\,\sigma/\sqrt{n}$$

Our estimate of the process standard deviation is

$$\sigma = \bar{s}/c_4$$

so that the control limits become:

$$\bar{\bar{x}} \pm 3(\bar{s}/c_4)/\sqrt{n} = \bar{\bar{x}} \pm 3\bar{s}/(c_4\sqrt{n})$$

Letting the constant $A_3 = 3/(c_4\sqrt{n})$, the control limits for $\bar{x}$ can be expressed as:

$$UCL(\bar{x}) = \bar{\bar{x}} + A_3\bar{s} \tag{10.24}$$

and

$$LCL(\bar{x}) = \bar{\bar{x}} - A_3\bar{s} \tag{10.25}$$

The value for A_3 depends on subgroup size and can be found in Table 8.

The zone boundaries are placed at one and two times the standard error on either side of the centerline. The standard error is expressed as $\bar{s}/(c_4\sqrt{n})$, so that the zone boundaries are:

$$\begin{matrix}\text{boundary between} \\ \text{lower zones A and B}\end{matrix} = \bar{\bar{x}} - 2\,\bar{s}/(c_4\sqrt{n}) \tag{10.26}$$

$$\begin{matrix}\text{boundary between} \\ \text{lower zones B and C}\end{matrix} = \bar{\bar{x}} - \bar{s}/(c_4\sqrt{n}) \tag{10.27}$$

$$\begin{matrix}\text{boundary between} \\ \text{upper zones B and C}\end{matrix} = \bar{\bar{x}} + \bar{s}/(c_4\sqrt{n}) \tag{10.28}$$

$$\begin{matrix}\text{boundary between} \\ \text{upper zones A and B}\end{matrix} = \bar{\bar{x}} + 2\,\bar{s}/(c_4\sqrt{n}) \tag{10.29}$$

x-Bar and s Charts: An Example

In a converting operation, a plastic film is combined with paper coming off a spooled reel. As the two come together they form a moving sheet that passes as a web over a series of rollers. The operation runs in a continuous feed, and the thickness of the plastic coating is an important product characteristic. Coating thickness is monitored by a highly automated piece of equipment that uses ten heads to take ten measurements across the web at half-hour intervals. Figure 10.7 shows a sequence of measurements taken over 20 time periods.

FIGURE 10.7 Plastic Coating Thickness

Head #	8:30	9:00	9:30	10:00	10:30	11:00	11:30
1	2.08	2.14	2.30	2.01	2.06	2.14	2.07
2	2.26	2.02	2.10	2.10	2.12	2.22	2.05
3	2.13	2.14	2.20	2.15	1.98	2.18	1.97
4	1.94	1.94	2.25	1.97	2.12	2.27	2.05
5	2.30	2.30	2.05	2.25	2.20	2.17	2.16
6	2.15	2.08	1.95	2.12	2.02	2.26	2.02
7	2.07	1.94	2.10	2.10	2.19	2.15	2.02
8	2.02	2.12	2.16	1.90	2.03	2.07	2.14
9	2.22	2.15	2.37	2.04	2.02	2.02	2.07
10	2.18	2.36	1.98	2.08	2.09	2.36	2.00
$\bar{x}$	2.14	2.12	2.15	2.07	2.08	2.18	2.06
s	.111	.137	.136	.098	.074	.099	.059

Head #	12:00	12:30	13:00	13:30	14:00	14:30	15:00
1	2.08	2.13	2.13	2.24	2.25	2.03	2.08
2	2.31	1.90	2.16	2.34	1.91	2.10	1.92
3	2.12	2.12	2.12	2.40	1.96	2.24	2.14
4	2.18	2.04	2.22	2.26	2.04	2.20	2.20
5	2.15	2.40	2.12	2.13	1.93	2.25	2.02
6	2.17	2.12	2.07	2.15	2.08	2.03	2.04
7	1.98	2.15	2.04	2.08	2.29	2.06	1.94
8	2.05	2.01	2.28	2.02	2.42	2.19	2.05
9	2.00	2.30	2.12	2.05	2.10	2.13	2.12
10	2.26	2.14	2.10	2.18	2.00	2.20	2.06
$\bar{x}$	2.13	2.13	2.14	2.19	2.10	2.14	2.06
s	.107	.141	.070	.125	.170	.084	.086

Head #	15:30	16:00	16:30	17:00	17:30	18:00	
1	2.04	1.92	2.12	1.98	2.08	2.22	
2	2.14	2.10	2.30	2.30	2.12	2.05	
3	2.18	2.13	2.01	2.31	2.11	1.93	
4	2.12	2.02	2.20	2.12	2.22	2.08	
5	2.00	1.93	2.11	2.08	2.00	2.15	
6	2.02	2.17	1.93	2.10	1.95	2.27	
7	2.05	2.24	2.02	2.15	2.15	1.95	
8	2.34	1.98	2.25	2.35	2.14	2.11	
9	2.12	2.34	2.05	2.12	2.28	2.12	
10	2.05	2.12	2.10	2.26	2.31	2.10	
$\bar{x}$	2.11	2.10	2.11	2.18	2.14	2.10	
s	.101	.136	.115	.121	.113	.106	

Arranging the data in this way helps determine whether there are any special causes of variation from measurement to measurement over time between the subgroups.

Values for $\bar{x}$ and s, the subgroup means and standard deviations, have been computed for each subgroup. The values for the averages of these subgroup means and standard deviations are computed using Equations 10.9 and 10.16 as:

$$\bar{\bar{x}} = 42.43/20 = 2.12$$

and

$$\bar{s} = 2.19/20 = 0.11$$

The s chart must be constructed first, and Equations 10.18 and 10.19 provide the upper and lower control limits:

$$UCL(s) = (1.716)\,(0.11) = 0.189$$

$$LCL(s) = (0.284)\,(0.11) = 0.031$$

The zone boundaries are computed using Equations 10.20, 10.21, 10.22, and 10.23.

boundary between
lower zones A and B $= 0.11 - (2/3)(0.11)(1.716 - 1) = 0.057$

boundary between
lower zones B and C $= 0.11 - (1/3)(0.11)(1.716 - 1) = 0.084$

boundary between
upper zones B and C $= 0.11 + (1/3)(0.11)(1.716 - 1) = 0.136$

boundary between
upper zones A and B $= 0.11 + (2/3)(0.11)(1.716 - 1) = 0.163$

The control chart is shown in Figure 10.8(a). As the control chart does not indicate a lack of control, the process variability appears stable. The x-bar portion of the chart may now be constructed.

It would not have been correct to analyze the x-bar portion before establishing the stability of the process standard deviation. If the standard deviation were not stable, the mean value of the subgroup standard deviations would be unreliable; the control limits on the x-bar chart would also be unreliable, as the control limits are based on the value for $\bar{s}$. The control chart might then fail to reveal patterns indicating a lack of control when they existed. This is why the estimate of the process standard deviation is not based on the standard deviation taken across all observations.

FIGURE 10.8 x-Bar and s Chart for Plastic Coating Thickness

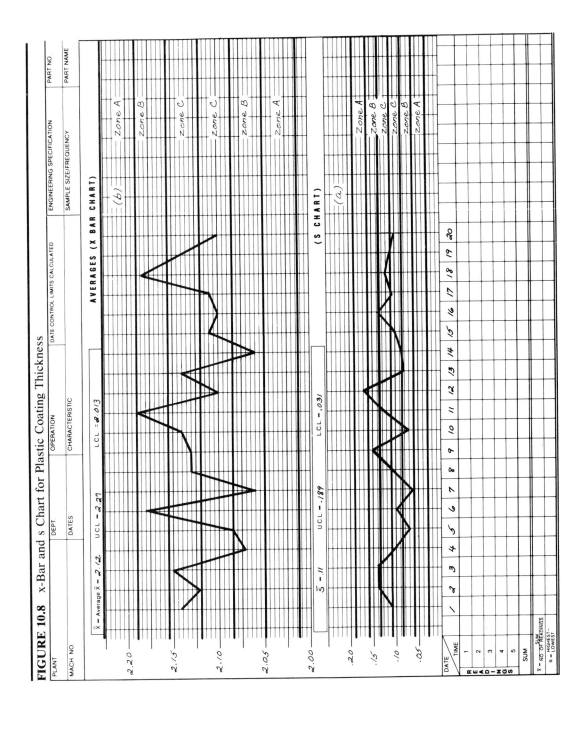

The average value for $\bar{x}$, $\bar{\bar{x}}$, is 2.12. Equations 10.24 and 10.25 are used to form the upper and lower control limits for the x-bar chart.

$$\text{UCL } (\bar{x}) = 2.12 + (0.975)(0.11) = 2.227$$

$$\text{LCL } (\bar{x}) = 2.12 - (0.975)(0.11) = 2.013$$

We then use Equations 10.26, 10.27, 10.28, and 10.29 to calculate the zone boundaries:

$$\begin{array}{l} \text{boundary between} \\ \text{lower zones A and B} \end{array} = 2.12 - 2[0.11/(0.9727)(3.16)] = 2.048$$

$$\begin{array}{l} \text{boundary between} \\ \text{lower zones B and C} \end{array} = 2.12 - [0.11/(0.9727)(3.16)] = 2.084$$

$$\begin{array}{l} \text{boundary between} \\ \text{upper zones B and C} \end{array} = 2.12 + [0.11/(0.9727)(3.16)] = 2.156$$

$$\begin{array}{l} \text{boundary between} \\ \text{upper zones A and B} \end{array} = 2.12 + 2[0.11/(0.9727)(3.16)] = 2.192$$

The x-bar chart appears in Figure 10.8(b) and completes the control chart. There are no indications of a lack of control, so the process can be considered to be stable and the output predictable with respect to time as long as conditions remain the same.

It is now appropriate to use some of the methods which will be described in Chapter 12, such as check sheets, Pareto analysis, or brainstorming, to attempt to reduce the common causes of variation in the never-ending quest to reduce variability. It is now also appropriate to rearrange the data to look for special sources of variation across the sheet; that is, to attempt to isolate variation on different sections of the sheet passing over the web at the measuring device. A rational subgrouping (as discussed in Chapter 8) to isolate this variation would group the data by measurement head. As measurements are taken every half hour, it might be apt to use all measurements taken on a given shift for each measuring head as a subgroup. This arrangement would yield subgroups with 16 measurements; each subgroup would represent a given head and shift. This rational subgrouping would address the question of differences between the measurement heads. It would also reveal differences between the shifts. The variation within each subgroup would be variation occurring from that subgroup's measuring head.

MEDIAN AND RANGE CHARTS

Median and range control charts employ the subgroup median, M_e, as a measure of the process location, and the subgroup range, R, as a measure of the process variability. A separate portion of this two-part control chart is constructed with each of these measures in a manner paralleling the construction of the x-bar and R, and x-bar and s control charts.

Conditions for Use

Resistance to control charting may have its origins in many corners. Difficulty with simple arithmetic is a frequently encountered problem. If proper training has not yet been instituted, or has been unsuccessful, some employees may be unwilling, unable, or both, to use control charts to help do a better job. Median and range charts present an alternative to x-bar and R or x-bar and s charts that may be initially more palatable to some employees. Nevertheless, x-bar and R charts and x-bar and s charts always contain more information from a statistical standpoint and are therefore more powerful than median and range control charts. Median and range control charts should always be replaced by either x-bar and R or x-bar and s control charts as soon as conditions permit.

Control charts for the subgroup medians and ranges involve little arithmetic. Education of those maintaining the charts may be less rigorous, and this may permit control charting to be used in circumstances where it might not otherwise be possible. Forms and graphs can be greatly simplified. The control limits to be used can be calculated by individuals other than those lacking the education or experience to do so. The charts can then be used with relatively simple tools for the identification of out-of-control points and/or patterns. Education of the work force is an important commitment on the part of management towards never-ending improvement. A logical progression towards more sophisticated methods of quality control cannot take place unless all personnel have been properly trained. Once the basic methods have been established with the work force, the transition to x-bar and R charts or x-bar and s charts will be easier and will meet with wider acceptance.

Subgroup Data

Simplicity is often the paramount reason for the use of median and range charts. For this reason, subgroup sizes should generally contain an odd number of elements—three, five, seven, or nine. Even numbers require calculation of the average of the two middle values to form the median, which may defeat the purpose of using this particular chart. Subgroup sizes of more than nine are also usually counterproductive because the size alone may be contrary to the goal of simplicity in the chart.

Because of the circumstances generally encountered, separate data collection forms for each subgroup are recommended for use with median charts. The forms should be clear and simple, with large spaces in which to enter the data. There should be no small lines to confuse the recorder, and the form should be tailored to the specific data being collected. That is, if the subgroup size is five, the form should have exactly five positions for data to be recorded. One form should be used for each subgroup. The form should then have a place for the employee to rank the data and then determine the median and the range for the subgroup. Figure 10.9 illustrates one possible recording chart. After the median and range have been found on the recording chart, they should be entered on the control chart.

Clear instructions and adequate training cannot be overstressed. Most commonly, this type of chart is valuable with minimally educated employees. Instructions should be clear; and process improvements resulting from the use of control charts should be highlighted for all those involved to see.

Constructing a Median and Range Control Chart

The sampling distribution of the subgroup medians has a mean value equal to the average of the subgroup medians, $\overline{M}_e$, and a standard error that can be estimated using a factor, A_6, which is based on the subgroup size, assumes that the process characteristic is stable and normally distributed, and can be found in Table 8.

$$\sigma_{M_e} = (A_6/3)\overline{R} \qquad (10.30)$$

The distribution of the subgroup ranges, discussed earlier in this chapter, has a mean value equal to $\overline{R}$ and a standard error as given in Equation 10.1 of $\sigma_R = d_3(\overline{R}/d_2)$, assuming the process characteristic is stable and normally distributed. These distributions can be used to construct the centerline and control limits for the median and range portions of the control chart.

Median and range control charts should be kept large and relatively easy to read. Here again the use of small scales and tight lines must be avoided. Chart scales must be large and constructed so that there is no chance that entries will fall outside the range of the entire chart. Scales should appear on both sides of the chart for ease of data entry.

Control limits, in cases where median charts are used, will typically be calculated by someone other than the recorder. We must first compute the average of the subgroup medians and ranges. Just as with the other variables control charts, the range portion must be constructed first so that its stability can be established. The subgroup ranges will, once again, form the basis for the process variability estimate for both the range and median portions. Hence, if the range is not stable, then control limits for the

FIGURE 10.9   Median Control Chart Data Recording Sheet

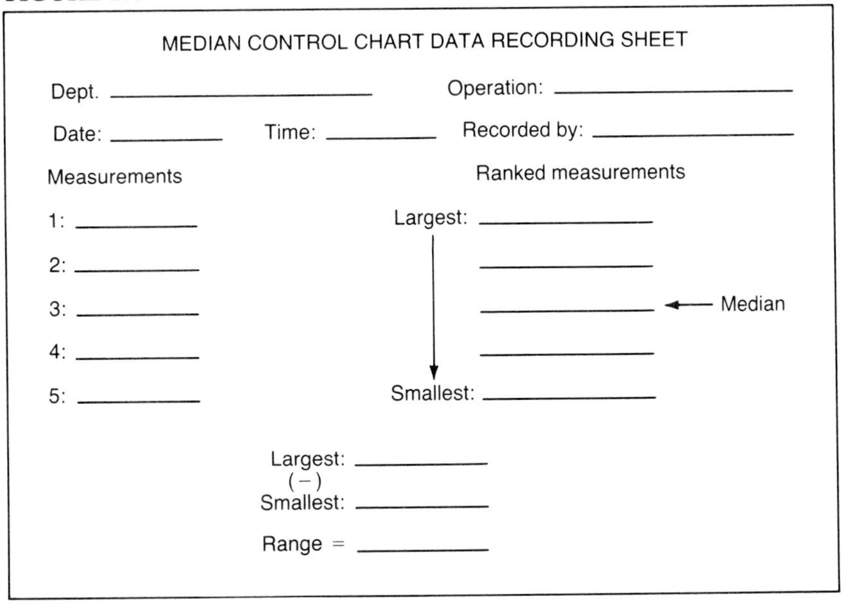

MEDIAN CONTROL CHART DATA RECORDING SHEET

Dept. _____   Operation: _____

Date: _____   Time: _____   Recorded by: _____

Measurements   Ranked measurements

1: _____   Largest: _____

2: _____   _____

3: _____   _____ ← Median

4: _____   _____

5: _____   Smallest: _____

Largest: _____
(−)
Smallest: _____

Range = _____

range and median portions will be unreliable. This may hide points or patterns indicating a lack of control.

Control limits for the range portion are identical to those used in the x-bar and R chart. We use Equation 10.2 for the centerline and Equations 10.3 and 10.4 to determine the upper and lower control limits. If there are no indications of a lack of control, the median portion of the control chart may be constructed. We use Equation 10.31 for the centerline for the control chart and Equations 10.32 and 10.33 to establish the control limits on the subgroup medians:

$$\text{centerline}(M_e) = \overline{M}_e = \Sigma M_e / k \qquad (10.31)$$

where k is the number of subgroups. Using Equation 10.30 we can create the upper and lower control limits by adding and subtracting three times the standard error from the centerline. Three times the standard error of the subgroup medians is $A_6\overline{R}$ so that:

$$\text{UCL}(M_e) = \overline{M}_e + A_6\overline{R} \qquad (10.32)$$

$$\text{LCL}(M_e) = \overline{M}_e - A_6\overline{R} \qquad (10.33)$$

Median and Range Charts: An Example

Consider the case of a manufacturer of low tension ceramic insulators. A mixture of various clays and water is crushed, milled, de-aired, pre-shaped, and then turned on a wheel to achieve the proper final shape. The manufacturer's customers have said that the diameter of the center hole is a critical dimension for the proper ultimate use of the insulator. Consistency in the diameter of the center holes has been identified as an important characteristic. The firm's customers, and hence the manufacturer, would like this dimension to be consistently on nominal with respect to time. To accomplish this goal the manufacturer must identify and remove any special causes of variation and proceed to improve the process by reducing the common variation.

The plant is located in an economically depressed area. The work force is poorly educated but is willing to learn and improve their processes. Median and range control charts are an effective way to introduce control charting in this situation. One particular operator is asked to help construct a control chart for the diameter of the first five ceramic insulator center holes produced in every two-hour period. Figure 10.10 illustrates results of the first 25 of these subgroups.

The average of the subgroup medians is found using Equation 10.31:

$$\overline{M}_e = 237.9/25 = 9.52$$

and, using Equation 10.2, we find that the average of the subgroup ranges is:

$$\overline{R} = 16.7/25 = 0.67$$

We calculate the control limits for the range portion using Equations 10.3 and 10.4:

$$UCL(R) = (2.114)\,(0.67) = 1.42$$

$$LCL(R) = (0)\,(0.67) = 0.00$$

These are shown in Figure 10.11(a). There are no indications of a lack of control so we may construct the median portion of the control chart.

The control limits for the median portion are calculated using Equations 10.32 and 10.33:

$$UCL(M_e) = 9.52 + (0.691)\,(0.67) = 9.98$$

$$LCL(M_e) = 9.52 - (0.691)\,(0.67) = 9.06$$

These values are shown in Figure 10.11(b).

Most often the median portion of the chart displays all of the individual measurements. Each point is entered as a dot, and the median is circled, as in Figure 10.11(b).

Notice that when more than one data point in a subgroup has the same

FIGURE 10.10   Diameters of Insulator Center Holes Measurement (mm)

Sub-group	Time	1	2	3	4	5	Median	Range
1	8:00 A.M.	9.2	9.6	9.4	9.6	9.0	9.4	0.6
2	10:00	9.8	9.9	9.1	9.5	9.4	9.5	0.8
3	12:00	9.9	9.9	9.4	9.2	9.9	9.9	0.7
4	2:00 P.M.	9.0	9.5	9.2	9.0	9.8	9.2	0.8
5	4:00	9.1	9.9	9.9	9.7	9.3	9.7	0.8
6	8:00 A.M.	9.1	9.6	9.1	9.7	9.4	9.4	0.6
7	10:00	9.0	9.8	9.8	9.8	9.3	9.8	0.8
8	12:00	9.9	9.1	9.8	9.2	9.9	9.8	0.8
9	2:00 P.M.	9.5	9.2	9.6	9.5	9.4	9.5	0.4
10	4:00	9.8	9.9	9.2	9.3	9.8	9.8	0.7
11	8:00 A.M.	9.7	9.7	9.9	9.4	9.8	9.7	0.5
12	10:00	9.5	9.5	9.4	9.1	9.4	9.4	0.4
13	12:00	9.7	9.2	9.1	9.7	9.1	9.2	0.6
14	2:00 P.M.	9.8	9.7	9.1	9.1	9.1	9.1	0.7
15	4:00	9.9	9.7	9.5	9.5	9.8	9.7	0.4
16	8:00 A.M.	9.7	9.4	9.4	9.3	9.1	9.4	0.6
17	10:00	9.0	9.1	9.8	9.8	9.5	9.5	0.8
18	12:00	9.6	9.3	9.0	9.3	9.8	9.3	0.8
19	2:00 P.M.	9.9	9.9	9.4	9.4	9.1	9.4	0.8
20	4:00	9.8	9.9	9.7	9.3	9.8	9.8	0.6
21	8:00 A.M.	9.4	9.2	9.8	9.8	9.9	9.8	0.7
22	10:00	9.3	9.3	9.1	9.5	9.6	9.3	0.5
23	12:00	9.0	9.4	9.7	9.5	9.6	9.5	0.7
24	2:00 P.M.	9.6	9.8	9.5	9.0	9.4	9.5	0.8
25	4:00	9.3	9.2	9.1	9.9	9.9	9.3	0.8
						Totals	237.9	16.7

numerical value, the entry is shown as multiple dots next to each other, enabling the reader to identify all points with the same numerical value in that subgroup. Additionally, zones A, B, and C have not been delineated. These should be introduced only when more sophisticated control charts are used.

The subgroups in this example have been selected in sets of five taken every two hours. They will reveal special sources of variation that create patterns in the data between the subgroups, if any exist. Here, the control chart will determine whether there are any differences in the subgroups with respect to time.

No points are out of control in the control chart in Figure 10.11, and there are no other indications of special sources of variation. The variation present from measurement to measurement appears to be common

FIGURE 10.11   Median and Range Chart for Center Hole Diameters

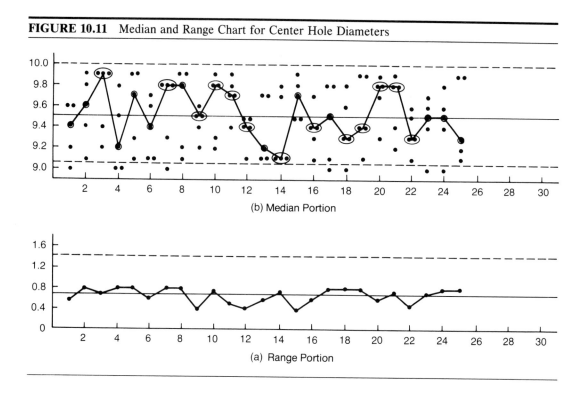

(b) Median Portion

(a) Range Portion

process variation. Efforts to reduce this variation should focus on process improvements, using the tools described in Chapter 12.

INDIVIDUALS CHARTS

In Chapter 9 we discussed individuals charts for attribute data. Variables data can also occur as single measurements. It is not uncommon to encounter a situation where only a single variable value can be periodically observed for control charting. Perhaps measurements must be taken at relatively long intervals, or the measurements are destructive and/or expensive; or perhaps they represent a single batch where only one measurement is appropriate, such as the total yield of a homogeneous chemical batch process. Whatever the case, there are circumstances when data must be taken as individual units that cannot conveniently be divided into subgroups.

As we saw in Chapter 9, individuals charts have two parts—one charting the process variability and the other charting the process average for the single measurements. The two parts are used in tandem in much the same way as the x-bar and R charts. Stability must first be established in the portion charting the variability because the estimate of the process variability provides the basis for the control limits of the portion charting the process average.

Single measurements of variables are considered a subgroup of size one. Hence, there is no variability within the subgroups themselves; and an estimate of the process variability must be made in some other way. In Chapter 9 we saw that for attribute data, an estimate of variability is based on the point-to-point variation in the sequence of single values. The same approach is used for variables data.

The point-to-point variation is measured by the moving range, the absolute value of the difference between each data point and the one that immediately preceded it.

$$R = |x_i - x_{i-1}| \tag{10.34}$$

An average of the moving ranges is used as the centerline for the moving range portion of the chart and as the basis of an estimate of the overall process variation.

$$\text{centerline(moving range)} = \overline{R} = \Sigma\,R/(k-1) \tag{10.35}$$

where k is the number of single measurements. As it is impossible to calculate the moving range for the first subgroup because none precedes it, there are only k − 1 range measurements; so the sum of the R values is divided by k − 1.

The sampling distribution of the subgroup ranges has an average given by Equation 7.17 of $\overline{R} = d_2\sigma$, and a standard error given by Equation 7.18 of $\sigma_R = d_3\sigma$, where σ is the process standard deviation from a stable and normally distributed process.

The estimate of the process variation is subsequently used to create the 3-sigma control limits for both the moving range portion and the single measurements portion of the control chart.

As we first saw in Chapter 8 and again earlier in this chapter, the control limits for a range are given by:

$$\text{UCL(moving range)} = D_4\overline{R} \tag{10.36}$$

$$\text{LCL(moving range)} = D_3\overline{R} \tag{10.37}$$

where D_3 and D_4 depend on subgroup size and are given in Table 8. In this case D_3 is 0.000 and D_4 is 3.267 because the subgroup size for the moving range portion is 2.

For the single measurements portion of the control chart the centerline is the average of the individual measurements. We find the control limits by adding and subtracting three times the standard deviation of the single measurements, estimated by $\overline{R}/d_2$.

$$\text{centerline(x)} = \overline{x} = \Sigma\,x/k \tag{10.38}$$

$$\text{UCL(x)} = \overline{x} + 3(\overline{R}/d_2)$$

Using the factor E_2 to represent $3/d_2$ the expression for the upper control limit becomes:

$$UCL(x) = \bar{x} + E_2\bar{R}$$

where E_2 depends on subgroup size and can be found in Table 8. In this case the subgroup size is 2, as we use two observations to calculate each moving range value. Hence, $E_2 = 2.66$, and

$$UCL(x) = \bar{x} + 2.66\,\bar{R} \qquad (10.39)$$

Similarly, the lower control limit is found using

$$LCL(x) = \bar{x} - 2.66\,\bar{R} \qquad (10.40)$$

Constructing an Individuals Chart for Variables Data: An Example

A chemical company produces 2,000-gallon batches of a liquid chemical product, A-744, once every two days. The product is a combination of six raw materials, of which three are liquids and three are powdered solids. Production takes place in a single tank, agitated as the ingredients are added and for several hours thereafter. Shipments of A-744 to the customer are made in bins as single lots when the batches are finished. The chemical company is concerned with the density of the finished product, which they measure in grams per cubic centimeter. As batches are constantly stirred during production, the density is assumed to be relatively uniform throughout each batch. Therefore, management decides that density will be measured by only one reading per batch.

During a 60-day period, 30 batches of A-744 are produced; Figure 10.12 shows the density readings for those batches.

Using Equation 10.34, we calculate the moving range by subtracting the previous observation from each observation, then taking the absolute value. Just as for the attribute version of the individuals chart, there is no moving range for the first observation, so there is one fewer moving range point than data points. The moving ranges are also shown in Figure 10.12.

Using Equation 10.35, the average of the 29 moving range values is:

$$\text{centerline(moving range)} = \bar{R} = 1.087/29 = 0.037$$

The control limits for the moving range portion of the control chart can be found using Equations 10.36 and 10.37:

$$UCL(\text{moving range}) = D_4\,\bar{R} = (3.267)\,(0.037) = 0.121$$

$$LCL(\text{moving range}) = D_3\,\bar{R} = (0.000)(0.037) = 0.000$$

The control chart for the moving range portion is shown in Figure 10.13(a). The moving range appears to be in a state of statistical control, so it is safe to use the average moving range value to construct the single measurements portion of the chart.

FIGURE 10.12 A-744 Batch Density

Date	Density	Moving Range	Date	Density	Moving Range
5/6	1.242	—	6/10	1.253	0.018
5/8	1.289	0.047	6/12	1.257	0.004
5/10	1.186	0.103	6/14	1.275	0.018
5/13	1.197	0.011	6/17	1.232	0.043
5/15	1.252	0.055	6/19	1.201	0.031
5/17	1.221	0.031	6/21	1.281	0.080
5/20	1.299	0.078	6/24	1.274	0.007
5/22	1.323	0.024	6/26	1.234	0.040
5/24	1.323	0.000	6/28	1.187	0.047
5/27	1.314	0.009	7/1	1.196	0.009
5/29	1.299	0.015	7/3	1.282	0.086
5/31	1.225	0.074	7/5	1.322	0.040
6/3	1.185	0.040	7/8	1.258	0.064
6/5	1.194	0.009	7/9	1.261	0.003
6/7	1.235	0.041	7/11	1.201	0.060
			Totals	37.498	1.087

Using Equation 10.38, the centerline for the control chart of the single measurements is:

$$\text{centerline}(x) = \bar{x} = 37.498/30 = 1.250$$

The control limits for the individual measurements portion are found using Equations 10.39 and 10.40:

$$\text{UCL}(x) = \bar{x} + 2.66\,\bar{R} = 1.250 + 2.66\,(0.037) = 1.348$$

$$\text{LCL}(x) = \bar{x} - 2.66\,\bar{R} = 1.250 - 2.66\,(0.037) = 1.152$$

Figure 10.13(b) illustrates the control chart for the single measurements portion. The process appears to be in a state of statistical control, for there are no points beyond the control limits and no other signs of any trends or patterns in the data.

Special Characteristics of Individuals Charts

Because each subgroup conists of only one value, and process variation is estimated on the basis of observation-to-observation changes, individuals charts have certain unique characteristics that distinguish them from other control charts. For example, in order for an individuals chart to be reliable, it is best to have at least 100 subgroups, whereas 25 will suffice for most other control chart forms. The 100 subgroups are necessary to

FIGURE 10.13 Single Measurement and Moving Range Chart for A–744 Density

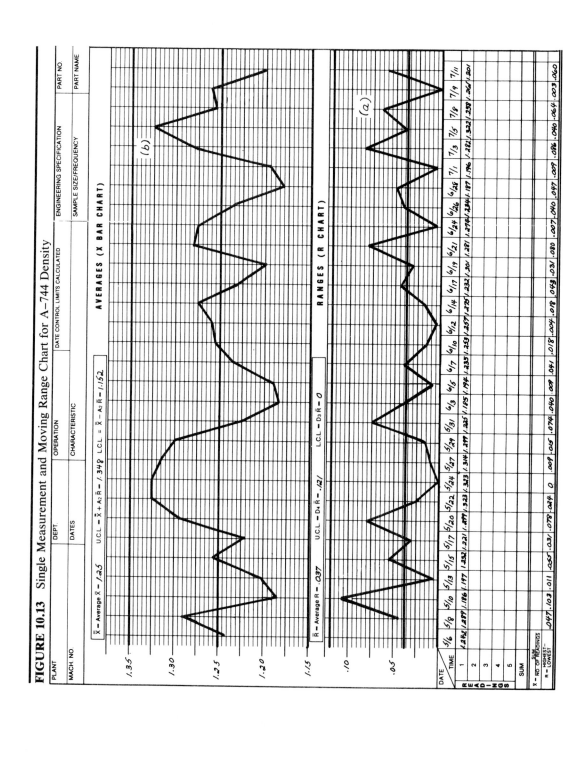

ascertain that the shape of the distribution of the output is approximately normal, which is a necessary condition for individuals charts.

Correlation in the Moving Range.   The moving ranges tend to be correlated. For example, a data point near the centerline followed by one in the upper zone A, followed by one in the lower zone C, will result in two large successive moving range points. As a consequence, large moving range values tend to be followed by other large moving range values, and small moving range values tend to be followed by small moving range values. Because of this, users must be cautious in applying rules for a lack of control dealing with patterns in the data. For example, using the rule designating the second of two out of three consecutive points in zone A or beyond as indicating a lack of control may tend to generate false alarms. For this reason, it is usually best to be conservative when applying the rules concerning patterns in the data other than points beyond the control limits that indicate a lack of control in moving range charts. For example, instead of 8 consecutive values above or below the centerline indicating a lack of control, we might require 10 or 12. Knowledge and experience are the best guides in establishing policy in this case.

Inflated Control Limits.   The control limits for individuals charts represent process limits because they are measurements on individual items. Hence, they are limits within which the single measurements will fall, provided that the variation in the process over the long run is approximately the same as the variation in the short run. Because the process variation is estimated using the moving range, the short-run (between consecutive subgroups) variation in the process provides the estimate of the process variation. Changes in the short-run or long-run variation may produce unreliable control limits.

One indication that the control limits are unreliable is the occurrence of at least two thirds of the data points below the centerline of the moving range portion of the control chart.[2] When this happens, control limits should be based on the median of the moving range values, M_{e_R}, as opposed to those based the average of the moving range values. If inflated control limits are suspected, control limits based on the median of the moving ranges should be calculated and compared to those based on the average moving range; the narrower of the two sets should be used. Control limits for the moving range portion, based on the median, can be computed using:

$$\text{centerline(median moving range)} = M_{e_R} \qquad (10.41)$$

$$\text{UCL(median moving range)} = D_6 M_{e_R}$$

$$\text{LCL(median moving range)} = D_5 M_{e_R}$$

where values for D_5 and D_6 depend on subgroup size, assume stability and a normal distribution of the process characteristic, and can be found in

Table 8. For the individuals chart the subgroup size is two, so the values used are $D_6 = 3.865$ and $D_5 = 0.000$. Hence:

$$\text{UCL(median moving range)} = 3.865 \, M_{e_R} \qquad (10.42)$$

$$\text{LCL(median moving range)} = 0.000 \qquad (10.43)$$

The standard error of the single measurements is estimated using M_{e_R}/d_4, where values for d_4 depend on subgroup size, assume stability and a normal distribution of the process characteristic, and can be found in Table 8. For subgroup size two, $d_4 = 0.954$. Hence, control limits for the single measurements portion are created by adding and subtracting three times $M_{e_R}/0.954$ from the centerline.

$$\text{UCL(x)} = \bar{x} + 3M_{e_R}/0.954 = \bar{x} + 3.145M_{e_R} \qquad (10.44)$$

Similarly,

$$\text{LCL(x)} = \bar{x} - 3.145M_{e_R} \qquad (10.45)$$

Individuals Charts with Inflated Control Limits: An Example. Consider the case of the manufacturer of chemicals discussed earlier. The chemical, A-744, is used by the manufacturer's customer as an ingredient in another process sensitive to the quantity of A-744. The manufacturer's customer is seeking to reduce costs by using the A-744 in whole bin lots. As the yield of the batches of A-744 can be expected to vary from its 2,000 gallon target, that yield is a likely candidate for the use of an individuals control chart. The yields of the 30 batches of A-744 are each carefully measured, yielding a sequence of 30 single values. The data and the computed moving ranges are shown in Figure 10.14.

The centerline and control limits for the moving range portion of the control chart are found using Equations 10.35, 10.36, and 10.37.

$$\bar{R} = 308.5/29 = 10.64$$

$$\text{UCL(moving range)} = D_4\,\bar{R} = (3.267)\,(10.64) = 34.76$$

$$\text{LCL(moving range)} = 0.00$$

The control chart for the moving range portion is shown in Figure 10.15(a). The moving range appears stable, so the average moving range value can be used to construct the single measurements portion of the chart.

Using Equation 10.38, the average of the 30 yields is:

$$\bar{x} = 60,049.0/30 = 2001.63$$

The control limits for the single measurements portion are found using Equations 10.39 and 10.40:

$$\text{UCL(x)} = \bar{x} + 2.66\,\bar{R} = 2001.63 + (2.66)\,(10.64) = 2029.93$$

$$\text{LCL(x)} = \bar{x} - 2.66\,\bar{R} = 2001.63 - (2.66)\,(10.64) = 1973.33$$

FIGURE 10.14 A-744 Batch Yields

Date	Yield	Moving Range	Date	Yield	Moving Range
5/6	1989.0	—	6/10	2002.3	4.9
5/8	1998.9	9.9	6/12	1999.5	2.8
5/10	2027.4	28.5	6/14	2000.8	1.3
5/13	2001.5	25.9	6/17	2022.4	21.6
5/15	1991.3	10.2	6/19	1998.3	24.1
5/17	2001.3	10.0	6/21	1999.8	1.5
5/20	1997.4	3.9	6/24	2000.9	1.1
5/22	1989.3	8.1	6/26	1994.3	6.6
5/24	1995.5	6.2	6/28	1998.7	4.4
5/27	2014.4	18.9	7/1	2013.5	14.8
5/29	1990.2	24.2	7/3	1998.1	15.4
5/31	1999.6	9.4	7/5	2002.5	4.4
6/3	2008.1	8.5	7/8	2000.2	2.3
6/5	1999.4	8.7	7/9	1996.1	4.1
6/7	1997.4	2.0	7/11	2020.9	24.8
			Totals	60049.0	308.5

The control chart for the single measurements portion is shown in Figure 10.15(b). This portion also appears to be in a state of statistical control. However, an experienced eye detects that more than two thirds of the data points (20 of the 29 moving ranges) are below the centerline, indicating that the control limits may be artificially inflated and therefore may be hiding indications of special sources of variation.

The median of the 29 moving range values is 8.5. This value can be used to calculate an alternate centerline and set of control limits using Equations 10.41, 10.42, and 10.43 for the moving range portion of the control chart:

$$\text{centerline(median moving range)} = 8.5$$

$$\text{UCL(median moving range)} = (3.865)(8.5) = 32.85$$

$$\text{LCL(median moving range)} = 0.00$$

The new control chart for the moving range portion is shown in Figure 10.16(a). The process still appears in a state of statistical control, so the median of the moving ranges may be used to construct a new single measurements portion of the chart.

Notice that it is essential to establish stability in the portion of the control chart dealing with variability (the moving range portion) before constructing the portion of the control chart dealing with the single measurements. This is because the control limits for the single measurements

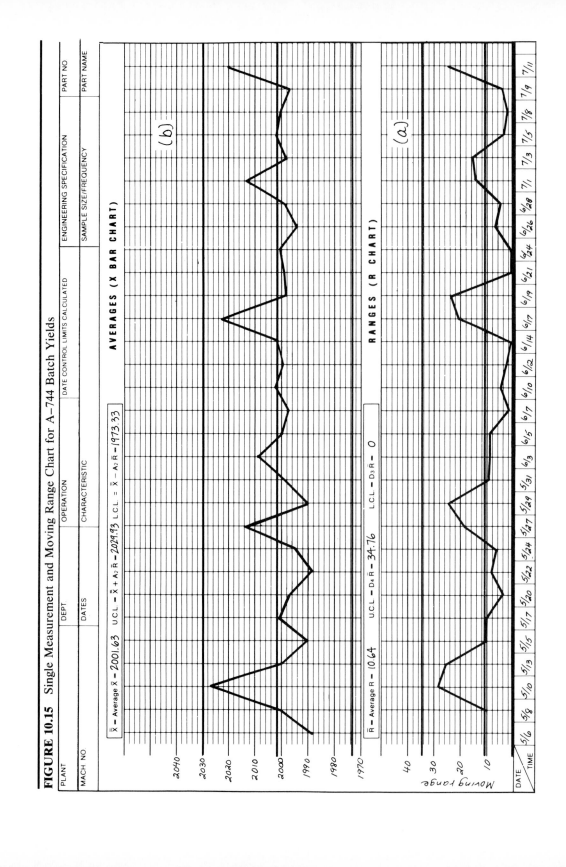

FIGURE 10.15 Single Measurement and Moving Range Chart for A–744 Batch Yields

FIGURE 10.16 Single Measurement and Moving Range Chart (Based on Median Moving Range Value) for A-744 Batch Yields

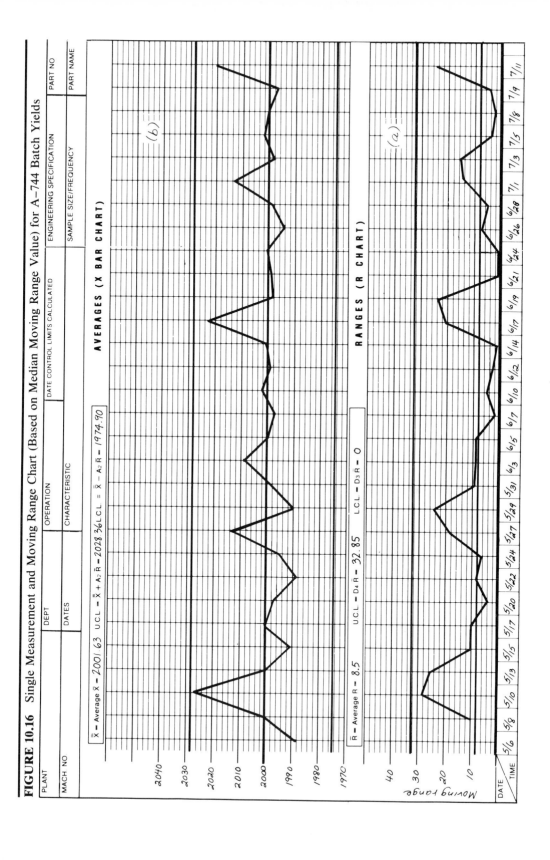

portion are based on the estimate of the process variability, generated by the moving range portion. A lack of stability in the moving range portion will produce unreliable estimates of the process variation, resulting in failure by the control chart to properly separate special and common variation.

The centerline remains at the average value, 2001.63, and the control limits for the single measurements portion are calculated using Equations 10.44 and 10.45:

$$\text{centerline}(x) = 2001.63$$

$$\text{UCL}(x) = 2001.63 + (3.145)\ (8.5) = 2028.36$$

$$\text{LCL}(x) = 2001.63 - (3.145)\ (8.5) = 1974.90$$

The control chart is shown in Figure 10.16(b). There are no indications of a lack of control. Because the control limits based on the median range are narrower, they are used in this case.

REVISING CONTROL LIMITS FOR VARIABLES CONTROL CHARTS

Frequent regular revision of control limits is undesirable and inappropriate. Control limits should be revised only for one of three reasons: a change in the process; when trial control limits have been used and are to be replaced with permanent control limits; and when points out of control have been removed from a data set.

Change in Process

Processes change for many reasons. For example, such things as technical improvements, new vendors, new machines, new machine settings, new operational definitions, or new operator instructions may induce process changes. Efforts toward the never-ending reduction of variation may precipitate the change. Whatever the cause, changes in the process itself change the variability and therefore necessitate recalculation of the control limits.

Trial Control Limits

When control charting is initiated for a process (either for a brand new process or an old process that is being charted for the first time), trial control limits are sometimes calculated from the first few subgroups.

After about 25 or 30 subgroups become available, these trial limits should be replaced with permanent control limits.

Removal of Out-of-Control Points

When out-of-control points used in the calculation of the control limits have been removed from a data set, the control limits must be recalculated. As the removed data values were used to calculate the process mean, range, standard deviation, or other statistics, the removal of these data points will precipitate changes in the centerline, control limits, and zone boundaries. A new centerline and new control limits and zone boundaries must be calculated. As we have seen, this may occasionally reveal other points that are out of control and will help to further identify areas requiring some special action or changes in the process.

SUMMARY

Variables control charts are a very important tool for process improvement. Variables control charts not only identify and differentiate between special and common causes of variation but also provide the data essential for process improvement.

x-bar and R charts use subgroups of from 2 to 10 items. When subgroup sizes are larger than 10, x-bar and s charts are used in place of x-bar and R charts. Both the x-bar and R charts and x-bar and s charts are important weapons in the war against variability. Variation of any kind must be continuously reduced if a firm is to battle its way upward on the spiral of never-ending process improvement.

Median and range control charts can serve as a means to introduce control charts to workers with little education. Even though they do not represent a long term solution approach, these charts can provide a needed introduction when other control charts are too complex.

Individuals charts are useful when, because of high cost or a lack of available data gathering opportunities, only one measurement is possible in a subgroup.

EXERCISES

10.1 A large hotel in a resort area has a housekeeping staff who clean and prepare all of the hotel's guest rooms daily. In an effort to improve service through reducing variation in the time required to clean and prepare a room, a series of measurements is taken of the times to service rooms in one section of the hotel. Cleaning times for five rooms selected each day for 25 consecutive days appear below:

Cleaning and Preparation Time

Room	Day 1	Day 2	Day 3	Day 4	Day 5	Day 6	Day 7	Day 8	Day 9	Day 10
1	15.6	15.0	16.4	14.2	16.4	14.9	17.9	14.0	17.6	14.6
2	14.3	14.8	15.1	14.8	16.3	17.2	17.9	17.7	16.5	14.0
3	17.7	16.8	15.7	17.3	17.6	17.2	14.7	16.9	15.3	14.7
4	14.3	16.9	17.3	15.0	17.9	15.3	17.0	14.0	14.5	16.9
5	15.0	17.4	16.6	16.4	14.9	14.1	14.5	14.9	15.1	14.2

Room	Day 11	Day 12	Day 13	Day 14	Day 15	Day 16	Day 17	Day 18	Day 19	Day 20
1	14.6	15.3	17.4	15.3	14.8	16.1	14.2	14.6	15.9	16.2
2	15.5	15.3	14.9	16.9	15.1	14.6	14.7	17.2	16.5	14.8
3	15.9	15.9	17.7	17.9	16.6	17.5	15.3	16.0	16.1	14.8
4	14.8	15.0	16.6	17.2	16.3	16.9	15.7	16.7	15.0	15.0
5	14.2	17.8	14.7	17.5	14.5	17.7	14.3	16.3	17.8	15.3

Room	Day 21	Day 22	Day 23	Day 24	Day 25
1	16.3	15.0	16.4	16.6	17.0
2	15.3	17.6	15.9	15.1	17.5
3	14.0	14.5	16.7	14.1	17.4
4	17.4	17.5	15.7	17.4	16.2
5	14.5	17.8	16.9	17.8	17.9

a. For each of the 25 subgroups, compute the mean and range.

b. Find the centerline values for both the x-bar and range portions of the control chart.

c. Estimate the standard deviation for the time to clean and prepare a room.

d. Find the values for the upper and lower control limits for both portions of the control chart.

e. Find the zone boundaries for all of the zones.

f. Determine whether there are any indications of a lack of control.

10.2 The drive-up window at a local bank is searching for ways to improve service. One of the tellers has decided to keep a control chart for the service time in minutes for the first four customers driving up to her window each hour for a three-day period. The results of her data collection appear below:

Drive-Up Teller Service Times

	Time							
Cust.	9 A.M.	10 A.M.	11 A.M.	12	1 P.M.	2 P.M.	3 P.M.	4 P.M.
1	1.4	3.8	3.6	4.3	4.0	1.3	0.9	4.7
2	2.3	5.2	2.5	1.2	5.2	1.1	4.4	5.1
3	1.9	1.9	0.8	3.0	2.7	4.9	5.1	0.9
4	5.1	4.8	2.9	1.5	0.3	2.3	4.6	4.7

	Time							
Cust.	9 A.M.	10 A.M.	11 A.M.	12	1 P.M.	2 P.M.	3 P.M.	4 P.M.
1	2.8	0.5	4.5	0.6	4.8	2.7	4.2	0.9
2	3.0	2.7	1.9	1.2	2.8	2.0	1.1	4.4
3	4.1	4.7	4.2	2.7	1.1	2.6	4.4	0.6
4	4.8	3.6	0.4	2.5	0.4	2.6	3.1	0.4

	Time							
Cust.	9 A.M.	10 A.M.	11 A.M.	12	1 P.M.	2 P.M.	3 P.M.	4 P.M.
1	0.3	3.5	5.2	2.9	3.3	4.0	2.8	0.6
2	2.4	3.4	0.3	1.9	3.7	3.3	0.7	2.1
3	5.0	4.6	2.4	0.8	3.8	5.0	1.6	3.3
4	0.9	3.3	3.9	0.3	2.1	2.8	4.6	2.7

a. For each of the 24 subgroups, compute the mean and the range.

b. Find the centerline values for both the x-bar and range portions of the control chart.

c. Estimate the standard deviation for the time to service a customer at the drive-up window.

d. Find the values for the upper and lower control limits for both portions of the control chart.

e. Find the zone boundaries for all zones.

f. Determine whether there are any indications of a lack of control.

10.3 A paper products manufacturer coats one particular paper product with wax. In an effort to control and stabilize the coating process, the employee running the coating machine collects data consisting of six measurements taken every 15 minutes during a one-day study period. The results of the data collection appear below.

Coating Thickness

				Time				
	8 A.M.	8:15	8:30	8:45	9 A.M.	9:15	9:30	9:45	10 A.M.
x̄	2.90	3.30	2.87	0.93	1.43	1.93	2.90	3.03	3.70
R	0.35	0.50	0.32	0.44	0.91	0.42	0.29	0.78	1.01

				Time				
	10:15	10:30	10:45	11 A.M.	11:15	11:30	11:45	12 NOON	12:15
x̄	3.70	3.00	3.20	2.80	2.90	3.53	3.90	3.73	3.23
R	1.09	0.95	0.56	0.42	0.55	0.16	1.04	0.90	0.86

				Time				
	12:30	12:45	1 P.M.	1:15	1:30	1:45	2 P.M.	2:15	2:30
x̄	2.73	3.00	3.33	2.50	2.13	1.97	2.70	3.10	2.63
R	0.19	0.84	0.20	0.71	0.90	0.92	1.00	0.56	0.88

			Time			
	2:45	3 P.M.	3:15	3:30	3:45	4 P.M.
x̄	2.63	3.40	3.20	2.20	2.03	3.23
R	0.89	1.06	0.66	0.93	0.59	0.91

a. Find the centerline values for both the x-bar and range portions of the control chart.

b. Find the values for the upper and lower control limits for both portions of the control chart.

c. Find the zone boundaries for the upper and lower A, B, and C zones.

d. Determine whether there are any indications of a lack of control.

10.4 Customers arriving at a restaurant arrive in groups and must wait to be seated by a hostess. Waiting time may be short if a table is available, or it may be long if the restaurant is crowded. The restaurant's hostess decides to study the waiting times and records the waiting time for the first eight people arriving from 7:00 P.M. each Friday, Saturday, and Sunday evening for a series of 10 consecutive weeks. She has computed the average and range for each of the subgroups. The results of the data collection and calculations are:

Waiting Times (minutes)

					Date					
	9/2	*9/3*	*9/4*	*9/9*	*9/10*	*9/11*	*9/16*	*9/17*	*9/18*	*9/23*
$\overline{x}$	15.5	23.0	18.0	11.5	12.0	12.3	15.5	13.5	21.5	12.2
R	5.1	6.2	2.6	2.4	4.8	2.6	2.8	4.0	4.8	1.8

					Date					
	9/24	*9/25*	*9/30*	*10/1*	*10/2*	*10/7*	*10/8*	*10/9*	*10/14*	*10/15*
$\overline{x}$	21.5	28.0	16.0	17.0	31.0	21.1	15.0	17.5	17.5	18.0
R	4.0	3.6	3.0	4.4	7.2	2.1	3.6	1.8	1.8	3.6

					Date					
	10/16	*10/21*	*10/22*	*10/23*	*10/28*	*10/29*	*10/30*	*11/4*	*11/5*	*11/6*
$\overline{x}$	17.5	20.0	26.0	17.0	19.0	20.0	16.0	18.0	18.5	19.5
R	3.6	4.8	5.6	3.4	3.4	3.6	3.2	3.8	3.4	3.6

a. Find the centerline values for both the x-bar and range portions of the control chart.

b. Find the values for the upper and lower control limits for both portions of the control chart.

c. Find the zone boundaries for the upper and lower A, B, and C zones.

d. Determine whether there are any indications of a lack of control.

10.5 The food services director of an airline wants to measure the weight of food left over on passenger food trays to measure the difference between product performance and customer expectations. Each day, 12 of the trays returned in the main cabin on a particular flight are set aside, placed into a special container, and returned to a laboratory for weighing. The average and standard deviation of each of these subgroups over a period of 25 days appear below.

Weight of Leftovers (grams)

					Day					
	1	2	3	4	5	6	7	8	9	10
$\bar{x}$	44.5	61.5	66.6	64.9	22.9	12.7	44.2	10.4	82.6	55.2
s	23.5	11.2	28.6	31.9	13.3	12.5	32.8	35.1	39.1	10.9

					Day					
	11	12	13	14	15	16	17	18	19	20
$\bar{x}$	92.9	25.5	92.9	27.5	84.8	82.2	24.7	25.8	79.7	19.5
s	35.5	38.3	27.5	21.3	22.5	35.2	20.7	19.4	14.7	23.5

			Day		
	21	22	23	24	25
$\bar{x}$	10.9	47.5	72.2	28.6	90.0
s	36.7	28.6	25.9	22.7	23.4

a. Find the centerline values for both the x-bar and s portions of the control chart.
b. Find the values for the upper and lower control limits for both portions of the control chart.
c. Find the zone boundaries for the upper and lower A, B, and C zones.
d. Determine whether there are any indications of a lack of control.

10.6 A tomato farmer has been using an x-bar and R chart to control the yield per plant on his highly mechanized farm. The yield is stable over time in both variability and location. The x-bar portion of the control chart has a centerline value of 10.50 pounds, and the upper control limit is 14.20 pounds.
a. Find the value of the lower control limit.
b. Find the value of the upper and lower A, B, and C zone boundaries.
c. Find the estimate of the value of the standard deviation of the yield of the tomato plants if the subgroup size is six.

10.7 A manufacturer of a special chemical fertilizer is concerned with the pH of the finished batches of product. A series of careful measurements reveals:

Batch	pH	Batch	pH	Batch	pH	Batch	pH
1	6.5	26	6.1	51	6.6	76	6.4
2	6.2	27	6.8	52	6.1	77	6.7
3	6.7	28	6.6	53	6.3	78	6.4
4	6.7	29	6.4	54	6.8	79	6.2
5	6.2	30	6.0	55	6.3	80	6.1
6	6.1	31	6.5	56	6.7	81	6.4
7	6.8	32	6.0	57	6.1	82	6.2
8	6.4	33	6.5	58	6.6	83	6.7
9	6.0	34	6.8	59	6.6	84	6.5
10	6.6	35	6.2	60	6.6	85	6.8
11	6.7	36	6.0	61	6.7	86	6.2
12	6.9	37	6.6	62	6.2	87	6.7
13	6.2	38	6.3	63	6.5	88	6.5
14	6.9	39	6.2	64	6.0	89	6.1
15	6.1	40	6.7	65	6.8	90	6.5
16	6.5	41	6.2	66	6.2	91	6.7
17	6.3	42	6.4	67	6.0	92	6.0
18	6.2	43	6.3	68	6.1	93	6.0
19	6.9	44	6.7	69	6.8	94	6.6
20	6.6	45	6.1	70	6.7	95	6.2
21	6.7	46	6.3	71	6.8	96	6.4
22	6.6	47	6.0	72	6.3	97	6.1
23	6.3	48	6.6	73	6.6	98	6.5
24	6.6	49	6.8	74	6.3	99	6.1
25	6.2	50	6.5	75	6.6	100	6.9

a. Find the moving range at each observation.
b. Construct the range portion of the control chart.
c. Are there any indications of a lack of statistical control?
d. Construct the single measurements portion of the control chart.
e. Are there any indications of a lack of control in this portion?
f. Do the control limits appear inflated? Explain.

10.8 A packaging operation is manned by unskilled and relatively uneducated workers. The workers' task is to shovel 30 kilograms of a granular product from a large pile into sacks that are then sealed and placed on pallets for shipping. The scale used by the workers is accurate, and an effort has been made to educate the workers about the need to measure each weight carefully. Given the conditions, initial efforts at process control will utilize a median and range control chart. A sequence of 25 subgroups, each consisting of the weights of five sacks, has been recorded.

Subgroup	Weight				
Number	1	2	3	4	5
1	35.4	35.6	34.8	34.7	34.8
2	36.0	35.6	34.9	34.8	35.9
3	35.2	35.0	35.0	35.4	35.1
4	34.8	35.8	35.2	35.0	34.9
5	34.2	35.0	36.1	34.9	35.1
6	36.0	35.0	35.2	34.8	34.9
7	36.1	34.9	34.5	35.0	35.1
8	35.1	35.0	35.6	34.9	36.2
9	35.0	35.6	36.1	34.8	35.6
10	35.4	35.8	36.0	34.2	36.0
11	35.2	35.3	35.2	35.9	34.8
12	35.9	36.0	35.1	35.1	35.6
13	35.2	35.6	35.0	34.9	35.0
14	35.2	35.6	35.8	35.0	35.1
15	34.9	34.8	35.0	35.2	34.9
16	35.2	35.3	35.2	35.6	35.1
17	35.6	35.8	35.2	35.4	34.9
18	35.2	35.6	35.4	35.6	35.2
19	34.7	34.9	35.6	35.2	35.0
20	35.0	35.1	35.6	35.0	35.1
21	35.6	35.0	35.8	35.2	34.6
22	34.9	35.1	35.6	35.0	35.2
23	35.9	36.0	35.2	36.0	35.2
24	35.2	35.1	35.4	34.9	35.1
25	35.2	35.2	35.1	35.6	34.9

a. Find the median and range for each of the subgroups.
b. Construct the range portion of the control chart.
c. Are there any indications of a lack of control in the range portion?
d. Construct the median portion of the control chart.
e. Are there any indications of a lack of control in the median portion?

NOTES

1. W. A. Shewhart, *Economic Control of Quality of Manufactured Product* (New York, N.Y.: D. Van Nostrand Co. Inc., 1931), p. 314.
2. Donald J. Wheeler and David S. Chambers, *Understanding Statistical Process Control* (Knoxville, Tenn.: Statistical Process Controls, Inc., 1986), p. 79.

Chapter 11

Out-of-Control Patterns

TYPES OF VARIATION THAT CREATE CONTROL CHART PATTERNS

The purpose of a control chart is to detect special (assignable or exogenous) causes of variation, or special disturbances. Ideally, once detected, special causes can be removed, leaving a process with only common (or endogenous) causes of variation. Common sources of variation are endogenous to the system and are not disturbances—they are the system.

Between and Within Group Variation

There are two types of special causes of variation: periodic disturbances and persistent disturbances. Periodic disturbances create special causes of variation that intermittently affect the process. The intermittent nature of these causes tends to affect sampled observations separated in time and, hence, in different subgroups. This is called *between group variation*. The effect of between group variation is to create control chart patterns in which subgroup statistics are beyond the control limits; in other words, its effect is to create control limits too narrow for the subgroup statistics.

A few examples of causes of variation that could generate between group variation are:

- Chaotic (unstable) functioning of automatic control devices.
- Operator carelessness in setting up machine runs (for example, timing or heat).
- Loose and wobbly braces for holding in place material to be worked on (for example, drilling, cutting, or sanding).
- Overadjustment of a machine.

Between group variation is demonstrated in the following example. One of the component parts of a certain machine tool is an eccentric cam in which a slot 1/2 inch deep is milled. The operator of the milling machine places five cams in the jig, tightens them in with an adjusting screw, and cuts five slots simultaneously.

During a study of this milling operation, it becomes apparent that the slots are not being held within tolerances. Both foreman and operator complain that the milling cutter has to be changed too often and that the slot goes out of tolerance before the cutter needs resharpening. A control chart analysis is made to reduce the amount of down time required for changing cutters and increase the efficiency of the operation.

As five slots are cut at one time, it seems logical to use groups of five simultaneous slots as the subgroup for the control chart. Measurements of slot depth therefore are made on each of five simultaneous slots; about 30 minutes elapse between inspections.

Figure 11.1 shows the x-bar chart for 18 samples of 5 each; the range portion of the chart (not shown) exhibited no indications of a lack of control. Either the process is very erratic, or the subgrouping is incorrect. A classification of the possible sources of special variation in the operation reveals the following:

Raw materials: variability in hardness of steel.

Dimensions of cam: variability from preceding operations.

Positioning of the milling jig: variability resulting from operator skill.

Wear in the cutting tool: variability caused by slot's growing shallower as cutter dulls.

As far as the chosen method of subgrouping is concerned, the cams are thoroughly scrambled or randomized before coming to the mills. The first and second sources are therefore included in the subgroups, because a

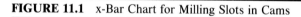

FIGURE 11.1   x-Bar Chart for Milling Slots in Cams

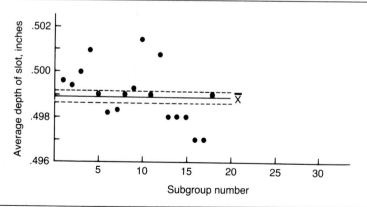

cam of any specified hardness or size would be as likely to appear in one subgroup as another. But the third and fourth causes are not included in the subgroups; their effect is *between* the subgroups. Positioning of the jig affects all five cams in the jig in the same way, but the next group of five might be positioned differently. Tool wear, a long-term directional effect, should appear as a trend between successive subgroups and would not affect the range within subgroups. Regarding the third cause, positioning of the jig, some variation between successive jig settings is unavoidable, for jig setting depends upon the manual skill of the operator; certainly excessive variations from this cause are undesirable.[1]

Persistent disturbances create special causes of variation that continually affect the process. The constant nature of these causes tends to affect all sampled observations, and hence sampled items both within and between subgroups. This is called *within group variation* and is the most difficult type of variation to identify and interpret. The effect of within group variation is to create control chart patterns in which subgroup statistics hug the centerline; in other words, to create control limits too wide for the subgroup statistics.

Several examples of causes of variation that could generate within group variation are:

Subcomponents used in final assemblies that come from two or more different sources.

Persistent differences in operators, where their work is mixed further down the line.

Variation in gauges, where measured items are mixed and used in later operations.

Within group variation is shown in the following extreme example. Shafts are cut to length by two machines, A and B. Each machine cuts 50 percent of all shafts in approximately the same amount of time. Machine A's shafts are all good, but machine B's shafts are all defective. Figure 11.2 presents a schematic of machines A and B.

As shafts are finished, they are placed on the conveyor belt and fall into a bin, which can hold 100 shafts. Once a bin is filled, a new one is placed at the end of the conveyor to take its place. Consequently, approximately 50 percent of the shafts in any bin are from machine A; the other 50 percent are from machine B.

Bins are then taken to the next operation. Employees at this next operation have started to complain about defective shafts. An inspection station is set up, as shown in Figure 11.3, and a control chart is constructed from 100 percent inspection of every 20th bin; these data appear in Figure 11.4.

FIGURE 11.2 Work Flow of Cut-to-Length Operation

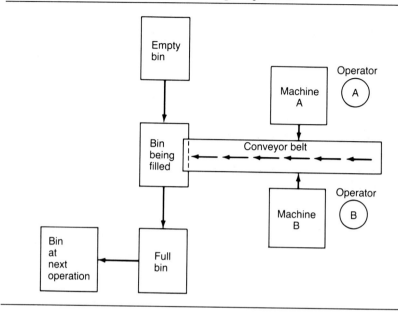

FIGURE 11.3 Work Flow of Cut-to-Length Operation with an Inspection Station

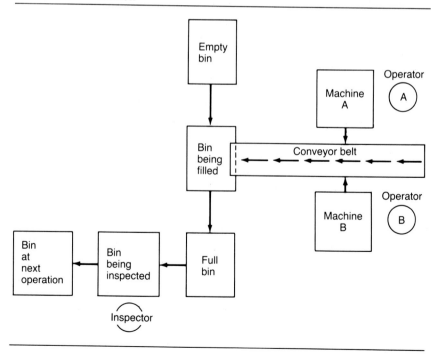

FIGURE 11.4 Cut-to-Length Data

Bin	Number of Items in Bin	Defective Items in Bin
1	100	48
2	100	53
3	100	46
4	100	47
5	100	50
6	100	53
7	100	48
8	100	53
9	100	47
10	100	49
11	100	53
12	100	47
13	100	51
14	100	49
15	100	48
Totals	1,500	742

$$p = .4947$$

$$UCL = .4947 + 3 \sqrt{\frac{.4947(.5053)}{100}} = .6447$$

$$LCL = .4947 - 3 \sqrt{\frac{.4947(.5053)}{100}} = .3447$$

An examination of the control chart in Figure 11.5 shows that the defective fraction hugs the centerline. Recall that one would expect approximately two thirds of all subgroup fractions to fall within one standard error of the mean; in this case, 100 percent fall in this region. Stated another way, it is extremely unlikely that a run of 13 or more points (in this case, 15) in a row would fall within a one-sigma band on either side of the mean. The process is abnormally quiet.

A novice to control chart interpretation might say that this process exhibits a large degree of stability and predictability, albeit at a very high defect rate; this is totally erroneous. The shaft-cutting process is plagued by within group variation. Each bin is made up of approximately 50 percent defective and 50 percent good shafts, resulting in large within group variation and small between group variation. As Figure 11.5 shows, this leads to computational procedures that generate control limits too wide for the subgroup statistics.

Three issues must be addressed in this shaft problem. First, the subgrouping should be made on a rational basis; that is, the sample should be

FIGURE 11.5 p Chart for Cut-to-Length Data

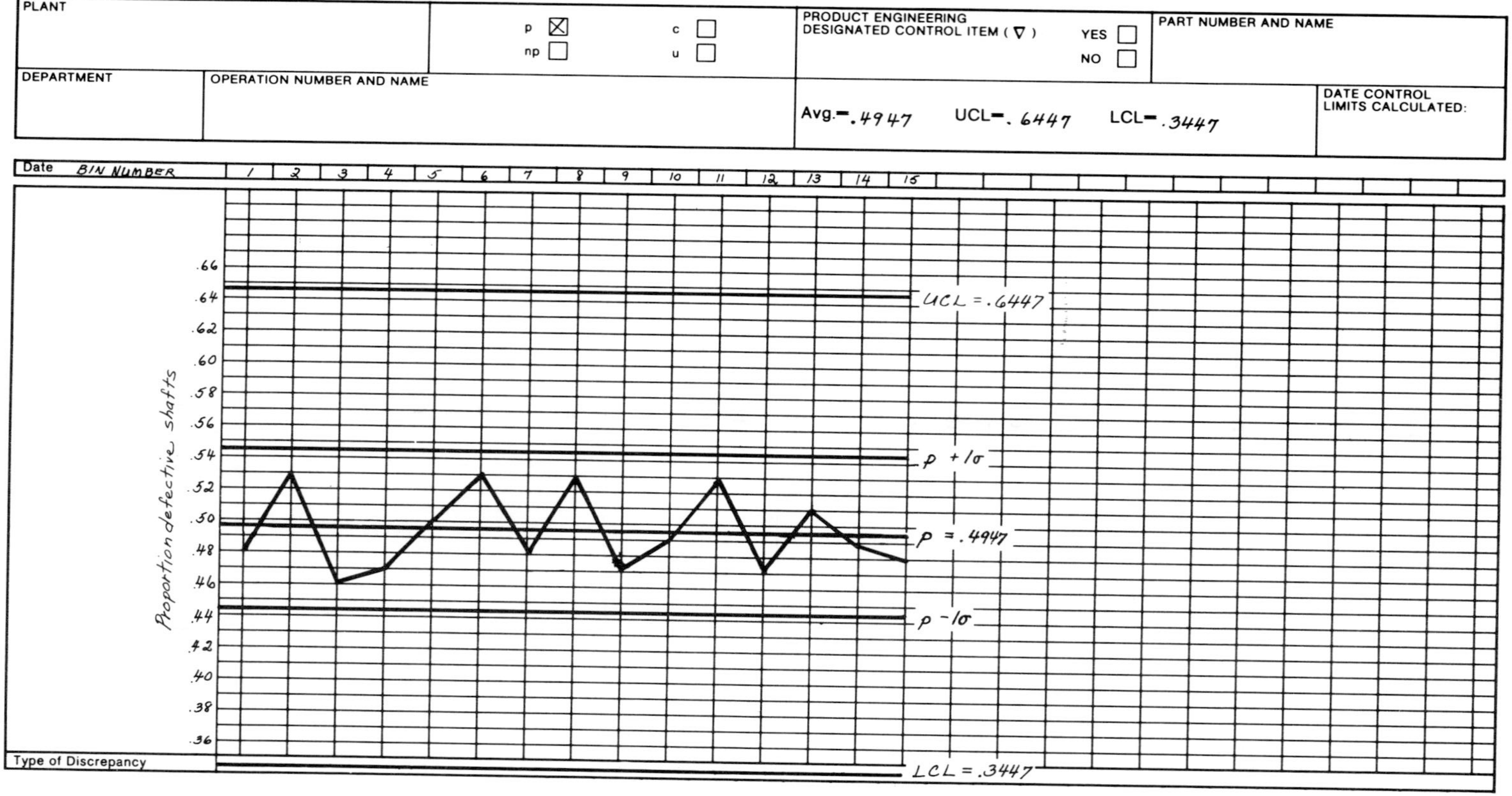

made separately from machines A and B. Second, the causes of machine B's defective output must be corrected. Last, both machines should be continually improved using statistical methods.

Distinguishing Within Group Variation from Common Variation

Within group special sources of variation are persistent, as are common sources of variation. However, the critical distinction is that within group special sources of variation are external (exogenous) disturbances to the process, while common sources of variation are internal (endogenous) to the process.

It is important to realize that both between and within group special sources of variation must be removed from a process before the process can be considered stable. As we have discussed, stability is essential to process improvement.

TYPES OF CONTROL CHART PATTERNS

Identifiable control chart patterns can occur as a consequence of the presence of between and/or within group causes of special variation in a process. Fifteen characteristic patterns have been identified by Western Electric Company engineers[2]: natural patterns; shift in level patterns (sudden shift in level, gradual shift in level, and trends); cycles; wild patterns (freaks and grouping/bunching); multi-universe patterns (mixtures, stable mixtures, associated with systematic variables and stratification, and unstable mixtures, associated with freaks and grouping/bunching); instability patterns; and relationship patterns (interaction and tendency of one chart to follow another). These patterns are useful because they can be compared to control charts in practice and used as diagnostic tools to detect special sources of variation.

Natural Patterns

A *natural pattern* is one that does not exhibit any points beyond the control limits, runs, or other nonrandom patterns and has most of the points near the centerline (approximately two thirds of the points within a one-sigma band of the centerline). Natural processes are not disturbed by either between group or within group special causes of variation. The process demonstrates a stable system of variation. Figures 11.6 and 11.7 illustrate a natural process.

FIGURE 11.6 Data from a Process Exhibiting a Natural Pattern

Day	Date	Subgroup Numbers				
		1	2	3	4	5
1	8/30	4.9	5.5	5.3	5.6	5.1
2	31	5.8	5.5	5.6	6.3	5.7
3	9/ 1	5.9	6.2	5.9	5.8	5.4
4	2	5.9	5.9	6.4	5.3	5.2
5	3	6.2	5.9	5.7	4.9	5.9
6	6	6.0	5.7	5.7	6.3	6.0
7	7	5.2	4.6	5.4	6.1	5.2
8	8	5.1	5.8	6.2	5.9	5.6
9	9	5.8	6.1	5.7	6.5	5.2
10	10	5.2	5.4	5.2	5.8	4.6
11	13	5.2	4.6	5.4	6.1	5.2
12	14	6.2	5.8	5.1	5.2	5.4
13	15	4.9	4.9	4.9	4.9	4.8
14	16	6.1	6.2	5.9	4.5	5.6
15	17	5.3	5.4	5.4	5.4	5.2
16	20	6.3	5.6	5.9	4.7	6.2
17	21	5.4	5.3	5.4	5.2	5.2
18	22	5.6	5.0	5.2	4.9	4.7
19	23	4.7	4.7	5.6	5.0	5.2
20	24	6.0	5.3	5.6	5.0	5.2
21	27	5.1	6.2	5.0	5.2	5.7
22	28	5.6	6.1	5.8	5.9	5.8
23	29	5.5	6.0	5.7	5.0	4.9
24	30	4.9	4.8	6.1	5.3	5.2
25	10/ 1	4.7	4.9	5.2	6.0	5.7
26	4	6.0	5.1	5.3	4.9	6.1
27	5	5.5	5.0	6.0	5.7	5.0
28	6	5.2	4.9	5.2	5.0	5.3
29	7	5.0	5.2	6.0	4.9	6.1
30	8	5.5	5.1	5.3	4.1	5.0

It is frequently necessary to create external disturbances (special sources of variation) to a natural process to create improvements—for example, to move a process's average toward nominal or to reduce the number of defects per unit. The purpose of these external disturbances is to alter the basic structure of the process.

Shift in Level Patterns

There are three types of *shift in level patterns:* sudden shift in level, gradual shift in level, and trends.

FIGURE 11.7 Control Chart for a Natural Process

PLANT	DEPT.	OPERATION	DATE CONTROL LIMITS CALCULATED	ENGINEERING SPECIFICATION	PART NO
156	Engine	Induction Harden + Quench		7.0 ± 3.5 mm	12 - 8963
MACH NO	DATES	CHARACTERISTIC		SAMPLE SIZE/FREQUENCY	PART NAME
3	8/30 - 10/8	Case hardness depth		9 every day	Cam shaft

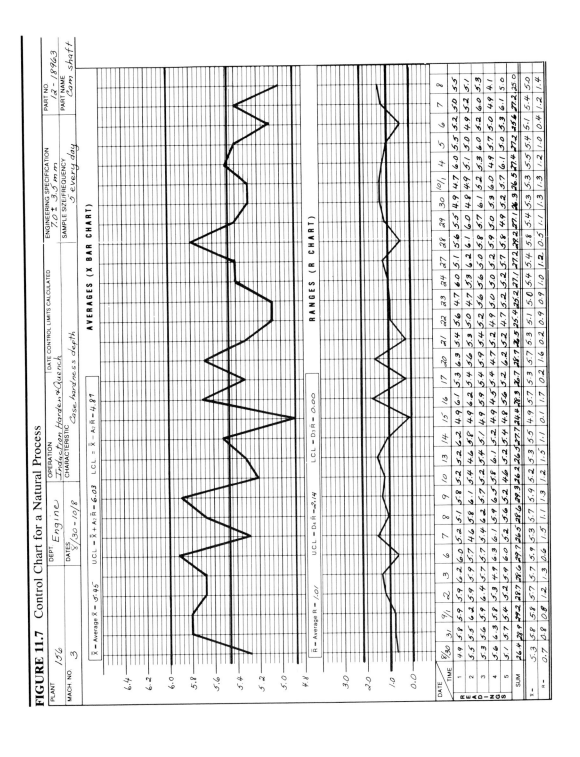

$\bar{X}$ – Average $\bar{X}$ = 5.45   UCL = $\bar{\bar{X}}$ + A₂$\bar{R}$ = 6.03   LCL = $\bar{\bar{X}}$ – A₂$\bar{R}$ = 4.87

AVERAGES (X BAR CHART)

$\bar{R}$ – Average R = 1.01   UCL = D₄$\bar{R}$ = 2.14   LCL = D₃$\bar{R}$ = 0.00

RANGES (R CHART)

DATE	8/30	31	9/1	2	3	6	7	8	9	10	13	14	15	16	17	20	21	22	23	24	27	28	29	30	10/1	4	5	6	7	8
TIME																														
R 1	4.9	5.8	5.9	6.2	6.2	6.0	5.2	5.1	5.8	5.2	5.2	5.2	4.9	6.1	5.3	6.3	5.4	5.6	4.7	6.0	5.1	5.6	5.5	4.9	4.7	6.0	5.5	5.2	5.0	5.5
E 2	5.5	5.5	6.2	5.9	5.9	5.7	4.6	5.8	6.1	5.4	4.6	5.8	4.9	6.2	5.4	5.6	5.3	5.0	4.7	5.3	6.2	6.1	6.0	4.8	4.9	5.1	5.0	4.9	5.2	5.1
A 3	5.3	5.6	5.9	6.4	5.7	5.7	5.4	6.2	5.7	5.2	5.4	5.1	4.9	5.9	5.4	5.9	5.4	5.2	5.6	5.6	6.2	5.8	5.7	6.1	5.2	6.0	5.2	6.0	5.3
D 4	5.6	6.3	5.8	5.3	4.9	6.3	5.9	5.9	6.3	5.8	6.1	5.2	4.9	4.5	5.5	4.7	5.2	4.9	5.0	5.0	5.2	5.9	5.0	5.3	6.0	4.9	5.7	5.0	4.9	4.1
S 5	5.1	5.7	5.2	5.9	6.0	6.0	5.2	5.6	5.2	4.6	5.2	5.4	4.8	5.6	5.4	6.2	5.2	4.7	5.2	5.7	5.8	4.9	5.3	5.2	5.7	6.1	5.3	5.3	6.1	5.0
SUM	26.4	28.9	29.2	28.7	28.6	29.2	26.5	28.6	29.3	26.2	26.5	27.7	24.4	28.3	26.7	28.7	26.5	25.4	25.2	27.1	27.2	29.2	27.1	27.2	26.5	27.4	27.2	25.6	27.2	25.0
$\bar{X}$ –	5.3	5.8	5.8	5.7	5.7	5.9	5.3	5.7	5.9	5.2	5.3	5.5	4.9	5.7	5.3	5.7	5.3	5.0	5.0	5.4	5.3	5.8	5.4	5.3	5.3	5.5	5.4	5.1	5.4	5.0
R –	0.7	0.8	0.8	1.2	1.3	0.6	1.5	1.1	1.3	1.2	1.5	1.1	0.1	1.7	0.2	1.6	0.2	0.9	0.9	1.0	1.2	0.5	1.1	1.3	1.3	1.2	1.0	1.2	0.4	1.4

Sudden Shift in Level. A *sudden shift in level pattern* involves a sudden rise or fall in the level of data on a control chart. This is one of the most easily detectable control chart patterns.

Sudden shifts on x-bar charts, or on charts for individuals, frequently result from a special source of variation, which first shifts the process's average to a new level, then has no further effect on the process. Sudden shifts on R charts can indicate the presence of some related variable affecting the process variability. For example, the addition of a new untrained worker to a trained and stable work force could increase the variability of output.

Sudden shifts on p charts can indicate such factors as a dramatic change in materials, methods, personnel, or operational definitions. Sudden shifts up indicate process degeneration, while sudden shifts down indicate process improvement.

To illustrate a sudden shift in level on a p chart, daily samples of 1,000 medium-sized ratchets are taken at random from a production process and tested for tight levers. The data appear in Figure 11.8, and Figure 11.9 illustrates the p chart. Variations are much larger than they should be, as shown by many out-of-control points. Some unpredictable factor is appar-

FIGURE 11.8   Ratchet Tight Lever Data

Date	Sample Size	Number Defectives	Fraction Defective
Oct. 3	1,000	25	.025
4	1,000	18	.018
5	1,000	16	.016
6	1,000	20	.020
7	1,000	33	.033
10	1,000	65	.065
11	1,000	30	.030
12	1,000	92	.092
13	1,000	45	.045
14	1,000	26	.026
17	1,000	17	.017
18	1,000	30	.030
19	1,000	8	.008
20	1,000	74	.074
21	1,000	41	.041
24	1,000	29	.029
25	1,000	28	.028
26	1,000	35	.035
27	1,000	90	.090
28	1,000	51	.051

FIGURE 11.8 *Concluded*

Date	Sample Size	Number Defectives	Fraction Defective
Nov. 2	1,000	53	.053
3	1,000	67	.067
4	1,000	34	.034
5	1,000	55	.055
6	1,000	24	.024
9	1,000	60	.060
10	1,000	81	.081
11	1,000	44	.044
12	1,000	50	.050
13	1,000	46	.046
16	1,000	12	.012
17	1,000	28	.028
18	1,000	40	.040
19	1,000	23	.023
20	1,000	29	.029
23	1,000	27	.027
24	1,000	65	.065
25	1,000	55	.055
26	1,000	69	.069
27	1,000	18	.018
30	1,000	51	.051
Dec. 1	1,000	47	.047
2	1,000	40	.040
3	1,000	52	.052
Totals	44,000	1,843	

$$\text{Average fraction defective} = p = \frac{1,843}{44,000} = 0.0419$$

$$p \pm 3 \sqrt{\frac{p(1 - p)}{n}} = .0419 \pm 3 \sqrt{\frac{.0419(1 - .0419)}{1000}}$$

$$= .061 \text{ and } .023$$

---

ently preventing consistent quality of work. A study of the assembly process soon reveals that the fixture used in welding the lever is designed so poorly that consistent welds cannot be obtained. The fixture is redesigned and more data collected, as shown in Figure 11.10. The redesigned fixture has eliminated most of the trouble and reduced the average percentage of tight levers from 4.2 to 0.6 percent. The old and new p charts, which are shown in Figure 11.11, demonstrate a sudden shift in level.

FIGURE 11.9 p Chart for Ratchet Tight Lever Data

Note in Figure 11.11 that the two points in the latter part of December are out of control as a result of a batch of faulty levers. Nevertheless, the 1.7 percent and 1.5 percent defective pieces found on these two days are fewer than were found most of the time on the old fixture.[3]

Gradual Shift in Level. *Gradual shifts in level* generally indicate that some portion of the process has been changed and that the effect of this change is a gradual shift in the average level of output from the process. For example, if new employees are put onto the work floor or new machines or maintenance procedures are set into motion, as they become integrated into the existing process they will continually and increasingly affect the average level of output. This type of pattern (in the constructive direction) is very common in the early stages of quality improvement efforts. Figure 11.12 depicts a gradual shift in level on a p chart.

Trends. *Trends* are steady changes, increasing or decreasing, in control chart level; they are gradual shifts in level that do not settle down. Trends can result from special sources of variation that gradually affect the process.

FIGURE 11.10   Redesigned Ratchet Tight Lever Data

Date	Number Defectives	Fraction Defective
Dec. 4	8	.008
5	10	.010
7	2	.002
8	5	.005
9	10	.010
10	6	.006
11	7	.007
14	6	.006
15	3	.003
16	1	.001
17	2	.002
18	3	.003
21	0	.000
22	4	.004
23	7	.007
24	12	.012
28	9	.009
29	17	.017
30	15	.015
31	10	.010
Jan. 3	8	.008
4	0	.000
7	6	.006
8	0	.000
9	3	.003
10	0	.000
11	1	.001
14	13	.013
15	12	.012
Total	180	

$$p = \frac{180}{29,000} = 0.0062$$

$$p \pm 3 \sqrt{\frac{p(1-p)}{n}} = .0062 \pm 3 \sqrt{\frac{.0062(1 - .0062)}{1000}}$$

$$= .0136 \text{ and } .0000$$

Trends on x-bar charts or individuals charts, result from disturbances (special sources of variation) that shift the process level up (or down) over time: tool wear; loosening of guide rails or holding devices; operator fatigue; and so on. Figure 11.13 shows trend on an x-bar chart for slot depth caused by tool wear in a cutting jig. Trends are relatively easy to

FIGURE 11.11 p Charts Comparing Ratchet Tight Lever Data before and after Fixture Redesign

detect, although those inexperienced in control chart diagnosis often see trends when they do not exist; thus caution is advised.

Cycles

Cycles are repeating waves of periodic low and high points on a control chart caused by special disturbances that appear and disappear with some degree of regularity, such as morning startups and periodic shifting of operators on x-bar charts; fluctuations in operator fatigue caused by coffee breaks and differences between shifts on R charts; and regular changes in inspectors on a p chart.

If a sampling frequency coincides with the cycle's pattern, sampling may reveal only high or low points; in this case, cycles will not show up on a control chart. The remedy is to sample more frequently to detect the cyclical pattern. Process knowledge is essential to detecting cycles. Figure 11.14 shows a control chart exhibiting cyclical special sources of variation.

Wild Patterns

There are two types of *wild patterns:* freaks and grouping/bunching. Both patterns are characterized by one or more subgroups that are very different from the main body of subgroups.

FIGURE 11.12 Control Chart for a Gradual Shift in Level

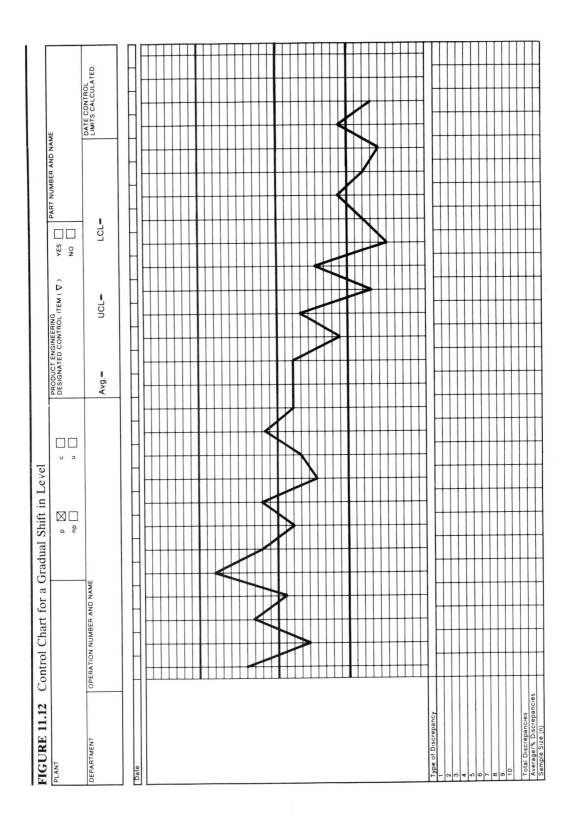

FIGURE 11.13 Trend on x-Bar Chart for Slot Depth Resulting from Tool Wear in Cutting Jig

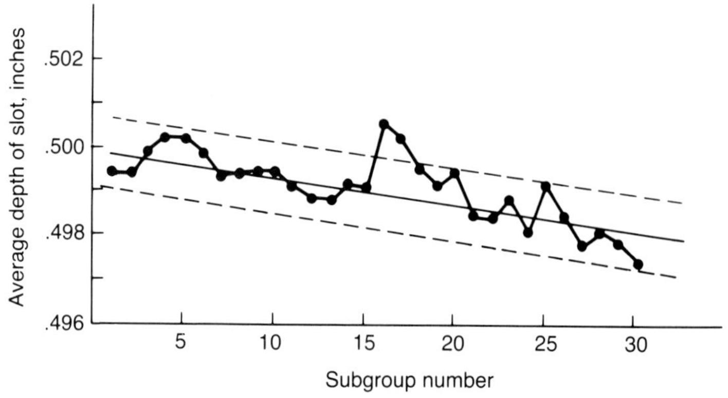

Freaks. *Freaks* can be caused by calculation errors or by external disturbances that can dramatically affect one subgroup. They show up on a control chart as points significantly beyond the control limits. Freaks are one of the easiest patterns to recognize, and it is usually relatively easy to determine the special cause of variation for the freak. Figure 11.15 shows an example of a freak.

Grouping/Bunching. *Grouping/bunching* is caused by the introduction into a process of a new system of disturbances that affect a "group" or "bunch" of points that are close together. Figure 11.16 illustrates grouping/bunching.

Multi-Universe Patterns

There are three *multi-universe patterns* and two groups of associated patterns. The three multi-universe patterns are mixtures, stable mixtures, and unstable mixtures. The first group of associated patterns are systematic variables and stratification, related to stable mixtures. The second group of associated patterns are freaks and grouping/bunching, related to unstable mixtures.

All multi-universe patterns are characterized by an absence of points near the centerline (large fluctuations) or by too many points near the centerline (small fluctuations).

Mixtures. *Mixture patterns* indicate the presence of two or more distributions for a quality characteristic; for example, two distributions of pipe

FIGURE 11.14 Control Chart for Cycle Pattern

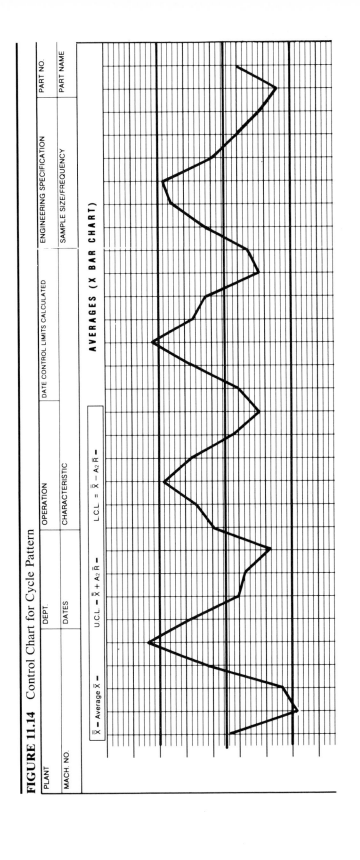

FIGURE 11.15 Control Chart for Freak Pattern

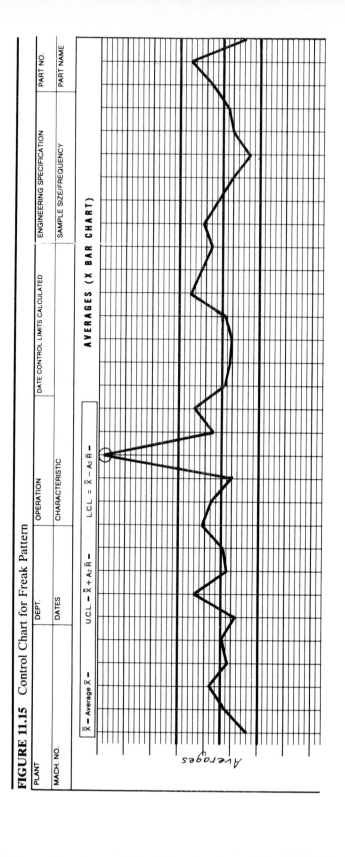

FIGURE 11.16 Control Chart for Grouping/Bunching Pattern

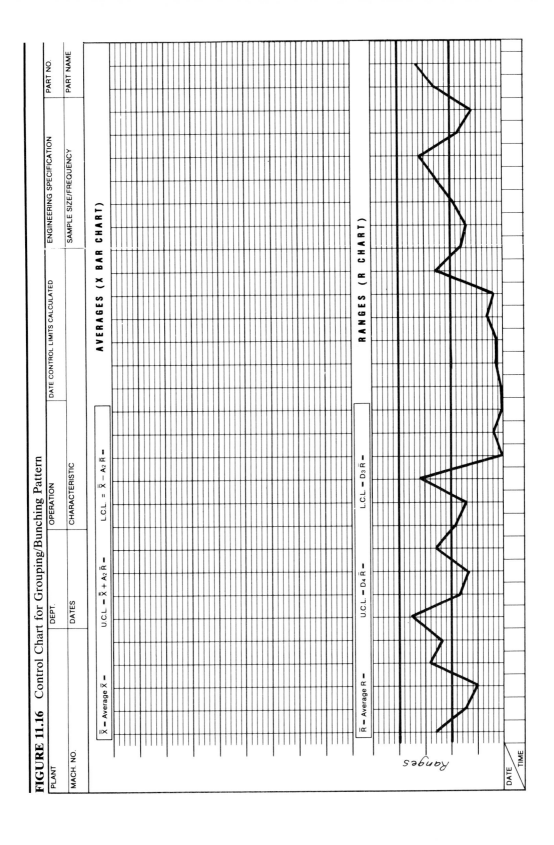

section diameter caused by numerous pipes coming from two different vendors. Mixtures become more apparent the greater the difference between the component distributions. There are two basic forms of mixtures: stable mixtures and unstable mixtures.

Stable Mixtures.   *Stable mixtures* indicate the presence of two or more distributions for a quality characteristic that does not change over time with respect to the proportion of items coming from each distribution and/ or the average for each distribution. For example, two vendors supply units to a buyer. Seventy percent of the incoming units are purchased from vendor A and weigh 10 pounds on average; the remaining 30 percent of incoming units are purchased from vendor B and weigh 12 pounds on average. As another example, samples are drawn consistently from two shifts or machines. These are stable mixture problems because the proportion of items coming from each distribution is stable over time.

Stable mixture patterns are characterized by an unusually high presence of control chart points near (or beyond) the upper and lower control limits or near the centerline.

Systematic Variables.   If samples are drawn separately from the component distributions, then the stable mixture pattern will appear on the control chart. This is a case of *systematic variables*. For example, if samples are alternatively drawn from two shifts that are widely different in output (two distributions), the points on a control chart will saw-tooth up and down—shift A high, shift B low, shift A high, shift B low, and so on. This is a systematic variable pattern. An example of the systematic variable form of a stable mixture is differences between tools or differences between shifts, where the data is systematically plotted to bring out the differences between tools or shifts. For example, suppose a box plant produces corrugated boxes. These boxes have many quality characteristics; in this example we will focus on "glued tab width." The glue tab is what forms the manufacturer's joint of a corrugated box, as shown in Figure 11.17. The glued tab width is important to insure proper strength of the final box.

Ten subgroups of the glued tab widths of four corrugated boxes were drawn from the outputs of shift A and shift B; Figure 11.18 presents the data and x-bar and R charts. The x-bar chart shows a classic saw-tooth pattern leading to the unusually high presence of control chart points beyond the control limits. The problem is the presence of a variable that systematically affects the process—in this case, shifts. The control charts in this example should be separated into the x-bar and R charts for shift A and shift B.

Stratification.   If samples are drawn from two or more distributions that have been combined, then the stable mixture pattern can create extremely small differences among statistics on x-bar, R, individuals, or p charts (an unusually high presence of control chart points near the cen-

---

FIGURE 11.17   Glued Tab Width

The glue tab is what forms the manufacturer's
joint of the box.
The carrier regulations specify that the
overlapped width of the joint must be a
minimum of 1¼". The glue tab is made
1⅜" wide so as to meet the minimum 1¼"
when the box blank is folded and glued.

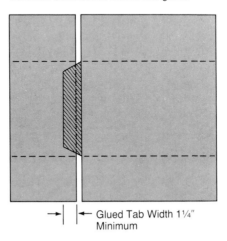

Glued Tab Width 1¼"
Minimum

---

terline). The small differences on x-bar, R, individuals, or p charts are
frequently interpreted by the novice control chart user as representing
unusually good control; nothing could be further from the truth.

An excellent example of *stratification* was shown in Figure 11.5. Note
the unusually high presence of control chart points near the average proportion defective. The control chart is not valid. In the presence of stratification, the control chart user must first correct errors in the methods of
sampling so that items from component distributions do not get mixed in
each sample.

Another illustration of the stratification-stable mixture pattern can be
shown via the box plant example discussed in Figures 11.17 and 11.18.
Suppose the data on the glued tab width had been collected and recorded
by a very concerned employee who wanted to make sure he obtained a
"representative" sample of output and consequently took two boxes
from each shift's output and grouped them together to form one day's
sample. The data and x-bar and R charts are shown in Figure 11.19. The
charts show unnaturally quiet patterns. The control charts in this example
are invalid because the concerned operator sampled simultaneously from
both shifts. The proper procedure would have been to sample separately
from each shift and have rational subgrouping.

FIGURE 11.18 Control Chart for Systematic Variable-Stable Mixture Pattern

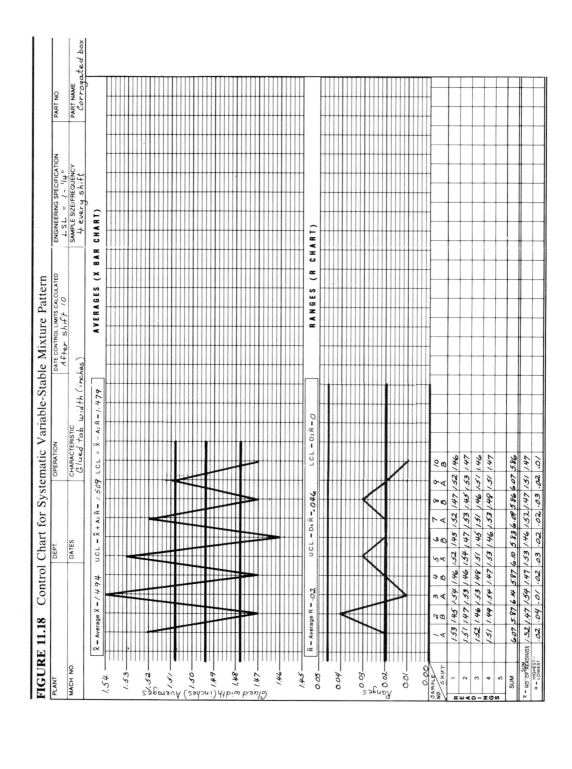

FIGURE 11.19 Control Chart for Stratification-Stable Mixture Pattern

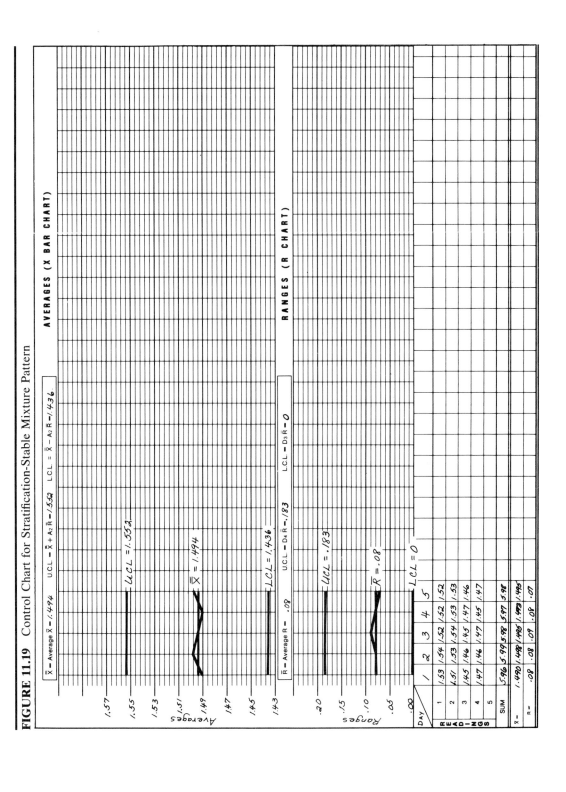

Unstable Mixtures.   *Unstable mixtures* indicate the presence of two or more distributions for a quality characteristic that changes over time with respect to the fraction of items coming from each distribution and/or the average for each distribution—for example, a buyer has two vendors for an item, and the fraction coming from each vendor and/or the quality characteristic averages for each vendor change over time. Figure 11.20 depicts an unstable mixture pattern.

Freaks and Grouping/Bunching.   The nature of an unstable mixture pattern implies that the multiple component distributions that make up the distribution of a quality characteristic are sporadically affected by special disturbances. This will cause a systematic variable effect; but this effect will occur unevenly, generating an unusually large number of control chart points near or beyond the control limits. However, those points will be in groups or bunches, depending upon the juxtaposition of the various component quality characteristic distributions. This pattern could also result in freaks.

Instability Patterns

Erratic points on a control chart exhibiting large swings up and down characterize a pattern called *instability*. Instability is a possible result of special variation when the control limits appear too narrow for the control chart; instability is characterized by large, erratic fluctuations in subgroup statistics. Instability is frequently associated with unstable mixtures.

Instability is caused either by one special disturbance that can sporadically affect the average or variability of a process or by two or more special disturbances, each of which can affect the average and/or variability of a process. These disturbances interact with each other and create complex process disturbances.

Simple instability occurs when one special disturbance creates a wide, bimodal or multimodal distribution in a quality characteristic or sporadically shifts the process average—for example, occasional lots of material from a supplier that are extremely good or bad, or sporadic adjustment of a machine. Both of the above situations create either wide, bimodal, or multimodal quality characteristic distributions or sporadically shift the quality characteristic's average. They create patterns in which control chart limits will appear too narrow for the subgroup statistics. For example, the erratic effect of the bad (or good) lot or the overadjustment of the machine create a situation comparable to the shift-to-shift differences found in the systematic variable pattern. Figure 11.21 depicts a pattern of instability.

FIGURE 11.20 Control Chart for Unstable Mixture Pattern

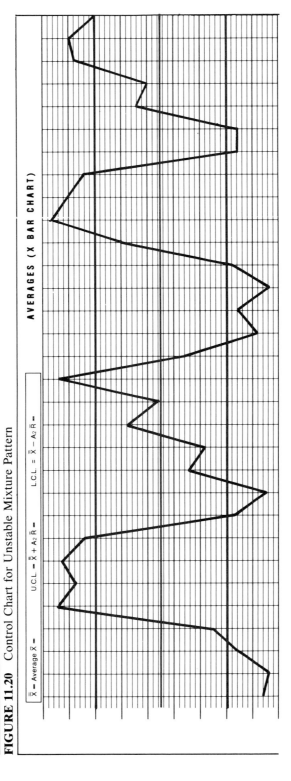

FIGURE 11.21 Control Chart for Simple Instability

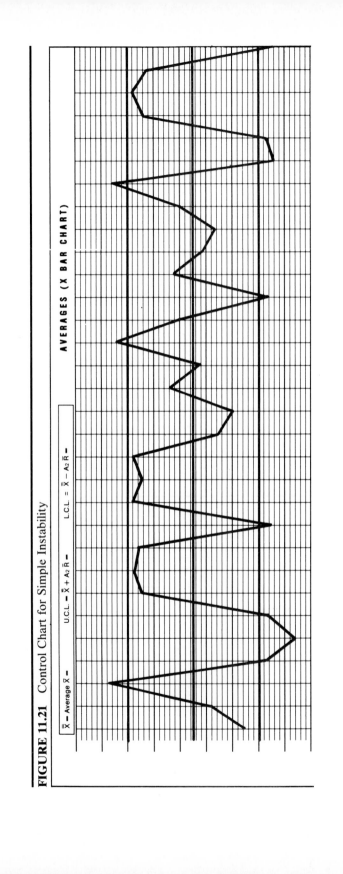

$\overline{\overline{X}}$ – Average $\overline{X}$ –   U.C.L. – $\overline{\overline{X}}$ + A$_2$ $\overline{R}$ –   L.C.L. = $\overline{\overline{X}}$ – A$_2$ $\overline{R}$ –

AVERAGES (X BAR CHART)

Relationship Patterns

There are two types of *relationship patterns:* interaction and tendency of one chart to follow another.

Interaction. *Interaction patterns* occur when one variable affects the behavior of another variable, or when two or more variables affect each other's behavior and create an effect that would not have been caused by either variable alone.

Interactions between variables are best investigated and understood through a set of statistical tools called experimental design. Interactions can also be investigated and understood through process capability studies.[4]

Interactions can be detected on x-bar, individuals, and p charts by changing the rational subgrouping of the data. For example, if the data in Figure 11.18 are broken into a shift A segment and a shift B segment (i.e., changing the rational subgrouping of the data), the interaction between glued tab width and shifts becomes apparent, as shown in Figure 11.22.

A run of low points on an R or s chart indicates that an interacting variable affecting the variability of the process has been temporarily removed or held at one level. This realization may lead to permanent removal of the interacting variable or continuous maintenance of the interacting variable at one level, consequently reducing the amount of process variability. Figure 11.23 illustrates this. The run of low points indicates the presence of an interacting variable that has temporarily reduced the variability of the process. It is a matter of expertise in the process being studied to identify and manipulate the interacting variable to aid process improvement.

Tendency of One Chart to Follow Another. These patterns may exist between two or more variables if the control charts for the variables tend to follow each other on a point-to-point basis. This type of pattern most frequently occurs when the control charts in question have been constructed from the same samples. For example, a sample of four items can be measured with respect to three different quality characteristics, and each measurement is plotted on its respective control chart. Figure 11.24 illustrates one chart following another.

In the case of x-bar and R charts, it is natural that the corresponding points come from the same samples. However, if the samples were randomly drawn from a normal distribution, there is no relationship between the corresponding points on x-bar and R charts; if the samples were randomly drawn from a skewed distribution, there is a relationship between the corresponding points on x-bar and R charts. Further, the greater the skewness of the distribution from which the samples are drawn, the greater the relationship between the corresponding points on

FIGURE 11.22 Control Chart for Interaction Pattern

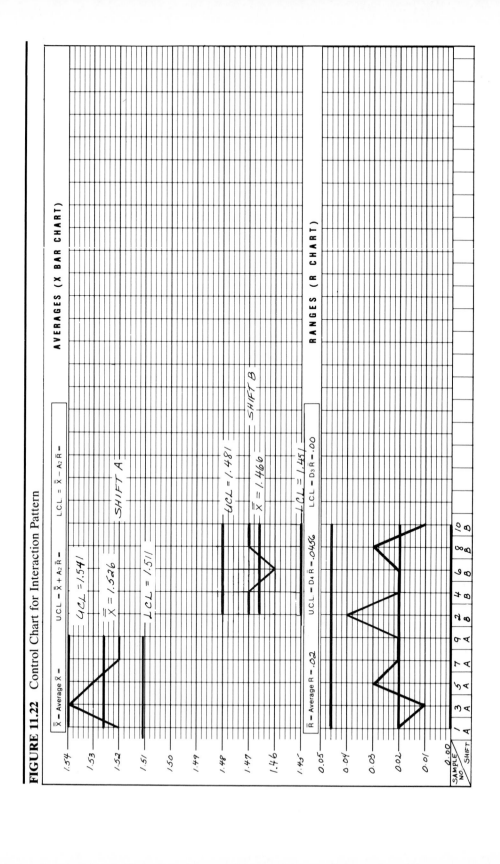

FIGURE 11.23 Control Chart Showing a Run of Low Variation: Interacting Variable

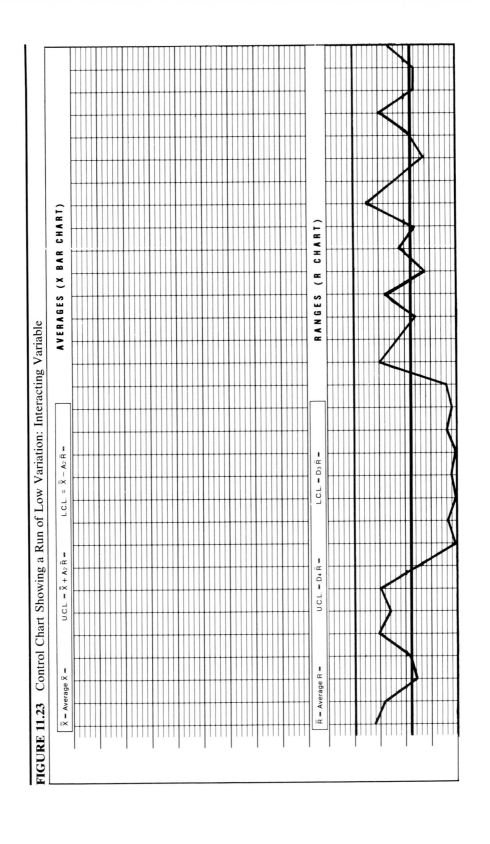

FIGURE 11.24 Control Chart for One Chart Following Another

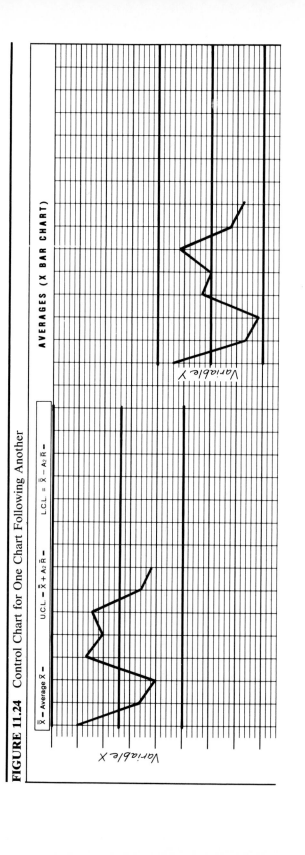

the x-bar and R chart. Consequently, point-to-point correspondence on x-bar and R charts is a sign of skewness in the conceptual population.

RULES OF THUMB FOR DETECTING OUT-OF-CONTROL PATTERNS

Natural control chart patterns exhibit four characteristics:

1. Most of the points are near the centerline.
2. A few of the points are near the control limits.
3. Rarely will a point exceed the control limits.
4. There are no runs among the points.

If one (or more) of these conditions is missing in a control chart pattern, the pattern will appear unnatural. Consequently, unnatural patterns will exhibit one or more of the following characteristics:

1. Absence of points near the centerline.
2. Absence of points near the control limits.
3. Points located beyond the control limits.
4. Runs or nonrandom patterns among the points.

For example, a stratification pattern would exhibit the absence of points near or beyond the control limits (too many points near the centerline); a grouping/bunching pattern would exhibit the absence of points near the centerline (too many points near or beyond the control limits); a freak pattern would exhibit points beyond the control limits; a systematic variable pattern would exhibit an unusually high number of runs up and down (high, low, high, low, high); and a gradual shift or trend pattern would exhibit an unusually long run of points up or down.

The rules of thumb for unnatural patterns are based on probability calculations that indicate, for a stable process, the proportion of points that will fall near the centerline, near the control limits, or beyond the control limits; or the expected number of runs above average, the expected number of runs below average, the length of a run up, or the length of a run down. The rules of thumb generally are applied beginning with the last point plotted and working backwards to consider as many points as are necessary to apply the rule. The probability calculations are only approximations because a stable process varies over time with respect to its parameters.

Recall the five out-of-control rules of thumb we introduced in Chapter 8. In the following section we will discuss these five rules, plus others, from a probability point of view.

There are one-sided and two-sided rules of thumb. One-sided rules of thumb refer to rules applied to either the upper or lower half of the control chart (both upper and lower sides can be considered, but they must be considered separately). The first four rules are one-sided rules of thumb. Two-sided rules of thumb are rules applied simultaneously to both sides of the control chart. Rule 5 is a two-sided rule of thumb.

One-Sided Rules of Thumb

These rules, introduced in Chapter 8 as Rules 1 through 4, are listed below and discussed in detail for normally distributed statistics, nonsymmetrically distributed statistics, and discrete statistics.

Normally Distributed Statistics.   Recall from Chapter 8 that for control charts whose statistics are continuous, stable, and normally distributed, four one-sided rules of thumb are defined.

Rule 1. A process exhibits a lack of control if any single value falls outside of a control limit; see Figure 8.7.

Rule 2. A process exhibits a lack of control if any two out of three consecutive points fall in zone A or beyond; see Figure 8.20.

Rule 3. A process exhibits a lack of control if four out of five consecutive points fall in zone B or beyond; see Figure 8.21.

Rule 4. A process exhibits a lack of control if eight or more consecutive points lie on one side of the centerline; see Figure 8.22.

The four rules above are subject to the following conditions: If a point falls on the centerline and the points before and after it are the same side of the centerline, consider the point on the centerline a separate run. If a point falls on the centerline and the points before and after it are on opposite sides of the centerline, group the point on the centerline with the shortest adjacent run.[5]

Probabilities.   The rationale for these one-sided rules of thumb is based on the normal distribution, requires a knowledge of the probability that a point will be in the different control chart zones, and assumes only one side of the control chart is being studied; these zones were introduced in Chapter 8, and are shown in Figure 8.19. The probabilities assume that the process is stable. The probability that a point will be beyond zone A is 0.00135. The probability that a point will be in zone A is 0.02135. The probability that a point will be in zone B is 0.1360. The probability that a point will be in zone C is 0.3413. The same probabilities apply to both sides of the control chart. Consequently, the one-sided probabilities for the rules of thumb can be computed.

Rule 1. The probability that a point will be beyond zone A is 0.00135.

Rule 2. The probability that two out of three consecutive points will be in zone A or beyond requires knowing the probability of being in zone A or beyond. This probability is the sum of 0.00135 and

0.02135, and equals 0.0227. Hence, using the binomial distribution with p = 0.0227, n = 3, and x = 2, the probability that two out of three points will be in zone A or beyond is 0.0015.

Rule 3. The probability that four out of five consecutive points will be in zone B or beyond requires knowing the probability of being in zone B or beyond. This probability is the sum of 0.00135, 0.02135, and 0.1360, and it equals 0.1587. Hence, using the binomial distribution with p = 0.1587, n = 5, and x = 4, the probability that four out of five points will be in zone B or beyond is 0.0027.

Rule 4. The probability that eight or more points will fall on one side of the centerline requires knowing the probability of being in zone C or beyond. This probability is 0.5000, the sum of 0.00135, 0.02135, 0.1360, and 0.3413. Hence, using the binomial distribution with p = 0.5, n = 8, and x = 8, the probability that eight successive points will fall on one side of the centerline is 0.0039.

As the probability that any one rule will occur by chance for a stable process is far lower than 1 in 100, when a particular pattern is observed, statisticians conclude that it is unlikely that the observed pattern occurred by chance and recommend searching for special sources of variation. The first four rules apply to either side of a control chart whose statistic is continuous, stable, and normally distributed.

Nonsymmetrically Distributed Statistics.   If the distribution of the control chart statistic is continuous, stable, and nonsymmetrical, then the probabilities associated with the first four rules must be recomputed, depending on the nature of the nonsymmetry. One interesting type of nonsymmetry is where the control chart statistic is drawn from a stable process in which the individual observations are distributed normally. This is interesting because range chart rules are based on this type of nonsymmetry. The rules for range charts based on subgroups of size five follow.

Rule 1. A process exhibits a lack of control if any single point falls outside of a control limit. The probability that a point will be beyond the upper control limit is 0.0046. The probability that a point will be below the lower control limit is less than 0.0001.

Rule 2. A process exhibits a lack of control if any two out of three consecutive points fall in zone A or beyond. The probability that this will occur in the upper half of the control chart is 0.0033 and in the lower half of the control chart is 0.0009.

Rule 3. A process exhibits a lack of control if four out of five consecutive points fall in zone B or beyond. The probability that this will occur in the upper half of the control chart is 0.0026 and in the lower half of the control chart is 0.0054.

Rule 4. A process exhibits a lack of control if eight or more consecutive points lie on one side of the centerline. The probability that this will occur in the upper half of the control chart is 0.0023 and in the lower half of the control chart is 0.0063. Some statisticians suggest that Rule 4 should be modified to a run of 12 below the centerline to keep the likelihood of the upper and lower cases of Rule 4 approximately the same. The probability that there will be a run of 12 or more below the centerline is 0.0005.

Again, the probability that any one rule will occur by chance is far lower than 1 in 100. Hence, these rules are considered to generate economical signals to search for special sources of variation.

Discrete Statistics. If the distribution of the control chart statistic is stable and the variable under study is discrete, then the probabilities associated with the first four rules must once again be recomputed, depending upon the underlying nature of the discrete distribution; this applies to p, np, c, and u charts. For example, p chart rules would be based on either the binomial or Poisson distributions, and c charts would be based on the Poisson distribution. No attempt will be made to develop a separate set of rules and associated probabilities for discrete variables because the distributions are often approximate and the probability calculations may be approximated by the continuous case. Hence, the previously stated rules apply.

One possible problem using rules of thumb with respect to discrete variables being control charted on c or u charts occurs when the process average for the discrete variable is small. As discussed in Chapter 9, this can create a situation where a zone can be entirely between two consecutive whole numbers, giving that zone a zero probability of occurring. For example, a firm makes radios with an average of 0.4 defects per unit, and the number of defects per unit is stable. Consequently, zones A, B, and C would range between the values shown in Figure 11.25. Note that zone B is entirely between one and two defects per unit, and hence has no probability of occurring. Consequently, if the average value of a discrete variable is small, only the first and fourth rules of thumb make sense:

Rule 1. A process exhibits a lack of control if any single value falls outside of a control limit. The probability that this will occur depends on the probability distribution of the statistic under study.

Rule 4. A process exhibits a lack of control if eight or more consecutive points lie on one side of the centerline. The probability that this will occur depends on the probability distribution of the statistic under study.

FIGURE 11.25   Zones for Number of Defects per Unit

$$\bar{c} + 3\sqrt{\bar{c}} \longrightarrow 2.30$$

A

$$\bar{c} + 2\sqrt{\bar{c}} \longrightarrow 1.66$$

B

$$\bar{c} + \sqrt{\bar{c}} \longrightarrow 1.03$$

C

$$\bar{c} \longrightarrow 0.40$$

C

$$\bar{c} - \sqrt{\bar{c}} \longrightarrow -0.23 \rightarrow 0$$

B

$$\bar{c} - 2\sqrt{\bar{c}} \longrightarrow -0.86 \rightarrow 0$$

A

$$\bar{c} - 3\sqrt{\bar{c}} \longrightarrow -1.50 \rightarrow 0$$

Two-Sided Rules of Thumb

These rules apply to the entire control chart. Rule 5 and Rule 6 are based on formal runs tests from a stable process where the variable under study can be either continuous or discrete. Rule 5 was discussed in Chapter 8.

Rule 5. A process exhibits a lack of control if eight or more consecutive points move upward in value or if eight or more consecutive points move downward in value. The run of eight points can be on either side of the centerline or cross over the centerline.

Rule 6. A process exhibits a lack of control if an unusually small number of runs above and below the centerline are present—in other words, a saw-tooth pattern. While this rule of thumb is largely a subjective decision on the part of the person analyzing the control chart, there are formal tests to determine if a saw-tooth pattern is present.[6]

The following rules are used to examine a process to determine if it is unusually quiet (low variability) or unusually noisy (high variability). These rules were not discussed in Chapter 8.

Normally Distributed Statistics. For a stable, continuous, symmetric, normally distributed statistic, such as x-bar, we have the following two-sided rule:

Rule 7. A process exhibits a lack of control if 13 consecutive points fall within zone C on either side of the centerline.

The probability of being in zone C on either side of the centerline is the sum of 0.3413 and 0.3413, and equals 0.6826. Hence, using the binomial distribution with p = 0.6826, n = 13, and x = 13, the probability that 13 consecutive points will be in either zone C is 0.007.

Nonsymmetrically Distributed Statistics. If the distribution of the control chart statistic is stable, continuous, and nonsymmetrical, another version of Rule 7 is needed, depending on the nature of the nonsymmetry. Again, an interesting case of this type is the distribution of the range based on samples of size five:

Rule 7. A process exhibits a lack of control if 13 consecutive points fall within zone C on either side of the centerline.

The probability of being in zone C on either side of the centerline when n = 5 is 0.6516. Hence, using the binomial distribution with p = 0.6516, n = 13, and x = 13, the probability that 13 consecutive points will be in either zone C is 0.0038.

Discrete Statistics. If the distribution of the control chart statistic is stable and the variable under study is discrete, another version of Rule 7 is required, depending on the underlying discrete distribution. No attempt will be made to revise the probability calculations for Rule 7 for discrete variables, because often, discrete variables may be treated as continuous, in which case Rule 7 applies. Again, if the mean of the discrete variable is small, this rule should not be used.

Probability of Error

It is not realistic to attempt to develop rules of thumb for all possible control chart patterns and situations. Consequently, knowledge, experience and expertise are required when analyzing control charts. It is also important to be aware that as the number of rules applied to a control chart increases, so does the probability of falsely searching for a special source of variation. If m independent rules of thumb are applied to a control chart, and if p_i is the probability that the ith test will falsely indicate a special source of variation, then the overall probability of falsely searching for a special source of variation, $p_{overall}$, is:

$$p_{overall} = 1 - \prod_{i=1}^{m} (1 - p_i) \tag{11.1}$$

Not all of the rules discussed are independent; hence, the $p_{overall}$ probability in Equation 11.1 must be used with discretion.

SUMMARY

In this chapter, we began by explaining the different types of special variation in an analytic study, between group variation and within group variation. Between group sources of variation are external sources of variation that affect a process periodically, while within group sources of variation are sources of variation that affect a process persistently.

Next, we presented and described 15 control chart patterns whose detection should be helpful to understanding and eliminating special sources of variation in a process: natural, sudden shift in level, gradual shift in level, trends, cycles, freaks, grouping or bunching, mixtures, stable mixtures with systematic variables, stable mixtures with stratification, unstable mixtures with freaks, unstable mixtures with grouping or bunching, instability, interaction, and tendency of one chart to follow another.

We then discussed several rules of thumb for detecting special sources of variation on control charts. For one-sided rules of thumb, regardless of the symmetry or continuity of the variable under study, the first four rules specified in Chapter 8 provide excellent guidelines for discovering special sources of variation. For two-sided rules of thumb, the fifth rule discussed in Chapter 8 provides another excellent guideline for discovering special sources of variation. In general, all rules specified must be applied cautiously in the context of the process being analyzed.

EXERCISES

11.1 a. Define special variation.
      b. Define common variation.
      c. Define the two types of special variation, between group variation and within group variation. Explain the difference between the two types of special variation. Give examples of each type of special variation.

11.2 Prepare a description of the 15 different types of control chart patterns. Give an example of a situation that would lead to each type of control chart pattern.

11.3 a. Describe the four characteristics of a stable control chart, or the natural control chart pattern.
      b. Describe the four characteristics of a chaotic, or unstable, control chart pattern.

11.4 Calculate the probability that a single value on an x-bar chart for a stable normally distributed process exceeds the upper control limit, given that the R chart is stable (Hint: Rule 1).

11.5 Calculate the probability that two out of three consecutive values on an x-bar chart for a stable normally distributed process fall in zone A or beyond, given that the R chart is stable (Hint: Rule 2).

11.6 Calculate the probability that four out of five consecutive values on an x-bar chart for a stable normally distributed process fall in zone B or beyond, given that the R chart is stable (Hint: Rule 3).

11.7 Calculate the probability that eight consecutive values are on one side of the process centerline on an x-bar chart for a stable normally distributed process, given that the R chart is stable (Hint: Rule 4).

11.8 Calculate the probability that 13 consecutive values fall within zone C on either side of the centerline on an x-bar chart for a stable normally distributed process, given that the R chart is stable (Hint: Rule 7).

NOTES

1. William B. Rice, *Control Charts in Factory Management* (New York: John Wiley & Sons, 1947), pp. 102–4.
2. AT&T, *Statistical Quality Control Handbook,* 10th printing, May 1984 (Indianapolis: AT&T, 1956), pp. 161–80.
3. Paraphrased from Rice, *Control Charts for Factory Management,* pp. 70–74.
4. AT&T, *Statistical Quality Control Handbook,* pp. 75–117.
5. Donald J. Wheeler and David S. Chambers, *Understanding Statistical Process Control* (Knoxville, Tenn.: Statistical Process Controls, Inc., 1986), p. 102.
6. Acheson Duncan, *Quality Control and Industrial Statistics;* 5th ed. (Homewood, Ill.: Richard D. Irwin, 1986) pp. 429–32.
7. Rice, *Control Charts in Factory Management,* p. 104.

Chapter 12

Diagnosing a Process

INTRODUCTION

As we have seen, statistical control charts are important aids in stabilizing and improving a process, and help decrease the difference between customer needs and process performance. However, other techniques can be used in conjunction with control charts to aid in process stabilization and improvement. The methods we will discuss are brainstorming, cause-and-effect diagrams, check sheets, Pareto analysis, and stratification.

BRAINSTORMING

Brainstorming is a way to elicit a large number of ideas from a group of people in a short period of time. The members of the group use their collective thinking power to generate ideas and unrestrained thoughts. Brainstorming is used for several purposes: to determine problems to work on; to find possible causes of a problem; to find solutions to a problem; and to find ways to implement solutions.[1] Brainstorming was fully developed and utilized by the ancient Greeks; it was known as *heuristics*.[2] Dr. Alex Osborn revived brainstorming in the 1940s in his work in advertising, after which the technique became very popular in industrial uses.[3]

Effective brainstorming should take place in a structured session. The group should be small in number, between 3 and 12 as a rule of thumb; having too large a group deters participation. The composition of the group should depend on the issue being examined; and it should include a variety of people, not all of whom should be technical experts in the particular area. The group leader should be experienced in brainstorming techniques. The leader's task is to keep the group focused, prevent distractions, keep ideas flowing, and record the outputs. The brainstorming session should be a closed-door meeting with no interruptions that might break the group's creative process or cause distractions. Seating should promote the free flow of ideas. A U-shape or circle arrangement is suggested. The leader should record the ideas so everyone can see them, preferably on a flip-chart, blackboard, or illuminated transparency.

Procedure

The following steps are recommended for a brainstorming session:

1. Select the topic or problem to be discussed.
2. Each group member makes a list of ideas on a piece of paper. This should take no longer than 10 minutes.
3. Each person reads one idea at a time from his/her list of ideas, sequentially, starting at the top of the list. As the ideas are read, they should be recorded and displayed by the group leader. The group members continue in this circular reading fashion until all the ideas on everyone's list are read.
4. If a member's next idea is a duplication, that member goes on to the subsequent idea on his/her list.
5. Members are free to pass on each go-round but should be encouraged to add something.
6. The leader then requests each group member, in turn, to think of any new ideas he/she had not thought of before. It is very likely that hearing others' ideas will result in more or related ideas. This is called *piggy-backing*. The leader continues asking each group member, in turn, for new ideas, until they cannot think of any more.
7. If the group reaches an impasse, the leader can ask for everyone's "wildest idea." An unrealistic idea can stimulate a valid one from someone else.

Rules

Certain rules should be observed by the participants to ensure a successful brainstorming session—otherwise, participation may be inhibited.

1. Do not criticize, by word or gesture, anyone's ideas.
2. Do not discuss any ideas during the session, except for clarification.
3. Do not hesitate to suggest an idea because it sounds silly. Many times a "dumb" idea can lead to the problem solution.
4. Only one idea should be suggested at a time by each team member.
5. Do not allow the group to be dominated by one or two people.
6. Do not let brainstorming become a gripe session.[4]

Aids to Better Brainstorming

A relaxed atmosphere in which people feel free to suggest any kind of idea enhances the brainstorming session. The following techniques may aid in improving brainstorming sessions by giving people ways to come up with new ideas.[5]

1. *Modification* is changing some aspect of an existing product or service. An example is lower priced movie tickets for senior citizens.
2. *Magnification* is enlarging a product or service, such as giant economy size packages.
3. *Minification* is altering a product or service so it becomes smaller or less complex. Examples are portable radios and televisions, electronic calculators, and no-frills airline travel.
4. *Substitution* is using a certain material or service in place of what has traditionally been employed. Examples are polyester instead of cotton, plastic in place of metal, and nurse-midwives instead of physicians.
5. *Rearrangement* is altering the configuration of basic elements in a product or service—for example, some housing developments use several floor plans but each have the same basic features.

An Example of Brainstorming

Consider a group of six people, one from each department of an organization, who meet to brainstorm about the problem of excessive employee absenteeism. They have already decided on the topic to be discussed, so they can proceed to making their lists of causes. After they have completed their lists, they read their ideas, sequentially, one at a time—and the designated leader records the ideas on a flip chart. The first person's list of possible causes of excessive employee absenteeism is:

1. Low morale.
2. No penalties for absence.
3. Boredom with job.
4. Personal problems.

The second person's list is:

1. Dislike of supervisor.
2. Drug problems.
3. Performance anxiety.
4. Anger over pay.
5. Work related accidents.

All of the other members have similar lists. After all have read their lists and the causes have been recorded, the leader requests any new ideas that have emerged. Piggybacking on one of the first person's causes—"personal problems"—might result in another cause, "family problems." Asking for wild ideas might generate a response such as "addiction to video games" or "rundown bathroom facilities."

After all of the ideas have emerged, each group member gets a copy of the list to study. The group meets again and evaluates the ideas. They rank them in order of importance and decide that low morale, drug problems, and boredom with job are the three most critical causes of absenteeism. They are then in the position to develop an action plan to deal with these causes.

CAUSE-AND-EFFECT DIAGRAMS

The *Cause-and-Effect (C&E) Diagram* is a structured problem-solving technique developed in 1943 by Professor Kaoru Ishikawa, President of the Musashi Institute of Technology in Tokyo.[6] He found that most plant personnel were overwhelmed by the number of factors that could influence a process. Consequently, Ishikawa developed and applied C&E diagrams to help plant personnel cope with the myriad of factors that affected their processes and to solve problems. Cause-and-Effect diagrams are also known as *Ishikawa Diagrams,* and they are widely used in manufacturing and service industries.

Cause-and-Effect analysis is used after brainstorming to organize the information generated in the brainstorming session. It includes gathering and organizing possible reasons or causes of a problem, selecting the most probable cause, and verifying possible causes until a valid cause-and-effect relationship, which leads to a solution, is established.

The technique consists of defining an occurrence or problem (effect). This is usually a result of a brainstorming session in which the team identifies something it wants to correct or change. Once the effect is defined, the factors that contribute to it (causes) are delineated. These are the possible reasons that the problem exists. While there may be only one or two actual causes of a problem, there are probably many potential causes that could appear on the C&E diagram. The relationships between all the contributing causes are illustrated on a C&E diagram.

A simple generic example of a C&E diagram is shown in Figure 12.1.

Procedure

As the purpose of using a C&E diagram is to understand the factors affecting a process, it is important that the diagram represents the perspective of many different people involved in the process rather than the views of only one or two individuals. Brainstorming sessions that allow

---

FIGURE 12.1   Generic Cause-and-Effect Diagram

---

for a broad base of process observation are ideally suited for drawing C&E diagrams. Guidelines for the session are that:

1. Everyone should be encouraged to participate.
2. No criticism should be made of any suggestion.
3. Suggestions do not have to be limited to factors in one's own work area.
4. A period of observation, between the time that the chart is started and the time that it is finished, may be helpful.
5. Concentration on how to eliminate the trouble, rather than getting involved with justifications of why the trouble has occurred, is desirable.

Constructing a C&E Diagram

The following steps are recommended for constructing a C&E Diagram.

1. *State the problem.*   Clearly define the problem or effect to avoid confusion and focus the discussion. It is far easier to delineate causes of a well-defined problem than to try to do so for a vaguely defined problem. After a problem has been clearly defined, write it on the right side of a flip chart and draw an arrow or pointer to it.

2. *Identify major causes.*   Identifying major causes is important because it gives the C&E process a structure. However, it is frequently difficult to identify major causes until many potential major causes have been studied. A common method of determining major causes is to use universal major causes (such as machines, methods, material, manpower, and environment) on a C&E diagram, as shown in Figure 12.1.

3. *Brainstorm sub-causes.*   The group should brainstorm to uncover all of the possible sub-causes that create the problem under study by the group. Sub-causes should be recorded on the diagram. It is important to remember that only causes of the problem should be suggested at this point, not solutions. Figure 12.2 shows an example of a C&E diagram for errors in producing airline tickets with major causes and sub-causes.[7] In this example, four major areas that may cause airline ticket errors are delineated. They are further broken down into sub-causes and sub-sub-causes during subsequent group brainstorming, until all possible causes of the problem under study have surfaced.

4. *Allow time to ponder the causes before evaluating them.*   Some questions to consider at this time are:

- Is this cause a variable or an attribute?
- Has the cause been operationally defined?
- Is there a control chart or running record for this cause?
- Does this cause interact with other causes?

This information is valuable because it can aid in better understanding a cause's impact on the problem under investigation.

FIGURE 12.2   Cause-and-Effect Diagram for Airline Ticket Errors

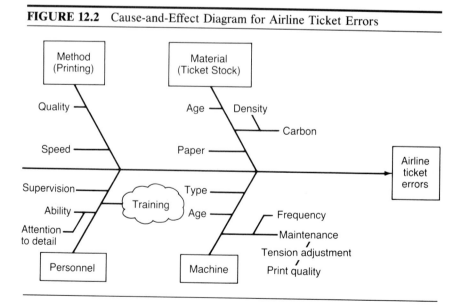

5. *Circle likely causes.* Causes of the problem under study should be evaluated and the most likely causes of the problem should be circled on the C&E diagram—for example, *training* is circled in Figure 12.2.

6. *Verify the cause.* The most likely cause of the problem under study should be analyzed by gathering data to see if it has a significant impact on the problem. If the most likely cause does not have a significant impact on the problem, the group should verify the next most likely choice to determine if it has a significant impact on the problem, and so on.

Root Cause Analysis

In some instances, the use of a C&E diagram does not result in finding the actual underlying causes of a problem. The group may become aware of this when discussing the cause or perhaps when a solution is implemented. At either point, the group can work on examining a cause that was identified in more depth; that is, the group could perform cause analysis when cause classification does not reach the root of the problem. Each of the causes originally identified can potentially be examined in a much more detailed manner by asking *who, what, where, when, why,* and *how* about each cause. In essence, each of the causes now becomes an effect (or problem) in a C&E diagram.

Returning to the example of the C&E diagram for airline ticket errors, we can illustrate the use of root cause analysis. Some of the causes identified were quality of printing, training of personnel, age of the ticket stock, and type of machine. Training of personnel is selected for further analysis. Figure 12.3 illustrates this concept. This procedure could be used to peel back the layers of a problem as one would peel off the layers of an onion—to get to the heart of a problem.

Process Analysis

Another form of a C&E diagram is for *process analysis*. It is used when a series of events (steps in a process) creates a problem and it is not clear which event or step is the major cause of the problem. Figure 12.4 illustrates a process analysis for making a chocolate mousse.[8] Note that each category or sub-process is examined for possible causes; after the causes from each step in the process are discovered, significant causes of the problem are selected and verified.

Summary

The C&E diagram should be used as a framework for collective efforts. If a process is stable, it will help organize efforts to improve the process. If a

FIGURE 12.3 Root Cause-and-Effect Diagram

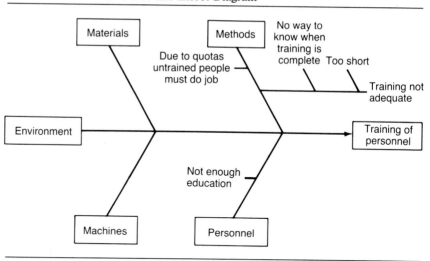

FIGURE 12.4 Cause-and-Effect Diagram of a Process for Making Chocolate Mousse

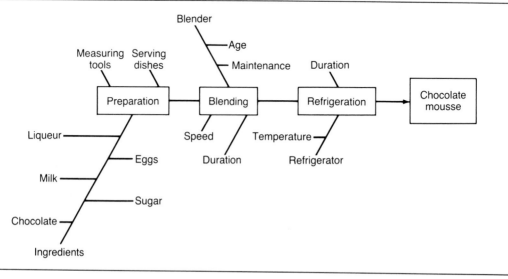

process is chaotic, the C&E diagram will help uncover areas that can help stabilize the process.

CHECK SHEETS

Check sheets are used for collecting or gathering data in a logical format (called *rational subgrouping,* discussed in Chapter 8). The data collected can be used in constructing a control chart; Pareto diagram (to be discussed in the next section); or histogram. Check sheets have several purposes, the most important of which is to enable the user(s) to gather and organize data in a format that permits efficient and easy analysis. The check sheet's design should facilitate data gathering.

Process improvement is aided by the determination of what data or information is needed to reduce the difference between customer needs and process performance. There are unlimited types of data that can be gathered in any organization, including information on the process, products, costs, vendors, inspection, customers, employees, administrative tasks, paperwork, sales, and personnel. Virtually any aspect of an organization can yield facts or data amenable to improvement of the extended process.

Types of Check Sheets

In designing a check sheet it is important to determine what the user(s) are attempting to learn by collecting the data and what action the user(s) will take given the results. This information will facilitate proper design of the check sheet to obtain the optimal benefit from the data. We will discuss three types of check sheets: attribute check sheets, variables check sheets, and defect location check sheets.[9] In reality, there are an unlimited number of formats for a check sheet because the user(s) can develop them based on the data needed to solve a particular problem and can be creative and invent a check sheet if the data are not amenable to an already established check sheet format.

Attribute Check Sheet. Gathering data about defects in a process is necessary to stabilize and improve the process. As there are many possible causes for any given defect, the logical way to collect data is to determine the number or percentage of defects generated by each cause. Based on the information collected, appropriate action can be taken to improve the process. Figure 12.5 is an example of an *attribute check sheet* for the causes of defects on corrugated board boxes.

FIGURE 12.5   Attribute Check Sheet of Defects in Corrugated Board Boxes

	Monday				Tuesday				
Type of Defect	8–10 A.M.	10–12 A.M.	12–2 P.M.	2–4 P.M.	8–10 A.M.	10–12 A.M.	12–2 P.M.	2–4 P.M.	Total
Smeared print	||		|||	|	|||		|		10
Box not glued properly	|		||	|||	||	|||	|	|||	15
Wrong symbol/ letter	|	||||	||	##|	||		||		16
Symbol/letter in wrong location	|||		##|	||	||	|||	##|	||	22
Total	7	4	12	11	9	6	9	5	63

This check sheet was created by tallying each type of defect during four two-hour time periods each day; it shows the types of defects and how many of each type occurred during each time period. Keeping track of these data for several days provides management with information upon which to base improvements to the process.

Variables Check Sheet.   Gathering data about a process also involves the collection of information about variables, such as size, length, weight, and diameter. These data are best represented by organizing the measurements into a frequency distribution on a *variables check sheet*.

Figure 12.6 is a variables check sheet showing the frequency distribution of the length of logs in a sample of 95 trees discussed in Chapter 5. Recall that the raw data, shown in Figure 5.4, could be displayed as a frequency distribution in Figure 5.6. These absolute frequencies appear in Figure 12.6. This type of check sheet is a simple way to examine the distribution of a process characteristic and its relationship to the specifi-

FIGURE 12.6   Variables Check Sheet for Length of Logs in a Sample of 95 Trees

400 but under 700	## ||||	9
		specification = 700
700 but under 1000'	## |||	8
1000 but under 1300	## ## ## ##	20
1300 but under 1600	## ## ## ## ## ## ##	35
1600 but under 1900	## ## ## |||	18
		specification = 1900
1900 but under 2200	##	5

FIGURE 12.7   Defect Location Check Sheet

Refrigerator Door Check Sheet

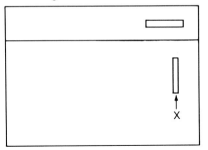

Examiner: John May

Date: July 18, 1988

Remarks: Bolt holding button of
handle is missing.
Handle loose.

Serial Number: HO 13988

cation limits (the boundaries of what is considered acceptable log lengths in Figure 12.6); the number and percentage of items outside the specification limit is easy to identify so that appropriate action can be taken to reduce the defectives.

Defect Location Check Sheet.   Another way to gather information about defects in a product is to use a *defect location check sheet*. This is a picture of a product or a portion of it on which an inspector indicates the location and nature of the defect. Figure 12.7 shows a defect location check sheet that could be used in collecting data regarding defects on the fronts of a model of refrigerator doors. It shows the location of a defect where the handle is attached to the refrigerator door. If this check sheet were used, and it was determined that most of the defects on the refrigerator doors occurred in the same location, further analysis could be performed. After investigation, the cause could be found and a plan to eliminate the problem could be implemented.

PARETO ANALYSIS

Pareto analysis is a tool used to identify and prioritize problems for solution. It is based on the work of Vilfredo Pareto, an Italian economist (1848–1923). Pareto focused attention on the concept of "the vital few versus the trivial many." The vital few are the few factors accounting for the largest part (percentage) of a total. The trivial many are the myriad of factors that account for the small remainder. Several other researchers have popularized this approach to prioritizing problem solving, most notably Joseph Juran[10] and Alan Lakelin[11] (a time-management specialist). Lakelin formulated the 80–20 rule based on an application of the Pareto principle. This rule says that approximately 80 percent of the value or costs come from 20 percent of the elements. For example, 80 percent of sales originate from 20 percent of customers, or 80 percent of phone calls are made by 20 percent of customers, or 80 percent of dollar inventory is accounted for by 20 percent of the items.

The Pareto diagram is a simple bar chart of the type discussed in Chapter 5, with the bars representing the frequency of each problem, arranged in descending order. That is, the tallest bars are on the left side of the chart descending to the right side. While Pareto analysis is commonly thought of as a problem-solving tool, it really helps determine what problems to solve, rather than how to solve them. The process of arranging the data, classifying it, and tabulating it helps determine the most important problem to be worked on.[12]

Constructing a Pareto Diagram

The following steps are recommended for constructing a Pareto diagram.[13] We illustrate the use of a Pareto diagram with an example concerning the sources of defective cards for a particular keypunch operator.

1. *Establish categories for the data being analyzed.* Data should be classified according to defects, products, work groups, size, and other appropriate categories; a check sheet listing these categories should be developed. In the example that follows, data on the type of defects for a keypunch operator will be organized and tabulated.

2. *Specify the time period during which data will be collected.* The time period to be studied will depend on the situation being analyzed. Three issues important in setting a time period to study are: 1) the selection of a convenient time, such as one week, one month, one quarter, one day, four hours; 2) the selection of a time period that is constant for all related diagrams for purposes of comparison; and 3) the selection of a time period that is relevant to the analysis, such as a specific season for a certain seasonal product. In the keypunch example, the time period is four months, January through April 1988.

In the example, the types of defects are recorded as they occur during the time period and are totaled, as shown in Figure 12.8.

3. *Construct a frequency table arranging the categories from the one with the largest number of observations to the one with the smallest number of observations.* The frequency table should contain a category column; a frequency column, indicating the number of observations per category with a total at the bottom of the column; a cumulative frequency column, indicating the number of observations in a particular category

FIGURE 12.8 Record of Defects for Keypunch Operator

Major causes of defective cards	Month 1/88	2/88	3/88	4/88	Total
Transposed numbers	7	10	6	5	28
Off-punched card	1		2		3
Wrong character	6	8	5	9	28
Data printed too lightly on card		1	1		2
Warped card	1	1		2	4
Torn card			1	1	2
Illegible source document			1		1
Total	15	20	16	17	68

Checksheet to determine the sources of operator 004's defective cards (1/88-4/88)

plus all frequencies in categories above it; a relative frequency column, indicating the percentage of observations within each category with a total at the bottom of the column; and a relative cumulative frequency column, indicating the cumulative percentage of observations in a particular category plus all categories above it in the frequency table.

The "other" category should be placed at the extreme right of the chart. If the "other" category accounts for as much as 50 percent of the total, the breakdown of categories should be reformulated. A rule of thumb is that the "other" bar should be smaller than the category with the largest number of observations.

The frequency table for the keypunch example appears in Figure 12.9. It shows that two types of defects (transposed numbers and wrong characters) are causing 82.4 percent of the total number of defective cards.

4. *Construct a Pareto diagram.* The steps required to construct the Pareto diagram are listed below.

a. Draw horizontal and vertical axes on graph paper and mark the vertical axis with the appropriate units, from zero up to the total number of observations in the frequency table.

b. Under the horizontal axis, write the most frequently occurring category on the far left, then the next most frequent to the right, continuing in decreasing order to the right. In the keypunch example, "trans-

FIGURE 12.9   Frequency Table of Defects for Keypunch Operator

Pareto analysis to determine major causes of defective cards for operator 004 (1-4/88)				
Major cause of defective cards	Freq.	Relative %	Cumulative Frequency	Cumulative %
Transposed numbers	28	41.2	28	41.2
Wrong character	28	41.2	56	82.4
Warped card	4	5.9	60	88.3
Off-punched card	3	4.4	63	92.7
Data printed too lightly on card	2	2.9	65	95.6
Torn card	2	2.9	67	98.5
Illegible source document	1	1.5	68	100.0
Total	68	100.0		

posed numbers'' and ''wrong character'' are the most frequently occurring defects and are positioned to the far left; ''illegible source document'' accounts for the fewest defective cards and appears at the far right of the chart.

c. Draw in the bars for each category. For some applications, this may provide enough information on which to base a decision; but often the percentage of change between the columns must be determined. Figure 12.10 displays the bars for the keypunch example.

d. Plot a cumulative line (cum line) on the Pareto diagram. Indicate an approximate cumulative percent scale on the right side of the chart and plot a cumulative line (''cum'' line) on the Pareto diagram. To plot the cum line, start at the lower left (zero) corner and move diagonally to the top right corner of the first column. In our example, the top of the line is now at the 28 level, as shown in Figure 12.11(a). Repeat the process, adding the number of observations in the second column. In our example, the line rests on the 56 level, as shown in Figure 12.11(b). The process is repeated for each column, until the line reaches the total number of observations level that includes 100 percent of the observations, as shown in Figure 12.11(c).

e. Title the chart and briefly describe its data sources. Without information on when and under what conditions the data were gathered, the Pareto diagram will not be useful.

FIGURE 12.10 Pareto Diagram of Defective Keypunch Cards for an Operator

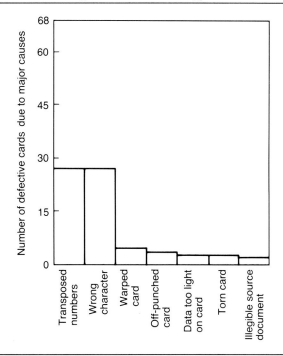

FIGURE 12.11 Cum Line Plotting

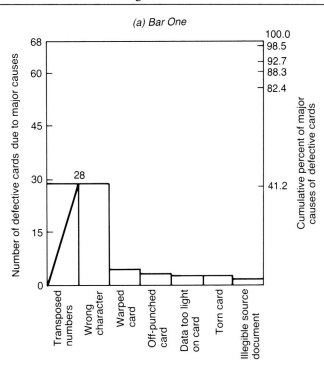

(a) Bar One

FIGURE 12.11 *Concluded*

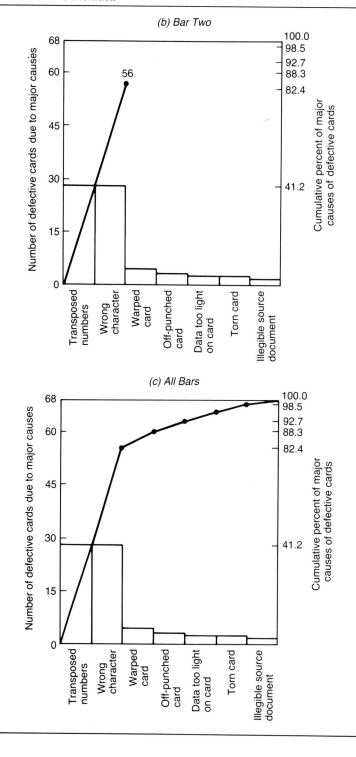

Cost Pareto Diagrams

Sometimes, Pareto diagrams can have more impact when problems or defects are represented in terms of their dollar costs. We can calculate the dollar cost for a particular type of defect by evaluating the unit cost each time the particular type of defect occurs, then multiplying that figure by the number of times that particular type of defect occurs. We then construct a Pareto diagram using the dollar cost of a defect as the vertical axis, rather than the number of defective units. A frequency table with costs per defect and a cost Pareto diagram for the keypunch example are shown in Figures 12.12 and 12.13. When the dollar cost of each defect is considered, a reordering of defect categories occurs.

This reordering occurs because of the high cost of some types of defects. For example, although there were only four "warped cards," the cost to the company for each warped card is $1 because of machine jams and downtime. Therefore, the dollar cost is $4, making it the most expensive problem. The 28 cards containing "transposed numbers," although high in number, drop down to a lower bar position in terms of dollar cost ($1.40), because the unit cost to correct a card containing transposed numbers is only $0.05 per card; only a replacement card is required to resolve the "transposed numbers" problem.

It is important to note that cost Pareto diagrams consider only visible and known costs; they do not consider unknown costs such as the cost of an unhappy customer receiving a card containing transposed numbers. This is an important limitation on the effective use of cost Pareto diagrams.

FIGURE 12.12 Frequency Table of Causes of Defective Keypunch Cards with Costs

Defect	Number of Defective Cards	Cost per Defective Card ($)	Dollar Cost for Type of Defect ($)
Transposed numbers	28	$0.05	$1.40
Wrong character	28	$0.05	$1.40
Warped card	4	$1.00	$4.00
Off-punched card	3	$0.05	$0.15
Data too light on card	2	$0.05	$0.10
Torn card	2	$1.00	$2.00
Illegible source document	1	$0.05	$0.05
Total	68		

FIGURE 12.13   Cost Pareto Diagram of Defective Keypunch Cards

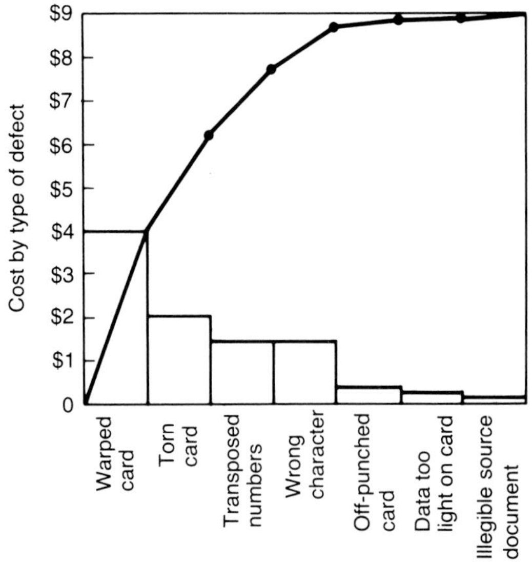

Use of Pareto Diagrams

Pursuing never-ending improvement depends upon cooperation between everyone concerned. The Pareto diagram is very useful in focusing a group's attention on a common problem and obtaining team cooperation.

As resources, manpower, and time are limited—and the need to improve is critical—it is extremely important to concentrate on the most important problems, those represented by the tallest bars on the Pareto diagram. It is often wiser to reduce a tall bar by half than to reduce a short bar to zero. If the tallest bar is reduced, it will be a considerable accomplishment, whereas a reduction in the short bars would result in less overall improvement.[14]

Pareto diagrams, as well as the other tools discussed in this book, can be used for improvement in all areas of organizations. They are an important tool for analyzing a problem, whether it be in production, administration, research, maintenance, or office work.

Use of Pareto Diagram for Root Cause Analysis

Pareto diagrams can be used to determine the root causes (underlying causes) of problems. For example, root cause analysis can be used to improve safety in a mill. First, a Pareto diagram is constructed to find out in which department the major injuries occur, as shown in Figure 12.14(a), assuming all departments are approximately the same size in terms of personnel. Clearly, from our illustration, the maintenance department accounts for the largest number of accidents. Another Pareto diagram is constructed to determine how the accidents to maintenance personnel occur, as shown in Figure 12.14(b). From this chart, it can be determined that injuries result chiefly from foreign objects entering eyes. Appropriate measures can now be taken to improve safety—for example, by insisting that maintenance personnel wear approved safety goggles.

FIGURE 12.14   Accident Pareto Diagram

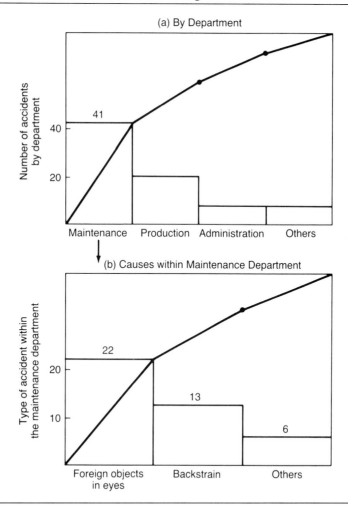

If the focus of the safety effort had been to determine the most serious injuries, perhaps those that disabled employees, the chart would be constructed differently, as shown in Figure 12.15. This is very much akin to cost Pareto diagrams. Here, although foreign objects in the eye may have accounted for the greatest number of accidents, they are not the most serious in terms of causing disability. Back strains account for the highest number of disabling injuries. The nature of the analysis that is performed can yield different priorities for improvement.

Use of a Pareto Diagram to Improve Chaotic Processes

Pareto diagrams can focus attention on the major problem on which a team concentrates its efforts. This is very effective when the ordering of problems remains stable over time—that is, if the process is stable. However, if the process is in a state of chaos, Pareto diagrams may not be effective. The following examples illustrate this.[15]

A Maintenance Nightmare. The foreman of a department considered one of the worst in terms of downtime in a factory realizes that he must try to improve the situation. The department has 72 machines, each with 36 spindles, for a total of 2,592 spindles that can break down. The foreman, trying to deal with a chaotic situation, constructs a rudimentary Pareto diagram, as shown in Figure 12.16. On this chart he lists the machines by number and creates two categories, spindle breakdowns, and other breakdowns. After collecting data, he sees that spindle breakdowns are the major source of problems and also identifies the machines that are experiencing the spindle breakdowns and therefore need an overhaul.

FIGURE 12.15 Pareto Diagram of Disability Accidents

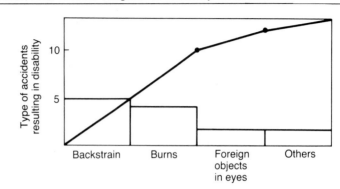

FIGURE 12.16 Foreman's Pareto Diagram

Other Breakdowns	Machine	Spindle Breakdowns
	1	xx
	2	x
x	3	xx
xxxx	4	x
x	5	xxxxxxx
	6	xxxx
x	7	x
	8	
xx	9	x
	10	xxxxxx
	.	
	.	
	.	
	72	

Thus, he can plan for the overhauls in order of priority and work them into his maintenance schedule. After three months, the number of breakdowns is reduced by about 70 percent.

In this case, the Pareto diagram clarified information that had been muddled in the chaotic day-to-day operations. This is an example of a successful application of a Pareto chart to an out-of-control process. It is successful because it tracks deterioration, which is a one-way process. The machines are not going to fix themselves—they will continue to deteriorate unless repairs are made. (In a sense, deterioration is stable instability; its effects will persist until a change is made.) The next example demonstrates what happens when Pareto diagrams are applied to an unstable process with changing special causes of variation.

Shifting Causes of Defects. A quality control specialist is interested in the causes of defects in rejected parts. She divides the causes into eight categories, each representing a different reason for rejection. Data are collected and used to create a Pareto diagram for one month. For this month foreign material is found to be the major cause for rejections, followed by damaged edges, others, stress, damaged mounts, bad trim, bubbles, and sinks. According to the Pareto diagram, foreign material and damaged edges account for 45 percent of the defective units, making these problems the logical starting points for process improvement.

Various distractions keep the quality control specialist from beginning work on process improvement after the first month. At the end of the second month she again prepares a Pareto diagram. This time stress and sinks are the two major problems. Yet in the first month, these had been the fourth and eighth largest reasons for rejection. The causes of defects have shifted over time, indicating process instability.

Pareto diagrams will not be effective if used on a chaotic process because the process is not ready for improvement. The process must first be

---

FIGURE 12.17 Pareto Diagram Showing Type and Number of Defects in a Corrugated Box Process before and after Improvement

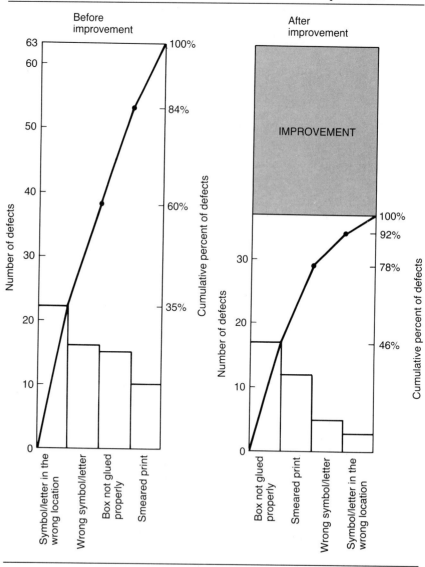

stabilized through the use of control charts, as discussed in Chapters 8 through 10.

Use of Pareto Diagrams to Gauge Improvement

Pareto diagrams can be very helpful in determining whether efforts towards process improvement are producing results. If effective actions have been taken, the order of the items on the horizontal axis will change. Figure 12.17 illustrates Pareto diagrams before and after improvements are implemented on a stable process producing corrugated board boxes. On the basis of the before-improvement chart, the major cause of a defective box is diagnosed as "symbol/letter in wrong location." After improvements are made to the process, "symbol/letter in wrong location" becomes the least frequent source of trouble, demonstrating the effect of the improvements through the use of Pareto diagrams. This is a powerful tool when used in this way because it can mobilize support for further process improvement and reinforce the continuation of current efforts.

STRATIFICATION

Stratification (not to be confused with the stratification control chart pattern discussed in Chapter 11) is a procedure used to describe the systematic subdivision of population or process data to obtain a detailed understanding of the structure of the population or process. Stratification can be used to break down a problem to discover its root causes and set into motion appropriate corrective actions, called *countermeasures*. Stratification is important to the proper functioning of the PDCA cycle. For example, the number of traffic accidents in Japan peaked in 1970. For each accident, the police officer present at the accident scene was required to complete an accident report, with information on the cause of the accident. The most common cause of accidents listed on the report, after analysis, was "careless driving." Unfortunately, this stratification did not yield the root cause of the accidents. Subsequently, a more in-depth stratification of the accident data—particularly the "careless driving" category—yielded specific locations where accidents occurred with high frequency. The determination of these locations gave the police and the Department of Highways the information they needed to set appropriate countermeasures to improve road conditions. These actions resulted in a drastic reduction in the number of traffic accidents in Japan. Proper stratification showed the root causes of the problem and led the way for the establishment of proper countermeasures.[16]

Tools Used for Stratification

Stratification and Pareto Diagrams. Figure 12.18 shows how stratification is used when performing root cause analysis with Pareto diagrams. 110 observations are made. We see that by breaking down a problem into its subcomponents (stratifying the 110 items into appropriate subcomponents A, B, etc.) and by breaking each subcomponent into its subcomponents (stratifying the 50 items in A into A_1 through A_6 and the 40 items in B into B_1 through B_4), we can focus on one or more of the root causes of a process or product problem, from which we can establish a countermeasure for resolving the problem.

In general, when all categories in a Pareto diagram are approximately the same size, as in Figure 12.19(a), stratifying on another product or

FIGURE 12.18 Pareto Diagrams with Stratification

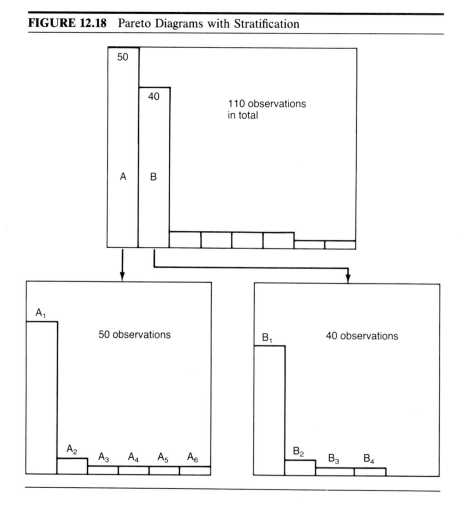

FIGURE 12.19 Pareto Diagram with Same-Size Categories

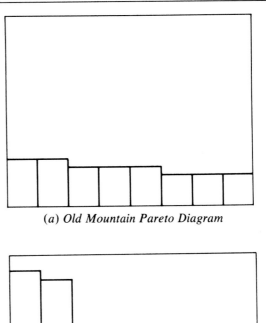

(a) Old Mountain Pareto Diagram

(b) New Mountain Pareto Diagram

process characteristic should be done until a Pareto diagram like the one shown in Figure 12.19(b) is found. Figure 12.19(a) is called an *old mountain stratification* (the mountain is worn flat) because no one category emerges as the obvious factor upon which to take process or product improvement action. Figure 12.19(b) is called a *new mountain stratification* (the mountain is young and has high peaks) because one or two categories emerge as the obvious starting point for process or product improvement action.

Stratification and Cause-and-Effect (C&E) Diagrams. Figure 12.20 shows how stratification is used when performing root cause analysis with C&E diagrams. We see that second tier C&E diagrams can be con-

FIGURE 12.20   Stratification and Cause-and-Effect Diagrams

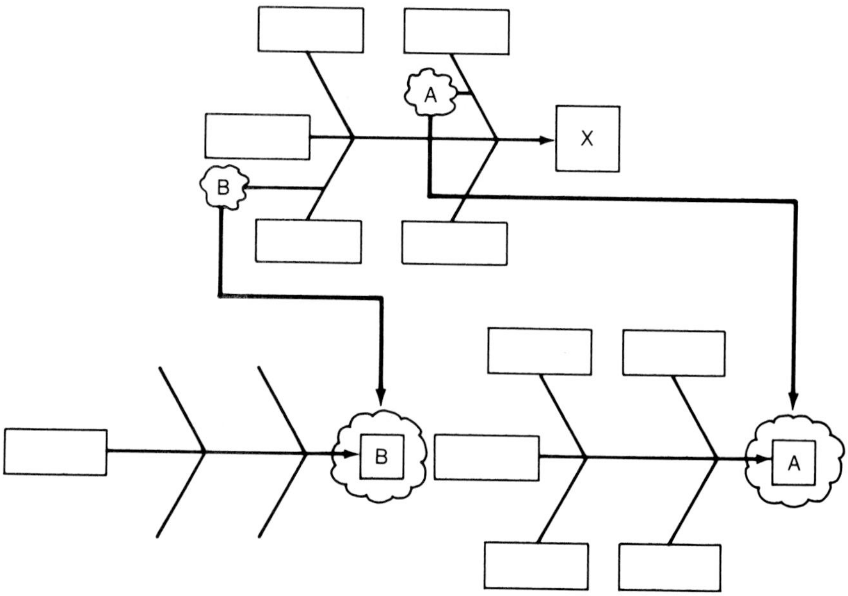

structed to study in depth any cause shown on a first-tier C&E diagram, and so on. For example, a C&E diagram used to study "problem X" generated two major possible causes for Problem X, A, and B. Consequently, both A and B are studied through their own C&E diagrams. This stratification could go on indefinitely until one or more of the root causes of Problem X are determined and appropriate countermeasures are set that should lead to process or product improvement.

Stratification with Pareto Diagrams and Cause-and-Effect Diagrams.
Figure 12.21(a) shows a Pareto diagram. Figure 12.21(b) shows a C&E diagram focusing exclusively on one of the bars in the Pareto diagram shown in Figure 12.21(a); this is the *correct* way to stratify a Pareto diagram to study in-depth the root causes of a problem. Figure 12.21(c) shows a C&E diagram focusing on all of the bars in the Pareto diagram shown in Figure 12.21(a); this is the *incorrect* way to stratify a Pareto diagram to study in-depth the root causes of a problem. A C&E diagram should be used to stratify one bar from a Pareto diagram at a time to get an in-depth understanding of the corresponding cause (bar) before any other cause (bar) is studied.

FIGURE 12.21   Stratification with Pareto Diagrams and Cause-and-Effect
Diagrams

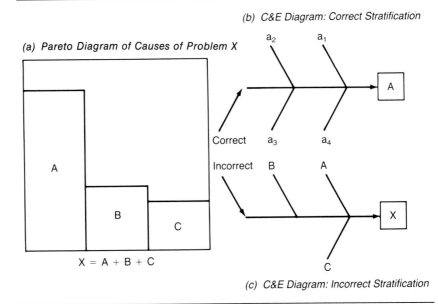

(b) *C&E Diagram: Correct Stratification*

(a) *Pareto Diagram of Causes of Problem X*

$X = A + B + C$

(c) *C&E Diagram: Incorrect Stratification*

Stratification with Control Charts, Pareto Diagrams, and C&E Diagrams.   Figure 12.22 shows how a Pareto diagram can be used to identify the common causes of variation from a stable process, and how these common causes of variation can be stratified through Pareto diagrams or C&E diagrams to determine root causes of process or product problems so that appropriate countermeasures can be established.

Other Combinations of Tools for Stratification.   Most of the tools and techniques presented in this text can be used in combination with each other to stratify data about a process or product problem, enabling an investigator to search for the root cause(s) of the problem. Once the root cause(s) of a product or process problem have been determined, appropriate countermeasures can be established that will result in process or product improvement(s).

Dangers and Pitfalls of Poor Stratification

Failure to perform meaningful stratification can result in the establishment of inappropriate countermeasures, which can then result in process or product deterioration. For example, analyzing a random sample of 100

FIGURE 12.22 Stratification with Control Charts, Pareto Diagrams, and
Cause-and-Effect Diagrams

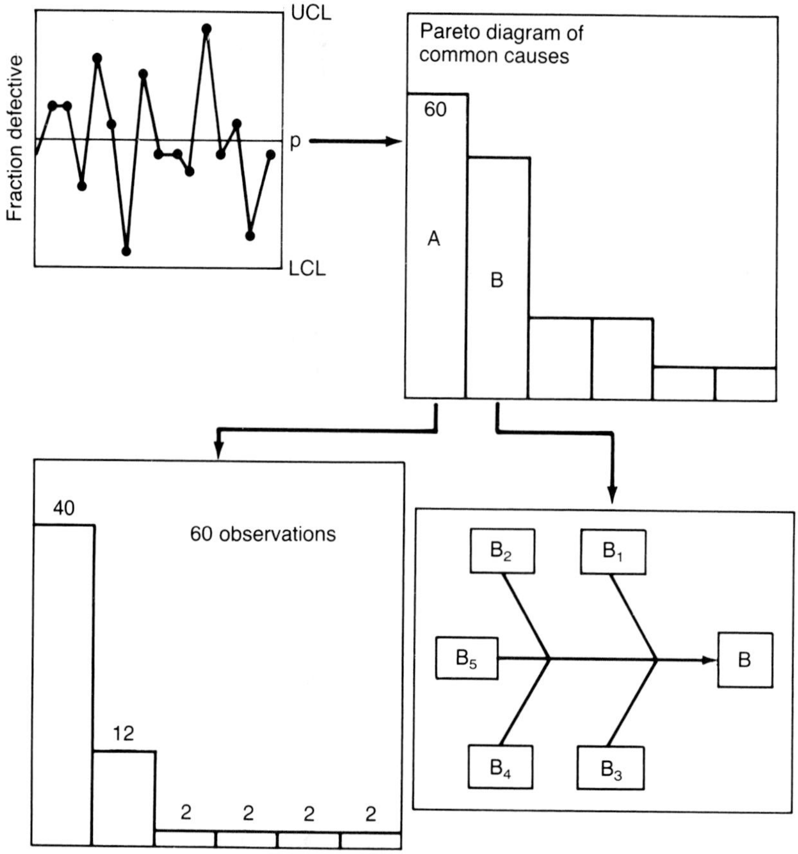

records of repairs performed in a factory by maintenance personnel in
which accidents occurred could lead to the isolation of alleged root causes
of accidents. Countermeasures designed to prevent future accidents
would be established based on these alleged root causes. However, this
analysis could lead to the determination of causes that are not unique to
repairs resulting in accidents but typical of both accident and nonaccident
repairs. Consequently, the alleged root causes would be ineffective in
establishing significant countermeasures to prevent future accidents. Alternatively, if a random sample of 50 records of repairs performed in a
factory by maintenance personnel in which accidents occurred were com-

pared with a random sample of 50 records of repairs performed in the same factory (under the same conditions, in which accidents did not occur), this comparison could lead to the isolation of real root causes of accidents. This alternative procedure would isolate real root causes because stratification focused attention on the differences between accident and nonaccident repairs.

Single Case Boring

Stratification is not possible when only a few data points are available—for example, if data exist on only four accident cases in a plant. In this situation, meaningful analysis is still possible; it is called *single case boring*. Single case boring is a procedure in which a few cases are studied in great depth. This procedure may yield information that could lead to conditional countermeasures, or countermeasures appropriate for the situation in which they were discovered—for example, a countermeasure established to prevent a particular type of accident, in a particular situation, by a particular individual.

SUMMARY

This chapter explains several tools and methods that are useful aids for stabilizing a process (taking action on special sources of variation) and for improving a process (taking action which will reduce common variation and center the process's average on nominal). These tools and methods are brainstorming, cause-and-effect diagrams, check sheets, Pareto diagrams, and stratification. Brainstorming is a way to bring forth a large number of ideas from a group about a process or product problem. A Cause-and-Effect diagram is a tool that can be used to organize the ideas about a problem collected during a brainstorming session and focus attention on one of the ideas as a possible solution to the problem being studied. A check sheet is a data collection form used to gather data logically, as in rational subgroups. A Pareto diagram is a tool used to identify and prioritize problems for solution. Finally, stratification describes the systematic subdivision of population or process data to obtain a detailed understanding of the structure of the population or process. Stratification can be used to break down a problem to discover its root causes and set into motion appropriate countermeasures. All of the above tools and methods, in conjunction with control charts, can be used in the PDCA cycle to relentlessly decrease the difference between customer needs and process performance.

EXERCISES

12.1 a. What is the purpose of brainstorming?
  b. List the seven steps required to conduct a brainstorming session.
  c. List the rules a group should follow to have an effective brainstorming session.

12.2 a. What is the function of a Cause-and-Effect diagram?
  b. Draw a picture of a generic Cause-and-Effect diagram.
  c. Construct a Cause-and-Effect diagram to understand the causes of being late to work.
  d. Explain the purpose of root cause analysis.
  e. Construct an example of a root cause analysis.
  f. Explain the purpose of process analysis.
  g. Construct an example of process analysis.

12.3 a. Explain the purpose of check sheets.
  b. Construct an attribute check sheet form to study the reasons for absenteeism in a factory.
  c. Construct a variables check sheet form to study the diameters of bearings produced in a factory.
  d. Construct a defect location check sheet to study blemishes on cordless telephones.

12.4 a. Explain the purpose of Pareto analysis.
  b. Given the following check sheet of causes of absenteeism in a factory for January 1988, construct a Pareto diagram.

Cause of Absenteeism	Number of Occurrences
Personal illness	28
Child's illness	46
Car broke down	4
Personal emergency	10
Day made long weekend	3
Other	12

12.5 a. Given the following subcauses of "child's illness" from the above Pareto diagram, perform a root cause analysis (stratify the data to determine the root cause of absenteeism resulting from "child's illness").

Cause of "Child's Illness" Absenteeism	Number of Occurrences
Had to take child to doctor	8
Could not leave child in day care	33
Other reasons	5

b. Suggest a possible countermeasure to rectify the above problem.

NOTES

1. Virgil Rehg, *Quality Circle Manual for Coordinators and Leaders* (Wright-Patterson AFB, Ohio), p. 19.
2. J. F. Beardsley & Associates, International, Inc., *Quality Circles: Member Manual,* 1977, p. 40.
3. Ibid.
4. Rehg, *Quality Circle Manual,* pp. 19–20.
5. Beardsley & Associates, *Quality Circles,* pp. 42–43.
6. Kaoru Ishikawa, *Guide to Quality Control,* 11th printing, 1983 (Tokyo: Asian Productivity Organization, 1976), pp. 26–28. Available through UNIPUB, Box 433, Murray Hill Station, New York, N.Y. 10157.
7. University of Miami School of Business Administration, *Methods for the Improvement of Quality and Productivity,* Class Project, 1986.
8. Ibid.
9. Ishikawa, pp. 29–35.
10. Beardsley & Associates, p. 88.
11. Ibid.
12. Ishikawa, p. 43.
13. Ibid., pp. 43–44.
14. Ibid., pp. 45–46.
15. Adapted from Donald J. Wheeler, and David S. Chambers, *Understanding Statistical Process Control* (Knoxville, Tenn.: Statistical Process Controls, Inc., 1986), pp. 294–97.
16. Speech by Dr. Noriaki Kano, Science University of Tokyo, given at Florida Power and Light Company, March 1987.

PART IV

Process Performance in Analytic Studies

This section is devoted to discussing process performance in analytic studies. Two aspects of process performance are treated, specifications and process capability studies.

Chapter 13 analyzes performance and technical specifications and includes examples of each type of specification. Technical specifications are subdivided into individual unit specifications; acceptable quality level (AQL) specifications; and distribution specifications. The significance of each type of technical specification to a process's performance is discussed, as is the fallacy of defining quality as conformance to specifications. Chapter 13 also introduces created dimensions, or dimensions created in the act of assembly. The statistical laws that govern the behavior of created dimensions are discussed.

Chapter 14 is concerned with process capability studies and quality improvement stories. Process capability studies provide important information about the condition of a process to facilitate process improvement actions. Two types of process capability studies are discussed, attribute studies and variables, or measurement, studies. In addition, process capability indices are presented as quantitative measures of the status of a process. Finally, Chapter 14 presents quality improvement (QI) stories. The QI story is an efficient format through which employees can present and publicize their quality improvement efforts. QI stories standardize quality control reports, avoid logical errors in analysis, and connect improvement efforts with organizational objectives. Further, QI stories can be used to enable employees to think in terms of the PDCA cycle in their daily work.

Chapter 13

Specifications

INTRODUCTION

The Deming cycle works to diminish the difference between process performance and customer needs. For this difference to be diminished, specifications for customer needs must be developed and revised, and process performance must be measured against these specifications.

If an organization is operating in a "defect detection" mode, specifications serve as vehicles to sort conforming and defective product or service. Unfortunately, when an organization operates in a defect detection mode, there is no feedback loop in the process to enable management to diminish the difference between customer needs (specifications) and process performance.

If an organization is operating in a "defect prevention" mode, specifications allow management to operationalize, measure, and diminish the proportion of product or service that fails to meet specifications. This is accomplished by creating a process history via attribute control charts. These attribute charts are then used in conjunction with other methods (as discussed in Chapter 12) as a basis for process stabilization and improvement. These improvements can take the form of a decrease in fraction defective, a decrease in number defective, or a decrease in number of defects per unit.

If an organization is operating in a "never-ending improvement" mode, specifications are important signposts to management for continuously reducing process variation within specifications and moving the process's mean to a level that creates customer satisfaction. In this mode of management, the Deming cycle is fully functional. Quality-of-performance studies lead to product or service redesign (revised specifications). Quality-of-conformance studies diminish the difference between revised specifications and process performance. Quality-of-performance studies check to see how the products or services are performing in the hands of the user.

This chapter discusses various types of specifications to better clarify their role in process improvement.

TYPES OF SPECIFICATIONS

Specifications fall into two broad categories: performance specifications and technical specifications.

Performance Specifications

Performance specifications address a need. For example, a vendor agrees to supply an air conditioning system sufficient to maintain a temperature range between 65 and 70 degrees Fahrenheit for a computer facility located in Room 325 of the customer's Spring River Plant. Performance is guaranteed for a period of no less than three years. Here, the customer's requirements are stated for the vendor.

Technical Specifications

Technical specifications describe performance at delivery. There are three types of technical specifications: individual unit specifications; acceptable quality level (AQL) specifications; and distribution specifications.

Individual Unit Specifications. *Individual unit specifications* state a boundary, or boundaries, that apply to individual units of a product or service. An individual unit of product or service is considered to conform to a specification if it is on or inside the boundary or boundaries.

Individual unit specifications are made up of two parts, which together form a third part. The first part of an individual unit specification is the *nominal value*. This is the desired value for process performance mandated by the needs of the customer. Ideally, if all quality characteristics were at nominal, products and services would perform as expected over their life cycle. The second part of an individual unit specification is a *tolerance*. A tolerance is an allowable departure from a nominal value established by design engineers that is deemed nonharmful to the functioning of the product or service over its life cycle. Tolerances are added and/or subtracted from nominal values. The third part of an individual unit specification is a *specification limit*. Specification limits are the boundaries created by adding and/or subtracting tolerances from a nominal value. It is possible to have two-sided specification limits:

USL = Nominal + Tolerance

LSL = Nominal − Tolerance

where USL is the upper specification limit and LSL is the lower specification limit; or one-sided specification limits, i.e., either USL or LSL.

An example of an individual unit specification and its three parts can be seen in the specification for the "case hardness depth" of a camshaft. A camshaft is considered to be conforming with respect to case hardness depth if each individual unit is between 7.0 mm ± 3.5 mm, (or 3.5 mm to 10.5 mm). The nominal value in the above specification is 7.0 mm; the two-sided tolerance is 3.5 mm; the lower specification limit is 3.5 mm (7.0 mm − 3.5 mm); and the upper specification limit is 10.5 mm (7.0 mm + 3.5 mm). As stated earlier, a camshaft is considered conforming with respect to case hardness depth if its case hardness depth is between 3.5 mm and 10.5 mm, inclusive.

From our earlier discussion of the philosophy of continuous reduction of variation, we saw that the goal of modern management should not be 100 percent conformance to specifications ("Zero Defects") but the never-ending reduction of process variation within specification limits so that all products/services are as close to nominal as possible. Specified tolerances become increasingly irrelevant as process variation is reduced so that the process's natural limits are well within the specification limits.

Acceptable Quality Level (AQL) Specifications. *Acceptable quality level (AQL) specifications* state a requirement that must be met by most of the individual units of product or service but allows a certain proportion of the units to exceed the requirements. For example, cam shafts shall be acceptable if no more than 3 percent of the units exceed the specification limits of 3.5 mm and 10.5 mm. This type of specification limit is frequently referred to as an Acceptable Quality Level (AQL). AQL specifications are much like individual unit specifications, except they have one added negative feature: they formally support the production of a certain percentage of defective product or service. This attitude toward the production of defective product/service is devastating to the extended process quality effort.

Distribution Specifications. *Distribution specifications* define an acceptable distribution for each product or service quality characteristic. The distribution is defined in terms of its mean, standard deviation, and shape. For example, the average case hardness depth of cam shafts shall be 7.0 mm and the individual units shall be distributed normally around this average with a dispersion not to exceed 3.5 mm on either side of the mean with probability 99.7 percent.

Stating that the process's quality characteristic dispersion should not exceed 3.5 mm on either side of the mean means that the process's standard deviation should not exceed 1.167 mm; that is, a process's capability is defined to be its mean plus or minus three times its standard deviation. Recall from Chapter 6 that if a stable process is operating according to a normal distribution, approximately 99.7 percent of its output will lie within three standard deviations on either side of its mean [7.0 mm ± 3(1.167 mm) = 7.0 mm ± 3.5 mm]. Consequently, for a maximum dispersion of 3.5 mm, three standard deviations would be 3.5 mm, and one standard deviation is 1.167 mm. Note that the mean and standard deviation, 7.0 mm and 1.167 mm, are simply directional goals for management. Management must use statistical methods to move the process average toward the nominal value of 7.0 mm and to decrease the process standard deviation as far below 1.167 mm as possible. Distribution requirements are stated in the language of the process and promote the never-ending improvement of quality.

PERFORMANCE SPECIFICATIONS AND TECHNICAL SPECIFICATIONS

Performance specifications are not commonly used in business; instead, technical specifications are used that are frequently passed on to purchasing departments (hence suppliers) by production and engineering departments. Unfortunately, this can cause major problems because technical specifications may not produce the desired performance specifications.

As an example, consider a hospital that serves medium (versus rare or well done) steak to patients who select steak for dinner.[1] The performance desired is patient satisfaction, within nutritional guidelines. However, performance specifications are not used; instead, a technical specification of five ounces of steak is substituted; it is assumed they are equivalent.

A hospital purchasing agent switches from meat vendor A to meat vendor B to secure a lower price, while still meeting the technical specification of five ounces. He does not discuss or inform the hospital nutritionist and kitchen staff of the switch in vendors. The hospital nutritionist begins receiving complaints from patients that the steak is tough and well done. She investigates and finds that vendor A's steaks were thick, while vendor B's are thin (but longer and wider). She realizes via statistical monitoring methods that the thinner steaks get hotter more quickly and hence cook faster, given the usual preparation regimen, as shown in Figure 13.1. She concludes: "If I had known that the steaks had been changed, I could have accommodated the change without creating patient dissatisfaction." The purchasing agent says: "I met the technical specifi-

FIGURE 13.1   Final Cooked Temperature of Five-Ounce Hospital Steaks

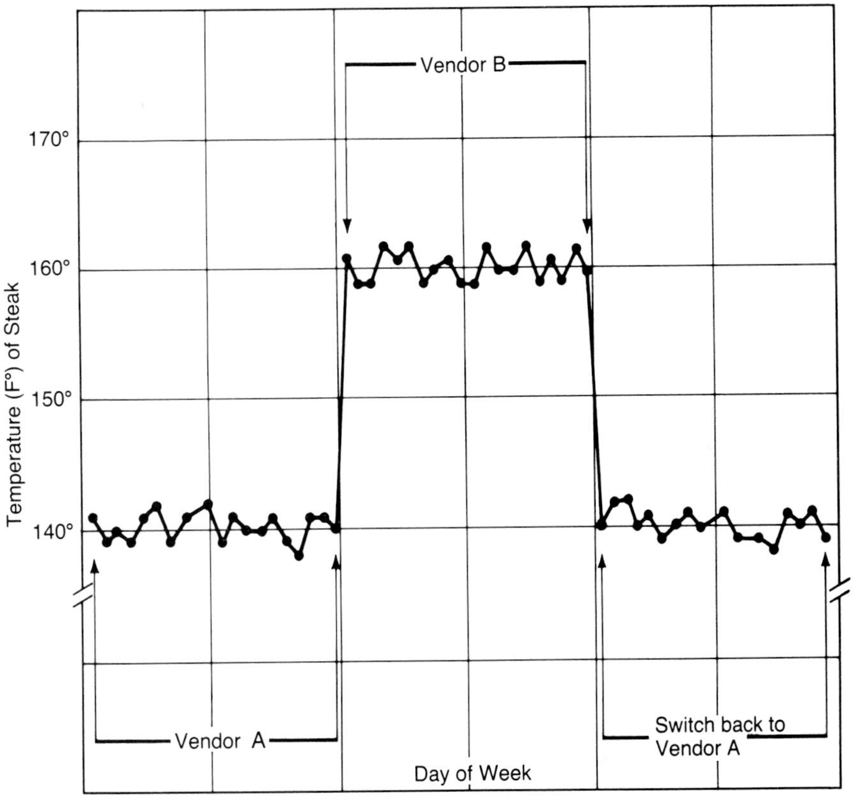

Note: Most patients can detect a 10°F
difference in steak temperature. The
difference in temperature between vendor A's
and vendor B's steak when subjected to the
hospital's cooking regimen is approximately 20°F.

cation of five ounces." The problem lies in assuming that technical specifications equal performance specifications; this is not necessarily true.

THE FALLACY THAT CONFORMANCE TO TECHNICAL SPECIFICATIONS DEFINES QUALITY

Mere conformance to specification limits is insufficient to achieve the level of quality required to compete effectively in today's marketplace. Management must constantly attempt to reduce process variation around a nominal value within specification limits to achieve the degree of uniformity required to produce products or services that function exactly as promised to the customer over their life cycle. The belief that there is no loss from products which are within specification limits regardless of the size of the deviation from the nominal value was discussed in Chapter 8 and will be discussed further in Chapter 15.

CREATED DIMENSIONS

When parts are assembled, new dimensions are created; these new dimensions have statistical distributions.[2] For example, if two boards are glued together to form a double thick board, the distribution of the thickness of the double-thick boards is a newly created dimension. Management must be able to control and reduce the variation of these created dimensions so that the final assemblies will perform perfectly for the customer over the product's life cycle. Understanding and controlling these created dimensions requires a working knowledge of the statistical rules of assemblies. If management does not pay attention to the statistical characteristics of created dimensions, these dimensions will fail to be within specification limits and will cause problems in production or service, will increase costs, and will lead to customer dissatisfaction. All information discussed about specifications earlier in this chapter applies to created dimensions.

Law of the Addition of Component Dimension Averages

If component parts are assembled so that the individual component dimensions are added to one another, the average dimension of the assembly will equal the sum of the individual component average dimensions. Figure 13.2 illustrates this concept.[3] If three component parts are glued together (assuming the glue takes no measurable dimensions), the average width of the assembled part is equal to the sum of the average individual part widths. That is,

$$\bar{X}_{assembly} = \bar{x}_1 + \bar{x}_2 + \bar{x}_3 \tag{13.1}$$

where

$\bar{X}_{assembly}$ = average width of the assembly

$\bar{x}_1$ = average width of part 1

$\bar{x}_2$ = average width of part 2

$\bar{x}_3$ = average width of part 3

FIGURE 13.2 Addition of Averages

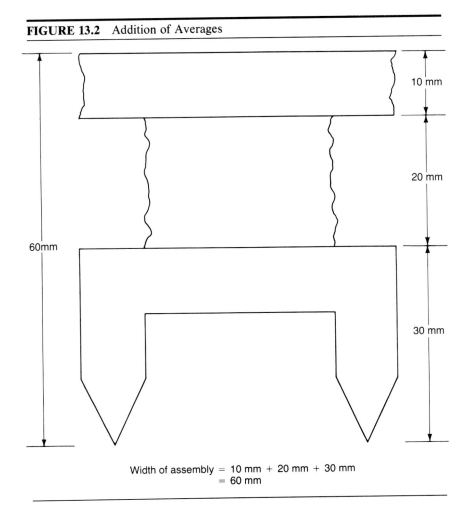

Width of assembly = 10 mm + 20 mm + 30 mm
= 60 mm

Part 1 has an average thickness of 10 mm, part 2 has an average thickness of 20 mm, and part 3 has an average thickness of 30 mm. Consequently, the average thickness of the final assembly is the sum of all three averages, or 60 mm (10 mm + 20 mm + 30 mm):

$$\bar{X}_{assembly} = \bar{X}_1 + \bar{X}_2 + \bar{X}_3$$

$$= 10 \text{ mm} + 20 \text{ mm} + 30 \text{ mm}$$

$$= 60 \text{ mm}$$

The above law holds only if the processes generating the components are in statistical control.

Law of the Differences of Component Dimensions of Averages

If component parts are assembled so that the individual component dimensions are subtracted from one another, the average dimension of the assembly will then equal the difference between the individual component average dimensions. Figure 13.3 illustrates this concept.[4] If a bolt is projected through a steel plate, the average length of the bolt projection through the steel plate equals the difference between the bolt's shank length and the width of the steel plate. That is,

$$\overline{X}_{\text{bolt projection}} = \overline{X}_s - \overline{X}_p \qquad (13.2)$$

where

$\overline{X}_{\text{bolt projection}}$ = average length of the bolt shank projection through the steel plate

$\overline{X}_s$ = average length of the bolt shank

$\overline{X}_p$ = average width of the steel plate

The bolt shank has an average length of 12 mm, and the steel plate has an average width of 8 mm. Consequently, the average bolt projection through the steel plate is 4 mm (12 mm − 8 mm):

$$\overline{X}_{\text{bolt projection}} = \overline{X}_s - \overline{X}_p$$

$$= 12 \text{ mm} - 8 \text{ mm}$$

$$= 4 \text{ mm}$$

FIGURE 13.3   Differences of Averages

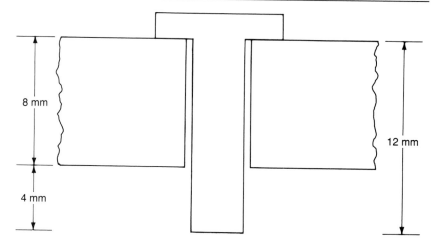

Bolt projection = 12 mm − 8 mm = 4 mm

Again, the above law holds only for component processes in statistical control.

Law of the Sums and Differences of Component Dimension Averages

If component parts are assembled so that the individual component parts are added and subtracted from one another, the average dimension of the assembly will then equal the algebraic sum of the individual component average dimensions. Figure 13.4 illustrates this concept.[5] If a bolt is screwed through a steel plate and washers are inserted on either side of the plate, the average length of the bolt projection through the steel plate and washers then equals the difference between the sum of the widths of the two washers and the steel plate, and the length of the bolt shank. That is,

$$\overline{X}_{\text{bolt projection}} = \overline{X}_s - (\overline{X}_{w1} + \overline{X}_p + \overline{X}_{w2}) \tag{13.3}$$

where

$\overline{X}_{\text{bolt projection}}$ = average length of the bolt shank projection through both washers and the steel plate

$\overline{X}_s$ = average length of the bolt shank

$\overline{X}_{w1}$ = average width of the top washer

$\overline{X}_{w2}$ = average width of the bottom washer

$\overline{X}_p$ = average width of the steel plate.

The bolt shank has an average length of 40 mm, the steel plate has an average thickness of 27 mm, the top washer has an average thickness of 3 mm, and the bottom washer has an average thickness of 4 mm. Hence, the average bolt projection through the steel plate and both washers is 6 mm [40 mm − (3 mm + 27 mm + 4 mm)]:

$$\overline{X}_{\text{bolt projection}} = \overline{X}_s - (\overline{X}_{w1} + \overline{X}_p + \overline{X}_{w2})$$

$$= 40 \text{ mm} - (3 \text{ mm} + 27 \text{ mm} + 4 \text{ mm})$$

$$= 6 \text{ mm}$$

Again, the above law holds only for component processes in statistical control.

Law of the Addition of Component Dimension Standard Deviations

If component parts are assembled at random (for example, so that each component part is drawn randomly from its own bin with no selection criteria), the standard deviation of the assembly will be the square root of the sum of the component variances, regardless of whether the compo-

FIGURE 13.4 Sums and Differences of Averages

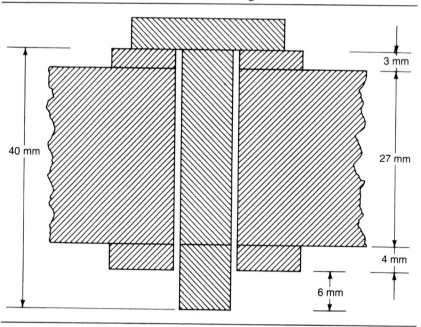

nents are added or subtracted from each other. This law applies to assemblies in which the component parts combine linearly and are statistically independent.

For example, consider again the bolt projection in Figure 13.3. Recall that $\bar{x}_s$ = 12 mm and $\bar{x}_p$ = 8 mm; consequently, we found from Equation 13.2 that $\bar{x}_{\text{bolt projection}}$ = 4 mm. Further, assume that the standard deviation of the bolt shank length, σ_s, is 0.010 mm and the standard deviation of the steel plate width, σ_p, is 0.008 mm. The standard deviation of the bolt projection thus is:

$$\sigma_{\text{bolt projection}} = [\sigma_s^2 + \sigma_p^2]^{1/2} \qquad (13.4)$$

$$= [(0.010)^2 + (0.008)^2]^{1/2}$$

$$= 0.0128 \text{ mm}$$

It is important to realize that the standard deviation of the bolt projection is not 0.018 mm, the sum of the individual component standard deviations. The square root of the sum of the individual component variances will always be less than the sum of the individual component standard deviations. This means that the assembly-to-assembly variation among random assemblies will be less than would be indicated by summing the individual components' unit-to-unit variations. Again, the above law holds only for component processes in statistical control.

Law of the Average for Created Areas and Volumes

If areas and volumes are created by the assembly of component parts, then the average area or volume of the assembly will equal the product of the individual component average dimensions if the component processes are stable and independent. Figure 13.5 illustrates this concept.[6] If a boxlike container is constructed with two short sides, two long sides, and two top/bottom sides, then the average internal volume of the boxlike container is equal to the product of the average length of the short side, the average length of the long side, and the average width of the sides. That is,

$$\bar{x}_v = (\bar{x}_s)(\bar{x}_l)(\bar{x}_w) \tag{13.5}$$

where

$\bar{x}_v$ = average internal volume of the constructed container

$\bar{x}_s$ = average length of short side

$\bar{x}_l$ = average length of long side

$\bar{x}_w$ = average width of the sides.

The average length of the short side is 3.0 mm, the average length of the long side is 8.0 mm, and the average width of the sides is 2.0 mm. Consequently, the average internal volume of the constructed container is 48 mm³ (3 mm × 8 mm × 2 mm).

$$\bar{x}_v = (\bar{x}_s)(\bar{x}_l)(\bar{x}_w)$$

$$= (3 \text{ mm})(8 \text{ mm})(2 \text{ mm})$$

$$= 48 \text{ mm}^3$$

Again, the above law holds only for component processes in statistical control.

FIGURE 13.5   Created Volume of a Container

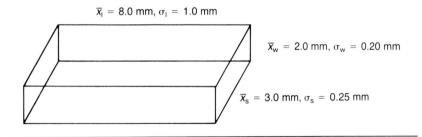

$\bar{x}_l = 8.0$ mm, $\sigma_l = 1.0$ mm

$\bar{x}_w = 2.0$ mm, $\sigma_w = 0.20$ mm

$\bar{x}_s = 3.0$ mm, $\sigma_s = 0.25$ mm

Law of the Standard Deviation for Created Areas and Volumes

If areas and volumes are created by the assembly of component parts, then the standard deviation of the created areas or volumes will equal:

$$\sigma_{area} = [\bar{x}_s^2 \sigma_l^2 + \bar{x}_l^2 \sigma_s^2 + \sigma_l^2 \sigma_s^2]^{1/2} \tag{13.6}$$

where

$\bar{x}_s$ = the average length of the short side

$\bar{x}_l$ = the average length of the long side

σ_s = the standard deviation of length of the short side

σ_l = the standard deviation of length of the long side

σ_{area} = the standard deviation of the created internal area

and

$$\sigma_{volume} = [\bar{x}_s^2 \bar{x}_w^2 \sigma_l^2 + \bar{x}_l^2 \bar{x}_w^2 \sigma_s^2 + \bar{x}_s^2 \bar{x}_l^2 \sigma_w^2$$
$$+ \bar{x}_s^2 \sigma_l^2 \sigma_s^2 + \bar{x}_l^2 \sigma_s^2 \sigma_w^2 + \bar{x}_w^2 \sigma_s^2 \sigma_l^2 \tag{13.7}$$
$$+ \sigma_l^2 \sigma_s^2 \sigma_w^2]^{1/2}$$

where

$\bar{x}_w$ = the average width

σ_w = the standard deviation of the width

and $\bar{x}_s$, $\bar{x}_l$, σ_s, σ_l = are defined as in Equation 13.6.

Equations 13.6 and 13.7 assume that the component processes are stable and independent. Figure 13.5 illustrates this concept.[7] If the means and standard deviations for the boxlike container's dimensions are as shown in Figure 13.5, then the standard deviation of the internal volume for the assembled container is:

$$\sigma_{volume} = [\bar{x}_s^2 \bar{x}_w^2 \sigma_l^2 + \bar{x}_l^2 \bar{x}_w^2 \sigma_s^2 + \bar{x}_s^2 \bar{x}_l^2 \sigma_w^2$$
$$+ \bar{x}_s^2 \sigma_l^2 \sigma_w^2 + \bar{x}_l^2 \sigma_s^2 \sigma_w^2 + \bar{x}_w^2 \sigma_s^2 \sigma_l^2$$
$$+ \sigma_l^2 \sigma_s^2 \sigma_w^2]^{1/2}$$
$$= [3^2 2^2 1^2 + 8^2 2^2 (.25)^2 + 3^2 8^2 (.20)^2$$
$$+ 3^2 1^2 (.20)^2 + 8^2 (.25)^2 (.20)^2 + 2^2 (.25)^2 (1^2)$$
$$+ (1^2)(.25)^2 (.20)^2]^{1/2}$$
$$= [36 + 16 + 23.04 + .36 + .16 + .25 + .0025]^{1/2}$$
$$= (75.8125)^{1/2}$$
$$= 8.71 \text{ mm}^3$$

Again, the above law holds only for component processes in statistical control.

SUMMARY

The quality consciousness of an organization can be better understood by examining the types of specifications it uses in production and service. If a firm uses individual unit and/or AQL specifications as guidelines to separate good product/service from bad, the firm is operating in a "defect detection" mode. If a firm uses individual unit specification as guidelines to determine the percentage of its output which is out-of-specification so that the difference between customer needs and process performance can be decreased, the firm has advanced to a "defect prevention" mode. Finally, if a firm uses distribution or performance specifications in its never-ending pursuit of total process improvement, it is operating in a "never-ending improvement" mode. The goals of never-ending improvement are the constant reduction of unit-to-unit variation and process movement toward nominal. Conformance to technical specifications ("Zero Defects") is not an acceptable form of quality consciousness.

This chapter introduced the statistical laws that govern product dimensions, created in the act of assembly: the law of the addition of component dimension averages, the law of the differences of component dimension averages, the law of the sums and differences of component dimension averages, the law of the addition of component dimension standard deviations, the law of the average for created areas and volumes, and the law of the standard deviation for created areas and volumes. These laws offer important insights into how to continually reduce the difference between customer needs (specifications for created dimensions) and process performance (for created dimensions).

It is necessary to understand the information in Chapter 13 to be able to determine a process's capability, or identity. The next chapter focuses on process capability studies.

EXERCISES

13.1 a. Discuss the significance of the phrase "defect detection" mode to the state of an organization's quality consciousness.
   b. Discuss the significance of the phrase "defect prevention" mode to the state of an organization's quality consciousness. Relate your discussion to the type of data an organization would use in its improvement efforts.
   c. Discuss the significance of the phrase "never-ending improvement" mode to the state of an organization's quality consciousness. Relate your discussion to the type of data an organization would use in its quality improvement efforts.

13.2 What is the basic function of a performance specification?

13.3 Explain the purpose and describe the construction of the three types of technical specifications: individual unit specifications, acceptable quality level specifications, and distribution specifications.

13.4 In an assembly operation steel sheet A is glued onto steel sheet B to create a double-sheet thickness. Assume that the glue has no discernable thickness and that the unit-to-unit variation in thickness for both types of steel sheets is stable over time. The resulting thickness of the combined steel sheets is the quality characteristic of interest. The following process statistics have been collected concerning both types of steel sheets:

Steel Sheet A	*Steel Sheet B*
mean = 2.50 inches	mean = 4.75 inches
std dev = 0.25 inches	std dev = 0.50 inches

a. Compute the mean of the double-sheet thickness.
b. Compute the standard deviation of the double-sheet thickness.

13.5 Rectangular sheets of material are produced in an assembly operation. The dimensions of the rectangular sheets are 9.0 inches in width by 14.0 inches in length. Assume that the unit-to-unit variation among the widths and lengths of the rectangular sheets are stable over time. The area of the sheets is the quality characteristic of interest. The following process statistics have been collected for the widths and lengths of the rectangular sheets:

Width	*Length*
mean = 9.0 inches	mean = 14.0 inches
std dev = 0.10 inches	std dev = 0.40 inches

a. Compute the mean area of the rectangular sheets.
b. Compute the standard deviation of the area of the rectangular sheets.

NOTES

1. A. Camp, University of Miami School of Business Administration, *Methods for the Improvement of Quality and Productivity,* Class Project, 1986.
2. AT&T, *Statistical Quality Control Handbook,* 10th printing, May 1984 (Indianapolis: AT&T, 1956), pp. 119–27.
3. Ibid.
4. Ibid.
5. Ibid.
6. Acheson Duncan, *Quality Control and Industrial Statistics,* 5th ed. (Homewood, Ill.: Richard D. Irwin, 1986), pp. 107–12.
7. Ibid.

Chapter 14

Process Capability and Improvement Studies

INTRODUCTION

Process capability studies determine whether a process is unstable, investigate any sources of instability, determine their causes, and take action to eliminate such sources of instability. After all sources of instability have been eliminated from a process, the natural behavior of the process is called its *process capability*. A process must have an established process capability before it can be improved. Consequently, a process capability study must be successfully completed before a process improvement study can have any chance for success.

Process improvement studies follow the Deming cycle of *Plan, Do, Check, Act*. First, managers construct a plan for process improvement, or a plan to decrease the difference between customer needs and process performance (Plan). Second, they test the plan's validity on a small experimental scale (Do). Third, they collect data to determine if the plan will lead to a decrease in the difference between customer needs and process performance (Check). Fourth, if the data collected about the plan show that the plan will achieve its objective(s), it is set into motion (Act); and the managers responsible for the plan return to the Plan phase of the Deming cycle to find other variables that will further reduce the difference between customer needs and process performance. If the data collected about the plan show that the plan will not achieve its objective(s), the managers responsible for the plan return to the Plan phase of the Deming cycle to find other variables that will reduce the difference between customer needs and process performance. The Deming cycle follows a never-ending path of process and quality improvement.

This chapter is divided into three sections: process capability studies, process improvement studies, and quality improvement stories. The quality improvement story is an effective format for quality control practitioners to present process capability and process improvement studies to management.

PROCESS CAPABILITY STUDIES

There are two types of process capability studies: attribute process capability studies and variables process capability studies. This chapter details both types.

Attribute Process Capability Studies

Attribute process capability studies determine a process's capability in terms of fraction of defective output or some other measure of process performance. The major tools used in attribute process capability studies are attribute control charts and the tools discussed in Chapter 12. The process capability for a p chart is p, or the average fraction of defective units generated by the process. The process capability for the np chart is np, or the average number of defective units generated by the process for a given subgroup size, n. The process capability for a c chart is $\bar{c}$, or the average number of defects per unit generated by the process for a given area of opportunity. Finally, the process capability for a u chart is $\bar{u}$, the average number of defects per unit generated by the process where the area of opportunity varies from subgroup to subgroup.

A shortcoming of this type of study is that it begins with a specification, but it is not specific about the reason for failure to meet that specification. The p chart does not indicate if defective units result from the process being off nominal and too close to the specification limit, or because the process has too much unit-to-unit variation, or because the process is not stable with respect to its mean and/or variance. Further, as p charts are relatively insensitive to shifts or trends in the process, problems can go undetected for so long that they cause defectives before they are checked. However, p charts are frequently based on readily available data, which explains their popularity and use.

Variables Process Capability Studies

Variables process capability studies determine a process's capability in terms of the distribution of process output or some other process quality characteristic. The major tools used in variables process capability studies are variables control charts and the tools discussed in Chapter 12. Variables control charts are used to stabilize a process so that it is possible to determine meaningful upper and lower natural limits. Natural limits are computed for stable processes by adding and subtracting three times the process's standard deviation to the process centerline. In general, for any variables control chart, the upper and lower natural limits are:

$$UNL = \bar{\bar{x}} + 3\sigma \tag{14.1}$$

$$LNL = \bar{\bar{x}} - 3\sigma \tag{14.2}$$

Specifically, for x-bar and R charts, the upper and lower natural limits are:

$$UNL = \bar{\bar{x}} + 3(\overline{R}/d_2) \tag{14.3}$$

$$LNL = \bar{\bar{x}} - 3(\overline{R}/d_2) \tag{14.4}$$

For x-bar and s charts, the upper and lower natural limits are:

$$UNL = \bar{\bar{x}} + 3(\bar{s}/c_4) \tag{14.5}$$

$$LNL = \bar{\bar{x}} - 3(\bar{s}/c_4) \tag{14.6}$$

For individuals charts, the upper and lower natural limits are:

$$UNL = \bar{\bar{x}} + 3(\overline{R}/d_2) \tag{14.7}$$

$$LNL = \bar{\bar{x}} - 3(\overline{R}/d_2) \tag{14.8}$$

Natural limits should not be confused with control limits, and as a rule, natural limits should not be shown on variables control charts because natural limits apply to individual units of output and control limits apply to subgroup statistics—and consequently the comparison would be misleading. One notable exception to the above rule is the individuals control chart for variables. In that case, the subgroups consist of individual units, and natural limits and control limits are the same.

Interpretation of the natural limits depends on the stability and shape of the process distribution under study. If the output distribution of a stable process is normal, then for the natural limits in Equations 14.1 through 14.8 we can say that approximately 99.7 percent of all process output will be between the natural limits. For example, if samples of four steel ingots are drawn from an ingot-producing process every hour, and the process is stable and distributed normally with a process average subgroup weight of 42.0 pounds ($\bar{\bar{x}} = 42.0$ lbs) and an average range of 0.6856 pounds, we can say the following about the process using Equations 14.3 and 14.4:

1. The process's upper natural limit is:

$$UNL = \bar{\bar{x}} + 3(\overline{R}/d_2) = 42.0 + 3(0.6856/2.059)$$

$$= 42.0 + 3(0.333) = 42.999$$

$$\cong 43.0 \text{ pounds}$$

2. The process's lower natural limit is:

$$\text{LNL} = \bar{\bar{x}} - 3(\bar{R}/d_2) = 42.0 - 3(0.6856/2.059)$$

$$= 42.0 - 3(0.333) = 41.001$$

$$\cong 41.0 \text{ pounds}$$

3. Approximately 99.7 percent of all steel ingots produced will weigh between 41.0 pounds and 43.0 pounds. This is what the steel ingot process is capable of producing; it is the process's identity.

If a stable process is not normally distributed, but skewed and unimodal, then for all natural limits in Equations 14.1 through 14.8 we can say, using the Camp-Meidel inequality in Equation 6.16, that at least 95.1 percent of all process output will lie between the natural limits.

For example, samples of five 20-foot rods are drawn from a process every day. The warpage of the rods is measured using an operationally defined method. The process is a stable and skewed unimodal process with an average subgroup warpage of one-half inch ($\bar{\bar{x}} = 0.50$ in) and an average range of 0.004652 inches. We can say the following about the process:

1. The process's upper natural limit is:

$$\text{UNL} = \bar{\bar{x}} + 3(\bar{R}/d_2) = 0.500 + 3(0.004652/2.326)$$

$$= 0.500 + 3(0.002)$$

$$= 0.506 \text{ in}$$

2. The process's lower natural limit is:

$$\text{LNL} = \bar{\bar{x}} - 3(\bar{R}/d_2) = 0.500 - 3(0.004652/2.326)$$

$$= 0.500 - 3(0.002)$$

$$= 0.494 \text{ in}$$

3. At least 95.1 percent of all 20-foot rods will have a warpage between 0.494 inches and 0.506 inches. Again, this is what the steel rod process is capable of producing with respect to warpage; it is the process's identity.

Finally, if a stable process's distribution is unknown, then for all the variables natural limits in Equations 14.1 through 14.8 we can say, using Tchebycheff's inequality in Equation 6.17, that at least 88.9 percent of all process output will lie between the natural limits. For example, if three samples of paper per day are drawn from a paper-producing process that is stable and generates an unknown shape distribution with an average subgroup basis weight of 69.0 pounds ($\bar{\bar{x}} = 69.0$ lb) and an average subgroup range of 0.423 pounds, we can say the following about the process:

1. The process's upper natural limit is:

$$UNL = \bar{\bar{x}} + 3(\bar{R}/d_2) = 69.0 + 3(0.423/1.693)$$

$$= 69.0 + 3(0.25)$$

$$= 69.75 \text{ pounds}$$

2. The process's lower natural limit is:

$$LNL = \bar{\bar{x}} - 3(\bar{R}/d_2) = 69.0 - 3(0.423/1.693)$$

$$= 69.0 - 3(0.25)$$

$$= 68.25 \text{ pounds}$$

3. At least 88.9 percent of all paper produced will have a basis weight between 68.25 pounds and 69.75 pounds. Again, this is the process's capability.

The disadvantage of variables process capability studies is that they frequently require that special data be collected. The advantages of variables process capability studies are that they provide information such as whether the process is centered on nominal, exhibiting too much unit-to-unit variation, or unstable with respect to its mean and/or variation—and furthermore, these studies are sensitive to shifts in the process and are helpful in detecting trends or shifts in the process before they cause trouble. Finally, variables process capability studies allow for the examination of specification limits to determine whether the specification limits were reasonable in the first place. We will discuss this further in the section on estimating the percentage out-of-specification later in this chapter.

Data Requirements for Process Capability Studies

Attribute Studies. Attribute process capability studies require a great deal of data. As a rule of thumb, the study should cover at least three or more distinct time periods, where each time period should contain 20 to 25 samples and each sample should have between 50 and 100 units. This rule of thumb is based on experience as well as statistical theory.

Variables Studies. Variables process capability studies require far less data than attribute studies. However, a variables study may be required for each quality characteristic that can cause a unit to be defective. As a rule of thumb, a variables study should cover at least three distinct time periods; the first period should contain about 50 samples of between three and five units each, and the second and third time periods should contain 25 samples of between three and five units each.

Addition of New Data onto a Process Capability Chart. After initial control limits have been calculated, the question arises as to what to do with additional data: should revised control limits be computed, or should the old control limits be extended across the control chart and new points plotted against the old limits? If the process has not changed significantly, new limits should not be calculated because they can cause action based on common variation.[1] In this case, the best procedure is to plot the new data against the old limits and search for a change in the data pattern. If the process has changed significantly, new limits should then be calculated from the additional data. These new limits allow for analysis of the process's new capability.

Process Capability Studies on Unstable Processes

Process statistics, such as the measures of location, dispersion, and shape discussed in Chapter 5, cannot be estimated from a process capability study performed on an unstable or chaotic process; nevertheless, useful information is still available. In such cases, the study often reveals information about the sources of special variation that affect the process and provides an opportunity to better understand the process.[2]

Process Capability Studies on Stable Processes

A process capability study on a stable process sets the stage for the estimation of the process's central tendency, $\bar{\bar{x}}$; standard deviation, σ; and shape (e.g., normal or nonnormal). These statistics allow comparisons between the process's performance and specifications, estimation of the fraction of output beyond specification limits, and use of centerlines, or process averages upon which to establish budgets and forecasts,[3] all required to fuel the Deming cycle. It is important to note that predicting a stable process's future behavior assumes that the process will remain stable. Unfortunately, it is impossible to know if this will be the case; hence, caution is advised.

Estimating the Fraction Out of Specification

A major function of a process capability study is to estimate the fraction of process output that will be out of specification. The procedures are different for attribute and variables process capability studies.

Attribute Process Capability Studies. The centerline on a stable attribute control chart should be used as an estimate of the "overall" process capability. However, there is one important proviso: an estimate of overall process capability is not specific as to the potential cause or causes of defective output. To identify these, we must separate out all possible sources of defects (such as operators, machines, and vendors) and perform individual process capability studies for each source. In such a case, x-bar and R charts are often more cost-effective, in terms of sample size and information, than attribute charts, if they can be used.

As an illustration of an attribute process capability study, because of a customer survey indicating dissatisfaction, a manager wants to determine the capability of the keypunching operation in her department, in terms of the proportion of defective cards produced.[4] She decides to take random

FIGURE 14.1 Attribute Process Capability Study on Keypunching Operation

Raw Data for Construction of Control Chart

Day	Cards Inspected	Number of Defective Cards	Fraction of Defective Cards
1	200	6	.030
2	200	6	.030
3	200	6	.030
4	200	5	.025
5	200	0	.000
6	200	0	.000
7	200	6	.030
8	200	14	.070
9	200	4	.020
10	200	0	.000
11	200	1	.005
12	200	8	.040
13	200	2	.010
14	200	4	.020
15	200	7	.035
16	200	1	.005
17	200	3	.015
18	200	1	.005
19	200	4	.020
20	200	0	.000
21	200	4	.020
22	200	15	.075
23	200	4	.020
24	200	1	.005
Totals	4,800	102	

samples of 200 cards from each day's output, inspect them for defects, and construct an initial p chart. Figure 14.1 shows the raw data, and Figure 14.2 shows the initial process control chart. The latter reveals that on days 8 (14 defective cards out of 200 inspected) and 22 (15 defective cards out of 200 inspected) something special happened, not attributable to the system, to cause defective cards to be keypunched.

The manager calls a meeting of the ten keypunch operators. The purpose of the meeting is to brainstorm for possible sources of the special causes of variation on days 8 and 22. The results of the brainstorming session are put onto the Cause-and-Effect diagram shown in Figure 14.3. The ten group members vote that their best guess for the problem on day 8 was a new untrained keypunch operator (see cause in Figure 14.3—circled in a cloud) who had been added to the work force, and that the one day it took the worker to acclimate to the new environment probably caused the unusually high number of keypunch errors. To ensure that this special cause will not be repeated, the manager institutes a one-day training program for all new employees. The ten group members also vote that their best guess for the problem on day 22 was that on the previous evening the department had run out of cards from the regular vendor, did

FIGURE 14.2 p Chart from Raw Data

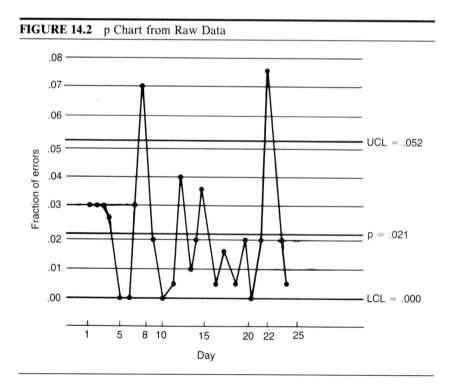

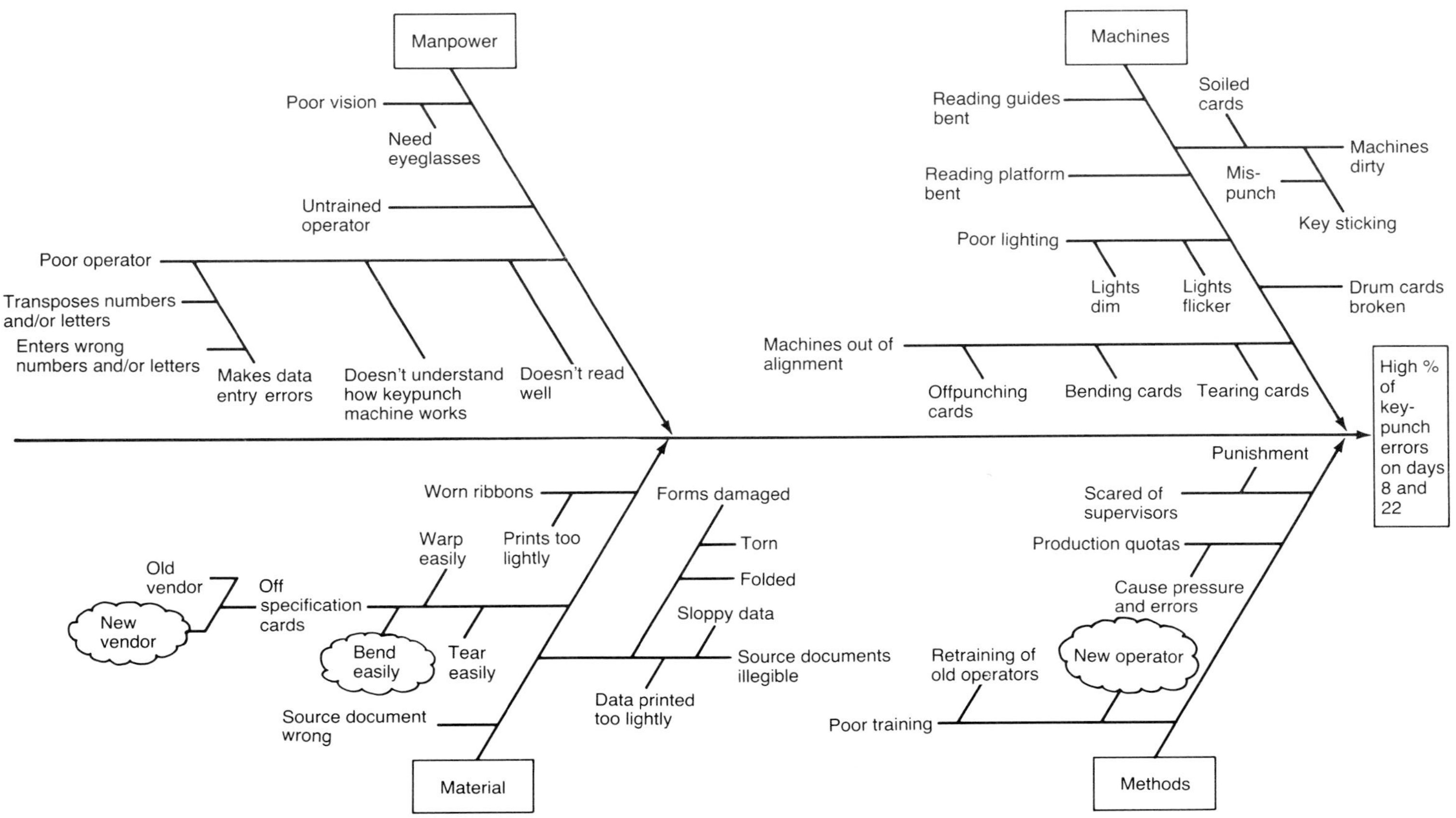

FIGURE 14.3 Cause-and-Effect Diagram to Determine Special Sources of Variation for All Operators on Days 8 and 22

not expect a new shipment until the morning of day 23 and consequently, purchased a one-day supply of cards from a new vendor. The operators found these cards were of inferior quality and bent easily, which caused the large number of defective cards. To correct this special cause of variation, the manager revises the firm's relationship with its regular card vendor and operationally defines acceptable quality for keypunch cards.

After eliminating the days for which special causes of variation are found, the manager recomputes the control chart statistics, using Equations 8.1 through 8.3:

$$\text{centerline(p)} = \text{average fraction of defective cards} = p$$

$$p = 73/4,400 = 0.01659 \cong 0.017$$

$$\text{UCL(p)} = 0.044$$

$$\text{LCL(p)} = 0.000$$

Figure 14.4 shows the revised control chart. The process appears stable. The centerline and control limits were extended out into the future for 25 days. Data from daily samples of 200 cards were collected for these 25 days and plotted with respect to the forecasted centerline and control limits. The process was found to be stable. The capability of the process is such that it will produce an average of 1.7 percent defective cards per day. Further, the percentage defective will rarely surpass 4.4 percent. Although the process's capability is now known, the manager is not satisfied with its capability and should not stop attempting further improvement. We will see how this is done in the section on process improvement studies later in this chapter.

Variables Process Capability Studies. The fraction of process output beyond specification limits can be estimated and predicted for stable processes. If the process is stable and its distribution is approximately normal, we can estimate the fraction of output beyond the specification limits as follows:

1. Estimate the process average, $\bar{\bar{x}}$.
2. Estimate the process standard deviation, σ, using $\bar{R}/d_2$ or $\bar{s}/c_4$, whichever is appropriate.
3. Calculate the standardized distance between the process average and the relevant specification limits. The standardized distance for the lower specification limit (LSL) is:

$$Z_{LSL} = \left[\frac{LSL - \bar{\bar{x}}}{\sigma} \right] \qquad (14.9)$$

where Z_{LSL} is the number of process standard deviation units between the lower specification limit and the process mean.

FIGURE 14.4   Revised p Chart Following Removal of Special Causes

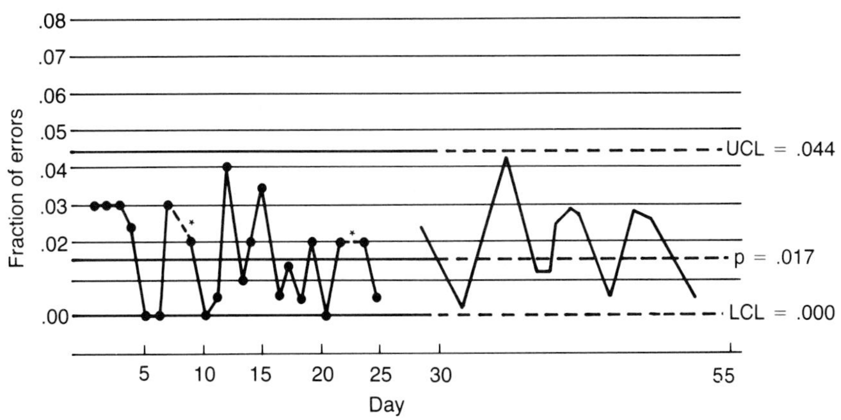

*Day 8 and day 22 removed.

The standardized distance for the upper specification limit (USL) is:

$$Z_{USL} = \left[\frac{USL - \bar{\bar{x}}}{\sigma} \right] \qquad (14.10)$$

where Z_{USL} is the number of process standard deviation units between the upper specification limit and the process mean.

4. Determine the probability that a unit of output will fall out of a relevant specification limit. The probability of a unit's falling below the lower specification unit or above the upper specification limit is obtained using Table 1, as discussed in Chapter 6 under the normal probability distribution. An example of this procedure is shown in the process capability study later in this chapter.

Engineers and managers should not assume that all well-behaved processes are normally and/or symmetrically distributed. Sometimes stable process distributions are skewed for sound engineering, physical, or mathematical reasons: for example, characteristics where the natural limit is zero, such as the proportion of an undesirable substance in a mixture or warpage; characteristics being measured near their physical limits, such as tensile strength or temperature of a liquid; or characteristics whose dimensions, when combined, will create areas or volumes. For skewed distributions, the fraction of output beyond the specification limits may be estimated.[5]

Process Capability Case Study

To illustrate a variables process capability study on a process whose characteristic of interest is normally distributed, consider an automobile manufacturer who wishes to purchase camshafts from a vendor.[6] The automobile manufacturer is concerned with the finish grind, diameters, and case hardness, as well as other quality characteristics of the camshaft. For illustrative purposes, this discussion will focus only on the case hardness depth of the camshafts.

The contract between the automobile manufacturer and camshaft vendor calls for camshafts that have an average case hardness depth of 7.0 mm and are distributed normally around the average with a dispersion not to exceed 3.5 mm; this is a distribution specification. Further, the contract requires that the vendor produce a process capability study demonstrating statistical control of his process. Consequently, the objective of management is to reduce camshaft-to-camshaft variation for case hardness depth and to move the process's average case hardness depth to the desired nominal of 7.0 mm.

FIGURE 14.5 Camshaft in an Engine

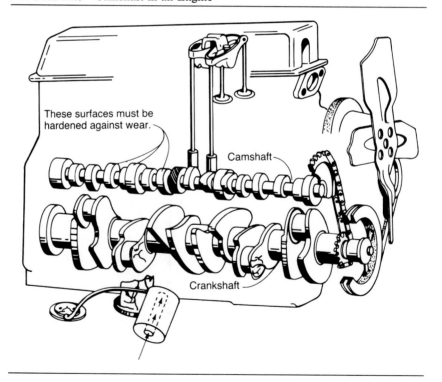

As background, a *camshaft* is a rod with elliptical lobes along its length. As the rod rotates, so do the elliptical lobes, and this ultimately causes intake and exhaust valves to open and close. The intake valves permit a mixture of fuel and air to enter the cylinders, where combustion takes place. The exhaust valves permit the waste gases to exit the cylinders after combustion. The surfaces of the elliptical lobes must be hardened (made brittle) to reduce wear, as seen in Figure 14.5. This hardening is called *case hardening* and is accomplished by immersing the camshaft in oil, placing electric bearing coils around the lobes, and passing electric current through the coils. This process heat treats the lobes and makes them hard (brittle), as shown in Figure 14.6. The depth to which the brittleness extends is called *case hardness depth*. The case hardness depth must be tightly controlled—for if the case hardness depth is too

FIGURE 14.6   Heat-Treating Camshaft Lobes

In the manufacturing process, the camshaft is immersed in oil, with coils around each surface to be hardened. Current is then passed through the coils, heat-treating the surfaces. The treated camshaft is later used in the finished assembly of the engine.

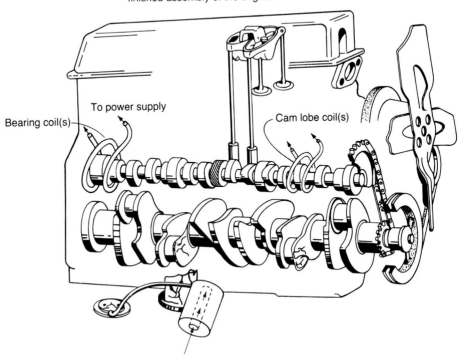

FIGURE 14.7 Process Capability Study of Camshaft

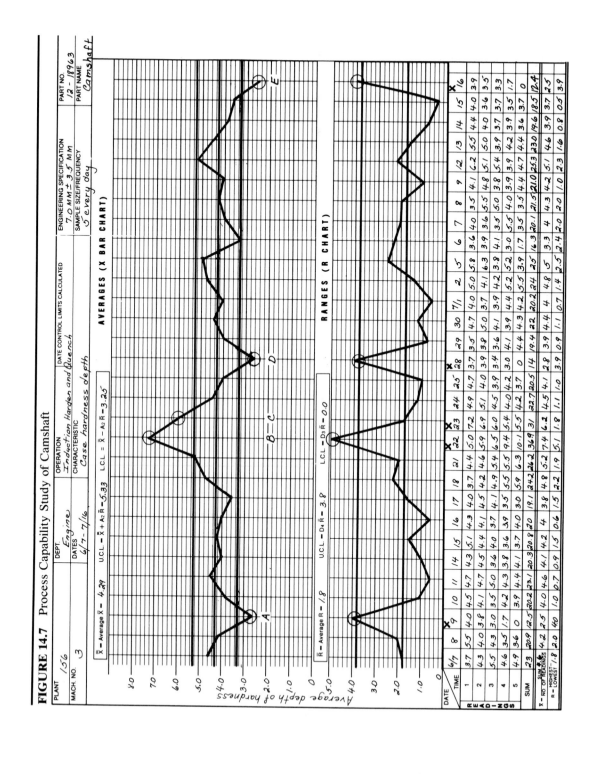

deep, the lobes will be too brittle and will break, and if the case hardness depth is too shallow, the lobes will be too soft and will bend.

Pursuant to the terms of the contract calling for a process capability study, a sample of five camshafts is drawn from the vendor's process every day. Each shaft is measured with respect to each of the relevant quality characteristics.

Figure 14.7 shows the initial data collected in the process capability study. These data reveal that the vendor's process is not in statistical control; this is indicated by points A through E on Figure 14.7 (June 9, 22, 23, 28, and July 16). Consequently, corrective action on the process is required by vendor management.

An engineer from the vendor's plant forms a brainstorming group comprised of the workers in the Induction Hardening and Quench Department. This is the department in which case hardening is performed on the camshafts. The purpose of the brainstorming group is to determine the causes for the out-of-control points on Figure 14.7. The results of the brainstorming sessions are shown on the cause-and-effect diagram in Figure 14.8. The group decides (votes) that the probable causes for out-of-

FIGURE 14.8   Cause-and-Effect Diagram to Diagnose Reasons for Out-of-Control Points

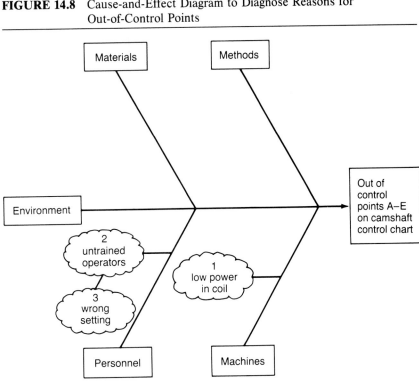

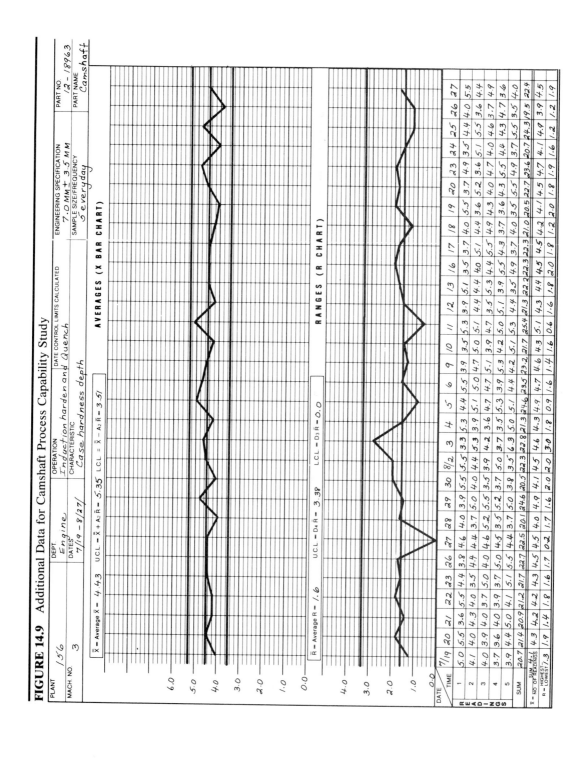

FIGURE 14.9 Additional Data for Camshaft Process Capability Study

control points were as follows:

1. Point A. Low power in the coil resulted in increased variability and less stable depth in the case hardness (see cloud 1 on Figure 14.8).
2. Point B. A temporary operator was used because the regular operator was sick (see cloud 2 on Figure 14.8).
3. Point C. The case hardness setting on the machine was incorrect (see cloud 3 on Figure 14.8).
4. Point D. Low power in the coil resulted in increased variability and less depth in case hardness (see cloud 1 on Figure 14.8).
5. Point E. Low power in the coil resulted in an out-of-control situation (see cloud 1 on Figure 14.8).

This analysis leads vendor management to take action on the process by repairing the voltage meter on the induction hardening machine (see points A, D, and E) and training all personnel in the proper operation of the machine (see points B and C).

After the above policies are instituted, the engineer collects 30 additional days of data, as shown in Figure 14.9. We see that the vendor's process is in statistical control. However, from the calculations shown in Figure 14.10, we see that approximately 8.9 percent of the camshafts will have a case hardness depth below the lower specification of 3.50 mm. This defect rate is not acceptable to the automobile manufacturers. At this point, the process capability of the camshaft vendor is a stable and known entity. The next steps are to further reduce camshaft-to-camshaft variation and better center the process average on nominal.

Process Capability Studies on Created Dimensions

In Chapter 13 we discussed created dimensions. In this section, we discuss how to determine the process capability for created dimensions and compute the fraction of created dimensions that will be beyond specification limits. For example, suppose that the process capability studies for the thickness of the sheet, pin, and both washer subcomponents of the assembly shown in Figure 13.4 were based on subgroups of five subcomponents and yielded the following statistics:

$$\bar{x}_s = 40 \text{ mm;} \quad \text{and} \quad \sigma_s = 0.0050$$

$$\bar{x}_{w1} = 3 \text{ mm;} \quad \text{and} \quad \sigma_{w1} = 0.0007$$

$$\bar{x}_{w2} = 4 \text{ mm;} \quad \text{and} \quad \sigma_{w2} = 0.0008$$

$$\bar{x}_p = 27 \text{ mm;} \quad \text{and} \quad \sigma_p = 0.0030$$

Further, assume that all component processes are independent, stable, normally distributed, and randomly assembled.

FIGURE 14.10   Fraction of Camshafts Out of Specification

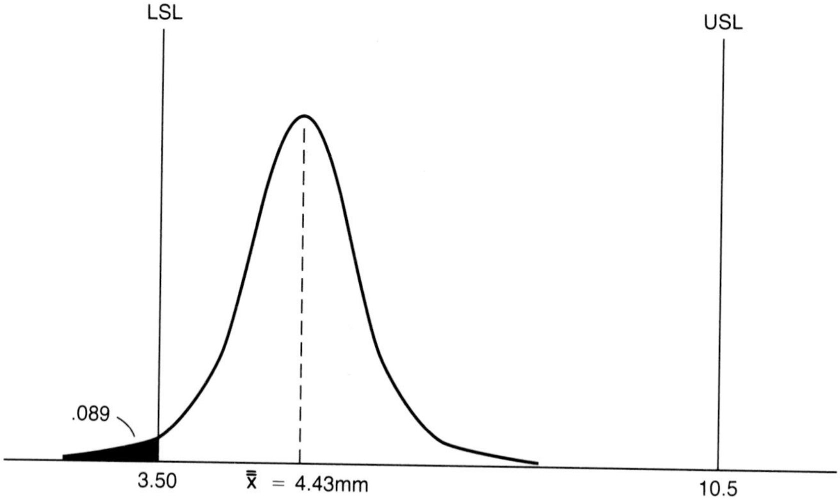

$$\sigma = \frac{\overline{R}}{d_2} = \frac{1.60}{2.326} = 0.688 \text{ mm} = \text{estimated process standard deviation}$$

$$Z_{LSL} = \frac{(3.50 - 4.43)}{0.688} = -1.35$$

The probability that a camshaft falls below the LSL is 8.9%.

The average bolt projection, from Equation 13.3, is

$$\overline{X}_{\text{bolt projection}} = 40 \text{ mm} - 3 \text{ mm} - 27 \text{ mm} - 4 \text{ mm} = 6 \text{ mm}$$

The standard deviation of the bolt projection, from Equation 13.4, is

$$\sigma_{\text{bolt projection}} = [(0.0050)^2 + (0.0007)^2 + (0.0008)^2 + (0.0030)^2]^{1/2}$$

$$= 0.00593 \cong .006$$

The process capability of the created bolt projection dimension is computed using Equations 14.1 and 14.2:

$$\text{UNL} = 6 \text{ mm} + 3(0.006) = 6 \text{ mm} + 0.018 \text{ mm} = 6.018 \text{ mm}$$

$$\text{LNL} = 6 \text{ mm} - 3(0.006) = 6 \text{ mm} - 0.018 \text{ mm} = 5.982 \text{ mm}$$

Consequently, 99.7 percent of all the bolt projections will be between 5.982 and 6.018 mm.

If a customer sets bolt projection specifications of 5.99 ± 0.02 mm, then the fraction of out-of-specification bolt projections is computed using Equations 14.9 and 14.10, as shown in Figure 14.11.

$$Z_{LSL} = \frac{\text{LSL} - \overline{\overline{x}}}{\sigma} = \frac{5.97 - 6.00}{.006} = -5.00$$

FIGURE 14.11   Fraction Out of Specification for Bolt Projection

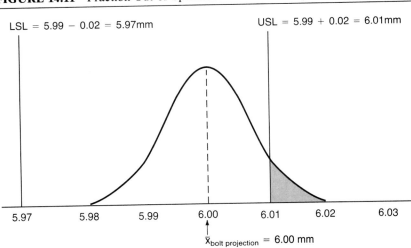

The lower specification limit is five process standard deviations below the process mean. Consequently, from Table 1, we see that no bolt projections will fall below the lower specification limit.

$$Z_{USL} = \frac{USL - \bar{\bar{x}}}{\sigma} = \frac{6.01 - 6.00}{.006} = +1.67$$

The upper specification limit is 1.67 process standard deviations above the process mean. Consequently, from Table 1, we see that 4.75 percent of the bolt projections will fall above the upper specifications limit. Thus, 4.75 percent of all bolt projections will be out of specification. This serious problem results from the process's not being centered on nominal and from unit-to-unit variation.

Finally, created dimensions should be control charted because a created dimension can be out of control while component dimensions are in control. Never-ending improvement cannot progress without reducing unit-to-unit variations and moving the process toward nominal for created dimensions.

The Relationships between Control Limits, Natural Limits, and Specification Limits for Variables Control Charts

Natural Limits and Control Limits.   *Natural limits* are used with respect to individual observations—consequently, on run charts. *Control limits* are used with respect to subgroup statistics—consequently, on control charts. The relationships between natural limits and control limits are shown in Figure 14.12. We see that for x-bar charts, if A_2 (x-bar and R chart) or A_3 (x-bar and s chart) is multiplied by the square root of the subgroup size $(\sqrt{n})$, and these new quantities $(A_2\sqrt{n})$ and $(A_3\sqrt{n})$ are added and subtracted from the process average $(\bar{\bar{x}})$, the control limits are transformed into natural limits. In the case of control limits for individuals charts, the control limits and the natural limits are identical because the subgroup size is one.

Natural Limits and Specification Limits. Natural limits and specification limits are comparable quantities for stable processes because they are both measured with respect to the individual units of output generated by the process under study. There are six basic relationships between natural limits and specification limits for stable processes.

Relationship 1. The process's natural limits are inside the specification limits, the process is centered on nominal, and the process's distribution is normal. This is illustrated in the run chart shown in Figure 14.13(a).

Relationship 2. The process's natural limits are inside the specification limits, the process is not centered on nominal, and the process's

---

FIGURE 14.12 Relationship between Control Limits and Natural Limits for Variables Location Charts

Chart Type	Control Limits	Natural Limits	Comments
$\bar{x}$ (with R chart)	$\bar{\bar{x}} \pm A_2\bar{R} =$ $\bar{\bar{x}} \pm 3\left(\dfrac{\bar{R}/d_2}{\sqrt{n}}\right)$ $\left(A_2 = \dfrac{3}{d_2\sqrt{n}}\right)$	$\bar{\bar{x}} \pm 3\,(\bar{R}/d_2)$	If A_2 and/or A_3 are multiplied by the square root of the subgroup size $(\sqrt{n})$, and this new quantity is added or subtracted from $\bar{\bar{x}}$, the new limits are the natural limits
$\bar{x}$ (with s chart)	$\bar{\bar{x}} \pm A_3\bar{s} =$ $\bar{\bar{x}} \pm 3\left(\dfrac{\bar{s}/c_4}{\sqrt{n}}\right)$ $\left(A_3 = \dfrac{3}{c_4\sqrt{n}}\right)$	$\bar{\bar{x}} \pm 3\,(\bar{s}/c_4)$	
x (individuals chart with moving range chart)	$\bar{\bar{x}} \pm E_2\bar{R} =$ $\bar{\bar{x}} \pm 3\,(\bar{R}/d_2)$ $(E_2 = 3/d_2)$	$\bar{\bar{x}} \pm 3\,(\bar{R}/d_2)$	Control limits and natural limits are equivalent

distribution is normal. This is illustrated in the run chart shown in Figure 14.13(b).

Relationship 3. The process's natural limits are outside the specification limits, the process is centered on nominal, and the process's distribution is normal. This is illustrated in the run chart shown in Figure 14.13(c).

Relationship 4. The process's natural limits are outside the specification limits, the process is not centered on nominal, and the process's distribution is normal. This is illustrated in the run chart shown in Figure 14.13(d).

Relationship 5. The process's natural limits are inside the specification limits, the process is not centered on nominal, and the process's distribution is non-normal. This is illustrated in the run chart shown in Figure 14.13(e).

FIGURE 14.13 Relationship between Natural Limits and Specification Limits

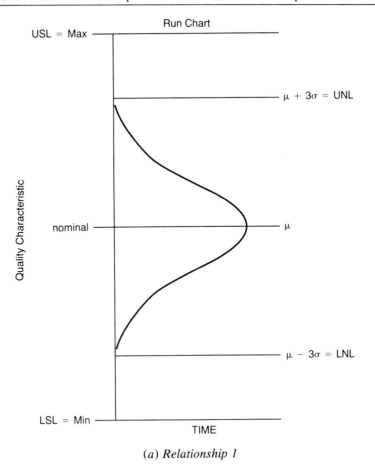

(a) Relationship 1

FIGURE 14.13 *Continued*

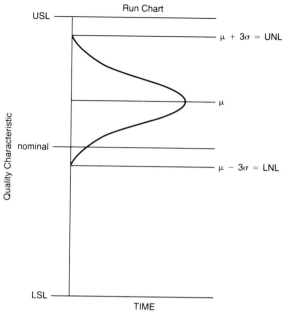

(b) Relationship 2

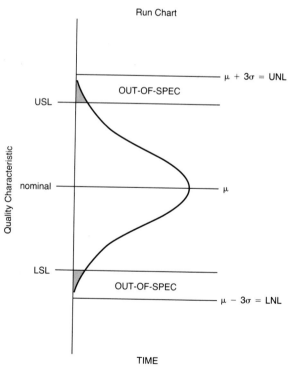

(c) Relationship 3

FIGURE 14.13 *Continued*

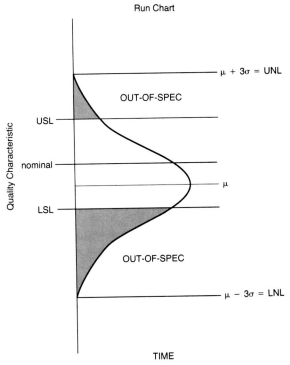

(*d*) *Relationship 4*

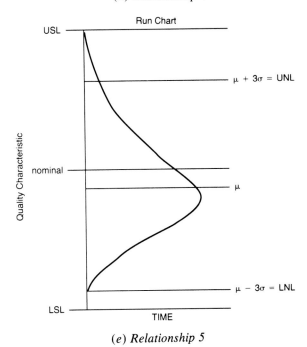

(*e*) *Relationship 5*

FIGURE 14.13   *Concluded*

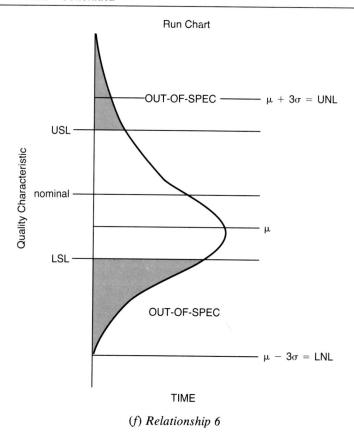

(*f*) *Relationship 6*

Relationship 6.   The process's natural limits are outside the specification limits, the process is not centered on nominal, and the process's distribution is non-normal. This is illustrated in the run chart shown in Figure 14.13(f).

In Chapter 8 we discussed the four states of a process.[7] Relationships 1 and 2 represent a process in its ideal state; given the same variation between relationships 1 and 2, relationship 1 is preferable. Relationships 3 and 4 represent a process in the threshold state; given the same variation between relationships 3 and 4, relationship 3 represents the more desirable situation. Relationship 5 represents a non-normal process in the ideal state. Relationship 6 represents a non-normal process in the threshold state.

Control Limits and Specification Limits.   In no case should specification limits be shown on x-bar charts. This is because while control limits apply to process statistics ($\bar{x}$), specification limits apply to individual units of process output (or some other quality characteristic). Nevertheless, specification limits are sometimes shown on control charts for individuals because in this special case the control limits are based on subgroups of size one—hence, on individual values.

Process Capability Indices for Variables Data

A common desire of many control chart users is to be able to state a process's ability to meet specifications in one summary statistic.[8] Such statistics are available and are called *process capability indices*. These indices are used to summarize internal processes as well as vendor processes.

Assumptions. All of the process capability indices we will discuss require three basic assumptions: process stability, variables data, and approximate normality of the process characteristic under study. It is common practice to assume that process capability indices are based on processes that are distributed normally. This ensures that all users of the capability index are "playing from the same score card."

Indices. Four process capability indices are commonly used: C_p, CPU, CPL, and C_{pk}. These indices are summarized in Figure 14.14.

C_p. The C_p index is used to summarize a process's ability to meet two-sided specification limits. C_p is computed as:

$$C_p = \frac{USL - LSL}{6\sigma} \tag{14.11}$$

In addition to the general assumptions stated in the prior section, the C_p index also assumes that the process average ($\bar{\bar{x}}$) is centered on the nominal value, m.

Recall that a process's capability is defined to be the range in which almost all of the output will fall; usually, this is described as plus or minus three standard deviations from the process's mean, or within an interval of six standard deviations (6σ). Consequently, if a process's USL = UNL = $\bar{\bar{x}} + 3\sigma$ and its LSL = LNL = $\bar{\bar{x}} - 3\sigma$, the process's capability is 1.0:

$$C_p = \frac{USL - LSL}{6\sigma} = \frac{(\bar{\bar{x}} + 3\sigma) - (\bar{\bar{x}} - 3\sigma)}{6\sigma} = \frac{6\sigma}{6\sigma} = 1.0$$

A process capability of 1.0 indicates that a process will generate approximately three out-of-specification units in 1,000, given the assumptions stated above.

FIGURE 14.14 Process Capability Indices

Index	Estimation Equation	Equation No.	Purpose	Assumptions about Process
C_p	$\dfrac{USL - LSL}{6\sigma}$	14.11	Summarize process potential to meet two-sided specification limits	1. Stable process 2. Normally distributed process 3. Variables data 4. Centered process (process average equals nominal)
CPU	$\dfrac{USL - \bar{\bar{x}}}{3\sigma}$	14.12	Summarize process potential to meet only a one-sided upper specification limit	1. Stable process 2. Normally distributed process 3. Variables data
CPL	$\dfrac{LSL - \bar{\bar{x}}}{3\sigma}$	14.13	Summarize process potential to meet only a one-sided lower specification limit	(same as CPU)
C_{pk}	$C_p - \dfrac{|m - \bar{\bar{x}}|}{3\sigma}$ where m = nominal value of the specification	14.14	1. Summarize process potential to meet two-sided specification limits. 2. $|m - \bar{\bar{x}}|/3\sigma$ is a penalty factor for the process's being off nominal. It is stated in terms of the number of natural limit units the process is off nominal.	(same as CPU)

For centered processes, given the assumptions above, there is a relationship between C_p, USL and LSL, and σ. These relationships are shown in Figure 14.15. Figure 14.15(a) shows a process with a $C_p = 1.0$. This indicates that the UNL = USL and the LNL = LSL; hence, 99.7 percent of the process's output will be within specification limits, and, as

$$C_p = \frac{USL - LSL}{6\sigma} = 1.0$$

then

$$\sigma = \frac{USL - LSL}{6}$$

FIGURE 14.15   Process Capability Indices

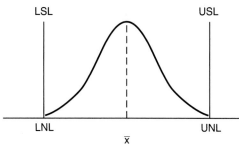

$$C_p = \frac{USL - LSL}{6\sigma} = 1.0$$

99.7% of output will be in-spec
$\sigma = (1/6)(USL - LSL)$

(a) C_p index is 1.0

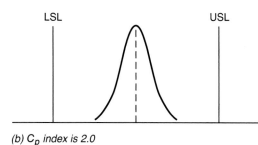

$$C_p = \frac{USL - LSL}{6\sigma} = 2.00$$

~100% of output will be in-spec
$\sigma = (1/12)(USL - LSL)$

(b) C_p index is 2.0

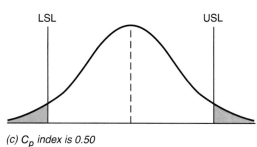

$$C_p = \frac{USL - LSL}{6\sigma} = 0.50$$

86.6% of output will be in-spec
$\sigma = (1/3)(USL - LSL)$

(c) C_p index is 0.50

Thus, we can say that if a process has a $C_p = 1.0$, the process standard deviation is one sixth of the distance between the upper and lower specification limits.

Figure 14.15(b) shows a process with a $C_p = 2.0$. This indicates that the UNL is half way between nominal and the USL and that the LNL is half way between the LSL and nominal; hence, more than 99.99 percent of the process's output will be within specification limits. As

$$C_p = \frac{USL - LSL}{6\sigma} = 2.0$$

then

$$\sigma = \frac{USL - LSL}{12}$$

Thus, we can say that if a process has a $C_p = 2.0$, the process standard deviation is one twelfth of the distance between the upper and lower specification limits.

Figure 14.15(c) shows a process with a $C_p = 0.50$. This indicates that the USL is half way between nominal and the UNL and that the LSL is half way between the LNL and nominal; hence, approximately 86.6 percent of the process's output will be within specification limits. As

$$C_p = \frac{USL - LSL}{6\sigma} = 0.50$$

then

$$\sigma = \frac{USL - LSL}{3}$$

Thus, we can say that if a process has a $C_p = 0.50$, the process standard deviation is one third of the distance between the upper and lower specification limits.

CPU. The CPU index is used to summarize a process's ability to meet a one-sided upper specification limit. In many situations process owners are concerned only that a process not exceed an upper specification limit. For example, for products that can warp in only one direction, there is no LSL for warpage; the lower the warpage the better. However, there is a USL for warpage, the value for warpage that will critically impair the product's ability to meet customer needs. It can also be used in situations where we want to examine one side of a two-sided specification limit.

CPU is computed as:

$$CPU = \frac{USL - \bar{\bar{x}}}{3\sigma} \tag{14.12}$$

The CPU index measures how far the process average ($\bar{\bar{x}}$) is from the upper specification limit in terms of natural tolerance limits (3σ). Natural tolerances, when added and subtracted from the process mean ($\bar{\bar{x}}$), yield the range in which a process is capable of operating—the process's capability, $\bar{\bar{x}} \pm 3\sigma = \bar{\bar{x}} \pm$ natural tolerance.

If a process USL = UNL ($= \bar{\bar{x}} + 3\sigma$), the CPU is 1.0:

$$CPU = \frac{USL - \bar{\bar{x}}}{3\sigma} = \frac{USL - \bar{\bar{x}}}{UNL - \bar{\bar{x}}} = 1.0$$

A CPU of 1.0 indicates that a process will generate approximately one and one half out-of-specification units in 1,000, assuming the process's output is stable and distributed normally.

If a process's UNL is greater than the USL, the CPU is less than one. As UNL − USL increases, the fraction of process output that is out of specification will increase geometrically. Conversely, if a process's UNL is less than the USL, then CPU is greater than one. As USL − UNL increases, the fraction of process output that is out of specification will decrease geometrically. To determine the fraction of process output that will be out of specification for a given value of CPU, we use Equation 14.10.

CPL. The CPL index is used to summarize a process's ability to meet a one-sided lower specification limit. The CPL operates exactly like the CPU. CPL is computed as:

$$CPL = \frac{LSL - \bar{\bar{x}}}{3\sigma} \qquad (14.13)$$

If a process's LNL is greater than the LSL, then CPL is less than one. As LNL − LSL increases, the fraction of process output that is out of specification will increase geometrically. Conversely, if a process's LNL is less than the LSL, then CPL is greater than one. As LSL − LNL increases, the fraction of process output that is out of specification will decrease geometrically. To determine the fraction of process output that will be out of specification for a given value of CPL, we use Equation 14.9.

C_{pk}. The C_{pk} index is used to summarize a process's ability to meet two-sided specification limits. The C_{pk} index uses the C_p index as a starting point for stating a process's capability, but it penalizes C_p if the process is not centered on nominal, m. C_{pk} is computed as:

$$C_{pk} = C_p - \left\{ \frac{|m - \bar{\bar{x}}|}{3\sigma} \right\} \qquad (14.14)$$

The term in brackets in Equation 14.14 is always positive, and hence lowers the value of C_p, which indicates that the process is less able to produce within specifications. The bracketed term is a measure of how many natural tolerance units (3σ) the process mean ($\bar{\bar{x}}$) is from nominal (m). The further off-center the process, the more C_p is penalized by the bracket factor. Hence, C_{pk} is a two-sided capability index that accounts for process centering.

A firm that exists in a defect detection mode will not know the process capability indices for its various processes. On the other hand, a firm operating in a defect prevention mode will know the values for its various processes and will be striving for C_p or C_{pk} approximately equal to 1.0. Finally, if a firm is pursuing never-ending improvement, it will be striving to move C_p and C_{pk} toward infinity. As C_p and C_{pk} become increasingly greater than one, the specification limits from which they were computed become increasingly irrelevant.

Limitations of Capability Indices. Several potential problems exist when using the C_p and C_{pk} indices. First, if a process is not stable, C_p and C_{pk} are meaningless statistics. Second, not all processes meet the assumption of normality. Hence, the naive user of capability indices will incorrectly assess the fraction of process output that will be out of specification. Last, experience shows that naive users of capability indices frequently confuse C_p and C_{pk}; they think they yield the same information about a process. Of course, this can result in a great deal of confusion.

An Example. Each of the process capability indices discussed earlier in this chapter is calculated using the camshaft example shown in Figures 14.7, 14.9, and 14.10. Figure 14.7 shows the camshaft operation out of control. After special sources of variation are removed from the process, it becomes stable, as shown in Figure 14.9. Further investigation reveals that the distribution of case hardness depth is approximately normally distributed. From Figure 14.9 we see that the average case hardness depth is 4.43 mm, the average range for case hardness depth is 1.60 mm, the upper specification limit is 10.5 mm, and the lower specification limit is 3.5 mm. The above figures are summarized as follows:

$$\bar{\bar{x}} = 4.43 \text{ mm}$$

$$\bar{R} = 1.60 \text{ mm; hence } \sigma = \bar{R}/d_2 = 1.60/2.326 = 0.688 \text{ mm}$$

$$\text{USL} = 10.5 \text{ mm}$$

$$\text{LSL} = 3.5 \text{ mm}$$

Given the above figures, the C_p, CPU, CPL, and C_{pk} can be computed and interpreted.

C_p. The computation of C_p is shown below.

$$C_p = \frac{\text{USL} - \text{LSL}}{6\sigma} = \frac{10.5 - 3.5}{6(0.688)} = 1.70$$

This C_p indicates an extremely capable process that will almost never produce out-of-specification product. Obviously, this is completely false; from Figure 14.10 we see that 8.9 percent of all camshafts will have an out-of-specification case hardness depth. The C_p index's failure to accurately state the process's capability results from the process's being 2.57 mm off-center ($|\bar{\bar{x}} - m| = |4.43 - 7.00| = 2.57$ mm) and the C_p's assumption that the process is centered.

CPU. The computation of CPU is shown below.

$$\text{CPU} = \frac{\text{USL} - \bar{\bar{x}}}{3\sigma} = \frac{10.5 - 4.43}{3(0.688)} = 2.94$$

The CPU accurately indicates that the process is operating well within the USL of 10.5 mm. We can calculate the fraction of camshafts that will be out of specification by using Equation 14.10.

$$Z_{USL} = \frac{USL - \bar{\bar{x}}}{\sigma} = \frac{10.5 - 4.43}{0.688} = 8.82$$

From Table 1, we see that the fraction of camshafts above the USL is 0.

CPL. The computation of CPL is shown below.

$$CPL = \frac{LSL - \bar{\bar{x}}}{3\sigma} = \frac{3.5 - 4.43}{3(0.688)} = -0.45$$

The CPL accurately indicates that the process is not operating within the LSL of 3.5 mm. We can calculate the fraction of camshafts that will be out of specification by using Equation 14.9:

$$Z_{LSL} = \frac{LSL - \bar{\bar{x}}}{\sigma} = \frac{3.5 - 4.43}{0.688} = -1.35$$

From Table 1 we see that 8.9 percent of the camshafts will be below the LSL.

C_{pk}. The computation of C_{pk} is shown below.

$$C_{pk} = C_p - \left\{ \frac{|m - \bar{\bar{x}}|}{3\sigma} \right\}$$

$$= \frac{USL - LSL}{6\sigma} - \frac{|m - \bar{\bar{x}}|}{3\sigma}$$

$$= \frac{10.5 - 3.5}{6(0.688)} - \frac{|7.0 - 4.43|}{3(0.688)}$$

$$= 1.70 - 1.25$$

$$= 0.45$$

This C_{pk} indicates a process that will produce a good deal of out-of-specification product. This is reasonable because 8.9 percent of the product will be out of specification.

In the end, C_p and C_{pk} can potentially cause more problems than they can provide benefits. Consequently, some practitioners recommend that they not be used; rather, Equations 14.9 and 14.10 should be used, for they clearly state the fraction of output that will be out of specification.

PROCESS IMPROVEMENT STUDIES

In Chapter 8 we discussed the Deming cycle as a vehicle for process improvements. The Deming cycle requires that managers understand and utilize both enumerative and analytic studies of design/redesign, conformance, and performance to pursue continuous and never-ending improvement of an organization's extended process. There are many tools and techniques available so that managers can conduct the previously mentioned studies; these tools and techniques are the subject of this book. In addition, successful process improvement endeavors require a managerial view that embraces the tools and techniques discussed in this book. The Deming philosophy is such a managerial view.

As with process capability studies, there are two types of process improvement studies: attribute improvement studies and variables improvement studies. In the pursuit of continuous and never-ending improvement, it is natural that attribute improvement studies lead to variables improvement studies.

Attribute Improvement Studies

Returning to the keypunching example discussed earlier in this chapter, recall that the percentage of defective cards was stabilized with an average 1.7 percent defective cards and would rarely go above 4.4 percent defective cards.

At this point the manager decides that to improve the process further she must study each keypunch operator individually. However, she must determine which operators to work with first. She makes a check sheet to record the number and fraction of defective cards by operator for December 1987, as shown in Figure 14.16. Next, she constructs a c chart for the

FIGURE 14.16 Check Sheet of Defective Cards by Operator (12/1/87–12/31/87)*

Operator	Tally	Frequency
001	II	2
002	III	3
003	I	1
004	ⅢⅢ ⅢⅢ ⅢⅢ IIII	19
005		0
006	II	2
007	I	1
008	III	3
009	ⅢⅢ ⅢⅢ ⅢⅢ II	17
010	II	2
Total		50

* All operators produced approximately the same number of cards during the period under study.

FIGURE 14.17   c Chart for Number of Defective Cards by Operator

(a) By Operator (12/1/87-12/31/87)

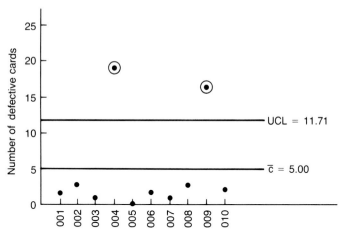

$\bar{c} = 50/10 = 5.00$
$3\sqrt{\bar{c}} = 6.71$
LCL = 0.00
UCL = 11.71
Operators 004 and 009 are out-of-control

*(b) By Operator - Without Operators
      004 and 009 (12/1/87-12/31/87)*

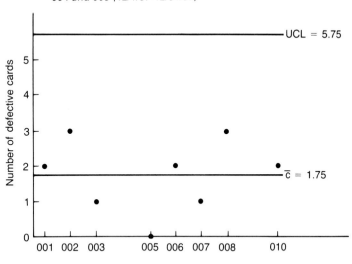

$\bar{c} = 14/8 = 1.75$
$3\sqrt{\bar{c}} = 4.00$
LCL = 0.00
UCL = 5.75
Operators are stable without operators 004 and 009.

FIGURE 14.18   Pareto Analysis of Defective Cards

(a) By Operator (12/1/87–12/31/87)

Operator	Frequency	Fraction	Cumulative Fraction
004	19	.38	.38
009	17	.34	.72
002	3	.06	.78
008	3	.06	.84
001	2	.04	.88
006	2	.04	.92
010	2	.04	.96
003	1	.02	.98
007	1	.02	1.00
005	0	.00	1.00
Totals	50	1.00	

(b) Pareto Diagram

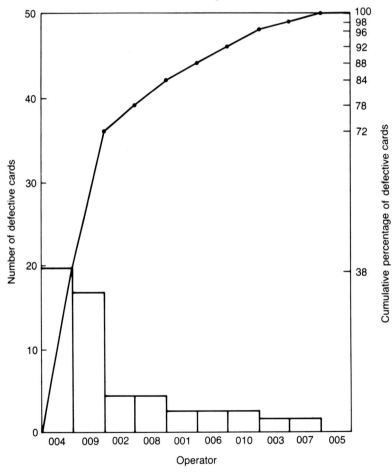

72% of all defective cards are produced by operators 004 and 009.

number of defective cards per operator for December 1987, as shown in Figure 14.17(a). From the c chart, she notes that operators 004 and 009 are out of control. She revises the c chart to determine if any other operators are out of control after having removed the impact of operators 004 and 009; she finds none, as shown in Figure 14.17(b). Next, she constructs a Pareto diagram for the number of defects per operator, as shown in Figures 14.18(a) and 14.18(b). From the Pareto diagram, she determines that 72 percent of all defective cards are produced by operators 004 and 009.

The manager decides to perform separate analyses for operators 004 and 009. She begins with operator 004 by setting up a check sheet, as shown in Figure 14.19, to determine the sources for operator 004's defects. The corresponding Pareto diagram, in Figures 14.20(a) and 14.20(b), shows 82 percent of operator 004's defects resulted from "transposed numbers" and "wrong character." Subsequently, the manager forms a brainstorming group composed of three select employees to do a Cause-and-Effect analysis of the above two problems. This is shown in Figure 14.21. The group members vote to attack both problems simultaneously. The Cause-and-Effect diagram leads the manager to send operator 004 to have her eyes checked. The optometrist finds that operator 004 is legally blind in her right eye. Eyeglasses correct her vision.

Next, the manager collects 24 more daily samples of 200 cards and constructs a p chart for the fraction of defective cards, as shown in Figure 14.22. From the p chart, the manager finds that operator 004's work is now stable, has an average defective rate of 0.8 percent (8 in 1000 cards), and rarely goes above 2.6 percent defective.

FIGURE 14.19   Check Sheet to Determine Defects for Operator 004 (1/88–4/88)

Major Causes of Defective Cards	*Month*				
	1/88	*2/88*	*3/88*	*4/88*	*Totals*
Transposed numbers	7	10	6	5	28
Off-punched cards	1		2		3
Wrong character	6	8	5	9	28
Data printed too lightly on cards		1	1		2
Warped cards	1	1		2	4
Torn cards			1	1	2
Illegible source document	—	—	1	—	1
Totals	15	20	16	17	68

FIGURE 14.20   Pareto Analysis to Determine Sources of Defects for Operator 004

(a) Causes (1/88–4/88)

Major Causes of Defective Cards	Frequency	Fraction	Cumulative Fraction
Transposed numbers	28	.412	.412
Wrong character	28	.412	.824
Warped cards	4	.059	.883
Off-punched cards	3	.044	.927
Data printed too lightly on cards	2	.029	.956
Torn cards	2	.029	.985
Illegible source document	1	.015	1.000
Totals	68	100.0	

(b) Pareto Diagram

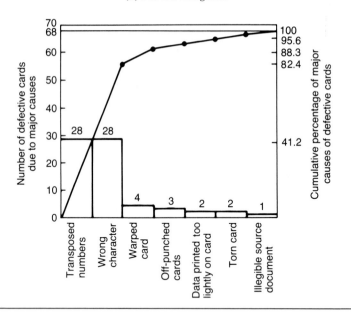

The manager realizes that if she wants to improve operator 004's performance further, she must switch from an attribute process improvement study to a variables process improvement study. Her next step is to plan her future courses of action, which will be: (1) study operator 009; and (2) review the entire department.

FIGURE 14.21 Cause-and-Effect Diagrams for Operator 004

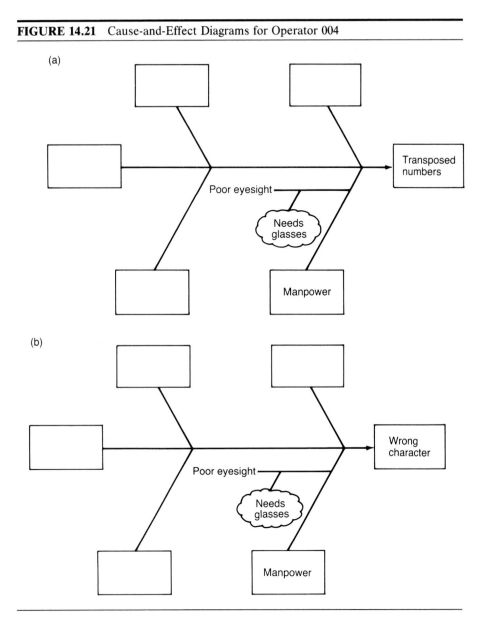

(a)

Transposed numbers

Poor eyesight

Needs glasses

Manpower

(b)

Wrong character

Poor eyesight

Needs glasses

Manpower

Variables Improvement Studies

Returning to the camshaft example discussed earlier in this chapter, recall that the case hardness depth of the camshafts was stable, with an average of 4.43 mm and a standard deviation of 0.688 mm.

FIGURE 14.22 p Chart for Operator 004 Following Fitting with Eye Glasses

At this point, the engineer assigned to study the induction quench-and-harden operation decides that to improve the process further the induction coil must be changed. The old coil is pitted and consequently emits an erratic electrical output, causing increased variability in case hardness depth between camshafts. The induction coil is changed on the evening of August 29, and 30 more days of data are collected (August 30–October 8) and control charted, as shown in Figure 14.23. The process is in statistical control, with an average case hardness depth of 5.45 mm and a standard deviation of 0.434 mm ($\overline{R}/d_2 = 1.01/2.326 = 0.434$). Note that the process has been shifted toward nominal (7.0 mm) and its unit-to-unit variation has been reduced.

Next, the process is checked to determine the percentage of camshafts that would be expected to be out of specification limits (3.5 mm to 10.5 mm), as shown in Figure 14.24. The necessary calculations are:

$$Z_{LSL} = \frac{LSL - \overline{\overline{x}}}{\sigma} = \frac{3.50 - 5.45}{0.434} = -4.49$$

$$P(Z_{LSL} < Z) \cong 0$$

$$Z_{USL} = \frac{USL - \overline{\overline{x}}}{\sigma} = \frac{10.5 - 5.45}{0.434} = 11.6$$

$$P(Z_{USL} > Z) \cong 0$$

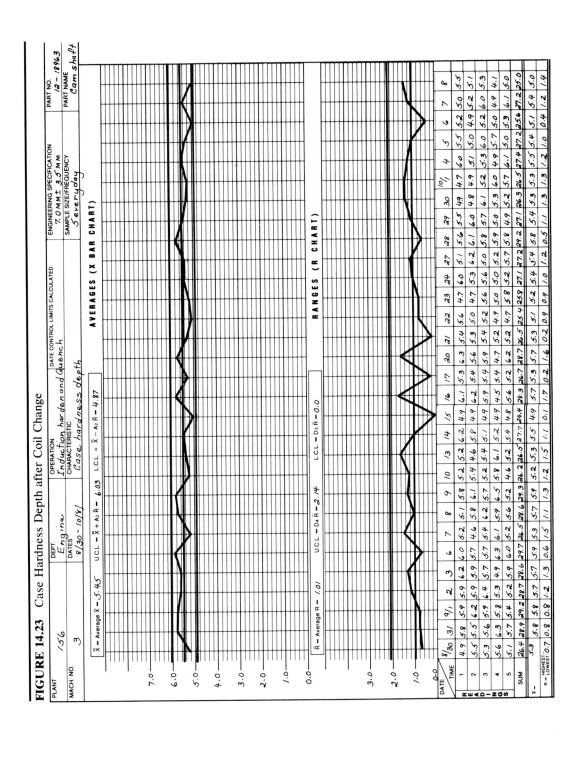

FIGURE 14.23 Case Hardness Depth after Coil Change

FIGURE 14.24 Fraction of Camshafts Out of Specification after Coil Redesign

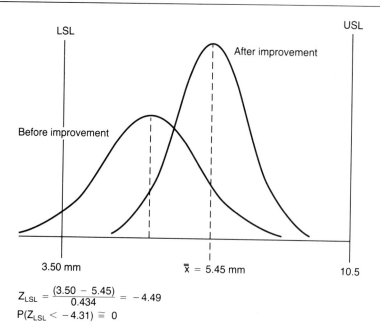

$$Z_{LSL} = \frac{(3.50 - 5.45)}{0.434} = -4.49$$

$$P(Z_{LSL} < -4.31) \cong 0$$

Hence, virtually no camshafts will be out of specification.

This example highlights the benefits of process improvement. The case hardness process was moved from chaos, to the threshold state, to the ideal state. The move from the threshold state (8.9 percent defective) to the ideal state (0.0 percent defective) resulted in:

1. Increased quality (8.9% − 0.0% = 8.9% more good output).
2. Increased productivity (8.9% − 0.0% = 8.9% more capacity for the same inputs.
3. Lower unit cost (it costs less to make good items than defective items because good items do not require rework).
4. Increased price flexibility resulting from lower unit costs.
5. Increased market share resulting from increased quality and price flexibility.
6. Increased profit resulting from lower unit costs and greater market share.
7. More secure jobs for all employees.

QUALITY IMPROVEMENT STORIES

Employees trying to improve quality have found that their ideas and recommendations are more persuasive when based on facts rather than opinions and guesses. The Quality Improvement (QI) story is an efficient format for employees to persuasively present process improvement studies to management. QI stories standardize quality control reports, avoid logical errors in analysis, and make the reports easy for all to understand. Further, QI stories can be used to facilitate the transition from the old to the new style of management.

Relationship between QI Stories and the PDCA Cycle

A seven-step procedure should be utilized to construct a QI story; these seven steps follow the Deming cycle of *Plan, Do, Check, and Act*. The Plan phase of the Deming cycle involves three steps: (1) selecting a theme for the QI story (obtaining all the background information necessary to understand the selected theme, including a process flow diagram; explaining the reason for selecting the theme; determining the organization and department objective(s) that should be influenced by the theme); (2) getting a full grasp of the present situation surrounding the theme; and (3) conducting an analysis of the present situation to identify appropriate action(s) (called *countermeasures*) to the process—that is, constructing a plan of action.

The Do phase of the Deming cycle involves a further step: (4) setting the appropriate countermeasures into action on a small scale so that the process improvement actions can be tested on an experimental basis.

The Check phase of the Deming cycle involves yet another step: (5) studying, creatively thinking about, collecting, and analyzing data concerning the effectiveness of the countermeasure(s) experimentally set into motion upon the targeted organization objective(s). Do the countermeasures reduce the difference between customer needs and process performance? Before and after comparisons of the effects of the experimental countermeasures on the targeted department and organization objectives must be presented.

The Act phase of the Deming cycle requires two final steps: (6) determining if the countermeasure(s) were effective in pursuing department and organization objectives. If not, we go back to the Plan stage to find other countermeasures that will be effective in pursuing departmental and organizational objectives; if the countermeasures *were* effective in pursuing department and organization objectives, we either go to the Plan stage to seek the optimal settings of the countermeasure(s) or formally establish revised standard operating procedures based on the data about the experimental countermeasures. Further actions must be taken to prevent backsliding for the countermeasures set into motion. This phase also includes: (7) identifying remaining process problems, establishing a plan for further actions, and reflecting on the positive and negative aspects of past countermeasures.

To summarize, the seven steps of a QI story are:

1. *Select a theme* for the QI story.
2. Get a full *grasp of the present situation* surrounding the theme.
3. Conduct an *analysis of the present situation* to identify appropriate countermeasures.
4. *Set the countermeasures into action* on a small experimental scale.
5. Study data concerning the *effectiveness of the countermeasures*.
6. Establish revised *standard operating procedures*.
7. Establish a *plan for future actions*.

The above steps should form the basis of QI story standard operating procedure; all organization personnel should rigidly adhere to the QI story standard operating procedure, and all sections of a QI story should be clearly numbered and labeled so that it can be related to one of the above seven steps, as shown in Figure 14.25.

Potential Difficulties

Two areas of potential difficulty when applying QI stories are qualitative (non-numerical) themes and exogenous problems. Themes that are difficult to describe with numerical values, such as "Improvement of Presidential Reviews," should be analyzed by focusing on the magnitude of the gap between actual performance and the desired performance.

FIGURE 14.25   Relationship between the QI Story and the PDCA Cycle

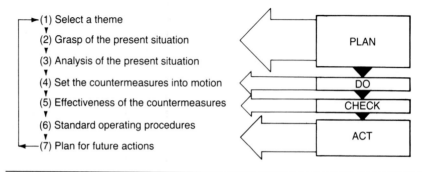

SOURCE: *Kaizen—The Keys to Japan's Competitive Success,* by Masaaki Imai. New York: Random House Business Division, 1986, p. 76.

If the primary cause of a problem is beyond the control of anyone in the organization (e.g., cold weather or no rain), we do not conclude that it is impossible to take countermeasures to remedy the exogenous problem; we attempt to determine why there are so many occurrences of the exogenous problem in area A versus area B, given that both areas have equal opportunities for the occurrence of the exogenous problem.

Pursuit of Objectives

Initially, QI stories will be selected because they are nearly complete resolutions to departmental problems and will not relate to organizational and departmental objectives. As employees gain experience with QI stories, they will want to select themes related to organizational and departmental objectives. Consequently, quantitative measures must be established that clearly recognize the relationship between organizations objectives and department objectives. For example, a purchasing department is aiming at "reduction in processing time for purchase orders." They claim this aim relates to the organizational objectives concerning "creating a great place to work free from system impediments." However, there is no quantitative measure that shows the relationship between "reduction in processing time for purchase orders" and "creating a great place to work free from system impediments." These quantitative relationships must be shown in the QI story. If QI story activities are not consistent with departmental objectives, and departmental objectives are not consistent with organizational objectives, then there is the distinct possibility that quality improvement efforts will not be in line with an organization's goals.

Quality Improvement Story Case Study

A QI story drawn from a data processing department is presented to demonstrate the role of QI stories in an organization's improvement efforts. The QI story is presented in QI story boards 1 through 14 in Figure 14.26.

As you can see, this QI story goes through two iterations of the PDCA cycle; nevertheless, a never-ending set of PDCA iterations will follow as the data processing department pursues continuous improvement in its daily work. The first iteration of the PDCA cycle focuses attention on all keypunch operators in the data processing department. In this iteration of the PDCA cycle, *Select a Theme* is presented in QI story board 1; this includes showing the background of theme selection and the reason for selecting the theme in relation to the organization's and department's objectives. *A Grasp of the Present Situation* is presented in QI story board 2. An *Analysis of the Present Situation,* shown in QI story board 3, is performed to determine appropriate countermeasures that pursue the theme and the organization and department objectives. *Set the Counter-measures into Motion* on a trial basis is presented in QI story board 4. The *Effectiveness of the Countermeasures* on the theme and the organization and department objectives are measured; this is shown in QI story board 5. *Standard Operating Procedure* is set that formalizes the countermea-sures and prevents backsliding; this is shown in QI story board 6. *A Plan for Future Actions* is presented in QI story board 7.

The second iteration of the PDCA cycle focuses attention on an individ-ual keypunch operator. In this iteration, *Select a Theme* is accomplished when the data processing manager realizes that future process improve-ments will require her to identify and train operators whose performance is out of control on the high side; see QI story board 8. In this iteration of the PDCA cycle, a *Grasp of the Present Situation* determined that key-punch operators 004 and 009 were out of control on the high side and why operator 004 was out of control on the high side; this is presented in QI story board 9. An *Analysis of the Present Situation,* which is shown in QI story board 10, determined the countermeasures necessary to improve operator 004's work. The manager *Set the Countermeasures into Motion;* this is shown in QI story board 11. The positive *Effectiveness of the*

FIGURE 14.26 Quality Improvement Story

QI STORY BOARD 1

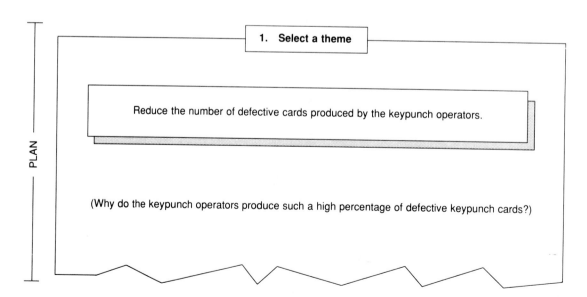

1. Select a theme

Reduce the number of defective cards produced by the keypunch operators.

(Why do the keypunch operators produce such a high percentage of defective keypunch cards?)

PLAN

FIGURE 14.26 *Continued*

PLAN

ORGANIZATIONAL OBJECTIVES

Background of theme selection (Relationship of the theme to organizational objectives)

The organization's mission mandates that every employee must base his/her decisions and actions on the following organizational objectives:
1. Pursuing continuous improvement in customer satisfaction.
2. Respecting and continuously improving all employees.
3. Establishing long-term and trusting relationships with suppliers.
4. Providing stockholders with a reasonable return.
5. Being a good corporate citizen.

DEPARTMENTAL OBJECTIVES

The data processing department's mission mandates that every employee must base his/her decisions and actions on the following departmental objectives:
1. Recognizing that customers are both internal and external to the organization and continuously strive to improve data processing services to all customers.
2. Identifying areas in which employees require improvement and establish necessary training programs to bring about the identified improvements.

The data processing department will achieve the first departmental objective by:
1. Keypunching all data exactly as it appears on the source document.
2. Pursuing continuous reduction in the amount of time it takes to process a keypunch job.

The manager of the data processing department realizes that the theme she selected to study is directly affected by the above objectives!

FIGURE 14.26 *Continued*

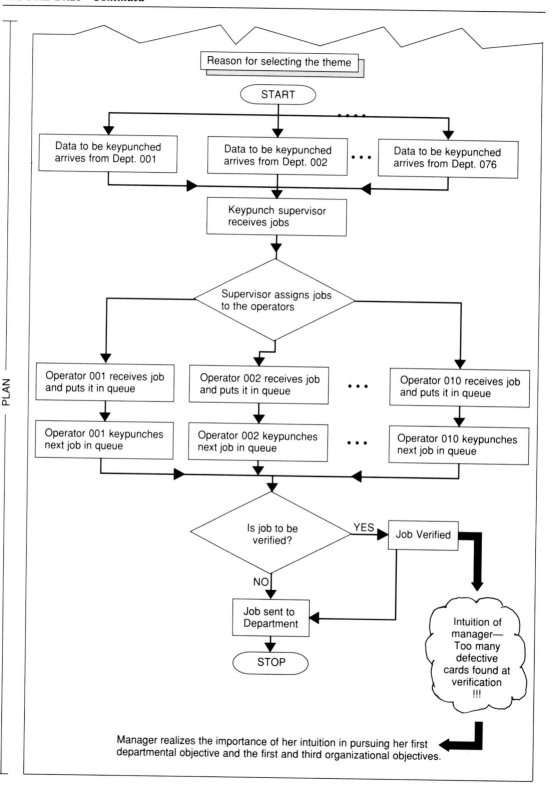

FIGURE 14.26 *Continued*

QI STORY BOARD 2

PLAN

> **2. Grasp of the present situation**

Manager's intuition leads her to conduct a survey to determine customer (other departments) satisfaction with her department's keypunching.

Manager constructs a list of her department's customers.

Administration
Production
Marketing

•
•
•

Manager constructs a questionnaire to determine customer satisfaction.

Department:_____

Supervisor: _____

(1) Do you feel that the error rate of the punched cards your department receives from the k.p. department is unsatisfactory [ ] satisfactory [ ] excellent [ ]

(2) Approximately what percent of the k.p. cards your department receives from our department contain errors attributable to our department? _____ %

FIGURE 14.26 *Continued*

PLAN

Questionnaires were sent to all of the departments and all of the departments responded. Analysis of the questionnaires yielded the following results:

Findings:

• Do you feel that the error rate of the k.p. cards your department receives from our department is:

unsatisfactory?	(72%)
satisfactory?	(20%)
excellent?	(8%)

• Approximately 2% of the k.p. cards received by the various departments contain errors attributable to the k.p. department.

FIGURE 14.26 *Continued*

PLAN

Due to customer dissatisfaction, the manager decided to collect data concerning the daily proportion of defective cards.

Day	Number of cards inspected	Number of defective cards	Proportion of defective cards
1	200	6	.03
2	200	6	.03
3	200	6	.03
4	200	5	.025
5	200	0	.0
6	200	0	.0
7	200	6	.03
8	200	14	.07
9	200	4	.02
10	200	0	.0
11	200	1	.005
12	200	8	.04
13	200	2	.01
14	200	4	.02
15	200	7	.035
16	200	1	.005
17	200	3	.015
18	200	1	.005
19	200	4	.02
20	200	0	.0
21	200	4	.02
22	200	15	.075
23	200	4	.02
24	200	1	.005
Total	4,800	102	

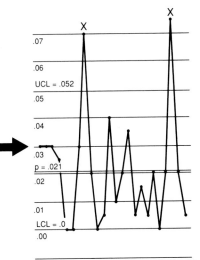

Finding:
The k.p. operation is in a state of chaos; it produces an unknown proportion of defective cards per day.

FIGURE 14.26 *Continued*

QI STORY BOARD 3

3. **Analysis of the present situation**

Process must be stabilized. Hence, causes for days 8 and 22 must be found and policy must be set to eliminate them from reoccurring.

The manager reviewed her daily comments concerning any unusual events that occurred on days 8 and 22.

Log sheet for k.p. operation	
Day	Comment

8	untrained operator used for a rush job

22	ran out of cards from usual vendor
23	
24	

Findings:

• Policy needed for training new cperators used for rush jobs.

• Inventory policy need to set safety stock level.

PLAN ESTABLISHED

PLAN

FIGURE 14.26 *Continued*

QI STORY BOARD 4

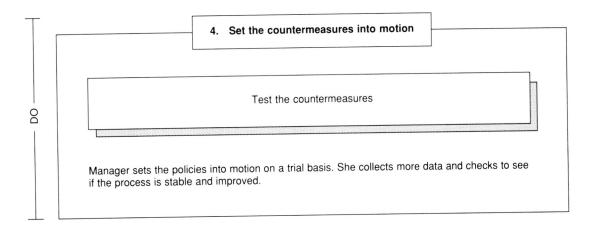

DO

> **4. Set the countermeasures into motion**
>
> Test the countermeasures
>
> Manager sets the policies into motion on a trial basis. She collects more data and checks to see if the process is stable and improved.

QI STORY BOARD 5

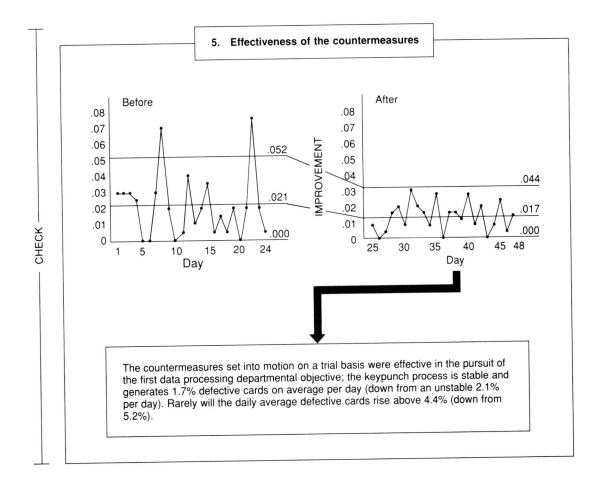

CHECK

> **5. Effectiveness of the countermeasures**
>
> The countermeasures set into motion on a trial basis were effective in the pursuit of the first data processing departmental objective; the keypunch process is stable and generates 1.7% defective cards on average per day (down from an unstable 2.1% per day). Rarely will the daily average defective cards rise above 4.4% (down from 5.2%).

FIGURE 14.26 *Continued*

QI STORY BOARD 6

ACT

6. Standard operating procedure

The manager establishes formal operating procedures, including appropriate training, for inventory policy and new operator skills development.

The manager decides that a random sample of 200 cards per month will be drawn from every keypunch operator's output. These samples will be analyzed so that appropriate actions can be taken to prevent any backsliding in areas that have been improved.

QI STORY BOARD 7

ACT

7. Plan for future actions

		When will future plans be carried out										Who will carry out plan
		12-87	1-88	2-88	3-88	4-88	5-88	6-88	7-88	8-88	9-88	
Phase 1	Work with operator 004	◄——————►										Manager and 004
Phase 2	Work with operator 009						◄—►					Manager and 009
Phase 3	Check progress of entire department								◄—►			Manager and 001-010
Phase 4	Survey customers to determine satisfaction with k.p.										◄►	Manager

FIGURE 14.26 *Continued*

QI STORY BOARD 8

PLAN

1. **Select a theme**

Manager realizes that to improve the k.p. process she must conduct a separate study for each operator.

QI STORY BOARD 9

PLAN

2. **Grasp of the present situation**

Checksheet of Defective cards by operator (All operators produced approximately the same number of cards during the period under study.) [12/1/87 - 12/31/87]

Operator	Tally	Frequency
001	I I	2
002	I I I	3
003	I	1
004	₩ ₩ ₩ IIII	19
005		0
006	I I	2
007	I	1
008	I I I	3
009	₩ ₩ ₩ II	17
010	I I	2
TOTAL		50

FIGURE 14.26 *Continued*

PLAN

$\bar{c}$-chart of number of
defective cards by
operator

Number of defective cards

X
X

UCL = 11.71
$\bar{c}$ = 5.00
LCL = 0

1 2 3 4 5 6 7 8 9 10

Operators 004 and 009 are
out of the k.p. system.

$\bar{c}$ - chart of the number of defective
cards by operator (excluding
operators 004 and 009

Number of defective cards

UCL = 5.73
$\bar{c}$ = 1.75
LCL = 0

1 2 3 4 5 6 7 8 9 10

Operators are stable without
operators 004 and 009.

What percentage of the departmental errors
are caused by operators 004 and 009?

Pareto Diagram of defective cards by
operator (12/1-31/87)

Number of defective cards

Cum. % of defective cards

100
98
96
92
88
84
78
72

38

4 9 2 8 1 6 10 3 7 5

Finding: 72% of all defective cards are
produced by operators 004 and 009!!!

Pareto Analysis of defective cards
by operator (12/1-31/87)

Operator	Freq.	%	Cum. %
4	19	38	38
9	17	34	72
2	3	6	78
8	3	6	84
1	2	4	88
6	2	4	92
10	2	4	96
3	1	2	98
7	1	2	100
5	0	0	100
Total	50	100	100

FIGURE 14.26 *Continued*

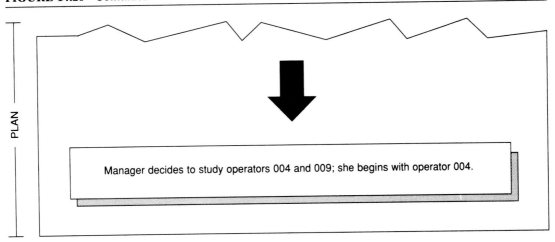

Manager decides to study operators 004 and 009; she begins with operator 004.

QI STORY BOARD 10

3. **Analysis of the present situation**

Checklist to determine the sources of operator 004's
defective cards (1/88-4/88)

Major causes of defective cards	Month				
	1/88	2/88	3/88	4/88	Total
Transposed numbers	7	10	6	5	28
Off-punched cards	1		2		3
Wrong charac.	6	8	5	9	28
Data printed too lightly on cards		1	1		2
Warped cards	1	1		2	4
Torn cards			1	1	2
Illegible source doc.			1		1
TOTAL	15	20	16	17	68

FIGURE 14.26 *Continued*

PLAN

Pareto Analysis to determine major causes of defective cards for operator 004 (1-4/88)

Major causes of defective cards	Freq.	%	Cumulative %
Transposed numbers	28	41.2	41.2
Wrong charac.	28	41.2	82.4
Warped cards	4	5.9	88.3
Off-punched cards	3	4.4	92.7
Data printed too lightly on cards	2	2.9	95.6
Torn cards	2	2.9	98.5
Illegible source doc.	1	1.5	100.0
TOTAL	68	100.0	

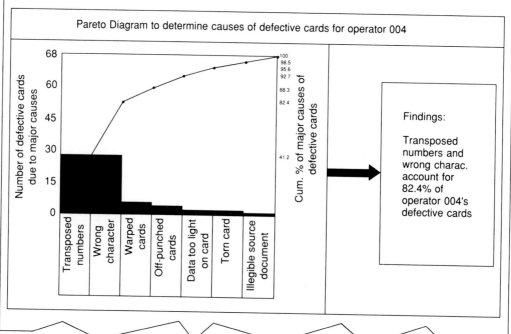

Pareto Diagram to determine causes of defective cards for operator 004

Findings:

Transposed numbers and wrong charac. account for 82.4% of operator 004's defective cards

FIGURE 14.26 *Continued*

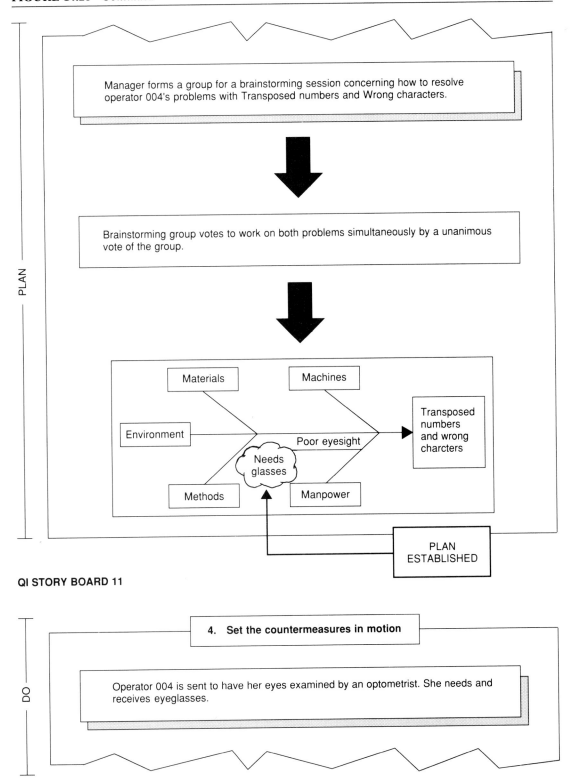

Manager forms a group for a brainstorming session concerning how to resolve operator 004's problems with Transposed numbers and Wrong characters.

Brainstorming group votes to work on both problems simultaneously by a unanimous vote of the group.

PLAN

Materials

Machines

Environment

Poor eyesight

Needs glasses

Transposed numbers and wrong charcters

Methods

Manpower

PLAN ESTABLISHED

QI STORY BOARD 11

4. **Set the countermeasures in motion**

DO

Operator 004 is sent to have her eyes examined by an optometrist. She needs and receives eyeglasses.

FIGURE 14.26 *Continued*

QI STORY BOARD 12

5. Effectiveness of the countermeasures

Manager collects 25 additional daily samples of 200 cards each to determine the effect of operator 004's eyeglasses on her defective card rate.

Day	# Defective	Total Cards	Proportion Defective
1	2	200	0.010
2	3	200	0.015
3	2	200	0.010
.	.	.	.
.	.	.	.
.	.	.	.
25	2	200	0.010
	40	5000	0.008

p charts comparing the proportion of defective cards produced by the "average" keypunch operator <u>before</u> improvement efforts with the proportion of defective cards produced by operator 004 <u>after</u> improvement efforts.

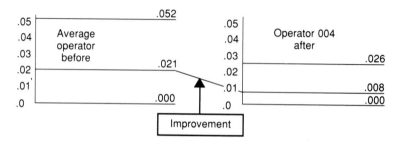

Finding: Operator 004 is stable and producing 8 defective cards per 1,000. Rarely will her defect rate go above 2.6 per 1,000. The countermeasure taken with operator 004 is effective in the pursuit of the first data processing department objective.

CHECK

FIGURE 14.26 *Concluded*

QI STORY BOARD 13

ACT

6. Standard operating procedure

The manager established a formal procedure by sending operator 004 for glasses.

The manager formally establishes a policy stating that all keypunch operators must have their eyes examined yearly and provide evidence of said examination. If any operator needs glasses, she will receive them. This policy should prevent backsliding in improvement efforts due to poor eyesioght

QI STORY BOARD 14

ACT

7. Plan for future action

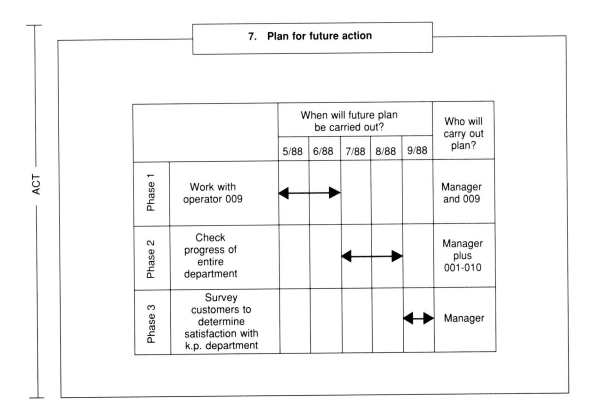

		When will future plan be carried out?					Who will carry out plan?
		5/88	6/88	7/88	8/88	9/88	
Phase 1	Work with operator 009	←→					Manager and 009
Phase 2	Check progress of entire department			←→			Manager plus 001-010
Phase 3	Survey customers to determine satisfaction with k.p. department					←→	Manager

Countermeasures on operator 004 and on the organization and department objectives was confirmed; see QI story board 12. *Standard Operating Procedure* is set, which formalizes the countermeasure to all operators and prevents backsliding; this is presented in QI story board 13. Finally, *A Plan for Future Actions* is specified in QI story board 14.

Improvement of the QI Story Process

The QI story process is subject to improvement efforts just as any other organizational process. These efforts must focus on reducing the differences in QI story objectives between QI story presenters and QI story reviewers. The following iterations of the PDCA cycle must be continuously performed to pursue improvement of the QI story process.

Plan.   A list (frame) must be constructed of QI story presenters and QI story reviewers for each level of QI story presentation in the organization. A survey instrument must be developed to gather data, for each level of QI story presentation, with respect to QI story variables to determine whether QI stories are being viewed more closely (more positive correlations) by QI story reviewers and presenters. Next, a sampling plan must be established that will be used to administer the survey; the sampling plan must consider nonresponse bias. A subplan has now been set with which to establish a plan for setting countermeasures to improve the QI story process.

The survey must be administered in strict accordance with the sampling plan. This is the Sub-Do phase of the Plan stage of the PDCA cycle focused on improvement of the QI story process.

The survey must be analyzed so that the correlations between QI story presenters and QI story reviewers, with respect to QI story variables, can be examined and countermeasures can be determined that would decrease the difference in QI story objectives between QI story presenters and reviewers (create strong positive correlations). This is the Sub-Check phase of the Plan stage of the PDCA cycle focused on improvement of the QI story process.

Countermeasures for improvement of the QI story process should be proposed that are a direct result of the survey. This is the Sub-Act phase of the Plan stage of the PDCA cycle focused on improvement of the QI story process.

Do.   The countermeasures based on the survey results should be set into motion on a trial basis. The trial might involve a sample of organization subcomponents across the entire organization, or just one pilot subcomponent from the organization.

Check. The correlations between QI story reviewers and presenters, with respect to QI story variables, must be examined to determine if the countermeasures set into motion on a trial basis made the correlations more positive and therefore decreased the difference in QI story objectives between presenters and reviewers.

Act. If the correlations between QI reviewers and presenters with respect to QI story survey variables become more positive, then formal changes must be established in QI story standard operating procedure. Of course, appropriate training in the new standard operating procedures must precede the revised standard operating procedures to prevent backsliding into old habits. Next, a return to the Plan phase of the PDCA cycle is required to search for new countermeasures by which to improve the QI story process.

QI stories can be used to get management to manage with data, instead of opinion and guesswork, and to think in terms of the PDCA cycle. QI stories take time to construct; however, the time spent is well worth the payoff of increased likelihood for process improvements.

SUMMARY

In this chapter we discussed process capability studies, process improvement studies, and quality improvement stories. Process capability studies determine if a process is unstable, investigate any sources of instability and determine their cause, and take action to eliminate these sources of instability. After all sources of instability have been eliminated, the natural behavior of the process is called its process capability.

We discussed the two types of process capability studies: attribute studies and variables studies. For each type of study, we considered data requirements; possible actions that can be taken on the process as a result of the process capability study; and estimating the fraction of a process output that will be out of specification. For variables process capability studies, we discussed estimating the fraction of output for a created dimension that will be out of specification; the relationship between control limits, natural limits, and specification limits; and process capability indices.

Also presented in this chapter were case studies of attribute and variables process improvement studies, followed by a discussion of quality improvement stories and the presentation of an attribute capability study shown in the context of a QI story.

EXERCISES

The ABC Company produces steel tubes. The steel tube process is a stable cut-to-length operation that generates tubes that are distributed normally with a mean of 12.00 inches and a standard deviation of 0.10 inches.

14.1 The XYZ Company wishes to buy tubes from the ABC Company. The XYZ Company requires steel tubes between 11.77 inches and 12.23 inches in length.
   a. Compute C_p.
   b. Compute CPU.
   c. Compute CPL.
   d. Compute C_{pk}.
   e. Compute Z_{LSL}.
   f. Compute Z_{USL}.
   g. Compare and contrast the above capability indices with respect to their ability to explain the capability of the ABC Company's steel tube process.
   h. Discuss the managerial implications of the capability indices that you computed in parts *a–f*.

14.2 The LMN Company wishes to buy tubes from the ABC Company. The LMN Company requires steel tubes 11.95 inches long, with a tolerance of 0.30 inches.
   a. Compute C_p.
   b. Compute CPU.
   c. Compute CPL.
   d. Compute C_{pk}.
   e. Compute Z_{LSL}.
   f. Compute Z_{USL}.
   g. Compare and contrast the above capability indices with respect to their ability to explain the capability of the ABC Company's steel tube process.
   h. Discuss the managerial implications of the capability indices you computed in parts *a–f*.

The Arco Company produces plastic containers. The plastic container process is a stable operation that generates containers with a mean volume of 12,500.00 cubic inches and a standard deviation of 10.00 cubic inches. (Note: It is not reasonable to assume that the distribution of volumes is normal.)

14.3 The Beta Company wishes to buy plastic containers from the Arco Company. The Beta Company requires plastic containers with a volume between 12,495.00 cubic inches and 12,545.00 cubic inches.
   a. Compute C_p.
   b. Compute CPU.
   c. Compute CPL.
   d. Compute C_{pk}.

e. Compute Z_{LSL}.

f. Compute Z_{USL}.

g. Compare and contrast the above capability indices with respect to their ability to explain the capability of the Beta Company's plastic container process.

h. Discuss the managerial implications of the capability indices that you computed in parts *a–f*.

14.4 The Largo Corporation wishes to buy plastic containers from the Arco Company. The Largo Corporation requires plastic containers with a volume of 12,495.00 cubic inches and a tolerance of 20.00 cubic inches.

a. Compute C_p.

b. Compute CPU.

c. Compute CPL.

d. Compute C_{pk}.

e. Compute Z_{LSL}.

f. Compute Z_{USL}.

g. Compare and contrast the above capability indices with respect to their ability to explain the capability of the Beta Company's plastic container process.

h. Discuss the managerial implications of the capability indices you computed in parts *a–f*.

14.5 How do you determine the capability of a process, given that the only information available comes from an attribute process capability study?

14.6 a. Discuss the purpose of a quality improvement (QI) story.

b. List the seven steps in a QI story.

c. Explain the relationship between the seven steps in a QI story and the four stages of the PDCA cycle.

14.7 Define *capability of a process* in statistical terms. Consider normally distributed, skewed, and other non-normally distributed processes in your definition.

14.8 a. Discuss the data requirements to conduct an attribute process capability study; consider the number of time periods and the number of subgroups per time period.

b. Discuss the data requirements to conduct a variables process capability study; consider the number of time periods and the number of subgroups per time period.

14.9 Discuss the value of estimating the fraction of process output that will be out of specification from a process that is not stable.

NOTES

1. AT&T, *Statistical Quality Control Handbook* 10th printing, May 1984 (Indianapolis: AT&T, 1956), pp. 34–37 and 45–73.
2. Ibid.
3. H. Gitlow and S. Gitlow, *The Deming Guide to Quality and Competitive Position* (Englewood Cliffs, N.J.: Prentice-Hall, 1987), p. 161.
4. H. Gitlow and P. Hertz, "Product Defects and Productivity," *Harvard Business Review,* September/October 1983, pp. 131–41.
5. AT&T, *Statistical Quality Control Handbook,* pp. 59, 135–36.
6. This example was modified from Operations Support Staffs, "Statistical Process Control Case Study," *Introduction to Ford's Operating Philosophy and Principles and Statistical Management Methods—Participant Notebook* (Ford Motor Company Publisher, September 1983), pp. 7.E.9.–7.E.18.
7. Donald J. Wheeler and David S. Chambers, *Understanding Statistical Process Control* (Knoxville, Tenn.: Statistical Process Controls, Inc., 1986), pp. 12–21.
8. V. Kane, "Process Capability Indices," *Journal of Quality Technology* 18, January 1986, pp. 41–52.

PART V

Process/Product Design

Chapter 15 presents some of the contributions Genichi Taguchi, a Japanese statistician, has made to the philosophy of quality improvement and statistics. In particular, the chapter considers the relationship between the quality of a manufactured product and the total loss created by that product to society; the necessity of continuous quality improvement and cost reduction for an organization's health in a competitive economy; the need for never-ending reduction of variation in product and/or process performance around nominal, or target, values; the relationship between society's loss due to performance variation and the deviation of the performance characteristic from its nominal value; the impact of product and process design on a product's quality and cost; the nonlinear effects between a product's and/or process's parameters and the product's desired performance characteristics; and the identification of product and/or process parameter settings that reduce performance variation.

Chapter 15

Taguchi Methods—Quality Improvement in Product and Process Design

INTRODUCTION

Traditionally, quality control activities have centered on control charts and process control; this is called *on-line* quality control. Dr. Genichi Taguchi, a Japanese statistician and Deming Prize winner, has extended quality improvement activities to include product and process design; this is called *off-line* quality control. Taguchi's methods provide a system to develop specifications, design those specifications into a product and/or process, and produce products that continuously surpass said specifications.

There are seven aspects[1] to off-line quality control:

1. The quality of a manufactured product is measured by the total loss created by that product to society.
2. Continuous quality improvement and cost reduction are necessary for an organization's health in a competitive economy.
3. Quality improvement requires the never-ending reduction of variation in product and/or process performance around nominal values.
4. Society's loss due to performance variation is frequently proportional to the square of the deviation of the performance characteristic from its nominal value.
5. Product and process design can have a significant impact on a product's quality and cost.
6. Performance variation can be reduced by exploiting the nonlinear effects between a product's and/or process's parameters and the product's desired performance characteristics.
7. Product and/or process parameter settings that reduce performance variation can be identified with statistically designed experiments.

In this chapter, we discuss these seven points. For a more detailed discussion, see Taguchi and Wu,[2] and Kackar.[3]

POINT 1: THE QUALITY OF A MANUFACTURED PRODUCT IS MEASURED BY THE TOTAL LOSS CREATED BY THAT PRODUCT TO SOCIETY

Taguchi defines *quality* in terms of the loss imparted to society from the time a product is shipped. Many people find his definition unusual because it presents quality in a negative fashion; the basis of Taguchi's words is that the smaller the loss caused to society by a product, the better the product's quality.

Viewing quality from a societal perspective is profound because it includes customers, manufacturers, and the community in the definition of quality. According to this perspective on quality, quality improvement saves society more resources than it costs, and it benefits everyone: customers, manufacturers, and the community. Hence, investment in quality improvement is worthwhile so long as it reduces the loss to society from the time a product is shipped.

The total loss to society to produce a product with given parameter values (nominal settings) consists of two component parts: (1) the production cost to the manufacturer of producing a product with given parameters, and (2) the inferior quality cost to the customer and community of producing a product with given parameters.

For example, an important characteristic of the vinyl sheets used to build vinyl houses for agricultural production is the thickness of the sheets.[4] In Figure 15.1, y is the thickness setting of the vinyl sheet. From the manufacturer's perspective, if the thickness of the vinyl sheet is increased, the production cost (including raw material cost, processing costs, and inventory costs) increases, as shown by curve C on Figure 15.1. From the customer's and community's perspective, if the thickness of the vinyl sheets is increased, the cost of inferior quality decreases because the sheets become more difficult to break and farmers must replace or repair them less frequently, as shown by curve Q on Figure 15.1. The total loss (cost) to society, L(y), is computed by summing the cost to manufacturers (C curve) and the cost to customers and society (Q curve) at every setting of vinyl sheet thickness, y, as shown by curve L(y) on Figure 15.1.

The optimal thickness setting is selected by picking the thickness setting, y, for which the loss to society, L(y), is a minimum; this is thickness setting $y = m$ on Figure 15.1. If the thickness setting is changed from $y = m$ to $y = m-e$, then a disproportionately large loss is created for customers; the cost to the customer and society increases from Q_1 to Q_2, or $\$(Q_2 - Q_1)$, while the cost to the manufacturer decreases from C_1 to C_2, or

FIGURE 15.1   Cost of Producing a Vinyl Sheet

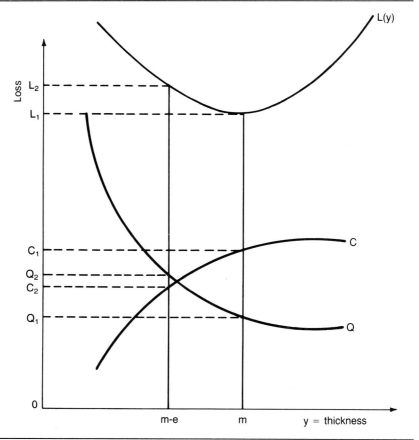

$(C_1 - C_2)$. As $(Q_2 - Q_1)$ is greater than $(C_1 - C_2)$, there is a net loss to society. This net loss is seen in that the total cost of producing the vinyl sheets has increased from L_1 to L_2. The thickness setting that minimizes the loss to society from the time the product is shipped is $y = m$.

POINT 2: CONTINUOUS QUALITY IMPROVEMENT AND COST REDUCTION ARE NECESSARY FOR AN ORGANIZATION'S HEALTH IN A COMPETITIVE ECONOMY

High quality and low cost are strategic factors in any plan for corporate health. Companies that realize the significance of these strategic factors know that quality can always be improved and costs can always be reduced. A major contention of Deming and Taguchi is that products and processes must be improved in a relentless and never-ending manner. The reasoning is demonstrated in Figure 15.2. Distribution A depicts unit-to-unit performance variation before a product and/or process is improved. The costs (losses to society) incurred using the system that produced distribution A are represented by the hatched area under distribution A beneath the total cost curve, L(y). Distribution B depicts unit-to-unit performance variation after a product and/or process is improved; the costs (losses to society) incurred using the improved system that produced distribution B are represented by the double-hatched area under distribution B beneath the total cost curve, L(y). The losses to society incurred under the system with lower unit-to-unit performance variation, system B, are clearly lower than the losses to society incurred with system A.

FIGURE 15.2 Loss and Variation Reduction

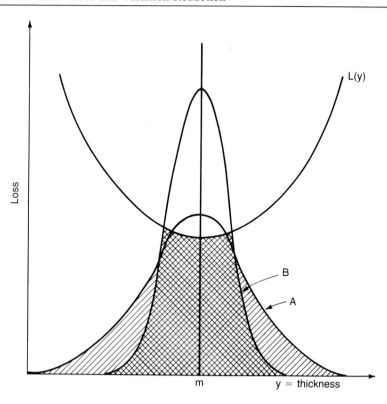

A = Loss incurred from performance variation *before* improvement.
B = Loss incurred from performance variation *after* improvement.

POINT 3: QUALITY IMPROVEMENT REQUIRES THE NEVER-ENDING REDUCTION OF VARIATION IN PRODUCT AND/OR PROCESS PERFORMANCE AROUND NOMINAL VALUES

Following the logic stated in Point 2, it is always economical to reduce unit-to-unit performance variation around nominal, even when we are within specification limits.

As an example illustrating this view, a car battery is charged by an alternator that has a voltage regulator that controls the charge to the battery. The alternator-voltage regulator assembly must put out a charge of 13.2 volts to keep the battery's charge at 12 volts. If the alternator produces a charge of fewer than 13.2 volts, eventually the electrolyte (acid) will turn into water and lose its charge, and the battery will die. If the alternator produces a charge of more than 13.2 volts, the battery plates will warp from excessive heat, the electrolyte will evaporate, and the battery will die.

This example demonstrates that *any* deviation from nominal causes a loss: just being within specification limits is not the minimum loss position.

POINT 4: SOCIETY'S LOSS DUE TO PERFORMANCE VARIATION IS FREQUENTLY PROPORTIONAL TO THE SQUARE OF THE DEVIATION OF THE PERFORMANCE CHARACTERISTIC FROM ITS NOMINAL VALUE

Any variation in a product's performance characteristic about its nominal value, at any randomly selected position in the product's life cycle, causes a loss to society, as discussed in the prior section. Again, let L(y) equal the total cost to society as a result of a product's having a value of y for a specified performance characteristic, given that the nominal value for the performance characteristic is m.

There are many possible forms for L(y); however, the two most common forms are shown in Equations 15.1 and 15.2:

$$L(y) = \begin{cases} A & \text{if} \quad y < LSL \text{ or } y > USL \\ 0 & \text{if} \quad LSL \leq y \leq USL \end{cases} \tag{15.1}$$

and

$$L(y) = \begin{cases} k(y - m)^2 & \text{if} \quad |y - m| > 0 \\ 0 & \text{if} \quad y - m = 0 \end{cases} \tag{15.2}$$

Equation 15.1 simply states that the loss caused by a product's characteristic value, y, deviating from nominal, m, is zero when y is within specification limits; when y is either lower than the lower specification limit or greater than the upper specification limit, the loss is a constant value, A, as shown in the shaded areas in Figure 15.3.

Equation 15.2 states that the loss caused by a product's characteristic value, y, deviating from nominal, m, is proportional to the squared distance between y and m, or $k(y - m)^2$, where k is a proportionality constant, or loss coefficient. If a product's characteristic value, y, is the same as the nominal value, m, then $y - m = 0$ and $L(y) = k(y - m)^2 = 0$. However, if a product's characteristic value is not the same as the nominal value, then $L(y) = k(y - m)^2$. In other words, the farther from nominal a product characteristic's value lies, the larger the loss.

The value of k is set by establishing the value of L(y) at a specification limit as A, so that $L(y) = A$ if $y = USL$ or $y = LSL$. Thus,

$$L(y) = k(y - m)^2$$

$$A = k_{USL}(USL - m)^2 \text{ and } A = k_{LSL}(LSL - m)^2$$

Then,

$$k_{USL} = \frac{A}{(USL - m)^2} \quad \text{and} \quad k_{LSL} = \frac{A}{(LSL - m)^2} \quad (15.3)$$

Using the loss function in Equation 15.2 is the same as stating that the minimum loss to society, L(y), occurs when $y = m$, or when a product's characteristic value is at nominal. This minimum-loss situation occurs with increasing frequency if a process's average is centered on nominal

FIGURE 15.3   Constant Loss Function for Performance Characteristic Outside Specification Limits

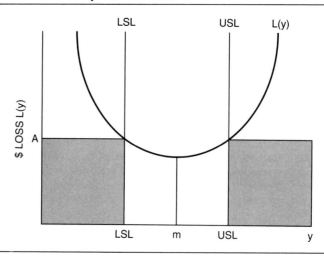

and its unit-to-unit variation is continuously reduced. In other words, when the average value of the product characteristic is m and its standard deviation about m approaches zero, the loss to society is minimized.

POINT 5: PRODUCT AND PROCESS DESIGN CAN HAVE A SIGNIFICANT IMPACT ON A PRODUCT'S QUALITY AND COST

The number of manufacturing imperfections in a product—hence the manufacturing cost of a product—is significantly affected by the product's design and the design of the process used to produce the product. Figure 15.4 shows the relationship between the number of manufacturing imperfections in a product and the degree of process control used to regulate the number of manufacturing imperfections in a product.

Curve A shows the relationship between the number of manufacturing imperfections and degree of process control for a given product/process

FIGURE 15.4   Manufacturing Imperfections for Various Degrees of Process Control

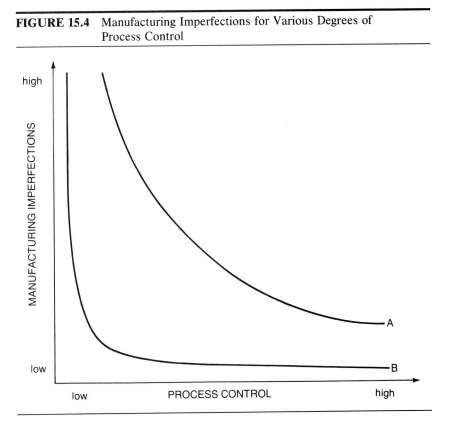

design configuration. Curve A indicates that greater process control results in fewer manufacturing imperfections, and vice versa.

Curve B shows the relationship between the number of manufacturing imperfections and degree of process control for an improved product/process design configuration. The product/process design configuration shown in curve B is superior to the product/process design configuration shown in curve A because the number of manufacturing imperfections is lower in curve B than in curve A for every level of process control. Curve B indicates that the number of manufacturing imperfections decreases as the amount of process control increases; however, the relationship is not linear. Hence, given an improved product/process design, the number of manufacturing imperfections can be dramatically reduced with a low level of process control. To state this another way, medium to high degrees of process control have little effect on reducing the number of manufacturing imperfections in the product shown in curve B because the number of manufacturing imperfections is already low as a result of the improved product/process design.

POINT 6: PERFORMANCE VARIATION CAN BE REDUCED BY EXPLOITING THE NONLINEAR EFFECTS BETWEEN A PRODUCT'S AND/OR PROCESS'S PARAMETERS AND THE PRODUCT'S DESIRED PERFORMANCE CHARACTERISTIC

Quality control activities must begin with *quality-of-design/redesign studies*. These studies lead to the development of product and process parameters, which create products that surpass the needs of customers. Product and process parameters must be stated in terms of specifications that have nominal values and tolerances around these nominal values.

Taguchi has developed a three-part procedure for constructing nominal values and tolerances for product and process parameters that will create products that surpass the needs of customers. The three-part procedure includes system design, parameter design, and allowance design.

System Design

System design involves using engineering and scientific knowledge to create an initial product prototype. The parameter settings of the initial product prototype's specifications are defined by a set of nominal values and their respective tolerances based on the results of quality-of-design/redesign and/or quality-of-performance studies. The parameter settings create a product prototype that considers the needs of customers and satisfies the requirements of manufacturability.

Parameter Design

Parameter design involves determining the specification settings for product and process parameters in terms of nominal values so that the final product will be less sensitive to sources of variation caused by environmental factors, product deterioration, and manufacturing variations.

Environmental Factors. *Environmental factors* are conditions that exist in the environment in which the product will be used by the customer, including human variations in operating the product.

Product Deterioration. *Product deterioration* consists of the changes in product parameters over time from wear and tear on the product during its life cycle.

Manufacturing Variations. *Manufacturing variations* are the manufacturing conditions that cause the production of product which deviates from its nominal values.

These three sources of product variation are usually common sources of variation because they are chronically present and affect the product's performance.

Let us consider the design of an electrical circuit to appreciate the purpose of parameter design. Kackar offers the following illustration.[5] Suppose the performance characteristic of interest is the output voltage of the electric circuit, y, and its target value is y_0. Assume that the output voltage of the circuit is largely determined by the gain of a transistor X in the circuit, and the circuit designer is at liberty to choose the nominal value of this gain. Suppose also that the effect of the transistor gain on the output voltage is nonlinear. This relationship is shown in Figure 15.5.

In order to obtain an output voltage of y_0, the circuit designer can select the nominal value of transistor gain to be x_0. If the actual transistor gain deviates from the nominal value x_0, the output voltage will deviate from y_0. The transistor gain can deviate from x_0 because of manufacturing imperfections in the transistor, deterioration during the circuit's life span, and environmental variables. If the distribution of transistor gain is as shown in Figure 15.5, the output voltage will have a large variation. One way of reducing the output variation is to use an expensive transistor whose gain has a very narrow distribution around x_0. Another way of reducing output variation is to select a different value of transistor gain. For instance, if the nominal transistor gain is x_1, the output voltage will have a much smaller variance. But the mean value y_1 associated with the transistor gain x_1 is far from the target value y_0. Now suppose there is another component in the circuit, such as a resistor, that has a linear effect on the output voltage, and the circuit designer is at liberty to choose the nominal value of this component. The circuit designer can then adjust this component to move the mean value of voltage from y_1 to y_0. Adjustment of the mean value of a performance characteristic to its target value

FIGURE 15.5   Effect of Transistor Gain on Output Voltage

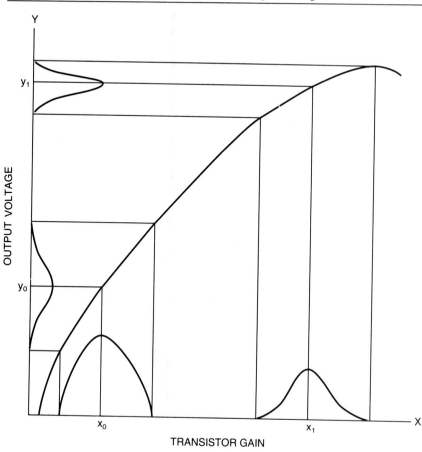

is usually a much easier engineering problem than the reduction of performance variation. When the circuit is designed in such a way that the nominal gain of transistor X is x_1, an inexpensive transistor having a wide distribution around x_1 can be used. Of course, this change would not necessarily improve circuit design if it were accompanied by an increase in the variance of another performance characteristic of the circuit.

The above example demonstrates that exploiting the nonlinear effects of product or process parameters on product performance characteristics can be an effective method for reducing the sensitivity of product performance to environmental factors, product deterioration, and manufacturing variations.

Allowance Design

In Chapter 13 we discussed tolerances and found that the amount of allowable tolerance around a nominal value is determined by balancing the customer's loss from increased performance variation resulting from wide tolerances and the manufacturer's loss from increased costs resulting from narrow tolerances. Allowance design involves establishing the allowable size of a tolerance around the nominal setting of product or process parameters determined in the parameter design stage.

Design engineers must resort to allowance design when the influences of environmental factors and product deterioration cannot be successfully reduced through parameter design. Allowance design should be performed after parameter design as it is less expensive to reduce performance variation through parameter design than it is to control performance variation through the establishment of tolerances.

The establishment of tolerances is not contradictory to the philosophy of never-ending improvement because engineers will continuously endeavor to produce a better design (by establishing nominal values which reduce performance variation), and production personnel will continuously strive to reduce variation around the nominal values established in the parameter design stage.

POINT 7: PRODUCT AND/OR PROCESS PARAMETER SETTINGS THAT REDUCE PERFORMANCE VARIATION CAN BE IDENTIFIED WITH STATISTICALLY DESIGNED EXPERIMENTS

Parameter design can be accomplished through a statistically designed experiment. The experiment requires that the variables affecting the performance variation of the product under study be classified into two categories: design parameter variables and noise variables.

Design Parameter Variables

Design parameter variables are variables that can be set at one of two or more possible parameter settings; for example, water level in a cooling system can be set at either a high (= 1) or a low (= 0) parameter setting. Design parameter variables represent the nominal values of product or process parameter settings that can be determined by a design engineer who has conducted and utilized the information from quality-of-design and quality-of-performance studies.

Noise Variables

Noise variables include all the factors that cause the product's performance characteristics to deviate from their nominal value or cause actual performance to differ from desired performance. The noise variables that most significantly affect the product's performance characteristics should be determined by design engineers and included in the parameter design experiment. The experimenter should systematically vary the levels of the most significant noise variables, or their surrogates, to determine their effects upon the product's performance characteristics.

In reality, it may not be possible to consider every noise variable in a parameter design experiment. It may be impossible to conduct an experiment with all known noise variables because of the large data requirements of the experiment; or the importance of a particular noise variable may be unknown to the person conducting the parameter design experiment. Consequently, design engineers must be wary of potential problems from unknown noise variables.

The purpose of a parameter design experiment is to determine the nominal values for the design parameter variables that yield the lowest impact on the product's performance characteristics by the noise variables. The nominal values of the parameter design variables are established by systematically varying their settings in conjunction with a selected combination of the settings of the noise variables, then comparing the resultant performance characteristics, as shown in Figure 15.6.

Selection of the parameter design variable settings and noise variable settings for a parameter design experiment is not a trivial task. Taguchi has developed a recommended procedure for performing a parameter design experiment, utilizing a design parameter variable matrix and a noise variable matrix.[6] The design variable matrix lists the design variables in its columns and the appropriate combinations of parameter design variable settings in its rows. The noise variable matrix lists the noise variables in its columns and the appropriate combinations of noise variable settings in its rows. The design parameter experiment consists of running tests for every row in the design parameter matrix (a given set of product design parameter nominal values) under the conditions specified in every row of the noise variable matrix (a given set of noise variable conditions); this is shown in Figure 15.6. The resulting performance characteristics are recorded for each specified combination of design parameter variables and noise variables.

For each configuration of the design parameter variables (a row in the design parameter matrix, representing a particular set of nominal values for the design parameter variables), all the test run settings (rows) of the noise variables shown in the noise variable matrix are used to compute a performance statistic. The performance statistic estimates the effects of

FIGURE 15.6   An Example of a Taguchi-Type Parameter Design Matrix

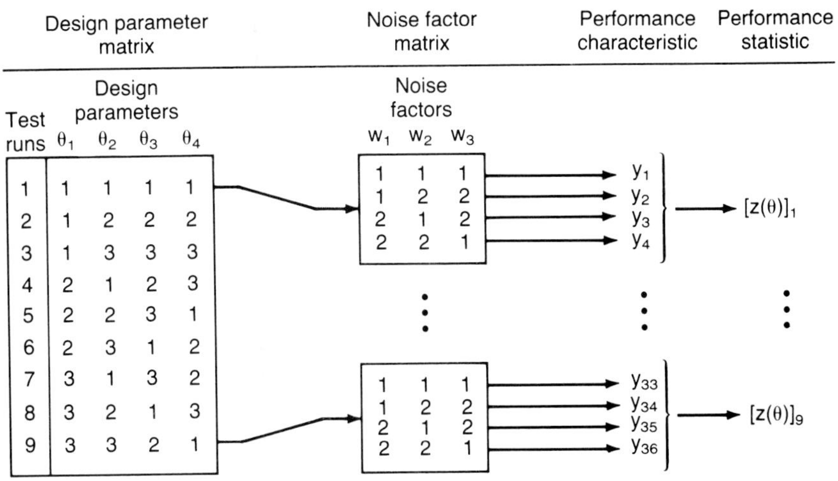

| Design parameter matrix | Noise factor matrix | Performance characteristic | Performance statistic |

SOURCE: Kackar, "Taguchi's Quality Philosophy: Analysis and Commentary," *Quality Progress,* December 1986, p. 27.

the noise variables on the performance characteristics for a given design (a particular set of nominal values for the design parameter variables). The setting of the parameter design variables that yield the best performance statistic is deemed the best product design.

Taguchi recommends using a performance statistic called a *signal-to-noise ratio.* Three types of signal-to-noise ratios exist for variables type performance statistics[7]:

1. The smaller the ratio the better the product design; for example, if friction is the performance characteristic under study, low friction would be desirable.
2. The larger the ratio the better the product design; for example, if adhesion is the performance characteristic under study, high adhesion would be desirable.
3. A nominal value is best; for example, if the gap size of a created dimension is the performance characteristic under study, and nominal is a gap size of 3 mm, a gap size of 3 mm is most desirable.

Signal-to-noise ratios can similarly be defined for attribute-type performance characteristics. Consequently, statistically designed experiments using a signal-to-noise ratio as a measure of product or process performance can be used to reduce variation by careful selection of parameter settings.

SUMMARY

Genichi Taguchi has made a large contribution to quality improvement philosophy and quality control statistics. His views on the definition of quality, quality loss functions, never-ending improvement, and parameter design experiments, to name a few of his contributions, have advanced the worldwide movement for quality improvement.

In this chapter we discussed several aspects of Taguchi's views of quality:

1. The quality of a manufactured product is measured by the total loss created by that product to society.
2. Continuous quality improvement and cost reduction are necessary for an organization's health in a competitive economy.
3. Quality improvement requires the never-ending reduction of variation in product and/or process performance around nominal values.
4. Society's loss due to performance variation is frequently proportional to the square of the deviation of the performance characteristic from its nominal value.
5. Product and process design can have a significant impact on a product's quality and cost.
6. Performance variation can be reduced by exploiting the nonlinear effects between a product's and/or process's parameters and the product's desired performance characteristics.
7. Product and/or process parameter settings that reduce performance variation can be identified with statistically designed experiments.

EXERCISES

15.1 Explain Taguchi's statement that the quality of a manufactured product is measured by the total loss created by that product to society.

15.2 a. Why is continuous quality improvement critical to any organization's health and competitive position?
    b. Explain the relationship between quality improvement and cost reduction for any organization.

15.3 Explain why being within specification limits is not enough to be competitive in today's world economy.

15.4 Explain why the concept of "zero defects" is flawed and will harm an organization in the long run.

15.5 Discuss two possible models for quantifying society's loss from poor quality: the linear loss function and the quadratic loss function. Explain the rationale behind each.

15.6 A firm produces steel rods with a length specification of 6.0 inches, plus or minus 0.10 inches. If a steel rod exceeds either specification limit it is melted down for stock to produce new rods. The cost of a rod's being out of specification is $0.75 per rod. Given the above information, construct a quadratic loss function.

15.7 Explain why a product's design, and the design of the process used to produce the product, has a significant impact on the product's quality and cost. Discuss the cost of process control versus the number of manufacturing imperfections in the product under question.

15.8 a. Explain the purpose of system design.
 b. Discuss the relevance of quality-of-design/redesign studies and quality-of-performance studies to system design.

15.9 a. Explain the purpose of parameter design.
 b. Define the term *design parameter variable*.
 c. Define the term *noise variable*.

15.10 a. Explain the purpose of allowance design.
 b. Explain why allowance design does not contradict the philosophy of continuous and never-ending improvement.

15.11 a. Explain the purpose of a signal-to-noise ratio.
 b. Describe the three types of signal-to-noise ratios for non-negative variables type data.

NOTES

1. This list of seven items has been paraphrased from R. Kackar, "Taguchi's Quality Philosophy: Analysis and Commentary," *Quality Progress,* December 1986, pp. 21–29.
2. G. Taguchi and Y. Wu, *Introduction to Off-Line Quality Control* (Nagoya, Japan: Central Japan Quality Control Association, 1980).
3. Kackar, "Taguchi's Quality Philosophy," pp. 21–29.
4. Abstracted from Taguchi and Wu, *Introduction to Off-Line Quality Control,* pp. 7–9.
5. Kackar, "Taguchi's Quality Philosophy," p. 26.
6. Genichi Taguchi has developed sets of matrices for performing parameter design experiments. The matrices are called *orthogonal design matrices.*
7. The measurement variable must be non-negative to use the performance statistics described.

PART VI

Inspection Policy

Chapter 16 discusses policies and procedures for inspection of incoming, intermediate, and final goods and services. Three possible options exist for inspection of goods and services: (1) no inspection, (2) 100 percent inspection, or (3) sampling inspection.

The first part of the chapter focuses on sampling inspection, commonly called acceptance sampling, as a method to determine whether to accept, reject, or screen goods and services. Three types of acceptance sampling plans are discussed: lot-by-lot plans, continuous plans, and special plans.

The second part of the chapter presents a theoretical argument against using acceptance sampling plans, followed by a discussion of the kp rule, an alternative inspection procedure that minimizes the total cost of inspection for incoming, intermediate, and final goods and services. The chapter ends with two mathematical proofs of the arguments made.

Inspection Policy

INSPECTING GOODS AND SERVICES

Goods or services enter an organization from a vendor or are passed on internally from one section of the organization to another, such as department to department, or operation to operation within a department. These goods or services move inter- or intra-organizationally in either discrete lots or in continuous flows and have certain customer-specified quality characteristics.

Organizations, or their subcomponents, must have some method for minimizing the total cost of inspection of incoming and intermediate goods or services plus the cost to repair and test these goods and services in process; or final goods or services that fail to meet specifications because of a defective good or service used in production. Three possible alternatives exist for inspection of goods or services: (1) no inspection (send items straight into use with no screening); (2) 100 percent inspection (screen all goods or services to weed out defectives); or (3) sampling inspection, also known as acceptance sampling (screen a sample of goods or services to determine if the remainder should be accepted, rejected, or screened). Historically, acceptance sampling has been considered useful if the inspection test is destructive (100 percent inspection will destroy all goods or services); the cost of 100 percent inspection is high; or too many units have to be inspected.

ACCEPTANCE SAMPLING

The purpose of acceptance sampling is to determine the disposition of goods or services (accept, reject, or screen). This is accomplished by selecting the disposition that minimizes the cost of inspection to achieve a desired level of quality (called the Acceptable Quality Level or AQL, as discussed in Chapter 13) or to financially penalize the vendor (external or internal) if his quality is poor. There are several types of acceptance sampling plans, shown in Figure 16.1.[1]

Lot-By-Lot Acceptance Sampling

Lot-by-lot acceptance sampling plans are used to inspect goods or services whenever the goods or services can conveniently be grouped into lots. All lot-by-lot acceptance plans are based on accepting, rejecting, or screening the remainder of the lot based on the number of defects found in the sample. Lot-by-lot plans exist for attribute data and variables data. Moreover, these types of plans can be broken down into *acceptance/ rejection plans* (plans in which a sample of items is drawn from the lot and the remainder of the lot is accepted or rejected based on an analysis of the sample) or *rectifying plans* (plans that call for either total or partial screening of remainders). The most common acceptance/rejection lot-by-lot acceptance sampling plan for variables used in American industry today is Military Standard 414.[2] This plan is used to control the fraction of incoming material that does not conform to specifications for variables data. The most common acceptance/rejection lot-by-lot acceptance sampling plan for attributes is Military Standard 105D.[3] This plan is used to constrain suppliers so that they will deliver at least an Acceptable Quality Level (AQL) of goods or services for attribute data.

FIGURE 16.1   Acceptance Sampling Plans

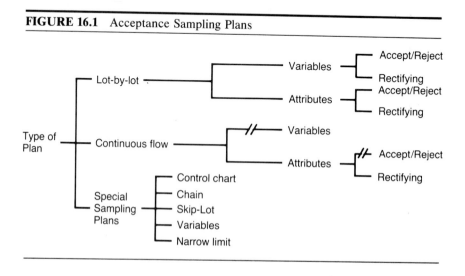

Continuous Flow Acceptance Sampling

Continuous flow acceptance sampling plans are used to inspect goods or services whenever the goods or services cannot be grouped into lots—for example, goods on conveyor belts or goods on a continuous moving line. Only rectifying attribute sampling plans exist for continuous flow processes. The most common acceptance sampling plan in this category used today is Military Standard 1235B.[4] All continuous flow acceptance plans are based on a *clear sampling,* the number of conforming units observed between the occurrence of two defective units. If the number of units between two defective units is greater than the number of units specified in the clearing sample, the units will be accepted and shipped; otherwise, they will be 100 percent inspected.

Special Sampling Plans

Other types of acceptance sampling plans have been developed that are of the lot-by-lot type but are applied to a series of lots considered as a group; they are called *special sampling plans.* Some of these plans are: control chart plans,[5] chain sampling plans,[6] skip-lot plans,[7] variables plans,[8] and narrow limit plans.[9]

Much attention is given to acceptance sampling plans in textbooks and courses. However, these plans do not minimize the total cost of inspection of: (1) incoming and intermediary goods or services plus the cost to repair and test these goods and services in process, or (2) final goods or services that fail to meet specifications because of a defective good or service that was used in production. They also place emphasis on inspection, not on process improvement in order to remove the need for inspection. Furthermore, there is a strong theoretical basis for not using acceptance sampling plans. A discussion follows.

A THEORETICAL INVALIDATION OF ACCEPTANCE SAMPLING

A discussion of the invalidity of acceptance sampling must consider the stability, or lack of stability, of the process's output undergoing inspection. Let us first consider the case of the stable process.

Stable Process

Suppose a lot of N independent items is drawn from a stable process that generates 100p percent defective output, and that x of the items are defective and N-x are conforming; recall from Chapter 7 that the number of defectives is binomially distributed with fraction defective p. This case is extremely common in stable processes. Then,

$$N = \text{total number of items in the lot,}$$
$$x = \text{number of defective items in the lot,}$$
$$N - x = \text{number of conforming items in the lot, and}$$
$$E(x/N) = p = \text{fraction defective items in the process.}$$

Suppose a sample of n items is drawn from the lot of N items (without replacement, due to the finite nature of the lot), such that r of the items are defective and n − r of the items are conforming. Then,

$$n = \text{total number of items in the sample,}$$
$$r = \text{number of defective items in the sample,}$$
$$n - r = \text{number of conforming items in the sample, and}$$
$$E(r/n) = p = \text{fraction defective items in the process.}$$

The selection of the sample from the lot creates a new entity we will call the *remainder,* or the rest of the lot. The remainder is composed of N − n items, such that x − r of the items are defective and (N − n) − (x − r) are conforming. Then,

$$N - n = \text{total number of items in the remainder,}$$
$$x - r = \text{number of defective items in the remainder,}$$
$$(N - n) - (x - r) = \text{number of conforming items in the remainder,}$$
$$\text{and}$$
$$E(x - r/N - n) = p = \text{fraction defective items in the process.}$$

Figure 16.2 illustrates the above sequence of item groupings.

If it can be shown that the number of defectives in the sample is independent of (not correlated with) the number of defectives in the remainder, then acceptance sampling plans that determine the disposition of a remainder (accept, reject, or screen) based on the number of defectives in a sample are invalid; this proof is given in Appendix 16.1 at the end of the chapter.

To state this another way, the number of defectives in the sample and in the remainder are both binomially distributed with the same mean fraction, p, and are independent. For example, if a lot of 1,000 fair coins was tossed repeatedly and a sample of 50 of the 1,000 was drawn each time for inspection, the fraction of heads in the sample and in the remainder would both be distributed around p = 0.5; but the number of heads in the samples would be independent of the number of heads in the remainders. In other words, the distribution of heads in the remainders associated with samples yielding 0 defectives would be the same as the distribution of heads in the remainders associated with samples yielding 50 defectives. As a direct result of this, acceptance sampling plans that determine the disposition of remainders based on samples are invalid for a stable pro-

FIGURE 16.2   Selection of Samples from Lots Drawn from a Process

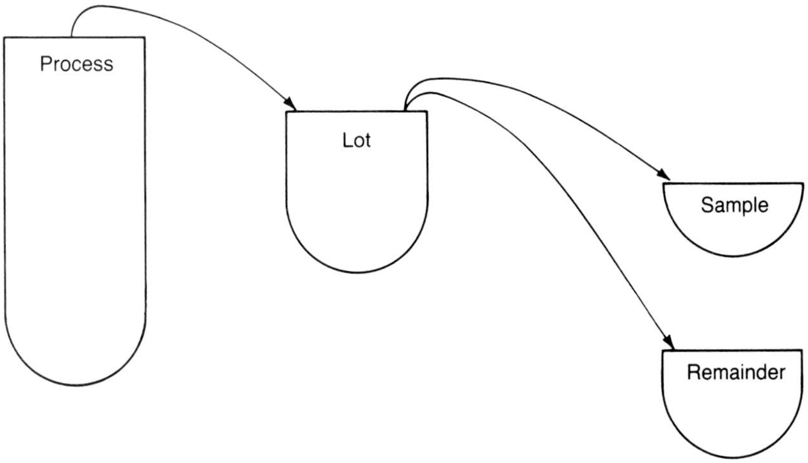

cess.[10] This is a shocking result to many.[11] An alternative to acceptance sampling from stable processes must be found. W. Edwards Deming offers, as an alternative, the *kp rule,* to be discussed later in this chapter.

Chaotic Process

Suppose a lot of N items is drawn from a chaotic or unknown process; x of the items are defective and N − x are conforming, and the process fraction defective p wanders from lot to lot (or day to day) and is not predictable. Then,

$$N = \text{total number of items in the lot,}$$
$$x = \text{number of defective items in the lot, and}$$
$$N - x = \text{number of conforming items in the lot.}$$

Suppose a sample of n items is drawn from the lot of N items (without replacement, as a result of the finite nature of the lot), such that r of the items are defective and n − r of the items are conforming. Then,

$$n = \text{total number of items in the sample,}$$
$$r = \text{number of defective items in the sample, and}$$
$$n - r = \text{number of conforming items in the sample.}$$

As with the stable process, the selection of the sample from the lot creates a remainder composed of N − n items, such that x − r of the items are defective and $(N - n) - (x - r)$ are conforming. Then,

$$N - n = \text{total number of items in the remainder,}$$
$$x - r = \text{number of defective items in the remainder, and}$$
$$(N - n) - (x - r) = \text{number of conforming items in the remainder.}$$

As before, Figure 16.2 represents the sequence of item groupings discussed in this section.

In this situation it can be shown that the number of defectives in the sample, r, is correlated with the number of defectives in the remainder, $x - r$. Thus, when p varies widely and unpredictably from lot to lot, the information from a sample provides insight into the remainder. Going back to the coin example, if lots of 1,000 biased coins were tossed repeatedly and samples of 50 of the 1,000 were drawn each time for inspection, the distribution of the fraction of heads in the samples and in the remainders would not be distributed around 0.5; rather, they would be related to the fraction of defectives in the lot and would be correlated. Consequently, the number of defectives in the sample and remainders would be correlated. Hence, acceptance sampling plans that determine the disposition of a remainder (accept, reject, or screen) based on the number of defectives in a sample are valid for a chaotic process.[12] The larger question of whether it is the most cost-effective plan given the chaotic nature of the process, will be discussed later in this chapter.

It is important to note that as processes are stabilized as a result of quality efforts, acceptance plans that are valid for chaotic processes—albeit at high cost—will no longer be effective on the stable process.

PLAN FOR MINIMUM AVERAGE TOTAL COST FOR TEST OF INCOMING MATERIALS AND FINAL PRODUCT FOR STABLE PROCESSES

The kp Rule

Given a stable process and the knowledge that acceptance sampling plans are not effective on such processes, we are left with only two of the inspection alternatives discussed at the beginning of this chapter: (1) no inspection or (2) 100 percent inspection. We discuss here Deming's kp rule, which specifies when to do no inspection and when to do 100 percent inspection, such that the total cost of incoming and intermediate materials, final products, and repairing and testing those products that fail will be minimized. The rule is derived in Appendix 16.2.

The assumptions for the use of the kp rule are listed below. These assumptions are not restrictive and are applicable to many common situations.[13]

1. All items are tested (inspected) before they move forward in the extended process. In other words, all nonconforming items are detected by a final inspection. When all items will not be subjected to a final inspection, the rule can be modified to reflect the possibility that a certain fraction of nonconforming parts, f, would be caught and the remaining fraction, 1 − f, would continue on into production or into the hands of customers.[14]

2. Inspection is completely reliable. If an item is defective, the item will fail inspection. "A defective part is one that by definition will cause the assembly to fail. If a part declared defective at the start will not cause trouble further down the line, or with the customer, then you have not yet defined what you mean by a defective part."[15]

3. The item vendor will give the buyer an extra supply of items, S, for replacement of any defective item found. The supplier adds the cost of these items onto his bill, either directly or indirectly. This cost is an overhead cost and would be present regardless of the inspection plan used. Hence, it need not be included in the cost function to be minimized.

The following notation is necessary to determine when to do 100 percent inspection and when to do no inspection. Let

p = the average incoming fraction of defective items in incoming lots of items. Recall that the process under study is stable and has a meaningful average incoming fraction of defective items, p, in incoming lots of items;

k_1 = the cost to initially inspect one item;

k_2 = the cost to dismantle, repair, reassemble, and test a good or service that fails because a defective item was used in its production.

If a process is stable around the fraction p, the kp rule states:

1. *If k_1/k_2 is greater than p, then 0 percent inspection* (no inspection minimizes the total cost). This occurs if the fraction of incoming defective items, p, is very low, the cost of inspecting an incoming item is high, and the cost of the defective item getting into production is low; therefore, no inspection is needed. The rationale is that there is little risk or penalty associated with incoming defective items.

2. *If k_1/k_2 is less than p, then 100 percent inspection* (100 percent inspection minimizes the total cost). This occurs if the fraction of incoming defective items, p, is high, the cost of inspecting an incoming item is low, and the cost of the defective item getting into production is high; therefore 100 percent inspection is needed. The rationale here is that there is great risk and a penalty attached to incoming defective items.

3. *If k_1/k_2 equals p, then either 0 percent or 100 percent inspection.* A decision must be made as to whether 0 percent or 100 percent inspection should be done in this case. In general, if p is not based on a substantial past history, it is vital to perform 100 percent inspection, for safety's sake.

To summarize, the kp rule will minimize the total cost of incoming and intermediary materials and final product for a stable process by proper selection of a 0 percent or 100 percent inspection policy. If the process under study is stable, then whether item i is defective is independent of whether any other item is defective. Hence, item i should be inspected, or not inspected, according to whether p is greater than or less than k_1/k_2. Recall that item i is a randomly selected item, and policy set for item i applies to any item. We can thus extend the policy for item i to all items in the lot. And consequently, either all or no items in the lot should be inspected, depending on whether p is greater than or less than the breakeven point, k_1/k_2.

It is important to note that 0 percent inspection does not mean zero information. Small samples should always be drawn from every lot—or on a skip-lot basis—for information about the process under study. This information should be recorded on control charts to facilitate process improvement.[16] The cost of these small samples is assumed to be a cost of doing business, and consequently, is not considered in the cost function to be minimized.

The kp rule is appropriate between any two points in the extended process, such as internally, in the vendor's processes, or between the firm and the vendor.

An Example of the kp Rule

An automobile manufacturer is deciding whether to purchase $25 million worth of equipment that would perform tests of engines purchased from vendors. The vendor's process is stable. The following figures have been determined:

- The inspection cost to screen out incoming defective engines is $50 per engine ($k_1$ = $50).
- The cost for corrective action if a defective engine gets into production is $500 per defective engine (k_2 = $500).
- On average, 1 in 150 incoming engines is defective (p = 1/150 = 0.0067).

Consequently, k_1/k_2 = 50/500 = 0.1. Note that 0.1 is greater than 0.0067. Therefore, k_1/k_2 is greater than p, and the correct course of action would be to do no initial inspection on incoming engines to achieve the minimum total cost.

If no engines are inspected, the automobile company would expect to incur the $500 cost in 1 out of 150 engines. This translates into an average corrective action cost of $3.33 per engine ($500 × 1/150). By eliminating initial inspection, the company would save $46.67 per engine ($50 − $3.33), on average. As the company purchases 4,000 engines per day, this translates into a daily savings of $186,680 (4,000 × $46.67), not including the savings of $25 million for testing equipment, interest on that money, and time freed up to work on improving quality! The next step in the pursuit of quality is for the automobile company to work with its engine vendor to reduce the fraction of defective engines.[17]

Exceptions to the kp Rule

Destructive Testing. The kp rule does not apply to destructive testing, in which an item is destroyed in the conduct of the test. The only solution in destructive testing is to achieve statistical control such that $p < k_1/k_2$, so that no inspection (other than routine small samples of the process) is the minimum cost policy. Note that achieving statistical control with $p < k_1/k_2$ is the best solution regardless of whether the test is destructive or non-destructive.[18]

Homogeneous Mixtures. The kp rule does not apply to homogeneous mixtures; for example, "a jigger of gin or whiskey. We accept the fact that it matters little whether we draw off a jigger from the top of the bottle or from the middle or from the bottom."[19] In this case, the sample is identical in composition to the remainder; hence, we can make judgments about the remainder from the sample.

Component Costs of k_1 and k_2

Some of the costs to consider when calculating k_1 and k_2 are shown in Figure 16.3(a) and 16.3(b).[20] The costs required to compute k_1 are usually known and can be calculated. A firm's financial personnel should be helpful in computing k_1. However, the costs required to compute k_2 are generally unknown and frequently difficult to compute; for example, the cost of customer dissatisfaction from recalls or lawsuits. A reasonable policy is to estimate k_2 without the more subjective costs. If $p > k_1/k_2$, there is no need to estimate the other components of k_2. On the other hand, if $p < k_1/k_2$ without including the subjective costs, an estimate of these missing costs may have to be made to determine more accurately the relationship between p and k_1/k_2.

ACHIEVING SUBSTANTIAL SAVINGS OVER 100 PERCENT INSPECTION IN THE COST OF INCOMING MATERIAL AND FINAL PRODUCT FOR CHAOTIC PROCESSES

Given a chaotic process and the knowledge that acceptance sampling plans are appropriate for such processes, we have the three inspection

---

FIGURE 16.3 Component Costs of k_1 and k_2

(*a*) *Some Inspection Costs k_1*

Capital Equipment
  Initial cost
  Depreciation (also considers residual value)
  Planned production volumes
  Cost of capital

Operating Costs
  Labor
  Rent, utilities, maintenance
  Piece cost (outside vendor quote)

(*b*) *Possible Detrimental Costs k_2*

The added costs of processing the nonconforming item further
The cost of sorting lots later to find a nonconforming item
The cost of repairing batches of assemblies later
The cost of lost production later if lots of parts or batches of
  assemblies must be guaranteed pending sorting and repair
Warranty costs
Cost of recalls
Law suits ($k_2 \rightarrow \infty$ for safety items)
Customer loyalty impinging upon future sales

---

alternatives discussed earlier: (1) no inspection, (2) 100 percent inspection, or (3) some form of acceptance sampling. The selection between alternatives (1) and (3) to achieve savings over 100 percent inspection (alternative (2)) depends upon the nature of the chaos in the process.

Mild Chaos

If the fraction defective in the process under study wanders in an unpredictable manner so that the fractions for the worst lots are below k_1/k_2, then no inspection (except for routine sampling of the process) should be performed, as Figure 16.4 shows. Significant effort, however, should be directed toward stabilizing the process from the information in the routine samples.

If the fraction defective in the process under study wanders in an unpredictable manner so that the fractions for the best lots are above k_1/k_2, then 100 percent inspection should be performed, as shown in Figure 16.5. Again, significant effort should be directed toward stabilizing the process by using the information from 100 percent inspection.

FIGURE 16.4   Mild Chaos with Low Fraction Defective

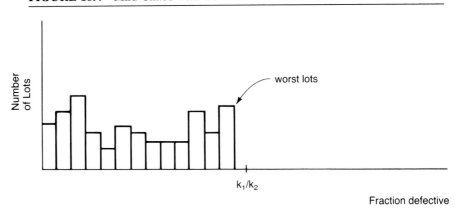

These two cases are considered mild chaos because the distribution of the fraction defective is chaotic within bounds. Of course, these bounds can disappear at any moment. These are not situations in which one should be lulled into thinking that the chaos will always stay within bounds; chaos is a wild beast that can run anywhere at anytime, including beyond any earlier boundary.

FIGURE 16.5   Mild Chaos with High Fraction Defective

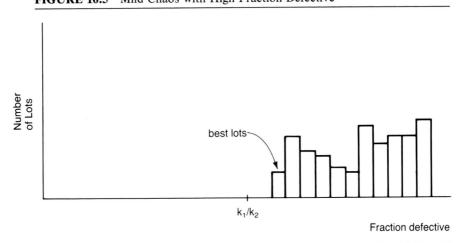

Severe Chaos

If the fraction defective in the process under study wanders in an unpredictable manner within a narrow range around k_1/k_2, the most practical plan is 100 percent inspection of all lots.[21] Any effort at acceptance sampling will not justify the cost of administering the plan.[22]

If the fraction defective in the process under study wanders in an unpredictable manner within a wide range[23] around k_1/k_2, Orsini[24] has devised a rule that yields substantial savings over 100 percent inspection:

If k_1/k_2 is less than 1/1,000, then inspect 100 percent of the incoming lots.

If k_1/k_2 is between 1/1,000 and 1/10, then test a sample of 200. If there are no defectives, then accept the remainder. Inspect the entire remainder if at least one defective item is found in the sample.

If k_1/k_2 is greater than 1/10, then do no inspection.

This rule is also helpful in working with a vendor to bring his or her process into control. A running record of the samples of 200 can be kept, and the number of defectives can be charted, sample by sample. Feedback to the vendor is extremely helpful in identifying problems.

Exceptions to Rules for Chaos

The rules for chaos are subject to the same exceptions as the kp rule for stable processes. Further, if items come into a firm from an unknown vendor, the optimal policy is to perform 100 percent inspection until enough information has been collected to construct a control chart(s) for the vendor's process/product. At that time, a selection of the best inspection plan can be made. 100 percent inspection should also be carried out for critical parts and safety items (for safety items k_2 is infinite).

SUMMARY

In this chapter we discussed several different types of acceptance sampling plans: lot-by-lot plans, continuous sampling plans, and special sampling plans. Lot-by-lot acceptance sampling plans are used to inspect goods or services whenever the goods or services can be conveniently grouped into lots. Continuous flow acceptance sampling plans are used to inspect goods or services whenever the goods or services cannot be grouped into lots—for example, goods on a conveyor belt. Special sampling plans are used for lots found in series. These acceptance sampling plans, do not however, minimize the average total cost of inspection of incoming, intermediate, and final goods and services. A theoretical argument invalidating acceptance sampling was presented.

Plans for minimizing the average total cost of testing incoming materials and final product, called the kp rule, were analyzed. One plan is used for stable processes, and the other plan is used for chaotic processes. Examples and exceptions to these plans were discussed. Both plans require the collection and control charting of either inspection data or routine samples to achieve process stability and pursue continuous and never-ending improvement.

Appendix 16.1

PROOF THAT THE NUMBER OF DEFECTIVES IN A SAMPLE IS INDEPENDENT OF THE NUMBER OF DEFECTIVES IN THE REMAINDER FOR LOTS DRAWN FROM A STABLE PROCESS

To prove: Given a stable process with 100p percent defective, the number of defectives in a sample of n items drawn from a lot of size N is independent of the number of defectives in the remaining N − n items.

$\quad$ *Proof:* $\qquad$ n = number of items in the sample

$\qquad\qquad$ N = number of items in the lot

$\qquad\qquad$ r = number of defective items in the sample

$\qquad\qquad$ x = number of defective items in the lot

$\qquad$ x − r = number of defective items in remaining N − n items

$\qquad$ f(x) = the probability of x defectives in a lot of N items drawn from a stable process with 100p percent defective

$\qquad$ f(r) = the probability of r defectives in a sample of n items drawn from a stable process with 100p percent defective

$\quad$ f(r|x) = the conditional probability of r defectives in a sample of n items drawn from a lot of N items given that the lot contains x defectives

f(r ∩ (x − r)) = the joint probability that there are r defectives in the sample of n items and x − r defectives in the remaining N − n items

We must show that f(r ∩ (x − r)) = f(r)f(x − r).

We know, from Equation 4.10, that the joint probability of x defectives in the lot and r defectives in the sample is given by:

$$f(x \cap r) = f(x)f(r|x)$$

The number of defectives, x, in a lot of size N, is binomially distributed, as given by Equation 6.3a:

$$f(x) = \binom{N}{x} p^x q^{N-x}$$

Similarly, the number of defectives, r, in a sample of size n, is binomially distributed:

$$f(r) = \binom{n}{r} p^r q^{n-r}$$

and the number of defectives, x − r, in the remaining N − n items, is also binomially distributed:

$$f(x - r) = \binom{N - n}{x - r} p^{x-r} q^{(N-n)-(x-r)}$$

The number of defectives, r, in a sample of size n, given a total of x defectives in a lot of size N, has a hypergeometric distribution, as given by:

$$f(r|x) = \frac{\binom{n}{r}\binom{N - n}{x - r}}{\binom{N}{x}}$$

Then

$$f(x \cap r) = \binom{N}{x} p^x q^{N-x} \frac{\binom{n}{r}\binom{N - n}{x - r}}{\binom{N}{x}}$$

This can be written as:

$$f(x \cap r) = p^{x-r+r} q^{N-x-n+n-r+r} \binom{n}{r}\binom{N - n}{x - r}$$

or, rearranging:

$$f(x \cap r) = \binom{N - n}{x - r} p^{x-r} q^{(N-n)-(x-r)} \binom{n}{r} p^r q^{n-r}$$

or,

$$f(x \cap r) = f(x - r)f(r)$$

This is the probability of x defectives in a lot of size N and r defectives in a sample of size n. But this is the same as the probability of x − r defectives in N − n items and r defectives in n items. So

$$f(x \cap r) = f((x - r) \cap r)$$

$$\text{Then } f((x - r) \cap r) = f(r)f(x - r)$$

But, from Equation 4.11, if the joint probability of two events is equal to the product of their unconditional probabilities, the events must be independent. Thus, the number of defectives in a sample of n items is independent of the number of defectives in the remaining N − n items.

Appendix 16.2

DERIVATION OF THE kp RULE FOR STABLE PROCESSES

Let

p = the average fraction of defective items in incoming lots of items;

k_1 = the cost to initially inspect one item;

k_2 = the cost to dismantle, repair, reassemble, and test a good or service that fails because a defective item was used in its production;

k = the average cost to test one or more items to find a conforming item from the supply, S, to replace a defective item found
  = $k_1/(1 - p)$

C_1 = the cost to initially inspect one item; and

C_2 = the cost to repair a failed good or service.

Further, let

$$x_i = 1 \text{ if item i is defective, and}$$

$$x_i = 0 \text{ if item i is conforming.}$$

Now, if one item, item i, is randomly drawn from a lot, then the probability that it is defective is p. The cost to initially inspect item i is

$$C_1 = k_1 + kx_i \quad \text{if we test item i, and}$$

$$C_1 = 0 \qquad \text{if we do not test item i.}$$

C_1 is composed of the cost to initially test one item plus the cost to replace the item if it is found to be defective; hence,

$$C_1 = k_1 + k \quad \text{if item i is tested and found to be defective,}$$

$$C_1 = k_1 \qquad \text{if item i is tested and found to conform, or}$$

$$C_1 = 0 \qquad \text{it item i is not tested.}$$

The cost to repair a failed good or service due to item i is

$$C_2 = (k_2 + k)x_i \quad \text{if we do not initially test item i, and}$$

$$C_2 = 0 \qquad \text{if we do initially test item i.}$$

C_2 is composed of the cost to repair a failed good or service if item i was not initially inspected; hence,

$C_2 = k_2 + k$ if item i was not initially inspected and item i is defective,

$C_2 = 0$ if item i was not initially inspected and item i is conforming, and

$C_2 = 0$ if item i was initially inspected.

C_1 and C_2 are mutually exclusive as they cannot occur simultaneously for a given item; if one is positive, the other is zero. The total cost for item i is

$$C = C_1 + C_2$$

Figure 16A.1 summarizes the cost structure for inspection versus no inspection for item i.[25]

FIGURE 16A.1   Cost Structure for Inspection Decision for Item i

Inspect the Item?	C_1	C_2	Total Cost $C = C_1 + C_2$
Yes	$k_1 + kx_i$	0	$k_1 + kx_i$
No	0	$(k_2 + k)x_i$	$(k_2 + k)x_i$

Figure 16A.2 extends the cost structure for inspection versus no inspection to the average cost per item over the lot. In this case x_i is replaced by p because

$$p = \sum_{i=1}^{N} [x_i/N]$$

None of the other elements in Figure 16A.2 is affected because they are constants.

FIGURE 16A.2   Cost Structure for Inspection Decision for Average Item

Inspect the Item?	C_1	C_2	Total Cost $C = C_1 + C_2$
Yes	$k_1 + kp$	0	$k_1 + kp$
No	0	$(k_2 + k)p$	$k_2p + kp$

Now, the breakeven point between inspection and no inspection can be determined by setting the total cost for inspection equal to the total cost for no inspection. That is, at the breakeven point: Cost (Inspect) = Cost (Do Not Inspect)

$$k_1 + kp = k_2p + kp$$

$$k_1 = k_2p$$

$$p = k_1/k_2$$

EXERCISES

16.1 Discuss the three possible alternatives for the inspection of goods or services.

16.2 a. Explain the purpose of acceptance sampling.
 b. Explain the purpose of lot-by-lot acceptance sampling plans. Describe the situations in which lot-by-lot acceptance sampling plans are used as a basis for action on a lot of goods.
 c. Explain the purpose of Military Standard 414.
 d. Explain the purpose of Military Standard 105D.
 e. Explain the purpose of continuous flow acceptance sampling plans.
 f. Briefly describe the operation of continuous flow acceptance sampling plans.

16.3 Explain why acceptance sampling plans are theoretically incorrect for stable processes and should not be used as a basis for action. Mathematically defend your explanation.

16.4 Explain why acceptance sampling is theoretically correct but not economical for chaotic processes and consequently should not be used as a basis for action.

16.5 a. Describe the kp rule for stable processes.
 b. Explain the assumptions required to use the kp rule.
 c. List several examples of k_1 inspection costs.
 d. List several examples of k_2 inspection costs.

16.6 Explain the term *mild chaos* and its significance to taking action on incoming or intermediary material, or final product.

16.7 A manufacturer of radios has a policy of inspecting every incoming radio speaker to ensure it conforms to specifications. What information would you need to question the wisdom of this inspection policy?

16.8 The production manager of the radio company discussed in Exercise 16.7 learned about the kp rule. He used past inspection data concerning the proportion of defective radio speakers purchased per day to construct a p chart with variable sample size. The p chart indicated that the incoming stream of radio speakers was stable with respect to the fraction of defective radio speakers purchased each day. The average fraction of defective radio speakers was found to be 0.002. Further study showed that it costs approximately $0.50 to inspect an incoming radio speaker and that it costs approximately $7.50 to repair a radio with a defective speaker before it leaves the factory.

a. Use the kp rule to determine if 0 percent of 100 percent inspection should be used for incoming radio speakers.

b. What should management do about the incoming radio speaker process, given your answer in part *a*?

16.9 Explain the term *severe chaos*. Describe an inspection rule that can be used when a process exhibits severe chaos.

NOTES

1. An excellent reference for details on acceptance sampling is Acheson Duncan, *Quality Control and Industrial Statistics,* 5th ed. (Homewood, Ill.: Richard D. Irwin, 1986), pp. 161–414.
2. Ibid., pp. 291–305.
3. Ibid., pp. 217–48.
4. Ibid., pp. 406–13.
5. Ibid., pp. 536–40.
6. Ibid., pp. 177–79.
7. Ibid., pp. 252–53.
8. Ibid., pp. 340–65.
9. Ibid., pp. 271–72.
10. A cautionary note: The above proof does not mean that statistical inference does not work—that is, that a random sample from a population or a stable process does not provide information about the sampling frame or conceptual frame. Rather, it only means that samples provide no information about remainders from stable processes.

    A proof that a random sample from a stable process provides information about the process p is shown below.

    $$\text{Recall:} \quad f(x)f(r|x) = f(r \cap (x - r)) = f(r)f(x - r) = f(r \cap x)$$

    $$\text{Hence:} \quad \sum_{x=0}^{N} f(r \cap x) = f(r) = \binom{n}{r} p^r (1 - p)^{n-r}$$

    This indicates that f(r) is the rth term of the binomial $(p + q)^n$, just as if the sample of n items was drawn (with replacement) directly from the process without the introduction of the lot between the process and the sample. This means that a random sample of n items from the process provides information on the process average via the estimation $p = r/n$.
11. Alexander M. Mood, "On the dependence of sampling inspection plans under population distributions," *Annals of Mathematical Statistics* 14, 1943, pp. 415–25. Also see W. Edwards Deming, *Some Theory of Sampling* (New York: John Wiley & Sons, 1950), p. 258.
12. This proof will not be shown.

13. W. E. Deming, *Quality, Productivity and Competitive Position,* (Cambridge, Mass: Massachusetts Institute of Technology, Center for Advanced Engineering Study, 1982, pp. 267–311. Joyce Orsini, "Simple Rule to Reduce Total Cost of Inspection and Correction of Product in State of Chaos," dissertation for the doctorate, Graduate School of Business Administration, New York University, 1982. G. P. Papadakis, "The Deming Inspection Criteria for Choosing Zero or 100 Percent Inspection," *Journal of Quality Technology* 17, July 1985, pp. 121–27.

14. Papadakis, "The Deming Inspection Criteria for Choosing Zero or 100 Percent Inspection," p. 123.

15. Deming, *Quality, Productivity and Competitive Position,* p. 268.

16. Ibid., p. 273.

17. Ibid., pp. 276–77.

18. Ibid., p. 274–75.

19. Ibid., p. 285.

20. Papadakis, "The Deming Inspection Criteria for Choosing Zero or 100 Percent Inspection," p. 124.

21. A narrow range is defined as a coefficient of variation of 0.3 or less:

$$\frac{nq}{p} \leq 0.3$$

22. Deming, *Quality, Productivity and Competitive Position,* p. 271.

23. A wide range is defined as a coefficient of variation of more than 0.3:

$$\frac{nq}{p} > 0.3$$

24. J. Orsini, "Simple Rule to Reduce Total Cost of Inspection and Correction of Product in a State of Chaos."

25. Deming, *Quality, Productivity and Competitive Position,* p. 302.

PART VII

Foundations of
Quality Revisited

This section presents the relationship between W. Edwards Deming's 14 points for management and the reduction of variation, as well as some current thoughts on statistical studies and statistical practice.

In Chapter 17, the purpose of statistics and management are shown to be identical—that is, the reduction of variation. A teaching aid, based upon an experiment utilizing a funnel, demonstrates the interconnectedness between management thought and statistical theory. Finally, the focal point of each of Deming's 14 points is shown to be the reduction of variation in the extended process.

Chapter 18 analyzes some current thinking about statistical studies and statistical practice and discusses differences in the theory of statistics for enumerative studies and analytic studies. The discussion also centers on the far-reaching implications of these differences. This chapter reproduces a paper written by Deming dealing with a code of professional conduct for the statistician. It includes the responsibilities of the statistician—and the subject matter expert (the client)—in conducting a statistical study.

Chapter 17

Deming's 14 Points and the Reduction of Variation

THE PURPOSE OF STATISTICS

The purpose of statistics is to study and understand process and product variation, interactions among product and process variables, and operational definitions—and ultimately to take action to reduce variation. As we discussed in Chapter 2, there are two types of statistical studies, enumerative studies and analytic studies. The purpose of an enumerative study is to survey the characteristics of a frame—one of those characteristics being variation—and to take action on the disposition of the material in the frame. The purpose of an analytic study is to examine the causes of variation in a process and to take action to reduce the variation of the process in the future and/or change the process's average.

THE PURPOSE OF MANAGEMENT

The purpose of management is to lead an organization in the direction of never-ending and continuous improvement. Never-ending improvement of an organization is pursued by continual reduction of variation of the organization's processes and products. This pursuit has both analytic and enumerative aspects.

AN AID TO UNDERSTANDING THE RELATIONSHIP BETWEEN STATISTICS AND MANAGEMENT

W. Edwards Deming has stated that "If anyone adjusts a stable process to try to compensate for a result that is undesirable, or for a result that is extra good, the output that follows will be worse than if he had left the process alone.''[1] This is called *overcontrol of the process,* or *tampering;* recall the discussion on overadjustment in Chapter 8. If management tampers with a process without profound knowledge of how to improve the process through statistical thinking, they will increase the variation of that process and reduce their ability to manage that process.

Deming illustrates this as follows: "A common example is to take action on the basis of a defective item, or on complaint of a customer. The result of his efforts to improve future output (only doing his best) will be to double the variance of the output, or even cause the system to explode. What is required for improvement is a fundamental change in the system, not tampering."[2]

The Funnel Experiment

Loss to an organization results from overcontrol of its processes, which include safety, training, hiring, supervision, union-management relations, policy formation, production, maintenance, shipping, purchasing, administration, and customer relations. This loss can be demonstrated by an experiment utilizing a funnel.[3] We will describe the apparatus and procedure for conducting the experiment, and we will demonstrate its relationship to management's pursuit of continual reduction of variation.

The materials needed to conduct the experiment, shown in Figure 17.1, are: (1) a funnel; (2) a marble that will fall through the funnel; (3) a flat surface (e.g., a table top); (4) a pencil; and (5) a holder for the funnel.

The following steps are required to perform the experiment: (1) designate a point on the flat surface as a target and consider this target to be the point of origin in a two dimensional space, where x and y represent the axes of the surface; hence, at the target $(x,y) = (0,0)$; (2) drop a marble through the funnel; (3) mark the spot where the marble comes to rest on the surface with a pencil; (4) drop the marble through the funnel again and mark the spot where the marble comes to rest on the surface; and (5) repeat step (4) through 50 drops.

A rule for adjustment of the funnel's position in relation to the target is needed to perform the fourth step. There are four possible rules, and the second rule can be done in two ways.

Rule 1. Set the funnel over the target at (0,0) and leave the funnel fixed through all 50 drops. This rule will produce a stable pattern of points on the surface; this pattern will approximate a circle, as shown in Figure 17.2.[4] Further, as we shall see, the variance of the diameters of all circles produced by repeated experimentation using rule 1 will be smaller than the variance resulting from any other rule used in the fourth step of the experiment.

Management's use of the first rule demonstrates an understanding of the distinction between special and common variation, and the different

FIGURE 17.1   Funnel Experiment Equipment

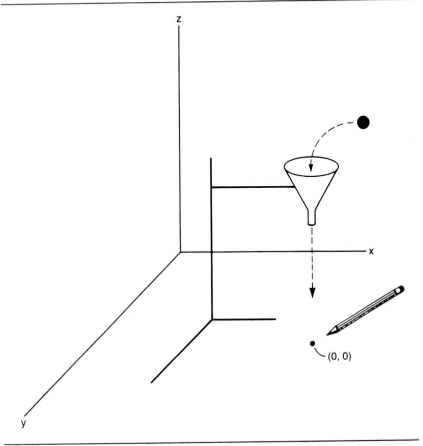

types of managerial action required for each type of variation. Rule 1 implies that the process is being managed by people who have knowledge of how to reduce variation.

   Rule 2. The funnel is set over the target at (0,0) prior to the initial drop. Let (x_k, y_k) represent the point where the kth marble dropped through the funnel comes to rest on the surface. Rule 2 states that the funnel should be moved the distance $(-x_k, -y_k)$ from its last resting point. In essence, this is an adjustment rule with a memory of the last resting point. This rule will produce a stable pattern of resting points on the surface, which will approximate a circle. However, the variance of the diameters of all circles produced by repeated experimentation using rule 2 will have double the variance of the circular pattern produced using rule 1, as shown in Figure 17.3.[5]

FIGURE 17.2   Rule 1

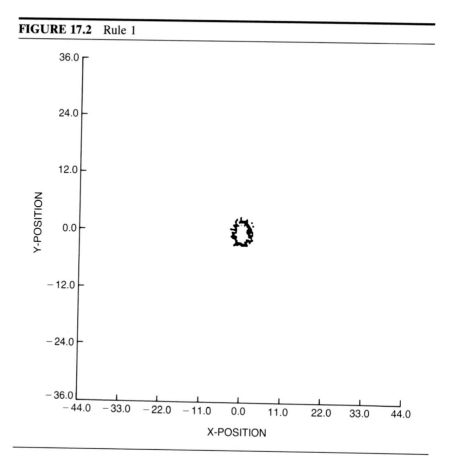

In terms of its application to management actions, rule 2 implies that the process is being tampered with by people with inadequate knowledge of how to manage the process to reduce its variation. It implies acting on common variation as if it were special variation. Rule 2 is commonly used as a method of "attempting" to make things better in a process. A list of examples follows.

1. Automatic process control. The automatic adjustment of a process to hold output within specified tolerance limits is an example of rule 2, assuming adjustments to the process are made from the last process measurement. This type of process adjustment procedure is frequently called *rule-based process control (RPC)*. The use of RPC is widespread in industry.

2. Operator adjustment. Operator adjustment to compensate for a unit of output's not being on target, or nominal, is an example of rule 2, assuming adjustments to the process are made from the last process

FIGURE 17.3   Rule 2

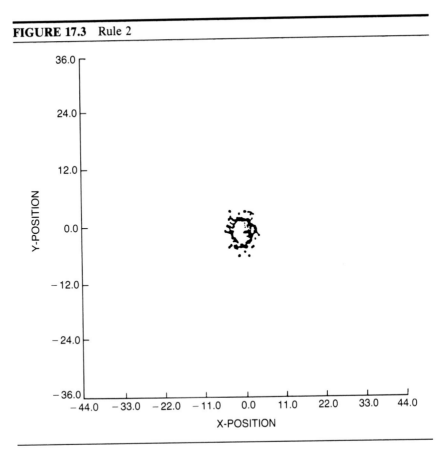

measurement.[6] This type of overcontrol frequently leads to a saw-tooth-type pattern on an x-bar chart.

3. Stock market. The stock market's reaction to good or bad news is often an overreaction to phenomena, which follows rule 2.

Rule 2a. A variant on rule 2 is often employed in industry. Rule 2a states if (x_k, y_k) is within a circle centered at $(0,0)$ with diameter d_{spec}, do not adjust the funnel. However, if (x_k, y_k) is outside the circle centered at $(0,0)$ with diameter d_{spec}, use the adjustment rule specified in rule 2. Rule 2a creates a "deadband" in which no process adjustment takes place. Research results obtained by Gitlow, Kang, and Kellogg demonstrate that rule 2a, with any size deadband, yields the same result of doubling the variation of the process as does rule 2, when compared with rule 1.[7]

Rule 3. The funnel is set over the target at $(0,0)$ prior to the initial drop. Let (x_k, y_k) represent the point where the kth marble dropped through the funnel comes to rest on the surface. Rule 3 states that the funnel should be moved a distance $(-x_k, -y_k)$ from the target $(0,0)$. In essence, this is an adjustment rule with no memory of the last resting point. This rule will produce an unstable and explosive pattern of resting points on the surface; as k increases without bound, the pattern will move farther and farther away from the target in some symmetrical pattern, such as the bow-tie shaped pattern shown in Figure 17.4.[8]

Rule 3, like rule 2, is commonly used as a method of "attempting" to make things better in the process. Rule 3 implies that the process is being tampered with by people with inadequate knowledge of how to manage the process to reduce its variation. It implies acting on common variation as if it were special variation. A list of examples of rule 3 follows.

1. Automatic process control. The automatic adjustment of a process to hold output within specified tolerance limits is an example of rule 3, assuming adjustments to the process are made from the target and not the last process measurement.

FIGURE 17.4 Rule 3

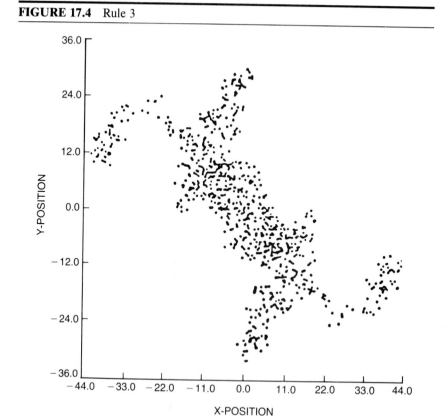

2. Operator adjustment. Operator adjustment to compensate for a unit of output's not being on target, or nominal, is an example of rule 3, assuming adjustments to the process are made from the target and not the last process measurement.
3. Setting the current period's goal based on last period's overage or underage. A sales quota policy that states that if you are short of this month's goal by $25,000, you must increase next month's goal by $25,000 is an example of rule 3.
4. Setting the inspection policy for the kth batch based on the k-1st batch's record. An AQL inspection policy that recommends tightened or loosened inspection for a batch of material based on the history of the prior batch is an example of rule 3. A better policy would be to use the kp rule, as discussed in Chapter 16, in conjunction with a variables control chart to estimate process capability and p.
5. Making up the previous period's shortage during the current period. A production policy that requires production personnel to make up any shortages from last month's production run in this month's production run is an example of rule 3.

Rule 4. The funnel is set over the target at (0,0) prior to the initial drop. Let (x_k, y_k) represent the point where the kth marble dropped through the funnel comes to rest on the surface. Rule 4 states that the funnel should be moved to the resting point, (x_k, y_k). In essence, this is an adjustment rule with no memory of either the last resting point or the position of the target at (0,0). This rule will produce an unstable and explosive pattern of resting points on the surface as k increases without bound, and it will eventually move farther and farther away from the target at (0,0) in one direction, as shown in Figure 17.5.[9]

Rule 4, like rule 2, is commonly used as a method of "attempting" to make things better in a process. Rule 4 implies that the process is being tampered with by people with inadequate knowledge of how to manage the process to reduce its variation. It implies acting on common variation as if it were special variation. A list of examples, many discussed by Deming in his management seminars, follows.

1. Make it like the last one. Using the last unit of output as the standard for the next unit of output will eventually produce material bearing no resemblance to the original piece; this is an example of rule 4. A possible solution to this problem is to use a master piece as a point of comparison. A common complaint is that "we sold the master (model) piece" so a standard is no longer available for comparison purposes.

   Another example of the above phenomenon is "a man who matches color from batch to batch for acceptance of material, without reference to the original swatch."[10] Two possible solutions to this problem are to use a standard color chip or to use the original color

---

FIGURE 17.5   Rule 4

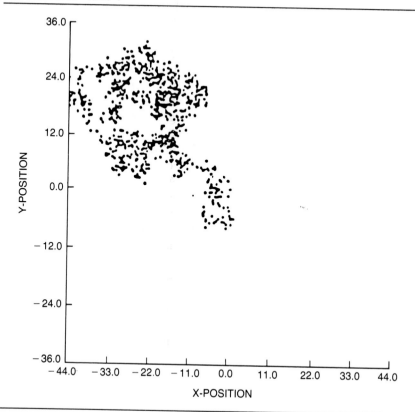

---

sample, assuming the color chip or the original color sample do not fade over time.

2. On-the-job training. Deming cites "A frightening example of rule 4 . . . where people on a job train a new worker. This worker is then ready in a few days to help to train a new worker. The methods taught deteriorate without limit. Who would know?"[11] Possible solutions to this problem are to formalize training with a video presentation or to utilize a master in the subject matter to do the training, to get a consistent and desired message to the trainees.

3. Budgeting. Setting the next period's budget as a percentage of the last period's budget is an example of rule 4.

4. Policy setting. Executives meeting to establish policy for an organization, without a mission statement to guide them and without profound knowledge of statistical thinking and variation, will set policies that become increasingly less consistent and more confusing, so that eventually their policies will be destructive to the organization.

5. The telephone game. The telephone game that small children play is an example of rule 4; the message gets continually more confusing, and after a time the current message bears no resemblance to the original message.

6. The grapevine. People who take action in their personal and professional lives based on information from the "grapevine" or "rumor mill" are using rule 4 as a basis for action; their next action is a function of only the most recent past action.

7. Engineering changes. Engineering changes to a product or process based on the latest version of a design without regard to the original design are made in accordance with rule 4. Eventually, the current design will bear no resemblance to the original design.

8. Policy surveys. An executive who changes policy based on the results of the latest employee survey, in a stream of employee surveys, is operating under rule 4; eventually the policy will have no bearing on its original intended purpose.

9. Adjusting work standards to reflect current performance. An organization that adjusts work standards to reflect current conditions is using rule 4. Work standards should be replaced with control charts that allow management to understand the capability of a process and take action to improve the process by reducing the variation in the process.

10. Collective bargaining. Union-management negotiations in which successive contracts are a reaction to current conditions is an example of rule 4.

Process Improvement

The experiment using the funnel illustrates that improvement of a system is accomplished not by overcontrol but by the reduction of variation. In the experiment, this means reducing the diameter of the circle created under rule 1, either by moving the funnel closer to the surface[12] or by straightening and lengthening the tube portion of the funnel to reduce the dispersion among the resting points. Note that both methods for improvement are system changes. In terms of an organization, the corresponding reduction of variation also involves system changes. As management is responsible for the system, only management can make the necessary changes to reduce this variation in the system.

We next discuss how the concept of reduction of variation relates to the 14 points discussed in Chapter 1.

DEMING'S 14 POINTS AND THE REDUCTION OF VARIATION

Deming's 14 points provide management with a guide for modifying its processes (the organizational counterparts of the five items and four rules required to perform the experiment) to create improvement and reduce variation (the organizational counterpart of reducing the diameter of the circle created with rule 1 in the experiment using the funnel). The 14 points are a road map for management to pursue their responsibilities for process improvement and the need to end management by the organizational counterparts of rules 2, 3, or 4.

Point 1: Create constancy of purpose toward improvement of product and service with a plan to become competitive, stay in business, and provide jobs.

Establishment of a mission statement is synonymous with setting a process's nominal or target level. Getting all employees (management, salaried, and hourly); members of the board of directors; and shareholders to interpret the mission statement uniformly and to behave in accordance to the common interpretation of the mission statement is a problem of reduction of variation.[13]

The mission statement should inspire employees, board members, and shareholders to think about innovation: what the consumer will want tomorrow. This will create uniformity in their concern for the customer and their commitment that there will be a tomorrow.

The mission statement should foster employee efforts for reducing variation. As employees focus on the reduction of variation to achieve customer satisfaction, a greater consistency in their actions will result.

An interesting example of this arose during a seminar at a Fortune 500 company. The company had a mission statement, and all employees received one half-day of training in the mission statement. The mission had four goals, including "create a great place to work" and "maximize profits." One new manager said he was concerned only with profit maximization, not creating a great place to work; his managerial actions reflected his view. His interpretation of the mission statement conflicted with other managers and created variability in management style. This created confusion, frustration, and a decrease in quality.

Point 2: Adopt the new philosophy. We are in a new economic age. We can no longer live with commonly accepted levels of delays, mistakes, defective material, and defective workmanship.

All people in an organization should embrace the view that the customer (internal and external) is the focus of all action. As everyone uniformly embraces customers as the main focus of the organization and their own jobs, variation in the way people view the organization—and in the way people interpret their job responsibilities—will be diminished.

The new philosophy requires that conformance to specifications be replaced by continuous and never-ending improvement of the organization's processes to reduce variation and reap the rewards, such as decrease in rework, increase in productivity, increase in quality, decrease in unit cost, increase in price flexibility, increase in competitive position, increase in market share, increase in profit, increase in job security, the creation of more jobs, and the increased availability of funds to fuel further process improvements. A corollary of the new philosophy is that tightening specifications does not improve quality and lower cost—it only creates more out-of-specification material.

Point 3: Cease dependence on mass inspection. Require, instead, statistical evidence that quality is built in to eliminate the need for inspection on a mass basis.

Defect detection is dependent upon mass inspection to sort conforming from defective material. Dependence on mass inspection does nothing to decrease variation. Moreover, inspection does not create a uniform product within specification limits—rather, product is bunched around specification limits; or at best, product is distributed within specification limits with large variance and tails truncated at the specification limits.

Some organizations that operate in a defect detection mode use automatic process control to hold processes within deadband limits; see the previous discussion of rule 2a. In fact, these systems double the variation of the process, at best—when compared to organizations operating under rule 1—rather than reduce the variation of the process.

Defect prevention can only reduce variation such that products are within specification. Unfortunately, as a result of entropy, a process that operates just within specification limits will eventually go out of specification limits. Further, defect prevention leaves employees with the impression that they are successfully accomplishing their jobs—with respect to the reduction of variation—if they achieve "Zero Defects." This creates an atmosphere of laxness that will ultimately take its toll.

Never-ending improvement works at continuously reducing variation within specification limits. Traditional loss functions, where loss is zero until a specification limit is reached and then becomes positive and constant, are inadequate for today's marketplace. Rather, a Taguchi quadratic loss function (as discussed in Chapter 15) should be used as it demonstrates the economic wisdom of continuous reduction of variation.

The kp rule facilitates the collection of process or product data such that variation can be continually reduced. This occurs when process data are collected with variables control charts, and estimates of the fraction of out-of-specification material, p, are calculated in a process capability study.

Point 4: End the practice of awarding business on the basis of price tag. Instead, depend on meaningful measures of quality, along with price. Move toward a single supplier for any one item on a long-term relationship of loyalty and trust.

Purchasing is a process that can be continuously improved through the never-ending reduction of variation in purchasing procedures, incoming materials, number of suppliers for any given item, and number of purchasing agents for any given item.

A one-time purchase from a supplier is an enumerative decision in which we are interested only in the frame and can be made without regard to the supplier's process. However, continuing purchases over time is an analytic decision and cannot be made without regard to the supplier's process. For the analytic decision, continuous reduction of variation in the supplier's process is critical. Long-term relationships with a supplier make sense if the supplier consistently meets the organization's needs and will continue to improve its ability to do so in the long run.

Multiple supplier processes, each of which has small variations, combine to create a process with large variation. This means an increase in the variability of inputs to the organization, which is counter to the reduction of variation. Consequently, reducing the supply base from many suppliers to one supplier is a rational action. This idea applies to both external and internal suppliers.

The number of purchasing agents who work with any given supplier should be reduced from several purchasing agents to one purchasing agent to decrease variation in purchasing practices.[14]

Point 5: Improve constantly and forever the system of production and service, to improve quality and productivity, and thus constantly decrease costs.

As we discussed in Chapter 1, we use the PDCA cycle to decrease the difference, or variation, between customer needs and process performance. The PDCA cycle is an analytic procedure useful for process improvement and the reduction of variation. In Chapter 3, we discussed using operational definitions to create uniformity in the interpretation of product, process, or job characteristics. Operational definitions allow for the reduction of an important source of variation, variation caused by multiple views of a product's, process's, or job's definition.

Never-ending improvement strives to continuously reduce variation within specification limits for operationally defined process and product characteristics. An example of an area that can benefit from the never-ending improvement resulting from the application of the PDCA cycle is union-management relations. Union-management relations are an analytic concern, not a one-time "sit down at the contract table" enumerative concern.

Point 6: Institute modern methods of training.

All employees in an organization should be trained in their job skills. That training should employ statistical methods to indicate when a state of statistical control is reached, and training is complete.

Everyone in an organization should similarly be trained in basic statistical methods and the organization should foster everyone's ability to understand variation. Statistical methods and statistical thinking will create a uniform way for employees to view their organization's processes. This will focus attention on the causes of variation and point to methods to remove those causes, reduce variation in a process, and create improvement in the organization. Statistical methods and statistical thinking allow management to separate systems problems for which only they are accountable from special problems that can be dealt with by all employees.

Point 7: Institute modern methods of supervision.

Managers should know how to lead their organizations, including employees and suppliers, toward continuous and never-ending improvement. To do this, each manager needs to know when people have special problems and how to continuously improve the system. Both functions require an understanding of variation, its two causes, and the appropriate managerial action for each type of variation.

Point 8: Drive out fear, so that everyone may work effectively for the company.

Fear stifles the desire to change and improve a process. Fear creates variability between an individual's or team's actions and the actions required to surpass customer needs and wants.

In a fear-filled environment, a system of statistically based management will not work. This is because people in the system will view statistics as vehicles for policing and judging rather than as vehicles for providing opportunities for improvement.

Point 9: Break down organizational barriers—everyone must work as a team to foresee and solve problems.

Barriers between departments result in multiple interpretations of a given message. This increases variability in the actions taken with respect to a given message. Operational definitions create a common language for communication between departments and consequently reduce variability in the actions taken on processes and products by different departments. "Operationally defining the ultimate customer's needs and expectations so that everyone understands how he contributes to the success of the organization is a solid step to breaking down barriers between departments."[15]

Point 10: Eliminate arbitrary numerical goals, posters, and slogans for the workforce which seek new levels of productivity without providing methods.

The system, and its variation, are the responsibility of management. Slogans and posters try to shift that responsibility to the worker. For example, a sign in a factory reading, "Safety is better than compensation!" is attempting to shift the burden for safety from factory management to the worker.

Point 11: Eliminate work standards and numerical quotas.

Work standards are negotiated values that have no bearing on a process or its capability. They create variability in performance by obscuring an employee's understanding of his job. Work standards create fears that undermine the smooth operation of the workplace and create an undesirable atmosphere, which is contrary to the Deming philosophy.

Work standards also place a cap on improvement. After the standard has been reached, employees stop working for fear of a new and higher standard. Thus, work standards create variability between actual performance and desired performance.

If a work standard is between a system's upper natural limit (UNL) and lower natural limit (LNL), there is a possibility that the standard can be met; but meeting the standard this way is simply a random lottery. If a work standard is above the UNL of the system, then there is little chance that the standard will be met unless management changes the system. Rather than focusing on the standard as a means to productivity, management should focus on stabilizing and improving the process to increase productivity.

The fallacy inherent in quotas and "management by objectives" (MBO) is illustrated when considering point 11 in terms of the experiment using the funnel. A quota is equivalent to declaring that an employee must achieve a given number of hits per time period within a circle centered at $(0,0)$ with diameter d_{spec}. This is equivalent to saying that an employee will be evaluated on performance that is solely a function of random chance. Try to imagine the frustration this can create as the results are really beyond the employee's control. MBO confounds common variation (like the random variation in the funnel experiment) and individual performance.

It is unfair and counterproductive to hold employees accountable for circumstances beyond their control. This kind of accountability leads to fear and frustration, often resulting in overcontrol, or the use of rules 2, or 3, or 4. As we have seen, this will only increase variation, thereby making it even more difficult to achieve the quota or MBO.

Point 12: Remove barriers that rob employees of their pride of workmanship.

Barrier 1: Managerial Ignorance Concerning Variation and Statistical Thinking. Management must be educated so they understand variation and its impact on decision making. By understanding the importance of not acting on common causes of variation as if they were special causes of variation, management can change their organization's systems to focus on continuous and never-ending improvement.

Barrier 2: Performance Appraisal Systems (PAS). The *performance appraisal system* (PAS) destroys teamwork by encouraging every person and every department to focus on individual goals—rather than on the organization's goals—to obtain a positive rating in their performance appraisal. This results in serious suboptimization with respect to the organization's goals. Simultaneous seeking of different, and possibly conflicting, goals creates variability in management's behavior, which creates confusion and fear as to exactly what everyone's job is. By eliminating the PAS, it becomes feasible to focus everyone's attention on the organization's goals.

PAS also reduce initiative or risk-taking because once an objective has been reached, effort stops. This is counter to the notion of continuous and never-ending improvement.

PAS can cause employees to lower their planned performance levels to increase the chances of meeting their objectives. Again, this creates variability between the desired objective and the negotiated objective, which will invariably result in lower productivity.

Further, PAS can foster banking of performance to create a cushion for the next PAS time period. This banking distorts the information upon which management makes decisions and subsequently creates variability between the desired objective and the actual situation.

PAS can increase variability in employee performance, resulting from actions such as rewarding everyone who is above average and penalizing everyone who is below average. In this type of situation, employees who are below average try to emulate employees who are above average. However, as the employees who are above average and those who are below average are part of the same system (only common variation is present), those who are below average are adjusting their behavior based on common variation. We know from the experiment using the funnel that this results in doubling performance variation at best, and exploding performance variation at worst.

The PAS assume that people are directly and solely responsible for their output. This assumption fails to consider the distinction between the effect of the individual and the effect of the system on output. In other words, it fails to appreciate the causes of variation and that most of the variation in performance can be attributed to the system, not the individual. The system is the responsibility of management.

PAS focuses on the short-term. The variability between where an organization wants to be after five years and where it will actually be after 20 quarters of short-term focus can be substantial.[16] Further, adjusting an organization's direction based upon short-term results is akin to rule 4 in the experiment using the funnel: what you do next depends solely on where you are now. This type of action will explode the variance of the system and create disastrous results.

The experiment using the funnel can be used to demonstrate the absurdity of PAS as shown in Figure 17.6:

1. Establish two concentric circles centered on the target of (0,0).
2. Name the area within the circle of smallest diameter (the innermost circle) "excellent performance"; name the area between the smallest circle and the next-larger circle "above average"; and so on, until the last area to be named—and name this area, the area between the second largest circle and the largest circle, "poor performance."
3. Allow each experimenter one drop of the marble, and record the area in which the marble comes to rest on the surface.
4. Label the experimenter in accordance with the name of the area in which his marble comes to rest on the surface.
5. Reward the experimenter accordingly: "excellent performance" gets high merit pay, while "poor performance" gets no cost-of-living pay and chastisement.
6. Observe what happens over time to the points where the marble comes to rest on the surface as a result of the above rating system.

The experimenters whose marbles fell in the outer circles will move to rules 2 or 3 or 4 out of frustration and fear, in their attempts to emulate the experimenters whose marbles came to rest on the innermost circle. Of course, we know that this is overcontrol based on common variation and will at best double the process variation (rule 2) and at worst explode the variation in the system (rules 3 or 4).

An Alternative to PAS. Instead of using PAS, we may view the development of *group objectives* (intra-departmental objectives, as opposed to inter-departmental objectives, which can be resolved by the department) as a process that management can control and improve through the reduction of variation.[17]

FIGURE 17.6   Performance Appraisal Systems

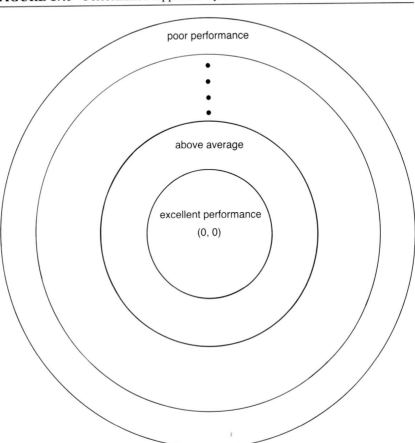

Variation in the pursuit of departmental objectives can be reduced by operationally defining group objectives. After group objectives have been defined, people from different departments can adopt the same group objectives because everyone participated in their development. This reduces the variability resulting from employees' search for different departmental objectives.

The *process* of developing group objectives focuses group attention on the sources of variation and the actions required to reduce that variation. Suppliers should meet with customers to determine what process or product characteristics they can provide to surpass customer needs. After customers and suppliers have agreed on the desired process or product characteristics, they can determine the possible areas, or sources, of improvement to satisfy customer needs. This can be accomplished, for example, by constructing a Cause-and-Effect diagram that indicates sources of variability in the characteristics requiring improvement and the possible resources to be used for reducing that variation. Next, it is important to determine which departments, and which employees within those departments, can effect the necessary improvements. After the appropriate individuals have been identified, they should form a group to develop mutual objectives that will lead to the desired improvement, which in turn will result in customer satisfaction.[18]

Improvement of the group objectives process requires an understanding of the sources of variability that can effect group objectives. After these sources of variability have been isolated, they must be eliminated to create an organization that can form group objectives consistently and efficiently. This can happen only when the *inputs* to the group objective process are stable with ever-decreasing variability—for example, when manpower does not vary with respect to objectives (individuals are all focused on the organization's objectives); when materials do not vary with respect to characteristics; when methods have been operationally defined and improved with statistical guidance; when machinery is managed on statistical signal; and when the environment creates uniformity in behavior.[19]

A Proposal that Fosters Consistency. Management must understand variation to recognize whether an employee is, or is not, operating in the system. An employee can exist in only one of three categories: in the system, out of the system on the negative side, and out of the system on the positive side. The latter two categories signify that the employee is in need of special attention. Further, if an employee shows improving or deteriorating performance for more than eight periods in a row (see Rule 5 in Chapter 8), that can be used as evidence that the employee is in need of special attention; this is an illustration of the institution of modern methods of supervision (Deming's seventh point in Chapter 1).

If all employees are within the system, then to reward or punish employees on the basis of performance (or merit) would be destructive. This is because the punished employees will try to emulate the rewarded employes, when in fact there is no difference between the two types of employees; they are both in the system and any difference only results from common variation. This type of reward/punish management will at best double the variation between employees, as illustrated by rule 2 of the experiment using the funnel. In this situation, any improvement must be a result of improvements in the system, which will benefit all employees.

If one or more employees are not within the system (special employees), then management must: (1) determine if the special employees form their own special system; and (2) determine if the rest of the employees are within the old revised system. Next, management must determine the sources of variation for the special employees and use this information to reduce variation, and create consistency and improvement.[20]

Barrier 3: Daily Production Reports. Daily production reports focus attention on yesterday's production, relative to the long-term production daily average, as a basis for action, without any acknowledgement of variation. This is an illustration of rules 2 and 3 of the experiment using the funnel. At best, this will double the variation in production (rule 2) and at worst, will explode the variation in the system of production into cycles of every-increasing variability (rule 3).

Daily production reports can create banking of production to meet tomorrow's quota. This reduces the accuracy of the information upon which management bases its actions, and hence, increases the variability of its decisions.

Daily production reports can create pressure to explain any variances from target production, when in fact variances may result from common variation. This leads to overcontrol and increases production variation.

A better method is to distribute production reports less frequently than every day—say, every week—and to include with the latest production information a control chart relative to the process's past history. In this way, management can see whether variation was caused by special or common causes.

Barrier 4: Cost of Quality. There are two types of figures that are important to decision making, visible figures (for example, break-even figures) and invisible figures (for example, the cost of an unhappy customer). If an organization is managed only on visible figures, decisions will be based on misleading information. This will result in greater variability in decision making, depending upon what visible figures are used.

Point 13: Institute a vigorous program of education and training.

As an organization improves, it will free up resources that can be used to educate and train its employees. These improvements will result in less variability in processes, products, and jobs, continuing the never-ending cycle of improvement.[21]

Management must use only attrition as a means to reduce the number of people in an organization. This will create a body of highly trained employees who are the resources that the organization should invest in through education and training. If management sporadically fires and hires employees depending on market conditions, it will create a body of employees whose performance is extremely variable as a result of a lack of training, a lack of common purpose, and a lack of commitment to the organization.

Point 14: Create a structure that will push the prior 13 points every day.

The statistical leader in an organization must work to reduce variability with respect to management's ability to use statistical thinking in their decision making, and in the application and teaching of statistical thinking and statistical methods in the organization. The statistical leader should strive to reduce variation between managers with respect to their views on the organization's mission and quality.

SUMMARY

This chapter explains the significance of the reduction of variation to each of Deming's 14 points.

The chapter begins with a discussion of statistics and management, and shows that they are both concerned with the never-ending reduction of variation. The discussion is facilitated through the use of an experiment that demonstrates that many types of management actions increase variation, rather than decrease it. Consequently, the experiment shows that management must begin to act based upon knowledge of variation.

The main portion of the chapter explains how each of Deming's 14 points helps an organization to continuously reduce variation in all of its processes.

EXERCISES

17.1 Explain the purpose of the experiment using the funnel.

17.2 Explain rule 1, and give an example of how it is used in an organization.

17.3 Explain rule 2, and give an example of how it is used in an organization.

17.4 Explain rule 2a, and give an example of how it is used in an organization.

17.5 Explain rule 3, and give an example of how it is used in an organization.

17.6 Explain rule 4, and give an example of how it is used in an organization.

NOTES

1. W. E. Deming, *Out of the Crisis* (Cambridge, Mass.: MIT Center for Advanced Engineering Study, 1986), p. 327.
2. Ibid., p. 327.
3. Ibid., pp. 327–32.
4. T. J. Boardman and H. Iyer, *The Funnel* (Fort Collins, Colo.: Colorado State University Press, 1986), p. 1.
5. Ibid., p. 1.
6. Deming, *Out of the Crisis,* see example 5, pp. 359–60.
7. H. Gitlow, K. Kang, and S. Kellogg, "Rule Based Process Control versus Statistical Process Control," *University of Miami Working Paper Series,* 1988.
8. Boardman and Iyer, *The Funnel,* p. 1.
9. Ibid., p. 1.
10. Deming, *Out of the Crisis,* p. 329.
11. Ibid., p. 330.
12. Private conversation with William Latzko, North Bergen, New Jersey.
13. William Scherkenbach, *The Deming Route to Quality and Productivity: Road Maps and Roadblocks* (Washington, D.C.: CeePress Books, 1986), pp. 133–34.
14. Ibid., pp. 133–36.
15. Ibid., p. 82.
16. Ibid., pp. 55–57.
17. Ibid., pp. 57–61.
18. Ibid., pp. 59–60.
19. Ibid., pp. 60–61.
20. Ibid., pp. 62–69.
21. Ibid., p. 126.

Chapter 18

Some Current Thinking about Statistical Studies and Practice

INTRODUCTION

This chapter discusses some current views relating to enumerative and analytic studies, as espoused most recently by W. Edwards Deming[1] and others.[2] We also discuss principles of professional conduct for the statistician.

ENUMERATIVE AND ANALYTIC STUDIES

In Chapter 2 we discussed the distinction between enumerative studies and analytic studies. We saw that the purpose of an enumerative study is to determine the disposition of material in a frame and that a statistical inference can be made only with respect to a frame. Consequently, an enumerative study conducted in Miami provides no basis for a statistical inference in Los Angeles or Spokane.

Also in Chapter 2, we saw that the purpose of an analytic study is to take action on a process to improve its performance in the future. As a statistical inference with a quantifiable degree of belief can be made only with respect to a frame—and it is impossible to get a frame of future events—statistical inferences with a quantifiable degree of belief cannot be made in analytic studies.

Figure 18.1 summarizes some of the important differences between enumerative and analytic studies.

FIGURE 18.1   Important Aspects of Enumerative and Analytic Studies

Aim	*Enumerative*	*Analytic*
Focus	Description of material in frame	Prediction of process output
Methods of access	Frame	Models of the process such as flowcharts and Cause-and-Effect diagrams (no frame)
Major sources of uncertainty	Sampling error	Extrapolation to the future (How well can I predict?)
Is the major source of uncertainty quantifiable?	Yes	No
Environment of the study	Static	Dynamic
Type of sample	Random	Judgment

We see that a frame is necessary in an enumerative study to take action on the disposition of material in the frame, at a given point in time; this is a static situation. A frame makes possible random sampling and quantification of a major source of uncertainty in an enumerative study; that is, sampling error. Consequently, the theory of estimation, including sampling distributions, standard errors, and confidence intervals, is completely applicable.

The major source of uncertainty in an analytic study is prediction. Unfortunately, current statistical theory provides no aid for quantifying the future effects of variables that will affect the process. There is no statistical theory for quantifying this type of uncertainty. Further, in Figure 18.1 we see that there is no frame for an analytic study; hence, there can be no random sampling and quantification of sampling error. The theory of estimation is not applicable. Because this is so, we are faced with a dilemma: how can we make predictions about the future behavior of a dynamic process without a quantifiable degree of belief? The best answer to the above question, given current knowledge, is that we must rely on a combination of expertise concerning the process under study, and statistical thinking and methods.

Errors in Analytic Studies

Two types of errors can occur in any study: a *type one error* occurs when action is taken on a process when it should have been left alone; or a *type two error* occurs when we fail to take action on a process when action is appropriate. It is impossible to calculate the probability of either type of error in an analytic study because it is impossible to know how things would have turned out if an alternative action, other than the one chosen, had been selected. Without the benefit of a quantifiable degree of belief of making type one and type two errors, we need a methodology for combining expertise with statistical notions.

Design of Analytic Studies

It is possible to increase the degree of belief in a prediction from an analytic study by considering: (1) sequential building of knowledge, (2) testing over a wide range of conditions, and (3) selection of units for the study.

Sequential Building of Knowledge.   The Deming philosophy requires that an organizational (team) focus be given to quality improvement efforts. In other words, improvement efforts should be based on an organizational belief that these efforts will decrease the difference between customer needs and process performance. Given the selection of a process for improvement efforts, the process must be described by documenting and defining it as discussed in Chapter 3. The PDCA cycle can then be used to decrease the difference between customer needs and process performance. The experiments performed in the iterations of the PDCA cycle, in combination with theory about the process from subject matter experts, may increase the degree of belief in predictions about the future behavior of the process under study. The degree of belief in predictions about the process is increased as sequential predictions about the process's future behavior come closer to the actual performance of the process.

It is important to note that there can be no significant prediction about the future behavior of a process without knowledge of that process. Further, knowledge of the process is continually improved and increased through successive iterations of the PDCA cycle. It is the predictive ability of knowledge that is the ultimate measure of its value.

Testing Over a Wide Range of Conditions.   The degree of belief in the predictive value of the knowledge gained from an analytic study is increased if the analytic study yields the same results over a wide range of conditions. For example, one of the contributions of Taguchi's off-line quality control methodology is that it facilitates the testing of various product configurations over a wide range of conditions. The conditions enter the experiment in the form of the noise matrix we discussed in Chapter 15. Another example of testing over a wide range of conditions is being able to predict the outcome of using a particular supplier's product—regardless of the batch from which it comes, or the time of year it is delivered, or how it is used.

We can summarize by saying that the degree of belief in the predictive power of the knowledge gained from an analytic study is increased as the same results are obtained over repeated trials under a wide range of conditions. Questions such as how wide a range of conditions is adequate to have a degree of belief sufficient enough to make a prediction, or, how close to actual conditions must the experimental conditions be to have a degree of belief high enough to make predictions are questions only an expert in the subject matter under study can answer. These questions cannot be answered by statistical theory—still, statistical methods can provide guidance for increasing the degree of belief in the knowledge gained from analytic studies.

Selection of the Units for the Study.   As there is no frame in an analytic study, there can be no random sample and quantifiable degree of uncertainty in a prediction about the future. Judgment samples are used to conduct analytic studies.[3] The judgment of a subject matter expert determines both the conditions under which a process will be studied and the measurements that will be taken for each set of conditions. Moreover, it is the expert who judges whether or not the results of an analytic study provide a sufficient degree of belief to take action on a process.

In analytic studies, judgment samples are almost always superior to random samples. For example, consider an analytic study being conducted to determine which of two machines is less sensitive to worker-to-worker variation. We need this information to purchase whichever machine yields more uniform output, regardless of operator. Suppose we have funds to include only 10 of 50 operators in the study. In this case, a random sample of the 10 operators would yield a quantifiable degree of uncertainty in the measurement variation, but the conditions under which the operators were studied will never be seen again; in the future, there will be new operators, new materials, and different training. For this study, a judgment sample is more appropriate. For example, a judgment sample might include the five most experienced operators and the five least experienced operators.

Suppose both machines perform best when used by an experienced operator and worst when used by an inexperienced operator. The degree of belief that the machine with the lower worker-to-worker variation is the machine to purchase will be greater if the variation is estimated from a judgment sample rather than from a sample of 10 randomly selected workers.

Analysis of Data from Analytic Studies

The purpose of an analytic study is to improve knowledge so that predictions can be made about the future behavior of a process. However, Deming states of analytic studies, "Analysis of variance, t-tests, confidence intervals, and other statistical techniques taught in the books, however interesting, are inappropriate because they provide no basis for prediction and because they bury the information contained in the order of production."[4] Thus, most of the techniques that are useful in enumerative studies are inappropriate for analytic studies.

The standard deviation of a stable process can be used to distinguish past common variation from past special variation. However, the calculation of the standard deviation based on the assumption of a stable process does not consider the most important source of uncertainty in an analytic study, prediction of the future behavior of the process.

A Stable Process. When a process is stable, it is easier to determine the effect that changes to the process have on the process's future behavior. Hence, a stable process provides a forum for a subject matter expert to conduct experiments to gain knowledge to predict the future behavior of the process.

Unfortunately, stability in the past does not guarantee stability in the future. The conditions that existed in the past that created the stable process may never be seen again. As a result, responsibility for prediction still rests with the subject matter expert.

Graphical Analysis. Because the distribution of a process that existed in the past cannot be used to predict the distribution of a future process, we need an approach to prediction that does not rely on past distribution statistics.

The general approach for conducting analytic studies relies on graphical techniques, such as control charts, which utilize both statistical knowledge and subject matter knowledge to learn about the process to predict its future behavior. Knowledge of the process should give meaning to control chart patterns and help distinguish common variation from special variation. For example, knowledge of a process might lead researchers to expect that when eight or more points in a row increase in value, tool wear is the likely special source of variation. This prediction about the process's behavior is a combination of subject matter knowledge and statistical knowledge.

Interactions between process variables can dramatically alter process behavior depending upon the process variables that are operational at any given time. For example, one machine might be better when used by an experienced operator, while another machine might be better when used by an inexperienced operator. Graphical techniques will allow a process, and its interactions, to be studied in a dynamic fashion over a wide range of conditions.

PRINCIPLES OF PROFESSIONAL PRACTICE

Every statistician, and every subject matter expert working with a statistician, should be familiar with the areas of responsibility for the statistician and his or her client. The remainder of this chapter is concerned with the following article by W. E. Deming, "Principles of Professional Statistical Practice."[5]

Purpose and Scope

1. The Statistician's Job. The statistician's responsibility, whether as a consultant on a part-time basis, or on regular salary, is to find problems that other people could not be expected to perceive. The statistician's tool is statistical theory (theory of probability). It is the statistician's use of statistical theory that distinguishes him from other experts. Statistical practice is a collaborative venture between statistician and experts in subject matter. Experts who work with a statistician need to understand the principles of professional statistical practice as much as the statistician needs them.

Challenges face statisticians today as never before. The whole world is talking about safety in mechanical and electrical devices (in automobiles, for example); safety in drugs; reliability; due care; pollution; poverty; nutrition; improvement of medical practice; improvement of agricultural practice; improvement in quality of product; breakdown of service; breakdown of equipment; tardy buses, trains and mail; need for greater output in industry and in agriculture; enrichment of jobs.

These problems cannot be understood and cannot even be stated, nor can the effect of any alleged solution be evaluated, without the aid of statistical theory and methods. One cannot even define operationally such adjectives as *reliable, safe, polluted, unemployed, on time* (arrivals), *equal* (in size), *round, random, tired, red, green,* or any other adjective, for use in business or in government, except in statistical terms. To have meaning for business or legal purposes, a standard (as of safety, or of performance or capability) must be defined in statistical terms.

2. Professional Practice. The main contribution of a statistician in any project or in any organization is to find the important problems, the problems that other people cannot be expected to perceive. An example is the 14 points that top management learned in Japan in 1950. The problems that walk in are important, but are not usually *the* important problems [4].

The statistician also has other obligations to the people with whom he works, examples being to help design studies, to construct an audit by which to discover faults in procedure before it is too late to make alterations, and at the end, to evaluate the statistical reliability of the results.

The statistician, by virtue of experience in studies of various kinds, will offer advice on procedures for experiments and surveys, even to forms that are to be filled out, and certainly in the supervision of the study. He knows by experience the dangers of (1) forms that are not clear, (2) procedures that are not clear, (3) contamination, (4) carelessness, and (5) failure of supervision.

Professional practice stems from an expanding body of theory and from principles of application. A professional man aims at recognition and respect for his practice, not for himself alone, but for his colleagues as well. A professional man takes orders, in technical matters, from standards set by his professional colleagues as unseen judges, never from an administrative superior. His usefulness and his profession will suffer impairment if he yields to convenience or to opportunity. A professional man feels an obligation to provide services that his client may never comprehend or appreciate.

A professional statistician will not follow methods that are indefensible, merely to please someone, or support inferences based on such methods. He ranks his reputation and profession as more important than convenient assent to interpretations not warranted by statistical theory. Statisticians can be trusted and respected public servants.

Their careers as expert witnesses will be shattered if they indicate concern over which side of the case the results seem to favor. "As a statistician, I couldn't care less" is the right attitude in a legal case or in any other matter.

Logical Basis for Division of Responsibilities

3. Some Limitations in Application of Statistical Theory. Knowledge of statistical theory is necessary but not sufficient for successful operation. Statistical theory does not provide a road map toward effective use of itself. The purpose of this article is to propose some principles of practice, and to explain their meaning in some of the situations that the statistician encounters.

Statisticians have no magic touch by which they may come in at the stage of tabulation and make something of nothing. Neither will their advice, however wise in the early stages of a study, ensure successful execution and conclusion. Many a study launched on the ways of elegant statistical design, later boggled in execution, ends up with results to which the theory of probability can contribute little.

Even though carried off with reasonable conformance to specifications, a study may fail through structural deficiencies in the method of investigation (questionnaire, type of test, technique of interviewing) to provide the information needed. Statisticians may reduce the risk of this kind of failure by pointing out to their clients in the early stages of the study the nature of the contributions that they themselves must put into it. (The word *client* will denote the expert or group of experts in a substantive field). The limitations of statistical theory serve as signposts to guide a logical division of responsibilities between statistician and client. We accordingly digress for a brief review of the power and the limitations of statistical theory.

We note first that statistical inferences (probabilities, estimates, confidence limits, fiducial limits, etc.), calculated by statistical theory from the results of a study, will relate only to the material, product, people, business establishments, and so on, that the frame was in effect drawn from, and only to the environment of the study, such as the method of investigation, the date, weather, rate of flow and levels of concentration of the components used in

tests of a production process, range of voltage, or of other stress specified for the tests (as of electrical equipment).

Empirical investigation consists of observations on material of some kind. The material may be people; it may be pigs, insects, physical material, industrial product, or records of transactions. The aim may be enumerative, which leads to the theory for the sampling of finite populations. The aim may be analytic, the study of the causes that make the material what it is, and which will produce other material in the future. A definable lot of material may be divisible into identifiable sampling units. A list of these sampling units, with identification of each unit, consistutes a *frame*. In some physical and astronomical investigations, the sampling unit is a unit of time. We need not detour here to describe nests of frames for multistage sampling. The important point is that without a frame there can be neither a complete coverage of a designated lot of material nor a sample of any designated part thereof. Stephan introduced the concept of the frame, but without giving it a name [7].

Objective statistical inferences in respect to the frame are the speciality of the statistician. In contrast, generalization to cover material not included in the frame, nor to ranges, conditions, and methods outside the scope of the experiment, however essential for application of the results of a study, are a matter of judgment and depend on knowledge of the subject matter [5].

For example, the universe in a study of consumer behavior might be all the female homemakers in a certain region. The frame might therefore be census blocks, tracts or other small districts, and the ultimate sampling unit might be a segment of area containing households. The study itself will, of course, reach only the people that can be found at home in the segments selected for the sample. The client, however, must reach generalizations and take action on the product or system of marketing with respect to all female homemakers, whether they be the kind that are almost always at home and easy to reach, or almost never at home and therefore in part omitted from the study. Moreover, the female homemakers on whom the client must take action belong to the future, next year, and the next. The frame only permits study of the past.

For another example, the universe might be the production process that will turn out next week's product. The frame for study might be part of last week's product. The universe might be traffic in future years, as in studies needed as a basis for estimating possible losses or gains as a result of a merger. The frame for this study might be records of last year's shipments.

Statistical theory alone could not decide, in a study of traffic that a railway is making, whether it would be important to show the movement of (for example) potatoes, separately from other agricultural products in the northwestern part of the United States; only he who must use the data can decide.

The frame for a comparison of two medical treatments might be patients or other people in Chicago with a specific ailment. A pathologist might, on his own judgment, without further studies, generalize the results to people that have this ailment anywhere in the world. Statistical theory provides no basis for such generalization.

No knowledge of statistical theory, however profound, provides by itself a basis for deciding whether a proposed frame would be satisfactory for a study. For example, statistical theory would not tell us, in a study of consumer research, whether to include in the frame people that live in trailer parks. The statistician places the responsibility for the answer to this question where it belongs, with the client: Are trailer parks in your problem? Would they be in it if this were a complete census? If yes, would the cost of including them be worthwhile?

Statistical theory will not of itself originate a substantive problem. Statistical theory cannot generate ideas for a questionnaire or for a test of hardness, nor can it specify what would be acceptable colors of dishes or of a carpet; nor does statistical theory originate ways to teach field workers or inspectors how to do their work properly or what characteristics of workmanship constitute a meaningful measure of performance. This is so in spite of the fact that statistical theory is essential for reliable comparisons of questionnaires and tests.

4. Contributions of Statistical Theory to Subject Matter. It is necessary, for statistical reliability of results, that the design of a survey or experiment fit into a theoretical model. Part of the statistician's job is to find a suitable model that he can manage, once the client has presented his case and explained why statistical information is needed and how the results of a study might be used. Statistical practice goes further than merely to try to find a suitable model (theory). Part of the job is to adjust the physical conditions to meet the model selected. Randomness, for example, in sampling a frame, is not just a word to assume; it must be created by use of random numbers for the selection of sampling units. Recalls to find people at home, or tracing illegible or missing information back to source documents, are specified so as to approach the prescribed probability of selection.

Statistical theory has in this way profoundly influenced the theory of knowledge. It has given form and direction to quantitative studies by molding them into the requirements of the theory of estimation, and other techniques of inference. The aim is, of course, to yield results that have meaning in terms that people can understand.

The statistician is often effective in assisting the substantive expert to improve accepted methods of interviewing, testing, coding, and other operations. Tests properly designed will show how alternative test or field procedures really perform in practice, so that rational changes or choices may be possible. It sometimes happens, for example, that a time-honored or committee-honored method of investigation, when put to statistically designed tests, shows alarming inherent variances between instruments, between investigators, or even between days for the same investigator. It may show a trend, or heavy intraclass correlation between units, introduced by the investigators. Once detected, such sources of variation may then be corrected or diminished by new rules, which are of course the responsibility of the substantive expert.

Statistical techniques provide a safe supervisory tool to help to reduce variability in the performance of worker and machine. The effectiveness of statistical controls, in all stages of a survey, for improving supervision, to achieve uniformity, to reduce errors, to gain speed, reduce costs, and to improve quality of performance in other ways is well known. A host of references could be given, but two will suffice [2, 3].

5. Statistical Theory as a Basis for Division of Responsibilities. We may now, with this background, see where the statistician's responsibilities lie.

In the first place, his specialized knowledge of statistical theory enables him to see which parts of a problem belong to substantive knowledge (sociology, transportation, chemistry of a manufacturing process, law), and which parts are statistical. The responsibility as logician in a study falls to him by default, and he must accept it. As logician, he will do well to designate, in the planning stages, which decisions will belong to the statistician and which to the substantive expert.

This matter of defining specifically the areas of responsibility is a unique problem faced by the statistician. Business managers and lawyers who engage an expert on corporate finance, or an expert on steam power plants, know pretty well what such experts are able to do and have an idea about how to work with them. It is different with statisticians. Many people are confused between the role of theoretical statisticians (those who use theory to guide their practice) and the popular idea that statisticians are skillful in compiling tables about people or trade, or who prophesy the economic outlook for the coming year and which way the stock market will drift. Others may know little about the contributions that statisticians can make to a study or how to work with them.

Allocation of responsibilities does not mean impervious compartments in which you do this and I'll do that. It means that there is a logical basis for allocation of responsibilities and that it is necessary for everyone involved in a study to know in advance what he or she will be accountable for.

A clear statement of responsibilities will be a joy to the client's lawyer in a legal case, especially at the time of cross-examination. It will show the kind of question that the statistician is answerable for, and what belongs to the substantive experts. Statisticians have something to learn about professional practice from law and medicine.

6. Assistance to the Client to Understand the Relationship. The statistician must direct the client's thoughts toward possible uses of the results of a statistical study of the entire frame without consideration of whether the entire frame will be studied or only a small sample of the sampling units in the frame. Once these matters are cleared, the statistician may outline one or more statistical plans and explain to the client in his own language what they mean in respect to possible choices of frame and environmental conditions, choices of sampling unit, skills, facilities and supervision required. The statistician will urge the client to foresee possible difficulties with definitions or with the proposed method of investigation. He may illustrate with rough calculations possible levels of precision to be expected from one or more statistical plans that appear to be feasible, along with rudimentary examples of the kinds of tables and inferences that might be forthcoming. These early stages are often the hardest part of the job.

The aim in statistical design is to hold accuracy and precision to sensible levels, with an economic balance between all the possible uncertainties that afflict data—built-in deficiencies of definition, errors of response, nonresponse, sampling variation, difficulties in coding, errors in processing, difficulties of interpretation.

Professional statistical practice requires experience, maturity, fortitude, and patience, to protect the client against himself or against his duly appointed experts in subject matter who may have in mind needless but costly tabulations in fine classes (five-year age groups, small areas, fine mileage brackets, fine gradations in voltage, etc.), or unattainable precision in differences between treatments, beyond the capacity of the skills and facilities available, or beyond the meaning inherent in the definitions.

These first steps, which depend heavily on guidance from the statistician, may lead to important modifications of the problem. Advance considerations of cost, and of the limitations of the inferences that appear to be possible, may even lead clients to abandon a study, at least until they feel most confident of the requirements. Protection of a client's bank account, and deliverance from more serious losses from decisions based on inconclusive or misleading results, or from misinterpretation, is one of the statistician's greatest services.

Joint effort does not imply joint responsibility. Divided responsibility for a decision in a statistical survey is as bad as divided responsibility in any venture—it means that no one is responsible.

Although they acquire superficial knowledge of the subject matter, the one thing that statisticians contribute to a problem, and which distinguishes them from other experts, is knowledge and ability in statistical theory.

Summary Statement of Reciprocal Obligations and Responsibilities

7. Responsibilities of the Client. The client will assume responsibility for those aspects of the problem that are substantive. Specifically, he will stand the ultimate responsibility for:

a. The type of statistical information to be obtained.
b. The methods of test, examination, questionnaire, or interview, by which to elicit the information from any unit selected from the frame.
c. The decision on whether a proposed frame is satisfactory.
d. Approval of the probability model proposed by the statistician (statistical procedures, scope, and limitations of the statistical inferences that may be possible from the results).
e. The decision on the classes and areas of tabulation (as these depend on the uses that the client intends to make of the data); the approximate level of statistical precision or protection that would be desirable in view of the purpose of the investigation, skills and time available, and costs.

The client will make proper arrangements for:

f. The actual work of preparing the frame for sampling, such as serializing and identifying sampling units at the various stages.
g. The selection of the sample according to procedures that the statistician will prescribe, and the preparations of these units for investigation.
h. The actual investigation; the training for this work, and the supervision thereof.
i. The rules for coding; the coding itself.
j. The processing, tabulations and computations, following procedures of estimation that the sampling plans prescribe.

The client or his representative has the obligation to report at once any departure from instructions, to permit the statistician to make a decision between a fresh start or an unbiased adjustment. The client will keep a record of the actual performance.

8. Responsibilities of the Statistician. The statistician owes an obligation to his own practice to forestall disappointment on the part of the client, who if he fails to understand at the start that he must exercise his own responsibilities in the planning stages, and in execution, may not realize in the end the fullest possibility of the investigation, or may discover too late that certain information that he had expected to get out of the study was not built into the design. The statistician's responsibility may be summarized as follows:

a. To formulate the client's problem in statistical terms (probability model), subject to approval of the client, so that a proposed statistical investigation may be helpful to the purpose.

b. To lay out a logical division of responsibilities for the client, and for the statistician, suitable to the investigation proposed.

c. To explain to the client the advantages and disadvantages of various frames and possible choices of sampling units, and of one or more feasible statistical plans of sampling or experimentation that seem to be feasible.

d. To explain to the client, in connection with the frame, that any objective inferences that one may draw by statistical theory from the results of an investigation can only refer to the material or system that the frame was drawn from, and only to the methods, levels, types, and ranges of stress presented for study. It is essential that the client understand the limitations of a proposed study, and of the statistical inferences to be drawn therefrom, so that he may have a chance to modify the content before it is too late.

e. To furnish statistical procedures for the investigation—selection, computation of estimates and standard errors, tests, audits, and controls as seem warranted for detection and evaluation of important possible departures from specifications, variances between investigators, nonresponse, and other persistent uncertainties not contained in the standard error; to follow the work and to know when to step in.

f. To assist the client (on request) with statistical methods of supervision, to help him to improve uniformity of performance of investigators, gain speed, reduce errors, reduce costs, and to produce a better record of just what took place.

g. To prepare a report on the statistical reliability of the results.

9. The Statistician's Report or Testimony. The statistician's report or testimony will deal with the statistical reliability of the results. The usual content will cover the following points:

a. A statement to explain what aspects of the study his responsibility included, and what it excluded. It will delimit the scope of the report.

b. A description of the frame, the sampling unit, how defined and identified, the material covered, and a statement in respect to conditions of the survey or experiment that might throw light on the usefulness of the results.

c. A statement concerning the effect of any gap between the frame and the universe for important conclusions likely to be drawn from the results. (A good rule is that the statistician should have before him a rough draft of the client's proposed conclusions.)

d. Evaluation (in an enumerative study) of the margin of uncertainty, for a specified probability level, attributable to random errors of various kinds, including the uncertainty introduced by sampling, and by small independent random variations in judgment, instruments, coding, transcription, and other processing.

e. Evaluation of the possible effects of other relevant sources of variation, examples being differences between investigators, between instruments, between days, between areas.

f. Effect of persistent drift and conditioning of instruments and of investigators; changes in technique.

g. Nonresponsive and illegible or missing entries.

h. Failure to select sampling units according to the procedure prescribed.

i. Failure to reach and to cover sampling units that were designated in the sampling table.

j. Inclusion of sampling units not designated for the sample but nevertheless covered and included in the results.

k. Any other important slips and departures from the procedure described.

l. Comparisons with other studies, if any are relevant.

In summary, a statement of statistical reliability attempts to present to readers all information that might help them to form their own opinions concerning the validity of conclusions likely to be drawn from the results.

The aim of a statistical report is to protect clients from seeing merely what they would like to see; to protect them from losses that could come from misuse of the results. A further aim is to forestall unwarranted claims of accuracy that a client's public might otherwise accept.

Any printed description of the statistical procedures that refer to this participation, or any evaluation of the statistical reliability of the results, must be prepared by the statistician as part of the engagement. If a client prints the statistician's report, it will be printed in full.

The statistician has an obligation to institute audits and controls as a part of the statistical procedures for a survey or experiment, to discover any departure from the procedures prescribed.

The statistician's full disclosure and discussion of all the blemishes and blunders that took place in the survey or experiment will build for him a reputation of trust. In a legal case, this disclosure of blunders and blemishes and their possible effects on conclusions to be drawn from the data are a joy to the lawyer.

A statistician does not recommend to a client any specific administrative action or policy. Use of the results that come from a survey or experiment are entirely up to the client. Statisticians, if they were to make recommendations for decision, would cease to be statisticians.

Actually, ways in which the results may throw light on foreseeable problems will be settled in advance, in the design, and there should be little need for a client or for anyone else to reopen a question. However, problems sometimes change character with time (as when a competitor of the client suddenly comes out with a new model), and open up considerations of statistical precision and significance of estimates that were not initially in view.

The statistician may describe in a professional or scientific meeting the statistical methods that he develops in an engagement. He will not publish actual data or substantive results or other information about a client's business without permission. In other words, the statistical methods belong to the statistician: the data to the client.

A statistician may at times perform a useful function by examining and reporting on a study in which he did not participate. A professional statistician will not write an opinion on another's procedures or inferences without adequate time for study and evaluation.

Supplemental Remarks

10. Necessity for the Statistician to Keep in Touch. A statistician, when he enters into a relationship to participate in a study, accepts certain responsibilities. He does not merely draft instructions and wait to be called. The people whom he serves are not statisticians, and thus cannot always know when they are in trouble. A statistician asks questions and will probe on his own account with the help of proper statistical design, to discover for himself whether the work is proceeding according to the intent of the instructions. He must expect to revise the instructions a number of times in the early stages. He will be accessible by mail, telephone, telegraph, or in person, to answer questions and to listen to suggestions.

He may, of course, arrange consultations with other statisticians on questions of theory or procedure. He may engage another statistician to take over certain duties. He may employ other skills at suitable levels to carry out independent tests or reinvestigation of certain units, to detect difficulties and departures from the prescribed procedure, or errors in transcription or calculation.

It must be firmly understood, however, that consultation or assistance is in no sense a partitioning of responsibility. The obligations of a statistician to his client, once entered into, may not be shared.

11. What Is an Engagement? Dangers of Informal Advice. It may seem at first thought that statisticians ought to be willing to give the world informally and impromptu the benefit of their knowledge and experience, without discussion or agreement concerning participation and relationships. Anyone who has received aid from a doctor of medicine who did his best without a chance to make a more thorough examination can appreciate how important the skills of a professional man can be, even under handicap.

On second thought, most statisticians can recall instances in which informal advice backfired. It is the same in any professional line. A statistician who tries to be helpful and give advice under adverse circumstances is in practice and has a client, whether he intended it so or not; and he will later find himself accountable for the advice. It is important to take special precaution under these circumstances to state the basis of understanding for any statements or recommendations, and to make clear that other conditions and circumstances could well lead to different statements.

12. When Do the Statistician's Obligations Come to a Close? A statistician should make it clear that his name may be identified with a study only as long as he is active in it and accountable for it. A statistical procedure, contrary to popular parlance, is not installed. One may install new furniture, a new carpet, or a new dean, but not a statistical procedure. Experience shows that a statistical procedure deteriorates rapidly when left completely to nonprofessional administration.

A statistician may draw up plans for a continuing study, such as for the annual inventory of materials in process in a group of manufacturing plants, or for a continuing national survey of consumers. He may nurse the job into running order, and conduct it through several performances. Experience shows, however, that if he steps out and leaves the work in nonstatistical hands, he will shortly find it to be unrecognizable. New people come on the job. They may think that they know better than their predecessor how to do the work; or they may not be aware that there ever were any rules or instructions, and make up their own.

What is even worse, perhaps, is that people that have been on a job a number of years think that they know it so well that they cannot go wrong. This type of fault will be observed, for example, in a national monthly sample in which households are to be revisited a number of times; when left entirely to non-statistical administration, it will develop faults. Some interviewers will put down their best guesses about the family, on the basis of the preceding month, without an actual interview. They will forget the exact wording of the question, or may think that they have something better. They will become lax about calling back on people not at home. Some of them may suppose that they are following literally the intent of the instructions, when in fact (as shown by a control), through a misunderstanding, they are doing something wrong. Or they may depart wilfully, thinking that they are thereby improving the design, on the supposition that the statistician did not really understand the circumstances. A common example is to substitute an average-looking sampling unit when the sampling unit designated is obviously unusual in some respect [1].

In the weighing and testing of physical product, people will in all sincerity substitute their judgment for the use of random numbers. Administration at the top will fail to rotate areas in the manner specified. Such deterioration may be predicted with confidence unless the statistician specifies statistical controls that provide detective devices and feedback.

13. The Single Consultation. It is wise to avoid a single consultation with a commercial concern unless satisfactory agenda are prepared in advance and satisfactory arrangements made for absorbing advice offered. This requirement, along with an understanding that there will be a fee for a consultation, acts as a shield against a hapless conference which somebody calls in the hope that something may turn up. It also shows that statisticians, as professional people, although eager to teach and explain statistical methods, are not on the lookout for chances to demonstrate what they might be able to accomplish.

Moreover, what may be intended as a single consultation often ends up with a request for a memorandum. This may be very difficult to write, especially in the absence of adequate time to study the problem. The precautions of informal advice apply here.

14. The Statistician's Obligation in a Muddle. Suppose that fieldwork is under way. Then the statistician discovers that the client or duly appointed representatives have disregarded the instructions for the preparation of the frame, or for the selection of the sample, or that the fieldwork seems to be falling apart.

"We went ahead and did so and so before your letter came, confident that you would approve," is a violation of relationship. That the statistician may approve the deed done does not mitigate the violation.

If it appears to the statistician that there is no chance that the study will yield results that he could take professional responsibility for, this fact must be made clear to the client, at a sufficiently high management level. It is a good idea for the statistician to explain at the outset that such situations, while extreme, have been known.

Statisticians should do all in their power to help clients to avoid such a catastrophe. There is always the possibility that a statistician may be partly to blame for not being sufficiently clear nor firm at the outset concerning his principles of participation, or for not being on hand at the right time to ask questions and to keep apprised of what is happening. Unfortunate circumstances may teach the statistician a lesson on participation.

15. Assistance in Interpretation of Nonprobability Samples. It may be humiliating, but statisticians must face the fact that many accepted laws of science have come from theory and experimentation without benefit of formal statistical design. Vaccination for prevention of smallpox is one; John Snow's discovery of the source of cholera in London is another [6]. So is the law $F = ma$ in physics; also Hooke's law, Boyle's law, Mendel's findings, Keppler's laws, Darwin's theory of evolution, the Stefan–Boltzmann law of radiation (first empirical, later established by physical theory). All this only means, as everyone knows, that there may well be a wealth of information in a nonprobability sample.

Perhaps the main contribution that the statistician can make in a nonprobability sample is to advise the experimenter against conclusions based on meaningless statistical calculations. The expert in the subject matter must take the responsibility for the effects of selectivity and confounding of treatments. The statistician may make a positive contribution by encouraging the expert in the subject matter to draw whatever conclusions he believes to be warranted, but to do so over his own signature, and not to attribute conclusions to statistical calculations or to the kind of help provided by a statistician.

16. A statistician will not agree to use of his name as advisor to a study, nor as a member of an advisory committee, unless this service carries with it explicit responsibilities for certain prescribed phases of the study.

17. A statistician may accept engagements from competitive firms. His aim is not to concentrate on the welfare of a particular client, but to raise the level of service of his profession.

18. A statistician will prescribe in every engagement whatever methods known to him seem to be most efficient and feasible under the circumstances. Thus, he may prescribe for firms that are competitive methods that are similar or even identical word for word in part or in entirety. Put another way, no client has a proprietary right in any procedures or techniques that a statistician prescribes.

19. A statistician will, to the best of his ability, at his request, lend technical assistance to another statistician. In rendering this assistance, he may provide copies of procedures that he has used, along with whatever modification seems advisable. He will not, in doing this, use confidential data.

References

[1] Bureau of the Census (1954). Measurement of Employment and Unemployment. Report of Special Advisory Committee on Employment Statistics, Washington, D.C.

[2] Bureau of the Census (1964). Evaluation and research program of the censuses of population and housing. 1960 series ER60. No. 1.

[3] Bureau of the Census (1963). The Current Population Survey and Re-interview Program. *Technical Paper No. 6.*

[4] Center for Advanced Engineering Study. (1983). *Management for Quality, Productivity, and Competitive Position.* Massachusetts Institute of Technology, Cambridge, Mass.

[5] Deming, W. E. (1960). *Sample Design in Business Research.* Wiley, New York, Chap. 3.

[6] Hill, A. B. (1953). Observation and experiment. *N. Eng. J. Med.* **248,** 995–1001.

[7] Stephan, F. F. (1936). Practical problems of sampling procedure. *Amer. Sociol. Rev.,* **1,** 569–80.

Bibliography

See the following works, as well as the references just given, for more information on the topic of principles of professional statistical practice.

Brown, T. H. (1952). The statistician and his conscience. *Amer. Statist.,* **6**(1), 14–18.

Burgess, R. W. (1947). Do we need a "Bureau of Standards" for statistics? *J. Marketing,* **11,** 281–282.

Chambers, S. P. (1965). Statistics and intellectual integrity. *J. R. Statist. Soc. A,* **128,** 1–16.

Court, A. T. (1952). Standards of statistical conduct in business and in government. *Amer. Statist.,* **6,** 6–14.

Deming, W. E. (1954). On the presentation of the results of samples as legal evidence. *J. Amer. Statist. Ass.,* **49,** 814–825.

Deming, W. E. (1954). On the Contributions of Standards of Sampling to Legal Evidence and Accounting. Current Business Studies, Society of Business Advisory Professions, Graduate School of Business Administration, New York University, New York.

Eisenhart, C. (1947). The role of a statistical consultant in a research organization. *Proc. Int. Statist. Conf.,* **3,** 309–313.

Freeman, W. W. K. (1952). Discussion of Theodore Brown's paper [see Brown, 1952]. *Amer. Statist.,* **6**(1), 18–20.

Freeman, W. W. K. (1963). Training of statisticians in diplomacy to maintain their integrity. *Amer. Statist.,* **17**(5), 16–20.

Gordon, R. A. et al. (1962). Measuring Employment and Unemployment Statistics. President's Committee to Appraise Employment and Unemployment Statistics. Superintendent of Documents, Washington, D.C.

Hansen, M. H. (1952). Statistical standards and the Census. *Amer. Statist.,* **6**(1), 7–10.

Hotelling, H. et al. (1948). The teaching of statistics (a report of the Institute of Mathematical Statistics Committee on the teaching of statistics). *Ann. Math. Statist.,* **19,** 95–115.

Jensen, A. et al. (1948). The life and works of A. K. Erlang. *Trans. Danish Acad. Tech. Sci.,* No. 2 (Copenhagen).

Molina, E. C. (1913). Computation formula for the probability of an event happening at least *c* times in *n* trials. *Amer. Math. Monthly,* **20,** 190–192.

Molina, E. C. (1925). The theory of probability and some applications to engineering problems. *J. Amer. Inst. Electr. Eng.,* **44,** 1–6.

Morton, J. E. (1952). Standards of statistical conduct in business and government. *Amer. Statist.,* **6**(1), 6–7.

Shewhart, W. A. (1931). *Economic Control of Quality of Manufactured Product.* D. Van Nostrand, New York.

Shewhart, W. A. (1939). *Statistical Method from the Viewpoint of Quality Control.* The Graduate School, Department of Agriculture, Washington, D.C.

Shewhart, W. A. (1958). Nature and origin of standards of quality. *Bell Syst. Tech. J.,* **37,** 1–2.

Zirkle, C. (1954). Citation of fraudulent data. *Science,* **120,** 189–190.

SUMMARY

This chapter presented some current thoughts on statistical studies and principles for statistical practice. These thoughts were related to the quantification of uncertainty in enumerative and analytic studies, and, in particular, to the impossibility of using enumerative methods to calculate the probabilities of the two types of errors that can occur in analytic studies. The discussion of the design of an analytic study included the importance of sequential building of knowledge to the development of an ability to predict the future with a comfortable degree of belief, the importance of testing a plan of action over a wide range of conditions to improve the degree of belief in a prediction based on the plan, and the importance of a judgment sample to increasing the degree of belief in a prediction from an analytic study. Some issues concerning the analysis of data from analytic studies include the importance of a stable process in gaining new knowledge about a process and the significance of graphical techniques for gaining knowledge from a process.

Finally, principles of professional practice were discussed and presented. These principles include the statistician's job, a description of what constitutes professional practice for a statistician, some limitations in application of statistical theory, a discussion of the contributions of statistical theory to subject matter, using statistical theory as a basis for the division of responsibilities between a statistician and client, the statistician's responsibility to assist the client in understanding the results of a statistical study, the responsibilities of the client and statistician, and the statistician's report or testimony.

NOTES

1. This material was presented at a seminar entitled, "Deming Seminar for Statisticians," March 28–30, 1988, New York University, New York.
2. Much of this chapter is paraphrased from Thomas W. Nolan, "Analytic Studies," a working paper from Associates in Process Management, 1988.
3. W. E. Deming, "On Probability as a Basis for Action," *The American Statistician* 29(4), November 1975, pp. 146–52.
4. W. E. Deming, *Out of the Crisis* (Cambridge, Mass.: MIT Center for Advanced Engineering Study, 1986), p. 132.
5. W. E. Deming, "Principles of Professional Statistical Practice," Reprinted with permission from Kotz-Johnson: *Encyclopedia of Statistical Science,* 7 (New York: John Wiley & Sons, 1986), pp. 184–93.

Statistical Tables

TABLE 1   Normal Curve Probabilities

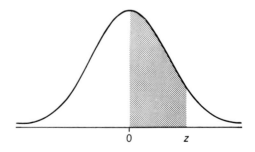

z	.00	.01	.02	.03	.04	.05	.06	.07	.08	.09
0.0	.0000	.0040	.0080	.0120	.0160	.0199	.0239	.0279	.0319	.0359
0.1	.0398	.0438	.0478	.0517	.0557	.0596	.0636	.0675	.0714	.0753
0.2	.0793	.0832	.0871	.0910	.0948	.0987	.1026	.1064	.1103	.1141
0.3	.1179	.1217	.1255	.1293	.1331	.1368	.1406	.1443	.1480	.1517
0.4	.1554	.1591	.1628	.1664	.1700	.1736	.1772	.1808	.1844	.1879
0.5	.1915	.1950	.1985	.2019	.2054	.2088	.2123	.2157	.2190	.2224
0.6	.2257	.2291	.2324	.2357	.2389	.2422	.2454	.2486	.2517	.2549
0.7	.2580	.2611	.2642	.2673	.2704	.2734	.2764	.2794	.2823	.2852
0.8	.2881	.2910	.2939	.2967	.2995	.3023	.3051	.3078	.3106	.3133
0.9	.3159	.3186	.3212	.3238	.3264	.3289	.3315	.3340	.3365	.3389
1.0	.3413	.3438	.3461	.3485	.3508	.3531	.3554	.3577	.3599	.3621
1.1	.3643	.3665	.3686	.3708	.3729	.3749	.3770	.3790	.3810	.3830
1.2	.3849	.3869	.3888	.3907	.3925	.3944	.3962	.3980	.3997	.4015
1.3	.4032	.4049	.4066	.4082	.4099	.4115	.4131	.4147	.4162	.4177
1.4	.4192	.4207	.4222	.4236	.4251	.4265	.4279	.4292	.4306	.4319
1.5	.4332	.4345	.4357	.4370	.4382	.4394	.4406	.4418	.4429	.4441
1.6	.4452	.4463	.4474	.4484	.4495	.4505	.4515	.4525	.4535	.4545
1.7	.4554	.4564	.4573	.4582	.4591	.4599	.4608	.4616	.4625	.4633
1.8	.4641	.4649	.4656	.4664	.4671	.4678	.4686	.4693	.4699	.4706
1.9	.4713	.4719	.4726	.4732	.4738	.4744	.4750	.4756	.4761	.4767
2.0	.4772	.4778	.4783	.4788	.4793	.4798	.4803	.4808	.4812	.4817
2.1	.4821	.4826	.4830	.4834	.4838	.4842	.4846	.4850	.4854	.4857
2.2	.4861	.4864	.4868	.4871	.4875	.4878	.4881	.4884	.4887	.4890
2.3	.4893	.4896	.4898	.4901	.4904	.4906	.4909	.4911	.4913	.4916
2.4	.4918	.4920	.4922	.4925	.4927	.4929	.4931	.4932	.4934	.4936
2.5	.4938	.4940	.4941	.4943	.4945	.4946	.4948	.4949	.4951	.4952
2.6	.4953	.4955	.4956	.4957	.4959	.4960	.4961	.4962	.4963	.4964
2.7	.4965	.4966	.4967	.4968	.4969	.4970	.4971	.4972	.4973	.4974
2.8	.4974	.4975	.4976	.4977	.4977	.4978	.4979	.4979	.4980	.4981
2.9	.4981	.4982	.4982	.4983	.4984	.4984	.4985	.4985	.4986	.4986
3.0	.4987	.4987	.4987	.4988	.4988	.4989	.4989	.4989	.4990	.4990

TABLE 2   Critical Values of the t-Distribution

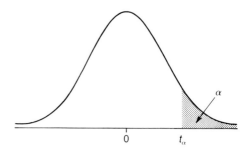

Degrees of Freedom	$t_{.4}$	$t_{.25}$	$t_{.1}$	$t_{.05}$	$t_{.025}$	$t_{.01}$	$t_{.005}$	$t_{.0025}$	$t_{.001}$	$t_{.0005}$
1	0.325	1.000	3.078	6.314	12.706	31.821	63.657	127.32	318.31	636.62
2	.289	.816	1.886	2.920	4.303	6.965	9.925	14.089	22.327	31.598
3	.277	.765	1.638	2.353	3.182	4.541	5.841	7.453	10.214	12.924
4	.271	.741	1.533	2.132	2.776	3.747	4.604	5.598	7.173	8.610
5	0.267	0.727	1.476	2.015	2.571	3.365	4.032	4.773	5.893	6.869
6	.265	.718	1.440	1.943	2.447	3.143	3.707	4.317	5.208	5.959
7	.263	.711	1.415	1.895	2.365	2.998	3.499	4.029	4.785	5.408
8	.262	.706	1.397	1.860	2.306	2.896	3.355	3.833	4.501	5.041
9	.261	.703	1.383	1.833	2.262	2.821	3.250	3.690	4.297	4.781
10	0.260	0.700	1.372	1.812	2.228	2.764	3.169	3.581	4.144	4.587
11	.260	.697	1.363	1.796	2.201	2.718	3.106	3.497	4.025	4.437
12	.259	.695	1.356	1.782	2.179	2.681	3.055	3.428	3.930	4.318
13	.259	.694	1.350	1.771	2.160	2.650	3.012	3.372	3.852	4.221
14	.258	.692	1.345	1.761	2.145	2.624	2.977	3.326	3.787	4.140
15	0.258	0.691	1.341	1.753	2.131	2.602	2.947	3.286	3.733	4.073
16	.258	.690	1.337	1.746	2.120	2.583	2.921	3.252	3.686	4.015
17	.257	.689	1.333	1.740	2.110	2.567	2.898	3.222	3.646	3.965
18	.257	.688	1.330	1.734	2.101	2.552	2.878	3.197	3.610	3.922
19	.257	.688	1.328	1.729	2.093	2.539	2.861	3.174	3.579	3.883
20	0.257	0.687	1.325	1.725	2.086	2.528	2.845	3.153	3.552	3.850
21	.257	.686	1.323	1.721	2.080	2.518	2.831	3.135	3.527	3.819
22	.256	.686	1.321	1.717	2.074	2.508	2.819	3.119	3.505	3.792
23	.256	.685	1.319	1.714	2.069	2.500	2.807	3.104	3.485	3.767
24	.256	.685	1.318	1.711	2.064	2.492	2.797	3.091	3.467	3.745
25	0.256	0.684	1.316	1.708	2.060	2.485	2.787	3.078	3.450	3.725
26	.256	.684	1.315	1.706	2.056	2.479	2.779	3.067	3.435	3.707
27	.256	.684	1.314	1.703	2.052	2.473	2.771	3.057	3.421	3.690
28	.256	.683	1.313	1.701	2.048	2.467	2.763	3.047	3.408	3.674
29	.256	.683	1.311	1.699	2.045	2.462	2.756	3.038	3.396	3.659
30	0.256	0.683	1.310	1.697	2.042	2.457	2.750	3.030	3.385	3.646
40	.255	.681	1.303	1.684	2.021	2.423	2.704	2.971	3.307	3.551
60	.254	.679	1.296	1.671	2.000	2.390	2.660	2.915	3.232	3.460
120	.254	.677	1.289	1.658	1.980	2.358	2.617	2.860	3.160	3.373
∞	.253	.674	1.282	1.645	1.960	2.326	2.576	2.807	3.090	3.291

SOURCE: *Biometrika Tables for Statisticians*, Vol. I, Table 12 (Cambridge, U.K.: Cambridge University Press, 1970), by permission of the *Biometrika* trustees.

TABLE 3 Critical Values of the Chi Square Distribution

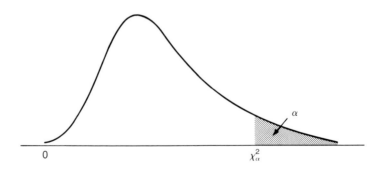

Degrees of Freedom	$\chi^2_{.995}$	$\chi^2_{.99}$	$\chi^2_{.975}$	$\chi^2_{.95}$	$\chi^2_{.05}$	$\chi^2_{.025}$	$\chi^2_{.01}$	$\chi^2_{.005}$
1	.0000393	.000157	.000982	.00393	3.841	5.024	6.635	7.879
2	.0100	.0201	.0506	.103	5.991	7.378	9.210	10.597
3	.0717	.115	.216	.352	7.815	9.348	11.345	12.838
4	.207	.297	.484	.711	9.488	11.143	13.277	14.860
5	.412	.554	.831	1.145	11.070	12.832	15.086	16.750
6	.676	.872	1.237	1.635	12.592	14.449	16.812	18.548
7	.989	1.239	1.690	2.167	14.067	16.013	18.475	20.278
8	1.344	1.646	2.180	2.733	15.507	17.535	20.090	21.955
9	1.735	2.088	2.700	3.325	16.919	19.023	21.666	23.589
10	2.156	2.558	3.247	3.940	18.307	20.483	23.209	25.188
11	2.603	3.053	3.816	4.575	19.675	21.920	24.725	26.757
12	3.074	3.571	4.404	5.226	21.026	23.337	26.217	28.300
13	3.565	4.107	5.009	5.892	22.362	24.736	27.688	29.819
14	4.075	4.660	5.629	6.571	23.685	26.119	29.141	31.319
15	4.601	5.229	6.262	7.261	24.996	27.488	30.578	32.801
16	5.142	5.812	6.908	7.962	26.296	28.845	32.000	34.267
17	5.697	6.408	7.564	8.672	27.587	30.191	33.409	35.718
18	6.265	7.015	8.231	9.390	28.869	31.526	34.805	37.156
19	6.844	7.633	8.907	10.117	30.144	32.852	36.191	38.582
20	7.434	8.260	9.591	10.851	31.410	34.170	37.566	39.997
21	8.034	8.897	10.283	11.591	32.671	35.479	38.932	41.401
22	8.643	9.542	10.982	12.338	33.924	36.781	40.289	42.796
23	9.260	10.196	11.689	13.091	35.172	38.076	41.638	44.181
24	9.886	10.856	12.401	13.848	36.415	39.364	42.980	45.558
25	10.520	11.524	13.120	14.611	37.652	40.646	44.314	46.928
26	11.160	12.198	13.844	15.379	38.885	41.923	45.642	48.290
27	11.808	12.879	14.573	16.151	40.113	43.194	46.963	49.645
28	12.461	13.565	15.308	16.928	41.337	44.461	48.278	50.993
29	13.121	14.256	16.047	17.708	42.557	45.722	49.588	52.336
30	13.787	14.953	16.791	18.493	43.773	46.979	50.892	53.672

TABLE 4 Binomial Coefficients

n	$\binom{n}{0}$	$\binom{n}{1}$	$\binom{n}{2}$	$\binom{n}{3}$	$\binom{n}{4}$	$\binom{n}{5}$	$\binom{n}{6}$	$\binom{n}{7}$	$\binom{n}{8}$	$\binom{n}{9}$	$\binom{n}{10}$
0	1										
1	1	1									
2	1	2	1								
3	1	3	3	1							
4	1	4	6	4	1						
5	1	5	10	10	5	1					
6	1	6	15	20	15	6	1				
7	1	7	21	35	35	21	7	1			
8	1	8	28	56	70	56	28	8	1		
9	1	9	36	84	126	126	84	36	9	1	
10	1	10	45	120	210	252	210	120	45	10	1
11	1	11	55	165	330	462	462	330	165	55	11
12	1	12	66	220	495	792	924	792	495	220	66
13	1	13	78	286	715	1287	1716	1716	1287	715	286
14	1	14	91	364	1001	2002	3003	3432	3003	2002	1001
15	1	15	105	455	1365	3003	5005	6435	6435	5005	3003
16	1	16	120	560	1820	4368	8008	11440	12870	11440	8008
17	1	17	136	680	2380	6188	12376	19448	24310	24310	19448
18	1	18	153	816	3060	8568	18564	31824	43758	48620	43758
19	1	19	171	969	3876	11628	27132	50388	75582	92378	92378
20	1	20	190	1140	4845	15504	38760	77520	125970	167960	184756

TABLE 5   Binomial Probabilities

						p					
n	x	.05	.10	.15	.20	.25	.30	.35	.40	.45	.50
1	0	.9500	.9000	.8500	.8000	.7500	.7000	.6500	.6000	.5500	.5000
	1	.0500	.1000	.1500	.2000	.2500	.3000	.3500	.4000	.4500	.5000
2	0	.9025	.8100	.7225	.6400	.5625	.4900	.4225	.3600	.3025	.2500
	1	.0950	.1800	.2550	.3200	.3750	.4200	.4550	.4800	.4950	.5000
	2	.0025	.0100	.0225	.0400	.0625	.0900	.1225	.1600	.2025	.2500
3	0	.8574	.7290	.6141	.5120	.4219	.3430	.2746	.2160	.1664	.1250
	1	.1354	.2430	.3251	.3840	.4219	.4410	.4436	.4320	.4084	.3750
	2	.0071	.0270	.0574	.0960	.1406	.1890	.2389	.2880	.3341	.3750
	3	.0001	.0010	.0034	.0080	.0156	.0270	.0429	.0640	.0911	.1250
4	0	.8145	.6561	.5220	.4096	.3164	.2401	.1785	.1296	.0915	.0625
	1	.1715	.2916	.3685	.4096	.4219	.4116	.3845	.3456	.2995	.2500
	2	.0135	.0486	.0975	.1536	.2109	.2646	.3105	.3456	.3675	.3750
	3	.0005	.0036	.0115	.0256	.0469	.0756	.1115	.1536	.2005	.2500
	4	.0000	.0001	.0005	.0016	.0039	.0081	.0150	.0256	.0410	.0625
5	0	.7738	.5905	.4437	.3277	.2373	.1681	.1160	.0778	.0503	.0312
	1	.2036	.3280	.3915	.4096	.3955	.3602	.3124	.2592	.2059	.1562
	2	.0214	.0729	.1382	.2048	.2637	.3087	.3364	.3456	.3369	.3125
	3	.0011	.0081	.0244	.0512	.0879	.1323	.1811	.2304	.2757	.3125
	4	.0000	.0004	.0022	.0064	.0146	.0284	.0488	.0768	.1128	.1562
	5	.0000	.0000	.0001	.0003	.0010	.0024	.0053	.0102	.0185	.0312
6	0	.7351	.5314	.3771	.2621	.1780	.1176	.0754	.0467	.0277	.0156
	1	.2321	.3543	.3993	.3932	.3560	.3025	.2437	.1866	.1359	.0938
	2	.0305	.0984	.1762	.2458	.2966	.3241	.3280	.3110	.2780	.2344
	3	.0021	.0146	.0415	.0819	.1318	.1852	.2355	.2765	.3032	.3125
	4	.0001	.0012	.0055	.0154	.0330	.0595	.0951	.1382	.1861	.2344
	5	.0000	.0001	.0004	.0015	.0044	.0102	.0205	.0369	.0609	.0938
	6	.0000	.0000	.0000	.0001	.0002	.0007	.0018	.0041	.0083	.0156
7	0	.6983	.4783	.3206	.2097	.1335	.0824	.0490	.0280	.0152	.0078
	1	.2573	.3720	.3960	.3670	.3115	.2471	.1848	.1306	.0872	.0547
	2	.0406	.1240	.2097	.2753	.3115	.3177	.2985	.2613	.2140	.1641
	3	.0036	.0230	.0617	.1147	.1730	.2269	.2679	.2903	.2918	.2734
	4	.0002	.0026	.0109	.0287	.0577	.0972	.1442	.1935	.2388	.2734
	5	.0000	.0002	.0012	.0043	.0115	.0250	.0466	.0774	.1172	.1641
	6	.0000	.0000	.0001	.0004	.0013	.0036	.0084	.0172	.0320	.0547
	7	.0000	.0000	.0000	.0000	.0001	.0002	.0006	.0016	.0037	.0078
8	0	.6634	.4305	.2725	.1678	.1001	.0576	.0319	.0168	.0084	.0039
	1	.2793	.3826	.3847	.3355	.2670	.1977	.1373	.0896	.0548	.0312
	2	.0515	.1488	.2376	.2936	.3115	.2965	.2587	.2090	.1569	.1094
	3	.0054	.0331	.0839	.1468	.2076	.2541	.2786	.2787	.2568	.2188
	4	.0004	.0046	.0185	.0459	.0865	.1361	.1875	.2322	.2627	.2734
	5	.0000	.0004	.0026	.0092	.0231	.0467	.0808	.1239	.1719	.2188
	6	.0000	.0000	.0002	.0011	.0038	.0100	.0217	.0413	.0703	.1094
	7	.0000	.0000	.0000	.0001	.0004	.0012	.0033	.0079	.0164	.0312
	8	.0000	.0000	.0000	.0000	.0000	.0001	.0002	.0007	.0017	.0039
9	0	.6302	.3874	.2316	.1342	.0751	.0404	.0207	.0101	.0046	.0020
	1	.2985	.3874	.3679	.3020	.2253	.1556	.1004	.0605	.0339	.0176
	2	.0629	.1722	.2597	.3020	.3003	.2668	.2162	.1612	.1110	.0703
	3	.0077	.0446	.1069	.1762	.2336	.2668	.2716	.2508	.2119	.1641
	4	.0006	.0074	.0283	.0661	.1168	.1715	.2194	.2508	.2600	.2461
	5	.0000	.0008	.0050	.0165	.0389	.0735	.1181	.1672	.2128	.2461
	6	.0000	.0001	.0006	.0028	.0087	.0210	.0424	.0743	.1160	.1641
	7	.0000	.0000	.0000	.0003	.0012	.0039	.0098	.0212	.0407	.0703
	8	.0000	.0000	.0000	.0000	.0001	.0004	.0013	.0035	.0083	.0176
	9	.0000	.0000	.0000	.0000	.0000	.0000	.0001	.0003	.0008	.0020

TABLE 5 Binomial Probabilities *(continued)*

n	x	.05	.10	.15	.20	.25	.30	.35	.40	.45	.50
10	0	.5987	.3487	.1969	.1074	.0563	.0282	.0135	.0060	.0025	.0010
	1	.3151	.3874	.3474	.2684	.1877	.1211	.0725	.0403	.0207	.0098
	2	.0746	.1937	.2759	.3020	.2816	.2335	.1757	.1209	.0763	.0439
	3	.0105	.0574	.1298	.2013	.2503	.2668	.2522	.2150	.1665	.1172
	4	.0010	.0112	.0401	.0881	.1460	.2001	.2377	.2508	.2384	.2051
	5	.0001	.0015	.0085	.0264	.0584	.1029	.1536	.2007	.2340	.2461
	6	.0000	.0001	.0012	.0055	.0162	.0368	.0689	.1115	.1596	.2051
	7	.0000	.0000	.0001	.0008	.0031	.0090	.0212	.0425	.0746	.1172
	8	.0000	.0000	.0000	.0001	.0004	.0014	.0043	.0106	.0229	.0439
	9	.0000	.0000	.0000	.0000	.0000	.0001	.0005	.0016	.0042	.0098
	10	.0000	.0000	.0000	.0000	.0000	.0000	.0000	.0001	.0003	.0010
11	0	.5688	.3138	.1673	.0859	.0422	.0198	.0088	.0036	.0014	.0005
	1	.3293	.3835	.3248	.2362	.1549	.0932	.0518	.0266	.0125	.0054
	2	.0867	.2131	.2866	.2953	.2581	.1998	.1395	.0887	.0513	.0269
	3	.0137	.0710	.1517	.2215	.2581	.2568	.2254	.1174	.1259	.0806
	4	.0014	.0158	.0536	.1107	.1721	.2201	.2428	.2365	.2060	.1611
	5	.0001	.0025	.0132	.0388	.0803	.1321	.1830	.2207	.2360	.2256
	6	.0000	.0003	.0023	.0097	.0268	.0566	.0985	.1471	.1931	.2256
	7	.0000	.0000	.0003	.0017	.0064	.0173	.0379	.0701	.1128	.1611
	8	.0000	.0000	.0000	.0002	.0011	.0037	.0102	.0234	.0462	.0806
	9	.0000	.0000	.0000	.0000	.0001	.0005	.0018	.0052	.0126	.0269
	10	.0000	.0000	.0000	.0000	.0000	.0000	.0002	.0007	.0021	.0054
	11	.0000	.0000	.0000	.0000	.0000	.0000	.0000	.0000	.0002	.0005
12	0	.5404	.2824	.1422	.0687	.0317	.0138	.0057	.0022	.0008	.0002
	1	.3413	.3766	.3012	.2062	.1267	.0712	.0368	.0174	.0075	.0029
	2	.0988	.2301	.2924	.2835	.2323	.1678	.1088	.0639	.0339	.0161
	3	.0173	.0853	.1720	.2362	.2581	.2397	.1954	.1419	.0923	.0537
	4	.0021	.0213	.0683	.1329	.1936	.2311	.2367	.2128	.1700	.1208
	5	.0002	.0038	.0193	.0532	.1032	.1585	.2039	.2270	.2225	.1934
	6	.0000	.0005	.0040	.0155	.0401	.0792	.1281	.1766	.2124	.2256
	7	.0000	.0000	.0006	.0033	.0115	.0291	.0591	.1009	.1489	.1934
	8	.0000	.0000	.0001	.0005	.0024	.0078	.0199	.0420	.0762	.1208
	9	.0000	.0000	.0000	.0001	.0004	.0015	.0048	.0125	.0277	.0537
	10	.0000	.0000	.0000	.0000	.0000	.0002	.0008	.0025	.0068	.0161
	11	.0000	.0000	.0000	.0000	.0000	.0000	.0001	.0003	.0010	.0029
	12	.0000	.0000	.0000	.0000	.0000	.0000	.0000	.0000	.0001	.0002
13	0	.5133	.2542	.1209	.0550	.0238	.0097	.0037	.0013	.0004	.0001
	1	.3512	.3672	.2774	.1787	.1029	.0540	.0259	.0113	.0045	.0016
	2	.1109	.2448	.2937	.2680	.2059	.1388	.0836	.0453	.0220	.0095
	3	.0214	.0997	.1900	.2457	.2517	.2181	.1651	.1107	.0660	.0349
	4	.0028	.0277	.0838	.1535	.2097	.2337	.2222	.1845	.1350	.0873
	5	.0003	.0055	.0266	.0691	.1258	.1803	.2154	.2214	.1989	.1571
	6	.0000	.0008	.0063	.0230	.0559	.1030	.1546	.1968	.2169	.2095
	7	.0000	.0001	.0011	.0058	.0186	.0442	.0833	.1312	.1775	.2095
	8	.0000	.0000	.0001	.0011	.0047	.0142	.0336	.0656	.1089	.1571
	9	.0000	.0000	.0000	.0001	.0009	.0034	.0101	.0243	.0495	.0873
	10	.0000	.0000	.0000	.0000	.0001	.0006	.0022	.0065	.0162	.0349
	11	.0000	.0000	.0000	.0000	.0000	.0001	.0003	.0012	.0036	.0095
	12	.0000	.0000	.0000	.0000	.0000	.0000	.0000	.0001	.0005	.0016
	13	.0000	.0000	.0000	.0000	.0000	.0000	.0000	.0000	.0000	.0001
14	0	.4877	.2288	.1028	.0440	.0178	.0068	.0024	.0008	.0002	.0001
	1	.3593	.3559	.2539	.1539	.0832	.0407	.0181	.0073	.0027	.0009
	2	.1229	.2570	.2912	.2501	.1802	.1134	.0634	.0317	.0141	.0056
	3	.0259	.1142	.2056	.2501	.2402	.1943	.1366	.0845	.0462	.0222
	4	.0037	.0349	.0998	.1720	.2202	.2290	.2022	.1549	.1040	.0611

TABLE 5   Binomial Probabilities *(continued)*

n	x	.05	.10	.15	.20	.25	.30	.35	.40	.45	.50
	5	.0004	.0078	.0352	.0860	.1468	.1963	.2178	.2066	.1701	.1222
	6	.0000	.0013	.0093	.0322	.0734	.1262	.1759	.2066	.2088	.1833
	7	.0000	.0002	.0019	.0092	.0280	.0618	.1082	.1574	.1952	.2095
	8	.0000	.0000	.0003	.0020	.0082	.0232	.0510	.0918	.1398	.1833
	9	.0000	.0000	.0000	.0003	.0018	.0066	.0183	.0408	.0762	.1222
	10	.0000	.0000	.0000	.0000	.0003	.0014	.0049	.0136	.0312	.0611
	11	.0000	.0000	.0000	.0000	.0000	.0002	.0010	.0033	.0093	.0222
	12	.0000	.0000	.0000	.0000	.0000	.0000	.0001	.0005	.0019	.0056
	13	.0000	.0000	.0000	.0000	.0000	.0000	.0000	.0001	.0002	.0009
	14	.0000	.0000	.0000	.0000	.0000	.0000	.0000	.0000	.0000	.0001
15	0	.4633	.2059	.0874	.0352	.0134	.0047	.0016	.0005	.0001	.0000
	1	.3658	.3432	.2312	.1319	.0668	.0305	.0126	.0047	.0016	.0005
	2	.1348	.2669	.2856	.2309	.1559	.0916	.0476	.0219	.0090	.0032
	3	.0307	.1285	.2184	.2501	.2252	.1700	.1110	.0634	.0318	.0139
	4	.0049	.0428	.1156	.1876	.2252	.2186	.1792	.1268	.0780	.0417
	5	.0006	.0105	.0449	.1032	.1651	.2061	.2123	.1859	.1404	.0916
	6	.0000	.0019	.0132	.0430	.0917	.1472	.1906	.2066	.1914	.1527
	7	.0000	.0003	.0030	.0138	.0393	.0811	.1319	.1771	.2013	.1964
	8	.0000	.0000	.0005	.0035	.0131	.0348	.0710	.1181	.1647	.1964
	9	.0000	.0000	.0001	.0007	.0034	.0116	.0298	.0612	.1048	.1527
	10	.0000	.0000	.0000	.0001	.0007	.0030	.0096	.0245	.0515	.0916
	11	.0000	.0000	.0000	.0000	.0001	.0006	.0024	.0074	.0191	.0417
	12	.0000	.0000	.0000	.0000	.0000	.0001	.0004	.0016	.0052	.0139
	13	.0000	.0000	.0000	.0000	.0000	.0000	.0001	.0003	.0010	.0032
	14	.0000	.0000	.0000	.0000	.0000	.0000	.0000	.0000	.0001	.0005
	15	.0000	.0000	.0000	.0000	.0000	.0000	.0000	.0000	.0000	.0000
16	0	.4401	.1853	.0743	.0281	.0100	.0033	.0010	.0003	.0001	.0000
	1	.3706	.3294	.2097	.1126	.0535	.0228	.0087	.0030	.0009	.0002
	2	.1463	.2745	.2775	.2111	.1336	.0732	.0353	.0150	.0056	.0018
	3	.0359	.1423	.2285	.2463	.2079	.1465	.0888	.0468	.0215	.0085
	4	.0061	.0514	.1311	.2001	.2252	.2040	.1553	.1014	.0572	.0278
	5	.0008	.0137	.0555	.1201	.1802	.2099	.2008	.1623	.1123	.0667
	6	.0001	.0028	.0180	.0550	.1101	.1649	.1982	.1983	.1684	.1222
	7	.0000	.0004	.0045	.0197	.0524	.1010	.1524	.1889	.1969	.1746
	8	.0000	.0001	.0009	.0055	.0197	.0487	.0923	.1417	.1812	.1964
	9	.0000	.0000	.0001	.0012	.0058	.0185	.0442	.0840	.1318	.1746
	10	.0000	.0000	.0000	.0002	.0014	.0056	.0167	.0392	.0755	.1222
	11	.0000	.0000	.0000	.0000	.0002	.0013	.0049	.0142	.0337	.0667
	12	.0000	.0000	.0000	.0000	.0000	.0002	.0011	.0040	.0115	.0278
	13	.0000	.0000	.0000	.0000	.0000	.0000	.0002	.0008	.0029	.0085
	14	.0000	.0000	.0000	.0000	.0000	.0000	.0000	.0001	.0005	.0018
	15	.0000	.0000	.0000	.0000	.0000	.0000	.0000	.0000	.0001	.0002
	16	.0000	.0000	.0000	.0000	.0000	.0000	.0000	.0000	.0000	.0000
17	0	.4181	.1668	.0631	.0225	.0075	.0023	.0007	.0002	.0000	.0000
	1	.3741	.3150	.1893	.0957	.0426	.0169	.0060	.0019	.0005	.0001
	2	.1575	.2800	.2673	.1914	.1136	.0581	.0260	.0102	.0035	.0010
	3	.0415	.1556	.2359	.2393	.1893	.1245	.0701	.0341	.0144	.0052
	4	.0076	.0605	.1457	.2093	.2209	.1868	.1320	.0796	.0411	.0182
	5	.0010	.0175	.0668	.1361	.1914	.2081	.1849	.1379	.0875	.0472
	6	.0001	.0039	.0236	.0680	.1276	.1784	.1991	.1839	.1432	.0944
	7	.0000	.0007	.0065	.0267	.0668	.1201	.1685	.1927	.1841	.1484
	8	.0000	.0001	.0014	.0084	.0279	.0644	.1134	.1606	.1883	.1855
	9	.0000	.0000	.0003	.0021	.0093	.0276	.0611	.1070	.1540	.1855

TABLE 5 Binomial Probabilities *(continued)*

n	x	.05	.10	.15	.20	.25	*p* .30	.35	.40	.45	.50
	10	.0000	.0000	.0000	.0004	.0025	.0095	.0263	.0571	.1008	.1484
	11	.0000	.0000	.0000	.0001	.0005	.0026	.0090	.0242	.0525	.0944
	12	.0000	.0000	.0000	.0000	.0001	.0006	.0024	.0081	.0215	.0472
	13	.0000	.0000	.0000	.0000	.0000	.0001	.0005	.0021	.0068	.0182
	14	.0000	.0000	.0000	.0000	.0000	.0000	.0001	.0004	.0016	.0052
	15	.0000	.0000	.0000	.0000	.0000	.0000	.0000	.0001	.0003	.0010
	16	.0000	.0000	.0000	.0000	.0000	.0000	.0000	.0000	.0000	.0001
	17	.0000	.0000	.0000	.0000	.0000	.0000	.0000	.0000	.0000	.0000
18	0	.3972	.1501	.0536	.0180	.0056	.0016	.0004	.0001	.0000	.0000
	1	.3763	.3002	.1704	.0811	.0338	.0126	.0042	.0012	.0003	.0001
	2	.1683	.2835	.2556	.1723	.0958	.0458	.0190	.0069	.0022	.0006
	3	.0473	.1680	.2406	.2297	.1704	.1046	.0547	.0246	.0095	.0031
	4	.0093	.0700	.1592	.2153	.2130	.1681	.1104	.0614	.0291	.0117
	5	.0014	.0218	.0787	.1507	.1988	.2017	.1664	.1146	.0666	.0327
	6	.0002	.0052	.0301	.0816	.1436	.1873	.1941	.1655	.1181	.0708
	7	.0000	.0010	.0091	.0350	.0820	.1376	.1792	.1892	.1657	.1214
	8	.0000	.0002	.0022	.0120	.0376	.0811	.1327	.1734	.1864	.1669
	9	.0000	.0000	.0004	.0033	.0139	.0386	.0794	.1284	.1694	.1855
	10	.0000	.0000	.0001	.0008	.0042	.0149	.0385	.0771	.1248	.1669
	11	.0000	.0000	.0000	.0001	.0010	.0046	.0151	.0374	.0742	.1214
	12	.0000	.0000	.0000	.0000	.0002	.0012	.0047	.0145	.0354	.0708
	13	.0000	.0000	.0000	.0000	.0000	.0002	.0012	.0045	.0134	.0327
	14	.0000	.0000	.0000	.0000	.0000	.0000	.0002	.0011	.0039	.0117
	15	.0000	.0000	.0000	.0000	.0000	.0000	.0000	.0002	.0009	.0031
	16	.0000	.0000	.0000	.0000	.0000	.0000	.0000	.0000	.0001	.0006
	17	.0000	.0000	.0000	.0000	.0000	.0000	.0000	.0000	.0000	.0001
	18	.0000	.0000	.0000	.0000	.0000	.0000	.0000	.0000	.0000	.0000
19	0	.3774	.1351	.0456	.0144	.0042	.0011	.0003	.0001	.0000	.0000
	1	.3774	.2852	.1529	.0685	.0268	.0093	.0029	.0008	.0002	.0000
	2	.1787	.2852	.2428	.1540	.0803	.0358	.0138	.0046	.0013	.0003
	3	.0533	.1796	.2428	.2182	.1517	.0869	.0422	.0175	.0062	.0018
	4	.0112	.0798	.1714	.2182	.2023	.1491	.0909	.0467	.0203	.0074
	5	.0018	.0266	.0907	.1636	.2023	.1916	.1468	.0933	.0497	.0222
	6	.0002	.0069	.0374	.0955	.1574	.1916	.1844	.1451	.0949	.0518
	7	.0000	.0014	.0122	.0443	.0974	.1525	.1844	.1797	.1443	.0961
	8	.0000	.0002	.0032	.0166	.0487	.0981	.1489	.1797	.1771	.1442
	9	.0000	.0000	.0007	.0051	.0198	.0514	.0980	.1464	.1771	.1762
	10	.0000	.0000	.0001	.0013	.0066	.0220	.0528	.0976	.1449	.1762
	11	.0000	.0000	.0000	.0003	.0018	.0077	.0233	.0532	.0970	.1442
	12	.0000	.0000	.0000	.0000	.0004	.0022	.0083	.0237	.0529	.0961
	13	.0000	.0000	.0000	.0000	.0001	.0005	.0024	.0085	.0233	.0518
	14	.0000	.0000	.0000	.0000	.0000	.0001	.0006	.0024	.0082	.0222
	15	.0000	.0000	.0000	.0000	.0000	.0000	.0001	.0005	.0022	.0074
	16	.0000	.0000	.0000	.0000	.0000	.0000	.0000	.0001	.0005	.0018
	17	.0000	.0000	.0000	.0000	.0000	.0000	.0000	.0000	.0001	.0003
	18	.0000	.0000	.0000	.0000	.0000	.0000	.0000	.0000	.0000	.0000
	19	.0000	.0000	.0000	.0000	.0000	.0000	.0000	.0000	.0000	.0000
20	0	.3585	.1216	.0388	.0115	.0032	.0008	.0002	.0000	.0000	.0000
	1	.3774	.2702	.1368	.0576	.0211	.0068	.0020	.0005	.0001	.0000
	2	.1887	.2852	.2293	.1369	.0669	.0278	.0100	.0031	.0008	.0002
	3	.0596	.1901	.2428	.2054	.1339	.0716	.0323	.0123	.0040	.0011
	4	.0133	.0898	.1821	.2182	.1897	.1304	.0738	.0350	.0139	.0046

TABLE 5   Binomial Probabilities *(concluded)*

n	x	.05	.10	.15	.20	.25	.30	.35	.40	.45	.50
	5	.0022	.0319	.1028	.1746	.2023	.1789	.1272	.0746	.0365	.0148
	6	.0003	.0089	.0454	.1091	.1686	.1916	.1712	.1244	.0746	.0370
	7	.0000	.0020	.0160	.0545	.1124	.1643	.1844	.1659	.1221	.0739
	8	.0000	.0004	.0046	.0222	.0609	.1144	.1614	.1797	.1623	.1201
	9	.0000	.0001	.0011	.0074	.0271	.0654	.1158	.1597	.1771	.1602
	10	.0000	.0000	.0002	.0020	.0099	.0308	.0686	.1171	.1593	.1762
	11	.0000	.0000	.0000	.0005	.0030	.0120	.0336	.0710	.1185	.1602
	12	.0000	.0000	.0000	.0001	.0008	.0039	.0136	.0355	.0727	.1201
	13	.0000	.0000	.0000	.0000	.0002	.0010	.0045	.0146	.0366	.0739
	14	.0000	.0000	.0000	.0000	.0000	.0002	.0012	.0049	.0150	.0370
	15	.0000	.0000	.0000	.0000	.0000	.0000	.0003	.0013	.0049	.0148
	16	.0000	.0000	.0000	.0000	.0000	.0000	.0000	.0003	.0013	.0046
	17	.0000	.0000	.0000	.0000	.0000	.0000	.0000	.0000	.0002	.0011
	18	.0000	.0000	.0000	.0000	.0000	.0000	.0000	.0000	.0000	.0002
	19	.0000	.0000	.0000	.0000	.0000	.0000	.0000	.0000	.0000	.0000
	20	.0000	.0000	.0000	.0000	.0000	.0000	.0000	.0000	.0000	.0000

TABLE 6 $e^{-\lambda}$

λ	$e^{-\lambda}$	λ	$e^{-\lambda}$	λ	$e^{-\lambda}$	λ	$e^{-\lambda}$
0.0	1.000	2.5	0.082	5.0	0.0067	7.5	0.00055
0.1	0.905	2.6	0.074	5.1	0.0061	7.6	0.00050
0.2	0.819	2.7	0.067	5.2	0.0055	7.7	0.00045
0.3	0.741	2.8	0.061	5.3	0.0050	7.8	0.00041
0.4	0.670	2.9	0.055	5.4	0.0045	7.9	0.00037
0.5	0.607	3.0	0.050	5.5	0.0041	8.0	0.00034
0.6	0.549	3.1	0.045	5.6	0.0037	8.1	0.00030
0.7	0.497	3.2	0.041	5.7	0.0033	8.2	0.00028
0.8	0.449	3.3	0.037	5.8	0.0030	8.3	0.00025
0.9	0.407	3.4	0.033	5.9	0.0027	8.4	0.00023
1.0	0.368	3.5	0.030	6.0	0.0025	8.5	0.00020
1.1	0.333	3.6	0.027	6.1	0.0022	8.6	0.00018
1.2	0.301	3.7	0.025	6.2	0.0020	8.7	0.00017
1.3	0.273	3.8	0.022	6.3	0.0018	8.8	0.00015
1.4	0.247	3.9	0.020	6.4	0.0017	8.9	0.00014
1.5	0.223	4.0	0.018	6.5	0.0015	9.0	0.00012
1.6	0.202	4.1	0.017	6.6	0.0014	9.1	0.00011
1.7	0.183	4.2	0.015	6.7	0.0012	9.2	0.00010
1.8	0.165	4.3	0.014	6.8	0.0011	9.3	0.00009
1.9	0.150	4.4	0.012	6.9	0.0010	9.4	0.00008
2.0	0.135	4.5	0.011	7.0	0.0009	9.5	0.00008
2.1	0.122	4.6	0.010	7.1	0.0008	9.6	0.00007
2.2	0.111	4.7	0.009	7.2	0.0007	9.7	0.00006
2.3	0.100	4.8	0.008	7.3	0.0007	9.8	0.00006
2.4	0.091	4.9	0.007	7.4	0.0006	9.9	0.00005

TABLE 7 Poisson Probabilities

					λ					
x	0.1	0.2	0.3	0.4	0.5	0.6	0.7	0.8	0.9	1.0
0	.9048	.8187	.7408	.6703	.6065	.5488	.4966	.4493	.4066	.3679
1	.0905	.1637	.2222	.2681	.3033	.3293	.3476	.3595	.3659	.3679
2	.0045	.0164	.0333	.0536	.0758	.0988	.1217	.1438	.1647	.1839
3	.0002	.0011	.0033	.0072	.0126	.0198	.0284	.3083	.0494	.0613
4	.0000	.0001	.0002	.0007	.0016	.0030	.0050	.0077	.0111	.0153
5	.0000	.0000	.0000	.0001	.0002	.0004	.0007	.0012	.0020	.0031
6	.0000	.0000	.0000	.0000	.0000	.0000	.0001	.0002	.0003	.0005
7	.0000	.0000	.0000	.0000	.0000	.0000	.0000	.0000	.0000	.0001

					λ					
x	1.1	1.2	1.3	1.4	1.5	1.6	1.7	1.8	1.9	2.0
0	.3329	.3012	.2725	.2466	.2231	.2019	.1827	.1653	.1496	.1353
1	.3662	.3614	.3543	.3452	.3347	.3230	.3106	.2975	.2842	.2707
2	.2014	.2169	.2303	.2417	.2510	.2584	.2640	.2678	.2700	.2707
3	.0738	.0867	.0998	.1128	.1255	.1378	.1496	.1607	.1710	.1804
4	.0203	.0260	.0324	.0395	.0471	.0551	.0636	.0723	.0812	.0902
5	.0045	.0062	.0084	.0111	.0141	.0176	.0216	.0260	.0309	.0361
6	.0008	.0012	.0018	.0026	.0035	.0047	.0061	.0078	.0098	.0120
7	.0001	.0002	.0003	.0005	.0008	.0011	.0015	.0020	.0027	.0034
8	.0000	.0000	.0001	.0001	.0001	.0002	.0003	.0005	.0006	.0009
9	.0000	.0000	.0000	.0000	.0000	.0000	.0001	.0001	.0001	.0002

					λ					
x	2.1	2.2	2.3	2.4	2.5	2.6	2.7	2.8	2.9	3.0
0	.1225	.1108	.1003	.0907	.0821	.0743	.0672	.0608	.0550	.0498
1	.2572	.2438	.2306	.2177	.2052	.1931	.1815	.1703	.1596	.1494
2	.2700	.2681	.2652	.2613	.2565	.2510	.2450	.2384	.2314	.2240
3	.1890	.1966	.2033	.2090	.2138	.2176	.2205	.2225	.2237	.2240
4	.0992	.1082	.1169	.1254	.1336	.1414	.1488	.1557	.1622	.1680
5	.0417	.0476	.0538	.0602	.0668	.0735	.0804	.0872	.0940	.1008
6	.0146	.0174	.0206	.0241	.0278	.0319	.0362	.0407	.0455	.0504
7	.0044	.0055	.0068	.0083	.0099	.0118	.0139	.0163	.0188	.0216
8	.0011	.0015	.0019	.0025	.0031	.0038	.0047	.0057	.0068	.0081
9	.0003	.0004	.0005	.0007	.0009	.0011	.0014	.0018	.0022	.0027
10	.0001	.0001	.0001	.0002	.0002	.0003	.0004	.0005	.0006	.0008
11	.0000	.0000	.0000	.0000	.0000	.0001	.0001	.0001	.0002	.0002
12	.0000	.0000	.0000	.0000	.0000	.0000	.0000	.0000	.0000	.0001

					λ					
x	3.1	3.2	3.3	3.4	3.5	3.6	3.7	3.8	3.9	4.0
0	.0450	.0408	.0369	.0334	.0302	.0273	.0247	.0224	.0202	.0183
1	.1397	.1304	.1217	.1135	.1057	.0984	.0915	.0850	.0789	.0733
2	.2165	.2087	.2008	.1929	.1850	.1771	.1692	.1615	.1539	.1465
3	.2237	.2226	.2209	.2186	.2158	.2125	.2087	.2046	.2001	.1954
4	.1734	.1781	.1823	.1858	.1888	.1912	.1931	.1944	.1951	.1954
5	.1075	.1140	.1203	.1264	.1322	.1377	.1429	.1477	.1522	.1563
6	.0555	.0608	.0662	.0716	.0771	.0826	.0881	.0936	.0989	.1042
7	.0246	.0278	.0312	.0348	.0385	.0425	.0466	.0508	.0551	.0595
8	.0095	.0111	.0129	.0148	.0169	.0191	.0215	.0241	.0269	.0298
9	.0033	.0040	.0047	.0056	.0066	.0076	.0089	.0102	.0116	.0132
10	.0010	.0013	.0016	.0019	.0023	.0028	.0033	.0039	.0045	.0053
11	.0003	.0004	.0005	.0006	.0007	.0009	.0011	.0013	.0016	.0019
12	.0001	.0001	.0001	.0002	.0002	.0003	.0003	.0004	.0005	.0006
13	.0000	.0000	.0000	.0000	.0001	.0001	.0001	.0001	.0002	.0002
14	.0000	.0000	.0000	.0000	.0000	.0000	.0000	.0000	.0000	.0001

TABLE 7 Poisson Probabilities *(continued)*

x	4.1	4.2	4.3	4.4	4.5	4.6	4.7	4.8	4.9	5.0
0	.0166	.0150	.0136	.0123	.0111	.0101	.0091	.0082	.0074	.0067
1	.0679	.0630	.0583	.0540	.0500	.0462	.0427	.0395	.0365	.0337
2	.1393	.1323	.1254	.1188	.1125	.1063	.1005	.0948	.0894	.0842
3	.1904	.1852	.1798	.1743	.1687	.1631	.1574	.1517	.1460	.1404
4	.1951	.1944	.1933	.1917	.1898	.1875	.1849	.1820	.1789	.1755
5	.1600	.1633	.1662	.1687	.1708	.1725	.1738	.1747	.1753	.1755
6	.1093	.1143	.1191	.1237	.1281	.1323	.1362	.1398	.1432	.1462
7	.0640	.0686	.0732	.0778	.0824	.0869	.0914	.0959	.1002	.1044
8	.0328	.0360	.0393	.0428	.0463	.0500	.0537	.0575	.0614	.0653
9	.0150	.0168	.0188	.0209	.0232	.0255	.0280	.0307	.0334	.0363
10	.0061	.0071	.0081	.0092	.0104	.0118	.0132	.0147	.0164	.0181
11	.0023	.0027	.0032	.0037	.0043	.0049	.0056	.0064	.0073	.0082
12	.0008	.0009	.0011	.0014	.0016	.0019	.0022	.0026	.0030	.0034
13	.0002	.0003	.0004	.0005	.0006	.0007	.0008	.0009	.0011	.0013
14	.0001	.0001	.0001	.0001	.0002	.0002	.0003	.0003	.0004	.0005
15	.0000	.0000	.0000	.0000	.0001	.0001	.0001	.0001	.0001	.0002

x	5.1	5.2	5.3	5.4	5.5	5.6	5.7	5.8	5.9	6.0
0	.0061	.0055	.0050	.0045	.0041	.0037	.0033	.0030	.0027	.0025
1	.0311	.0287	.0265	.0244	.0225	.0207	.0191	.0176	.0162	.0149
2	.0793	.0746	.0701	.0659	.0618	.0580	.0544	.0509	.0477	.0446
3	.1348	.1293	.1239	.1185	.1133	.1082	.1033	.0985	.0938	.0892
4	.1719	.1681	.1641	.1600	.1558	.1515	.1472	.1428	.1383	.1339
5	.1753	.1748	.1740	.1728	.1714	.1697	.1678	.1656	.1632	.1606
6	.1490	.1515	.1537	.1555	.1571	.1584	.1594	.1601	.1605	.1606
7	.1086	.1125	.1163	.1200	.1234	.1267	.1298	.1326	.1353	.1377
8	.0692	.0731	.0771	.0810	.0849	.0887	.0925	.0962	.0998	.1033
9	.0392	.0423	.0454	.0486	.0519	.0552	.0586	.0620	.0654	.0688
10	.0200	.0220	.0241	.0262	.0285	.0309	.0334	.0359	.0386	.0413
11	.0093	.0104	.0116	.0129	.0143	.0157	.0173	.0190	.0207	.0225
12	.0039	.0045	.0051	.0058	.0065	.0073	.0082	.0092	.0102	.0113
13	.0015	.0018	.0021	.0024	.0028	.0032	.0036	.0041	.0046	.0052
14	.0006	.0007	.0008	.0009	.0011	.0013	.0015	.0017	.0019	.0022
15	.0002	.0002	.0003	.0003	.0004	.0005	.0006	.0007	.0008	.0009
16	.0001	.0001	.0001	.0001	.0001	.0002	.0002	.0002	.0003	.0003
17	.0000	.0000	.0000	.0000	.0000	.0001	.0001	.0001	.0001	.0001

TABLE 7 Poisson Probabilities *(continued)*

x	6.1	6.2	6.3	6.4	6.5	6.6	6.7	6.8	6.9	7.0
0	.0022	.0020	.0018	.0017	.0015	.0014	.0012	.0011	.0010	.0009
1	.0137	.0126	.0116	.0106	.0098	.0090	.0082	.0076	.0070	.0064
2	.0417	.0390	.0364	.0340	.0318	.0296	.0276	.0258	.0240	.0223
3	.0848	.0806	.0765	.0726	.0688	.0652	.0617	.0584	.0552	.0521
4	.1294	.1249	.1205	.1162	.1118	.1076	.1034	.0992	.0952	.0912
5	.1579	.1549	.1519	.1487	.1454	.1420	.1385	.1349	.1314	.1277
6	.1605	.1601	.1595	.1586	.1575	.1562	.1546	.1529	.1511	.1490
7	.1399	.1418	.1435	.1450	.1462	.1472	.1480	.1486	.1489	.1490
8	.1066	.1099	.1130	.1160	.1188	.1215	.1240	.1263	.1284	.1304
9	.0723	.0757	.0791	.0825	.0858	.0891	.0923	.0954	.0985	.1014
10	.0441	.0469	.0498	.0528	.0558	.0588	.0618	.0649	.0679	.0710
11	.0245	.0265	.0285	.0307	.0330	.0353	.0377	.0401	.0426	.0452
12	.0124	.0137	.0150	.0164	.0179	.0194	.0210	.0227	.0245	.0264
13	.0058	.0065	.0073	.0081	.0089	.0098	.0108	.0119	.0130	.0142
14	.0025	.0029	.0033	.0037	.0041	.0046	.0052	.0058	.0064	.0071
15	.0010	.0012	.0014	.0016	.0018	.0020	.0023	.0026	.0029	.0033
16	.0004	.0005	.0005	.0006	.0007	.0008	.0010	.0011	.0013	.0014
17	.0001	.0002	.0002	.0002	.0003	.0003	.0004	.0004	.0005	.0006
18	.0000	.0001	.0001	.0001	.0001	.0001	.0001	.0002	.0002	.0002
19	.0000	.0000	.0000	.0000	.0000	.0000	.0000	.0001	.0001	.0001

x	7.1	7.2	7.3	7.4	7.5	7.6	7.7	7.8	7.9	8.0
0	.0008	.0007	.0007	.0006	.0006	.0005	.0005	.0004	.0004	.0003
1	.0059	.0054	.0049	.0045	.0041	.0038	.0035	.0032	.0029	.0027
2	.0208	.0194	.0180	.0167	.0156	.0145	.0134	.0125	.0116	.0107
3	.0492	.0464	.0438	.0413	.0389	.0366	.0345	.0324	.0305	.0286
4	.0874	.0836	.0799	.0764	.0729	.0696	.0663	.0632	.0602	.0573
5	.1241	.1204	.1167	.1130	.1094	.1057	.1021	.0986	.0951	.0916
6	.1468	.1445	.1420	.1394	.1367	.1339	.1311	.1282	.1252	.1221
7	.1489	.1486	.1481	.1474	.1465	.1454	.1442	.1428	.1413	.1396
8	.1321	.1337	.1351	.1363	.1373	.1382	.1388	.1392	.1395	.1396
9	.1042	.1070	.1096	.1121	.1144	.1167	.1187	.1207	.1224	.1241
10	.0740	.0770	.0800	.0829	.0858	.0887	.0914	.0941	.0967	.0993
11	.0478	.0504	.0531	.0558	.0585	.0613	.0640	.0667	.0695	.0722
12	.0283	.0303	.0323	.0344	.0366	.0388	.0411	.0434	.0457	.0481
13	.0154	.0168	.0181	.0196	.0211	.0227	.0243	.0260	.0278	.0296
14	.0078	.0086	.0095	.0104	.0113	.0123	.0134	.0145	.0157	.0169
15	.0037	.0041	.0046	.0051	.0057	.0062	.0069	.0075	.0083	.0090
16	.0016	.0019	.0021	.0024	.0026	.0030	.0033	.0037	.0041	.0045
17	.0007	.0008	.0009	.0010	.0012	.0013	.0015	.0017	.0019	.0021
18	.0003	.0003	.0004	.0004	.0005	.0006	.0006	.0007	.0008	.0009
19	.0001	.0001	.0001	.0002	.0002	.0002	.0003	.0003	.0003	.0004
20	.0000	.0000	.0001	.0001	.0001	.0001	.0001	.0001	.0001	.0002
21	.0000	.0000	.0000	.0000	.0000	.0000	.0000	.0000	.0001	.0001

In both subtables the column group is headed by λ.

TABLE 7 Poisson Probabilities *(concluded)*

x	8.1	8.2	8.3	8.4	8.5	λ 8.6	8.7	8.8	8.9	9.0
0	.0003	.0003	.0002	.0002	.0002	.0002	.0002	.0002	.0001	.0001
1	.0025	.0023	.0021	.0019	.0017	.0016	.0014	.0013	.0012	.0011
2	.0100	.0092	.0092	.0086	.0079	.0068	.0063	.0058	.0054	.0050
3	.0269	.0252	.0237	.0222	.0208	.0195	.0183	.0171	.0160	.0150
4	.0544	.0517	.0491	.0466	.0443	.0420	.0398	.0377	.0357	.0337
5	.0882	.0849	.0816	.0784	.0752	.0722	.0692	.0663	.0635	.0607
6	.1191	.1160	.1128	.1097	.1066	.1034	.1003	.0972	.0941	.0911
7	.1378	.1358	.1338	.1317	.1294	.1271	.1247	.1222	.1197	.1171
8	.1395	.1392	.1388	.1382	.1375	.1366	.1356	.1344	.1332	.1318
9	.1256	.1269	.1280	.1290	.1299	.1306	.1311	.1315	.1317	.1318
10	.1017	.1040	.1063	.1084	.1104	.1123	.1140	.1157	.1172	.1186
11	.0749	.0776	.0802	.0828	.0853	.0878	.0902	.0925	.0948	.0970
12	.0505	.0530	.0555	.0579	.0604	.0629	.0654	.0679	.0703	.0728
13	.0315	.0334	.0354	.0374	.0395	.0416	.0438	.0459	.0481	.0504
14	.0182	.0196	.0210	.0225	.0240	.0256	.0272	.0289	.0306	.0324
15	.0098	.0107	.0116	.0126	.0136	.0147	.0158	.0169	.0182	.0194
16	.0050	.0055	.0060	.0066	.0072	.0079	.0086	.0093	.0101	.0109
17	.0024	.0026	.0029	.0033	.0036	.0040	.0044	.0048	.0053	.0058
18	.0011	.0012	.0014	.0015	.0017	.0019	.0021	.0024	.0026	.0029
19	.0005	.0005	.0006	.0007	.0008	.0009	.0010	.0011	.0012	.0014
20	.0002	.0002	.0002	.0003	.0003	.0004	.0004	.0005	.0005	.0006
21	.0001	.0001	.0001	.0001	.0001	.0002	.0002	.0002	.0002	.0003
22	.0000	.0000	.0000	.0000	.0001	.0001	.0001	.0001	.0001	.0001

x	9.1	9.2	9.3	9.4	9.5	λ 9.6	9.7	9.8	9.9	10
0	.0001	.0001	.0001	.0001	.0001	.0001	.0001	.0001	.0001	.0000
1	.0010	.0009	.0009	.0008	.0007	.0007	.0006	.0005	.0005	.0005
2	.0046	.0043	.0040	.0037	.0034	.0031	.0029	.0027	.0025	.0023
3	.0140	.0131	.0123	.0115	.0107	.0100	.0093	.0087	.0081	.0076
4	.0319	.0302	.0285	.0269	.0254	.0240	.0226	.0213	.0201	.0189
5	.0581	.0555	.0530	.0506	.0483	.0460	.0439	.0418	.0398	.0378
6	.0881	.0851	.0822	.0793	.0764	.0736	.0709	.0682	.0656	.0631
7	.1145	.1118	.1091	.1064	.1037	.1010	.0982	.0955	.0928	.0901
8	.1302	.1286	.1269	.1251	.1232	.1212	.1191	.1170	.1148	.1126
9	.1317	.1315	.1311	.1306	.1300	.1293	.1284	.1274	.1263	.1251
10	.1198	.1210	.1219	.1228	.1235	.1241	.1245	.1249	.1250	.1251
11	.0991	.1012	.1031	.1049	.1067	.1083	.1098	.1112	.1125	.1137
12	.0752	.0776	.0799	.0822	.0844	.0866	.0888	.0908	.0928	.0948
13	.0526	.0549	.0572	.0594	.0617	.0640	.0662	.0685	.0707	.0729
14	.0342	.0361	.0380	.0399	.0419	.0439	.0459	.0479	.0500	.0521
15	.0208	.0221	.0235	.0250	.0265	.0281	.0297	.0313	.0330	.0347
16	.0118	.0127	.0137	.0147	.0157	.0168	.0180	.0192	.0204	.0217
17	.0063	.0069	.0075	.0081	.0088	.0095	.0103	.0111	.0119	.0128
18	.0032	.0035	.0039	.0042	.0046	.0051	.0055	.0060	.0065	.0071
19	.0015	.0017	.0019	.0021	.0023	.0026	.0028	.0031	.0034	.0037
20	.0007	.0008	.0009	.0010	.0011	.0012	.0014	.0015	.0017	.0019
21	.0003	.0003	.0004	.0004	.0005	.0006	.0006	.0007	.0008	.0009
22	.0001	.0001	.0002	.0002	.0002	.0002	.0003	.0003	.0004	.0004
23	.0000	.0001	.0001	.0001	.0001	.0001	.0001	.0001	.0002	.0002
24	.0000	.0000	.0000	.0000	.0000	.0000	.0000	.0001	.0001	.0001

TABLE 8 Control Chart Constants

Number of Observations in Subgroup, n	A_2	A_3	A_6	B_3	B_4	c_4	d_2	d_3	d_4	D_3	D_4	D_5	D_6	E_2
2	1.880	2.659		0.000	3.267	0.7979	1.128	0.853	0.954	0.000	3.267	0.000	3.865	2.660
3	1.023	1.954	1.187	0.000	2.568	0.8862	1.693	0.888	1.588	0.000	2.574	0.000	2.745	1.772
4	0.729	1.628		0.000	2.266	0.9213	2.059	0.880	1.978	0.000	2.282	0.000	2.375	1.457
5	0.577	1.427	0.691	0.000	2.089	0.9400	2.326	0.864	2.257	0.000	2.114	0.000	2.179	1.290
6	0.483	1.287		0.030	1.970	0.9515	2.534	0.848	2.472	0.000	2.004	0.000	2.055	1.184
7	0.419	1.182	0.509	0.118	1.882	0.9594	2.704	0.833	2.645	0.076	1.924	0.078	1.967	1.109
8	0.373	1.099		0.185	1.815	0.9650	2.847	0.820	2.791	0.136	1.864	0.139	1.901	1.054
9	0.337	1.032	0.412	0.239	1.761	0.9693	2.970	0.808	2.915	0.184	1.816	0.187	1.850	1.054
10	0.308	0.975		0.284	1.716	0.9727	3.078	0.797	3.024	0.223	1.777	0.227	1.809	1.010
11	0.285	0.927	0.350	0.321	1.679	0.9754	3.173	0.787	3.121	0.256	1.744			0.975
12	0.266	0.886		0.354	1.646	0.9776	3.258	0.778	3.207	0.283	1.717			
13	0.249	0.850		0.382	1.618	0.9794	3.336	0.770	3.285	0.307	1.693			
14	0.235	0.817		0.406	1.594	0.9810	3.407	0.762	3.356	0.328	1.672			
15	0.223	0.789		0.428	1.572	0.9823	3.472	0.755	3.422	0.347	1.653			
16	0.212	0.763		0.448	1.552	0.9835	3.532	0.749	3.482	0.363	1.637			
17	0.203	0.739		0.466	1.534	0.9845	3.588	0.743	3.538	0.378	1.622			
18	0.194	0.718		0.482	1.518	0.9854	3.640	0.738	3.591	0.391	1.608			
19	0.187	0.698		0.497	1.503	0.9862	3.689	0.733	3.640	0.403	1.597			
20	0.180	0.680		0.510	1.490	0.9869	3.735	0.729	3.686	0.415	1.585			
21	0.173	0.663		0.523	1.477	0.9876	3.778	0.724	3.730	0.425	1.575			
22	0.167	0.647		0.534	1.466	0.9882	3.819	0.720	3.771	0.434	1.566			
23	0.162	0.633		0.545	1.455	0.9887	3.858	0.716	3.811	0.443	1.557			
24	0.157	0.619		0.555	1.445	0.9892	3.895	0.712	3.847	0.451	1.548			
25	0.153	0.606		0.565	1.435	0.9896	3.931	0.709	3.883	0.459	1.541			
More than 25	$3/\sqrt{n}$			$1 - 3/\sqrt{2n}$	$1 + 3/\sqrt{2n}$									

SOURCE: A_2, A_3, B_3, B_4, c_4, d_2, d_3, D_3, D_4, E_2 reprinted with permission from ASTM Manual on the Presentation of Data and Control Chart Analysis (Philadelphia, Penn.: ASTM 1976), pp. 134–36. Copyright ASTM.

A_6, d_4, D_5, D_6 reprinted with permission from D. J. Wheeler and D. S. Chambers, Understanding Statistical Process Control (Knoxville: Statistical Process Controls, Inc., 1986), pp. 307, 309, 312.

TABLE 9 2,500 Four-Digit Random Numbers

5347	8111	9803	1221	5952	4023	4057	3935	4321	6925
9734	7032	5811	9196	2624	4464	8328	9739	9282	7757
6602	3827	7452	7111	8489	1395	9889	9231	6578	5964
9977	7572	0317	4311	8308	8198	1453	2616	2489	2055
3017	4897	9215	3841	4243	2663	8390	4472	6921	6911
8187	8333	1498	9993	1321	3017	4796	9379	8669	9885
1983	9063	7186	9505	5553	6090	8410	5534	4847	6379
0933	3343	5386	5276	1880	2582	9619	6651	7831	9701
3115	5829	4082	4133	2109	9388	4919	4487	4718	8142
6761	5251	0303	8169	1710	6498	6083	8531	4781	0807
6194	4879	1160	8304	2225	1183	0434	9554	2036	5593
0481	6489	9634	7906	2699	4396	6348	9357	8075	9658
0576	3960	5614	2551	8615	7865	0218	2971	0433	1567
7326	5687	4079	1394	9628	9018	4711	6680	6184	4468
5490	0997	7658	0264	3579	4453	6442	3544	2831	9900
4258	3633	6006	0404	2967	1634	4859	2554	6317	7522
2726	2740	9752	2333	3645	3369	2367	4588	4151	0475
4984	1144	6668	3605	3200	7860	3692	5996	6819	6258
2931	4046	2707	6923	5142	5851	4992	0390	2659	3306
3046	2785	6779	1683	7427	0579	0290	6349	0078	3509
2870	8408	6553	4425	3386	8253	9839	2638	0283	3683
1318	5065	9487	2825	7854	5528	3359	6196	5172	1421
6079	7663	3015	4029	9947	2833	1536	4248	6031	4277
1348	4691	6468	0741	7784	0190	4779	6579	4423	7723
3491	9450	3937	3418	5750	2251	0406	9451	4461	1048
2810	0481	8517	8649	3569	0348	5731	6317	7190	7118
5923	4502	0117	0884	8192	7149	9540	3404	0485	6591
8743	8275	7109	3683	5358	2598	4600	4284	8168	2145
2904	0130	5534	6573	7871	4364	4624	5320	9486	4871
6203	7188	9450	1526	6143	1036	4205	6825	1438	7943
3885	8004	5997	7336	5287	4767	4102	8229	2643	8737
4066	4332	8737	8641	9584	2559	5413	9418	4230	0736
4058	9008	3772	0866	3725	2031	5331	5098	3290	3209
7823	8655	5027	2043	0024	0230	7102	4993	2324	0086
9824	6747	7145	6954	0116	0332	6701	9254	9797	5272
6997	7855	6543	3262	2831	6181	1459	7972	5569	9134
3984	2307	4081	0371	2189	9635	9680	2459	2620	2600
6288	8727	9989	9996	3437	4255	1167	9960	9801	4886
5613	6492	2945	5296	8662	6242	3016	7618	9531	3926
9080	5602	4899	6456	6746	6018	1297	0384	6258	9385
0966	4467	7476	3335	6730	8054	9765	1134	7877	4501
3475	5040	7663	1276	3222	3454	1810	5351	1452	7212
1215	7332	7419	2666	7808	5363	5230	0000	0570	6353
6938	0773	9445	7642	1612	0930	6741	6858	8793	3884
9335	6456	4376	4504	4493	6997	1696	0827	6775	6029
3887	3554	9956	8540	0491	6254	7840	0101	8618	2207
5831	6029	7239	6966	1247	9305	0205	2980	6364	1279
8356	1022	9947	7472	2207	1023	2157	2032	2131	5712
2806	9115	4056	3370	6451	0706	6437	2633	7965	3114
0573	7555	9316	8092	5587	5410	3480	8315	0453	8136

SOURCE: Reprinted from *A Million Random Digits with 100,000 Normal Deviates* by the RAND Corporation (New York: The Free Press, 1955). Copyright 1955 and 1983 by The RAND Corporation.

TABLE 9   2,500 Four-Digit Random Numbers *(continued)*

2668	7422	4354	4569	9446	8212	3737	2396	6892	3766
6067	7516	2451	1510	0201	1437	6518	1063	6442	6674
4541	9863	8312	9855	0995	6025	4207	4093	9799	9308
6987	4802	8975	2847	4413	5997	9106	2876	8596	7717
0376	8636	9953	4418	2388	8997	1196	5158	1803	5623
8468	5763	3232	1986	7134	4200	9699	8437	2799	2145
9151	4967	3255	8518	2802	8815	6289	9549	2942	3813
1073	4930	1830	2224	2246	1000	9315	6698	4491	3046
5487	1967	5836	2090	3832	0002	9844	3742	2289	3763
4896	4957	6536	7430	6208	3929	1030	2317	7421	3227
9143	7911	0368	0541	2302	5473	9155	0625	1870	1890
9256	2956	4747	6280	7342	0453	8639	1216	5964	9772
4173	1219	7744	9241	6354	4211	8497	1245	3313	4846
2525	7811	5417	7824	0922	8752	3537	9069	5417	0856
9165	1156	6603	2852	8370	0995	7661	8811	7835	5087
0014	8474	6322	5053	5015	6043	0482	4957	8904	1616
5325	7320	8406	5962	6100	3854	0575	0617	8019	2646
2558	1748	5671	4974	7073	3273	6036	1410	5257	3939
0117	1218	0688	2756	7545	5426	3856	8905	9691	8890
8353	1554	4083	2029	8857	4781	9654	7946	7866	2535
1990	9886	3280	6109	9158	3034	8490	6404	6775	8763
9651	7870	2555	3518	2906	4900	2984	6894	5050	4586
9941	5617	1984	2435	5184	0379	7212	5795	0836	4319
7769	5785	9321	2734	2890	3105	6581	2163	4938	7540
3224	8379	9952	0515	2724	4826	6215	6246	9704	1651
1287	7275	6646	1378	6433	0005	7332	0392	1319	1946
6389	4191	4548	5546	6651	8248	7469	0786	0972	7649
1625	4327	2654	4129	3509	3217	7062	6640	0105	4422
7555	3020	4181	7498	4022	9122	6423	7301	8310	9204
4177	1844	3468	1389	3884	6900	1036	8412	0881	6678
0927	0124	8176	0680	1056	1008	1748	0547	8227	0690
8505	1781	7155	3635	9751	5414	5113	8316	2737	6860
8022	8757	6275	1485	3635	2330	7045	2106	6381	2986
8390	8802	5674	2559	7934	4788	7791	5202	8430	0289
3630	5783	7762	0223	5328	7731	4010	3845	9221	5427
9154	6388	6053	9633	2080	7269	0894	0287	7489	2259
1441	3381	7823	8767	9647	4445	2509	2929	5067	0779
8246	0778	0993	6687	7212	9968	8432	1453	0841	4595
2730	3984	0563	9636	7202	0127	9283	4009	3177	4182
9196	8276	0233	0879	3385	2184	1739	5375	5807	4849
5928	9610	9161	0748	3794	9683	1544	1209	3669	5831
1042	9600	7122	2135	7868	5596	3551	9480	2342	0449
6552	4103	7957	0510	5958	0211	3344	5678	1840	3627
5968	4307	9327	3197	0876	8480	5066	1852	8323	5060
4445	1018	4356	4653	9302	0761	1291	6093	5340	1840
8727	8201	5980	7859	6055	1403	1209	9547	4273	0857
9415	9311	4996	2775	8509	7767	6930	6632	7781	2279
2648	7639	9128	0341	6875	8957	6646	9783	6668	0317
3707	3454	8829	6863	1297	5089	1002	2722	0578	7753
8383	8957	5595	9395	3036	4767	8300	3505	0710	6307

TABLE 9 2,500 Four-Digit Random Numbers *(continued)*

5503	8121	9056	8194	1124	8451	1228	8986	0076	7615
2552	9953	4323	4878	4922	0696	3156	2145	8819	0631
8542	7274	9724	6638	0013	0566	9644	3738	5767	2791
6121	4839	4734	3041	3939	9136	5620	7920	0533	3119
2023	0314	5885	1165	2841	1282	5893	3050	6598	2667
9577	8320	5614	5595	8978	6442	0844	4570	8036	6026
0760	1734	0114	8330	9695	6502	3171	8901	7955	4975
0064	1745	7874	3900	3602	9880	7266	5448	6826	3882
6295	8316	6150	3155	8059	4789	7236	7272	0839	3367
7935	1027	8193	2634	0806	6781	0665	8791	7416	8551
4833	6983	5904	8217	9201	5844	6959	5620	9570	8621
0584	0843	7983	5095	3205	3291	1584	1391	4136	8011
2585	0220	0730	5994	7138	7615	1126	3878	6154	2260
2527	1615	8232	7071	9808	3863	9195	4990	7625	3397
7300	2905	1760	4929	4767	9044	6891	0567	2382	8489
8131	9443	2266	0658	3814	0014	1749	5111	6145	6579
1002	4471	5983	8072	6371	6788	2510	4534	5574	6761
8467	5280	8912	3769	2089	8233	2262	0614	0577	0354
2929	5816	2185	3373	9405	8880	5460	0038	6634	6923
5177	9407	7063	4128	9058	8768	1396	5562	2367	3510
4216	5625	6077	5167	3603	7727	8521	1481	9075	2367
7835	6704	2249	5152	3116	3045	2760	4442	9638	2677
0955	5134	3386	8901	7341	8153	7739	3044	9774	1815
1577	6312	3484	0566	0615	4897	5569	6181	9176	2082
1323	9905	9375	3673	4428	4432	1572	3750	4726	1333
5058	0357	3847	7323	6761	7278	7817	1871	9909	6411
9948	5733	1063	7490	9067	1964	6990	6095	1796	3721
5467	3952	7378	4886	6983	6279	6520	6918	0557	7474
9934	7154	1024	7603	3170	7686	8890	6957	2764	0033
3549	4023	3486	5535	1284	6809	5264	3273	6701	4678
9817	2538	0384	2392	4795	1035	7011	1117	6329	9990
0267	8615	5686	0259	0164	4220	7995	3776	8234	7195
3693	4287	8163	7995	0706	4162	9680	9238	8886	6858
5685	1277	2430	7366	8426	2466	1668	0223	6602	6413
0546	2889	1427	2377	8859	1708	3388	8878	3901	5711
1502	2023	6338	7112	0662	0741	9498	3232	7942	7038
9561	0803	8146	9106	8885	5658	0122	2809	1972	7146
0902	4037	0573	5512	7429	4919	3166	4260	3036	9642
8143	9995	5246	6766	9732	6980	2124	6592	1262	9289
2143	5933	5862	9482	6548	0964	4101	8510	1611	3207
9583	7614	1163	8028	1778	9793	1282	7389	6600	2752
9981	4463	4374	9979	8682	1211	3170	0502	2815	0420
7721	3114	5054	1160	5093	0249	0918	9587	8584	7195
1326	0260	7983	6605	8027	0853	2867	3753	7053	8235
4428	7173	2662	5469	1490	5213	8111	7454	7885	3199
7052	4595	7963	5737	0505	3196	3337	1323	8566	8661
8838	1122	2508	7146	0981	4600	1906	6898	1831	7417
8316	7399	1720	7944	6409	4979	1193	4486	8697	3453
5021	7172	3385	4514	0569	2993	1282	0159	0845	5282
9768	2934	6774	8064	1362	2394	4939	8368	3730	9535

TABLE 9   2,500 Four-Digit Random Numbers *(continued)*

1236	2389	3150	9072	1871	8914	5859	9942	2284	0826
3889	3023	3423	2257	7442	2273	2693	4060	1078	8012
8078	5541	3977	9331	1827	2114	5208	7809	8563	8114
0239	7758	0885	2356	3354	4579	1097	4472	2478	0969
7372	7018	6911	7188	8014	7287	3898	2340	6395	4475
6138	1722	5523	1896	3900	9350	1827	4981	5280	6967
3916	4428	1497	9749	2597	3360	6014	3003	7767	4929
8090	7448	3988	1988	3731	0420	4967	3959	0105	4399
0905	6567	6366	3403	0657	8783	2812	4888	5048	5573
3342	2422	3204	6008	2041	8504	5357	3255	6409	5232
7265	6947	7364	7153	5545	1957	1555	2057	1212	5003
0414	3209	8358	6182	3548	3273	6340	9149	3719	0276
8522	1419	5221	6074	2441	5785	3188	5126	8229	7355
5488	0357	9167	5950	0861	3379	2901	8519	6226	2868
3325	5151	8203	4523	3935	3322	5946	6554	7680	1698
7597	1595	3240	8208	0221	5714	3352	4719	9452	7325
9063	7531	3538	3445	4924	1146	2510	7148	8988	9970
6506	1549	9334	3356	1942	6682	0304	9736	0815	4748
6442	0742	8223	9781	3957	0776	6584	2998	1553	9011
2717	1738	7696	7511	4558	9990	4716	5536	2566	2540
3221	3009	8727	5689	1562	3259	8066	0808	1942	8071
5420	5804	7235	8982	0270	1681	8998	3738	4403	5936
5928	6696	8484	7154	6755	3386	8301	6621	6937	2390
8387	5816	0122	9555	2219	6590	3878	0135	4748	2817
8331	5708	0336	8001	3960	4069	5643	6405	0249	5088
6454	2950	1335	7864	9262	1935	6047	5733	5213	0711
3926	0007	5548	0152	7656	2257	2032	8462	3018	4390
2976	0567	2819	6551	1195	7859	6390	2134	1921	9028
0631	0299	0146	2773	9028	1769	6451	3955	3469	0321
9754	4760	5765	5910	2185	4444	0797	5429	8467	7875
8296	8571	1161	9772	5351	5378	9894	3840	7093	1131
7687	3472	1252	9064	1692	1366	1742	8448	6830	8524
8739	7888	8723	9208	9563	6684	2290	6498	8695	5470
7404	1273	5961	3369	1259	4489	6798	7297	8979	1058
4789	4141	6643	7004	5079	4592	9656	6795	5636	4472
8777	7169	6414	5436	9211	3403	5906	6205	6204	3352
9697	6314	7221	8004	1199	4769	9562	7299	2904	8589
4382	1328	7781	8169	2993	7075	0202	3237	0055	8668
5720	8396	4009	3923	6595	5991	9141	5557	8842	4557
4906	7217	8093	0601	9032	6368	0793	9958	4901	2645
9425	8427	9579	1347	8013	2633	5516	7341	4076	4517
6814	8138	8238	1867	4045	9282	3004	3741	4342	4513
1220	9780	3361	2886	4164	1673	8886	3263	4198	8461
8831	8970	2611	1241	1943	6566	6098	5976	1141	1825
5672	8035	2961	6305	1525	4468	6468	4235	5102	7768
0713	1232	0107	1930	8704	5892	2845	8106	9397	6665
2118	6455	5561	3608	2433	8439	1602	1220	7755	7566
0215	1225	8873	4391	0365	2109	6080	6324	2684	3581
9095	8523	3277	0730	3618	4742	1968	3318	4138	0324
8010	9130	1285	4129	0032	1501	1957	9113	1272	9260

TABLE 9 2,500 Four-Digit Random Numbers *(concluded)*

9263	7824	1926	9545	5349	2389	3770	7986	7647	6641
7944	7873	7154	4484	2610	6731	0070	3498	6675	9972
5965	7196	2738	5000	0535	9403	2928	1854	5242	0608
3152	4958	7661	3978	1353	4808	5948	6068	8467	5301
0634	7693	9037	5139	5588	7101	0920	7915	2444	3024
2870	5170	9445	4839	7378	0643	8664	6923	5766	8018
6810	8926	9473	9576	7502	4846	6554	9658	1891	1639
9993	9070	9362	6633	3339	9526	9534	5176	9161	3323
9154	7319	3444	6351	8383	9941	5882	4045	6926	4856
4210	0278	7392	5629	7267	1224	2527	3667	2131	7576
1713	2758	2529	2838	5135	6166	3789	0536	4414	4267
2829	1428	5452	2161	9532	3817	6057	0808	9499	7846
0933	5671	5133	0628	7534	0881	8271	5739	2525	3033
3129	0420	9371	5128	0575	7939	8739	5177	3307	9706
3614	1556	2759	4208	9928	5964	1522	9607	0996	0537
2955	1843	1363	0552	0279	8101	4902	7903	5091	0939
2350	2264	6308	0819	8942	6780	5513	5470	3294	6452
5788	8584	6796	0783	1131	0154	4853	1714	0855	6745
5533	7126	8847	0433	6391	3639	1119	9247	7054	2977
1008	1007	5598	6468	6823	2046	8938	9380	0079	9594
3410	8127	6609	8887	3781	7214	6714	5078	2138	1670
5336	4494	6043	2283	1413	9659	2329	5620	9267	1592
8297	6615	8473	1943	5579	6922	2866	1367	9931	7687
5482	8467	2289	0809	1432	8703	4289	2112	3071	4848
2546	5909	2743	8942	8075	8992	1909	6773	8036	0879
6760	6021	4147	8495	4013	0254	0957	4568	5016	1560
4492	7092	6129	5113	4759	8673	3556	7664	1821	6344
3317	3097	9813	9582	4978	1330	3608	8076	3398	6862
8468	8544	0620	1765	5133	0287	3501	6757	6157	2074
7188	5645	3656	0939	9695	3550	1755	3521	6910	0167
0047	0222	7472	1472	4021	2135	0859	4562	8398	6374
2599	3888	6836	5956	4127	6974	4070	3799	0343	1887
9288	5317	9919	9380	5698	5308	1530	5052	5590	4302
2513	2681	0709	1567	6068	0441	2450	3789	6718	6282
8463	7188	1299	8302	8248	9033	9195	7457	0353	9012
3400	9232	1279	6145	4812	7427	2836	6656	7522	3590
5377	4574	0573	8616	4276	7017	9731	7389	8860	1999
5931	9788	7280	5496	6085	1193	3526	7160	5557	6771
2047	6655	5070	2699	0985	5259	1406	3021	1989	1929
8618	8493	2545	2604	0222	5201	2182	5059	5167	6541
2145	6800	7271	4026	6128	1317	6381	4897	5173	5411
9806	6837	8008	2413	7235	9542	1180	2974	8164	8661
0178	6442	1443	9457	7515	9457	6139	9619	0322	3225
6246	0484	4327	6870	0127	0543	2295	1894	9905	4169
9432	3108	8415	9293	9998	8950	9158	0280	6947	6827
0579	4398	2157	0990	7022	1979	5157	3643	3349	7988
1039	1428	5218	0972	2578	3856	5479	0489	5901	8925
3517	5698	2554	5973	6471	5263	3110	6238	4948	1140
2563	8961	7588	9825	0212	7209	5718	5588	0932	7346
1646	4828	9425	4577	4515	6886	1138	1178	2269	4198

References

AT&T. *Statistical Quality Control Handbook*. Indianapolis, Ind.: AT&T, 1956. 10th printing, May 1984.

J. F. Beardsley & Associates, International, Inc. *Quality Circles: Member Manual*. San Jose, Calif.: J. F. Beardsley, 1977.

Boardman, T. J., and H. Iyer. *The Funnel*. Fort Collins, Colo.: Colorado State University Press, 1986.

Deming, W. E. *Some Theory of Sampling*. New York: John Wiley & Sons, 1950.

————. "On the Distinction between Enumerative and Analytic Surveys." *Journal of the American Statistical Association* 48, 1953, pp. 244–55.

————. "On Some Statistical Aids toward Economic Production." *Interfaces* 5, August 1975, pp. 1–15.

————. "On Probability as a Basis for Action." *The American Statistician* 29(4), November 1975, pp. 146–52.

————. "On the Use of Judgment-Samples." *Reports of Statistical Applications* 23, March 1976, pp. 25–31.

————. *Quality, Productivity, and Competitive Position*. Cambridge, Mass.: MIT Center for Advanced Engineering Study, 1982.

————. *Out of the Crisis*. Cambridge, Mass.: MIT Center for Advanced Engineering Study, 1986.

————. "Principles of Professional Statistical Practice." *Encyclopedia of Statistical Sciences* 7. Edited by Kotz-Johnson. New York: John Wiley & Sons, 1986.

Duncan, A. *Quality Control and Industrial Statistics*. 5th ed. Homewood, Ill.: Richard D. Irwin, 1986.

Feigenbaum, A. V. "Total Quality Control." *Harvard Business Review,* November 1956.

Fitzgerald, J. M., and A. F. Fitzgerald. *Fundamentals of Systems Analysis*. New York: John Wiley & Sons, 1973.

Ford Motor Company. "Statistical Process Control Case Study." *Introduction to*

Ford's Operating Philosophy and Principles and Statistical Management Methods—Participant Notebook, September 1983, pp 7.E.9.–7.E.18.

————. *Continuing Process Control and Process Capability Improvement,* February 1984.

Gitlow, H. "Definition of Quality." *Proceedings—Case Study Seminar—Dr. Deming's Management Methods: How They Are Being Implemented in the U.S. and Abroad.* Andover, Mass.: G.O.A.L., November 1984, pp. 4–18.

———— and S. Gitlow. *The Deming Guide to Quality and Competitive Position.* Englewood Cliffs, N.J.: Prentice-Hall, 1987.

———— and P. Hertz. "Product Defects and Productivity." *Harvard Business Review,* September–October 1983, pp. 131–41.

————; K. Kang; and S. Kellogg. "Rule Based Process Control versus Statistical Process Control." University of Miami Working Paper Series, 1988.

———— and R. Oppenheim. *Stat City: Understanding Statistics Through Realistic Applications.* 2nd ed. Homewood, Ill.: Richard D. Irwin, 1986.

Golomski, W. A. "Quality Control—History in the Making." *Quality Progress,* July 1976, pp. 16–18.

Grant, E. L., and R. S. Leavenworth. *Statistical Quality Control.* 5th ed. New York: McGraw-Hill, 1980.

Harrington, H. J. "Quality's Footprints in Time." IBM Technical Report TR 02.1064. San Jose, Calif., September 20, 1983.

Imai, M. *KAIZEN—The Keys to Japan's Competitive Success.* New York: Random House, 1986.

Ishikawa, K. *Guide to Quality Control.* Tokyo: Asian Productivity Organization, 1976. 11th printing, 1983. (Available through UNIPUB, Box 433, Murray Hill Station, New York, N.Y., 10157.)

————. *What Is Total Quality Control? The Japanese Way.* Englewood Cliffs, N.J.: Prentice Hall, 1985.

Juran, J. *Quality Control Handbook.* 3rd ed. New York: McGraw-Hill, 1979.

Kackar, R. "Taguchi's Quality Philosophy: Analysis and Commentary." *Quality Progress,* December 1986, pp. 21–29.

Kane, E. J. "IBM's Quality Focus on the Business Process." *Quality Progress,* April 1986, pp. 26–33.

Kane, V. "Process Capability Indices." *Journal of Quality Technology* 18, January 1986, pp. 41–52.

Melan, E. H. "Process Management in Service and Administrative Operations." *Quality Progress,* June 1985, pp. 52–59.

Mood, A. M. "On the Dependence of Sampling Inspection Plans under Population Distributions." *Annals of Mathematical Statistics* 14, 1943, pp. 415–25.

Nolan, T. W. "Analytic Studies." Working paper, Associates in Process Management, 1988.

Orsini, J. "Simple Rule to Reduce Total Cost of Inspection and Correction of Product in State of Chaos." Ph.D. dissertation, Graduate School of Business Administration, New York University, 1982.

Papadakis, G. P. "The Deming Inspection Criteria for Choosing Zero or 100 Percent Inspection." *Journal of Quality Technology* 17(3), July 1985, pp. 121–27.

Rehg, V. *Quality Circle Manual for Coordinators and Leaders*. Wright Patterson AFB, Ohio.

Rice, W. B. *Control Charts in Factory Management*. New York: John Wiley & Sons, 1947.

Scherkenbach, W. *The Deming Route to Quality and Productivity: Road Maps and Roadblocks*. Washington, D.C.: Ceepress Books, 1986.

Shewhart, W. A. *Economic Control of Quality of Manufactured Product*. New York: D. Van Nostrand Co., Inc., 1931.

Silver, G. A., and J. B. Silver. *Introduction to Systems Analysis*. Englewood Cliffs, N.J.: Prentice-Hall, 1976.

Taguchi, G., and Y. Wu. *Introduction to Off-Line Quality Control*. Nagoya, Japan: Central Japan Quality Control Association, 1980.

Wheeler, D. J., and D. S. Chambers. *Understanding Statistical Process Control*. Knoxville, Tenn.: Statistical Process Controls, Inc., 1986.

Index

Index